T0310086

CIRCULATING TUMOR CELLS

Isolation and Analysis

CHEMICAL ANALYSIS

A Series of Monographs on Analytical Chemistry and Applications

VOLUME 184

A complete list of the titles in this series appears at the end of this volume.

CIRCULATING TUMOR CELLS

Isolation and Analysis

Edited by

Z. HUGH FAN

Published by John Wiley & Sons, Inc., Hoboken, New Jersey
Published simultaneously in Canada

For general information on our other products and services or for technical support, please contact our Customer Care Department within the United States at (800) 762-2974, outside the United States at (317) 572-3993 or fax (317) 572-4002.

Wiley also publishes its books in a variety of electronic formats. Some content that appears in print may not be available in electronic formats. For more information about Wiley products, visit our web site at www.wiley.com.

Library of Congress Cataloging-in-Publication Data

Names: Fan, Z. Hugh, editor.
Title: Circulating tumor cells : isolation and analysis / edited by Z. Hugh Fan.
Description: Hoboken, New Jersey : John Wiley & Sons, Inc., [2016] | Includes
bibliographical references and index.
Identifiers: LCCN 2015044832 | ISBN 9781118915530 (cloth)
Subjects: LCSH: Tumor markers.
Classification: LCC RC270.3.T84 C57 2016 | DDC 616.99/4–dc23 LC record available at http://lccn.loc.gov/2015044832

Set in 10/12pt, TimesLTStd by SPi Global, Chennai, India.

Printed in the United States of America

10 9 8 7 6 5 4 3 2 1

CONTENTS

List of Contributors **xv**

Foreword **xxi**

Preface **xxv**

PART I INTRODUCTION **1**

1 Circulating Tumor Cells and Historic Perspectives **3**
Jonathan W. Uhr
1.1 Early Studies on Cancer Dormancy Led to the Development of a Sensitive Assay for CTCs (1970–1998) 3
1.2 Modern Era for Counting CTCs: 1998–2007 6
1.3 Proof of Malignancy of CTCs 7
1.4 New Experiments Involving CTCs 7
1.5 Clinical Cancer Dormancy 8
1.6 Human Epidermal Growth Factor Receptor 2 (HER2) Gene Amplification can be Acquired as Breast Cancer Progresses 10
1.7 uPAR and HER2 Co-amplification 11
1.8 Epithelial–Mesenchymal Transition (EMT) 12
1.9 New Instruments to Capture CTCs 14
1.10 Genotypic Analyses 15
1.11 Conclusions 18
References 20

2 Introduction to Microfluidics **33**

Kangfu Chen and Z. Hugh Fan

2.1 Introduction 33
2.1.1 Brief History 33
2.1.2 Fluids 34
2.1.3 Microfluidics 34
2.1.4 Advantages of Microfluidics 35
2.2 Scaling Law 36
2.2.1 Laminar Flow 36
2.2.2 Flow Rate 37
2.2.3 Diffusion 38
2.3 Device Fabrication 39
2.3.1 Materials 39
2.3.2 Fabrication Methods 40
2.3.2.1 Photolithography 40
2.3.2.2 Etching 41
2.3.2.3 Bonding 42
2.3.2.4 Molding 42
2.4 Functional Components in Microfluidic Devices 43
2.4.1 Micropumps 43
2.4.1.1 Mechanical Pumps 43
2.4.1.2 Nonmechanical Pumps 44
2.4.2 Microvalves 44
2.4.3 Micromixers 45
2.4.4 Other Components 46
2.5 Concluding Remarks 46
References 47

PART II ISOLATION METHODS **51**

3 Ensemble-decision Aliquot Ranking (eDAR) for CTC Isolation and Analysis **53**

Mengxia Zhao, Perry G. Schiro, and Daniel T. Chiu

3.1 Overview of eDAR 53
3.2 Individual Components and Analytical Performance of eDAR 55
3.2.1 Aliquot Ranking 55
3.2.2 Aliquot Sorting 59
3.2.2.1 Active Sorting Scheme Based on an On-Chip Solenoid 59
3.2.2.2 Active Sorting Scheme Based on an Off-Chip Solenoid 60
3.2.3 On-Chip Purification 63
3.2.3.1 Purification via Integrated Planar Filter 63
3.2.3.2 Purification via Microfabricated Slits 64

3.2.4 Secondary Labeling and the Enumeration of CTCs 65
3.2.5 Characterization and Analytical Performance of eDAR 67
3.3 Application and Downstream Analyses of eDAR 69
3.3.1 Enumeration of CTCs from Cancer Patients using eDAR 69
3.3.2 Downstream Analysis of CTCs Isolated by eDAR 71
3.3.3 Automated High-Throughput Counting of CTCs – A "Simplified" Version of eDAR 75
3.4 Conclusion and Perspective 80
References 81

4 Sinusoidal Microchannels with High Aspect Ratios for CTC Selection and Analysis 85

Joshua M. Jackson, Małgorzata A. Witek, and Steven A. Soper

4.1 Introduction 85
4.1.1 Brief Perspective 85
4.1.2 Overview of CTC Selection Modalities and Assay Metrics 87
4.2 Parallel Arrays of High-Aspect-Ratio, Sinusoidal Microchannels for CTC Selection 90
4.2.1 Production of CTC *In Vitro* Diagnostic Devices in Thermoplastics 91
4.2.2 Activation of High-Aspect-Ratio Microchannels for Efficient Ab Immobilization 94
4.2.3 CTC Selection in Sinusoidal Microchannels from a Fluid Dynamics Perspective 99
4.2.3.1 Centrifugal Forces in Curved Microchannels 100
4.2.3.2 Transient Dynamics of CTC-Ab Binding 104
4.2.4 Parallel Arrays for High-Throughput Sample Processing 107
4.3 Clinical Applications of Sinusoidal CTC Microchip 114
4.4 Conclusion 118
Acknowledgments 119
References 119

5 Cell Separation using Inertial Microfluidics 127

Nivedita Nivedita and Ian Papautsky

5.1 Introduction 127
5.2 Device Fabrication and System Setup 128
5.3 Inertial Focusing in Microfluidics 129
5.4 Cancer Cell Separation in Straight Microchannels 132
5.5 Cancer Cell Separation in Spiral Microchannels 136
5.6 Conclusions 142
References 142

6 **Morphological Characteristics of CTCs and the Potential for Deformability-Based Separation** **147**
Simon P. Duffy and Hongshen Ma
6.1 Introduction 147
6.2 Limitations of Antibody-based CTC Separation Methods 148
6.3 Morphological and Biophysical Differences Between CTCs and Hematological Cells 149
6.4 Historical and Recent Methods in CTC Separation Based on Biophysical Properties 153
6.5 Microfluidic Ratchet for Deformability-Based Separation of CTCs 155
6.5.1 Microfluidic Ratchet Mechanism 155
6.5.2 Design of the Microfluidic Ratchet 156
6.5.3 Validation of the Microfluidic Ratchet Mechanism 158
6.5.4 Viability of Cells Enriched by the Microfluidic Ratchet Mechanism 159
6.6 Resettable Cell Trap for Deformability-based Separation of CTCs 160
6.6.1 Resettable Cell Trap Mechanism 160
6.6.2 Validation of the Resettable Cell Trap Device 163
6.6.3 Application of the Resettable Cell Trap for CTC Enrichment 164
6.7 Summary 165
References 166

7 **Microfabricated Filter Membranes for Capture and Characterization of Circulating Tumor Cells (CTCs)** **173**
Zheng Ao, Richard J. Cote, Ram H. Datar, and Anthony Williams
7.1 Introduction 173
7.2 Size-based Enrichment of Circulating Tumor Cells 174
7.3 Comparison Between Size-based CTC Isolation and Affinity-based Isolation 177
7.4 Characterization of CTCs Captured by Microfilters 178
7.4.1 Genomic Analysis of CTCs Enriched by Microfabricated Filter Membrane 178
7.4.2 Gene Expression Analysis of CTC Enriched by Microfabricated Filter Membrane 179
7.4.3 Functional Characterization of CTCs Enriched by Microfabricated Filter Membrane 180
7.5 Conclusion 180
References 181

8 Miniaturized Nuclear Magnetic Resonance Platform for Rare Cell Detection and Profiling 183

Sangmoo Jeong, Changwook Min, Huilin Shao, Cesar M. Castro, Ralph Weissleder, and Hakho Lee

8.1 Introduction 183
8.2 μNMR Technology 184
8.2.1 Magnetic Nanoparticles with High Transverse Relaxivity 185
8.2.2 Bioorthogonal Strategy for Efficient MNP Labeling 188
8.2.3 Miniaturized NMR Probe 188
8.3 Clinical Application of μNMR for CTC Detection and Profiling 191
8.3.1 Quad-Marker Assay Integrated with μNMR 192
8.3.2 Comparison of Biomarkers in CTC and Bulk Tumor Cell 192
8.4 Conclusion 196
References 196

9 Nanovelcro Cell-Affinity Assay for Detecting and Characterizing Circulating Tumor Cells 201

Millicent Lin, Anna Fong, Sharon Chen, Yang Zhang, Jie-fu Chen, Paulina Do, Morgan Fong, Shang-Fu Chen, Pauline Yang, An-Jou Liang, Qingyu Li, Min Song, Shuang Hou, and Hsian-Rong Tseng

9.1 Introduction 202
9.1.1 Circulating Tumor Cells 202
9.1.2 Current CTC Capture Methods 202
9.1.3 The Evolution of NanoVelcro Cell-Affinity Assays 205
9.1.4 Nanostructured Substrates for Cell Biology 206
9.2 Proof-of-Concept Demonstration of NanoVelcro Cell-Affinity Substrates 207
9.2.1 Stationary NanoVelcro CTC Assay 207
9.2.2 General Applicability of NanoVelcro CTC Substrates 207
9.3 First-Generation NanoVelcro Chips for CTC Enumeration 209
9.3.1 Device Configuration of First-Generation NanoVelcro Chips 209
9.3.2 Clinical Utility of First-Generation NanoVelcro Chips 212
9.3.3 An Alternative Capture Agent, Aptamer 213
9.4 Second-Generation NanoVelcro-LMD Technology for Single CTC Isolation 214
9.4.1 Preparation of PLGA NanoVelcro Chips 215
9.4.2 NanoVelcro-LMD Technology and Mutational Analysis 215
9.4.3 NanoVelcro-LCM Technology and Whole Exome Sequencing 215

9.5 Third-Generation Thermoresponsive NanoVelcro Chips 219
9.6 Conclusions and Future Perspectives 220
Acknowledgment 221
References 221

10 Acoustophoresis in Tumor Cell Enrichment 227
Per Augustsson, Cecilia Magnusson, Hans Lilja, and Thomas Laurell
10.1 Introduction 227
10.1.1 Background 227
10.1.2 System Specification 229
10.2 Factors Determining Acoustophoresis Cell Separation 230
10.2.1 Acoustic Field 231
10.2.2 Acoustic Radiation Force 231
10.2.3 Trajectory of a Cell 232
10.2.4 The Microchannel Flow Profile 233
10.3 Acoustophoresis System for Separating Cells 234
10.3.1 Acoustophoresis Chip 234
10.3.2 Actuation of Ultrasound 234
10.3.3 Flow System 234
10.3.4 Sample Preparation and Analysis 236
10.3.4.1 Blood and Cancer Cell Preparation 236
10.3.4.2 Cell Labeling 236
10.3.5 Device Testing 237
10.3.5.1 Varying the Flow Rate 237
10.3.5.2 Cancer Cell Number 237
10.3.5.3 Cancer Cell Diversity 237
10.4 Acoustophoresis Platform for Clinical Sample Processing 239
10.4.1 The Acoustophoresis Chip 239
10.4.2 Flow System 241
10.4.2.1 Pressure-Driven Flow 241
10.4.2.2 Operation of the Flow System for Cell Separation 241
10.4.3 Temperature Control System 242
10.4.4 Software Interface 242
10.4.5 System Calibration using Microbeads 243
10.4.6 Cell Separations 244
10.5 Unperturbed Cell Survival and Phenotype after Microchip Acoustophoresis 244
10.6 Summary 246
References 246

11 Photoacoustic Flow Cytometry for Detection and Capture of Circulating Melanoma Cells 249
John A. Viator, Benjamin S. Goldschmidt, Kiran Bhattacharyya, and Kyle Rood

11.1 Introduction 249
11.1.1 Biomedical Photoacoustics 251
11.1.2 Photoacoustic Flow Cytometry 251
11.2 Current Methods for Detection and Capture of CMCs 254
11.2.1 Two-Phase Flow for Cell Capture 255
11.2.2 Photoacoustic Flow Cytometer 256
11.2.3 Blood Sample Preparation 256
11.2.4 Capture Process 258
11.2.5 Results of CMC Capture Study 259
11.3 Discussion 259
11.3.1 Extension to Nonpigmented CTCs 260
11.4 Future Work 261
References 262

12 Selectin-Mediated Targeting of Circulating Tumor Cells for Isolation and Treatment 267

Jocelyn R. Marshall and Michael R. King

12.1 Introduction 267
12.1.1 Selectin Adhesion 267
12.1.2 Circulating Tumor Cells 270
12.2 CTC Capture by E-selectin 271
12.3 Applications for E-selectin in Cancer Diagnosis and Treatment 273
12.3.1 E-Selectin Capture for Drug Efficacy Testing 273
12.3.2 E-Selectin for use in Targeted Cancer Therapy 274
12.4 Conclusions 278
References 279

13 Aptamer-Enabled Tumor Cell Isolation 287

Jinling Zhang and Z. Hugh Fan

13.1 Introduction 287
13.2 Aptamers and their Biomedical Applications 288
13.2.1 Identification of Aptamers 288
13.2.2 Aptamers versus Antibodies 289
13.2.3 List of Aptamers 290
13.3 Aptamer-based Tumor Cell Isolation 290
13.3.1 Aptamers with Microfluidics 290
13.3.1.1 Device Designs 290
13.3.1.2 Surface Functionalization 293
13.3.1.3 Tumor Cell Isolation 293
13.3.1.4 Instrument Setup 293
13.3.2 Aptamers with Nanoparticles 294
13.3.3 Aptamers with Innovative Schemes 295
13.3.4 Aptamers for CTC Isolation 296
13.4 Conclusion and Outlook 297
References 297

14 Depletion of Normal Cells for CTC Enrichment **301**
Jeffrey J. Chalmers, Maryam B. Lustberg, Clayton Deighan, Kyoung-Joo Jenny Park, Yongqi Wu, and Peter Amaya
14.1 Introduction 301
14.2 Estimates of Number and Type of Cells in Blood 302
14.3 Summary of Examples of Negative Depletion 303
14.3.1 Removal of RBCs 303
14.3.2 Removal of Normal Nucleated Cells 304
14.4 Types of Cells Observed After Depletion of Normal Cells 305
14.5 Incomplete Depletion of Normal Cells 305
14.6 Conclusion 310
References 311

PART III POST-ISOLATION ANALYSIS AND CLINICAL TRANSLATION **313**

15 Tumor Heterogeneity and Single-cell Analysis of CTCs **315**
Evelyn K. Sigal and Stefanie S. Jeffrey
15.1 Introduction 315
15.2 Tumor Heterogeneity 316
15.3 Single-Cell Analysis of CTCs and CTC Heterogeneity 318
15.4 Gene Expression Analysis 319
15.5 Mutational Analysis 321
15.6 Conclusion: Clinical Implications and Future Perspectives 323
References 324

16 Single-Cell Molecular Profiles and Biophysical Assessment of Circulating Tumor Cells **329**
Devalingam Mahalingam, Pawel Osmulski, Chiou-Miin Wang, Aaron M. Horning, Anna D. Louie, Chun-Lin Lin, Maria E. Gaczynska, and Chun-Liang Chen
16.1 Introduction 329
16.2 Methods 331
16.2.1 Single-Cell Molecular Profiling 331
16.2.1.1 High-Throughput Single-Cell qRT-PCR using Microfluidic BioMark™ HD 332
16.2.1.2 Single-Cell Transcriptome Analysis using Gene Expression Microarray 332
16.2.1.3 Single-Cell RNA-Seq 333
16.2.2 Probing Cellular Biophysical Properties of Single Cells 333
16.2.2.1 The Instruments for Biophysical Assessment of Single Cells 333
16.2.2.2 The Biophysical Parameters of Single Cells 336

16.3 CTC Applications 336
16.3.1 Single-cell Molecular Profiling of CTCs 336
16.3.2 Analysis of Nanomechanical Phenotypes of Single CTCs 341
16.4 Conclusions 342
References 343

17 Directing Circulating Tumor Cell Technologies Into Clinical Practice 351
Benjamin P. Casavant, David Kosoff, and Joshua M. Lang
17.1 Introduction 351
17.2 Defining Biomarkers 352
17.2.1 Prognostic CTC Biomarkers 353
17.2.2 Predictive CTC Biomarkers 355
17.2.3 Pharmacodynamic CTC Biomarkers 355
17.2.4 Diagnostic CTC Biomarkers 356
17.2.5 Surrogate CTC Biomarkers 356
17.3 The Technology 356
17.3.1 Translated Technologies 357
17.4 Translating Technology 357
17.4.1 The Technology Side 358
17.4.2 The Clinic Side 360
17.5 Conclusions 360
References 361

PART IV COMMERCIALIZATION 365

18 DEPArray™ Technology for Single CTC Analysis 367
Farideh Z. Bischoff, Gianni Medoro, and Nicolò Manaresi
18.1 Challenges in Molecular Profiling of CTCs 367
18.2 DEPArray™ Technology Solution 368
18.3 DEPArray™ for Single Tumor Cell Analysis 369
18.4 Clinical Significance in Single CTC Profiling 373
18.5 Conclusion 374
References 374

19 CELLSEARCH® Instrument, Features, and Usage 377
Denis A. Smirnov, Brad W. Foulk, Mark C. Connelly, and Robert T. McCormack
19.1 Introduction 377
19.2 Principles of CELLSEARCH® 379
19.3 EpCAM Density and CTC Capture 380

19.4 Clinical Applications of CELLSEARCH® CTCs 383
19.4.1 CTC Enumeration 383
19.4.2 Expanding Enumeration 384
19.4.3 CTC Enumeration and Clinical Utility 385
19.4.4 Characterization of CTCs using CELLSEARCH® 387
19.5 Beyond EpCAM Capture 390
19.6 Discussion 391
References 394

PART V GLOSSARY **401**

Circulating Tumor Cell Glossary **403**
Jose I. Varillas and Z. Hugh Fan

Index **423**

LIST OF CONTRIBUTORS

Peter Amaya William G. Lowrie Department of Chemical and Biomolecular Engineering, The Ohio State University, Columbus, OH, USA

Zheng Ao Sheila and David Fuente Graduate Program in Cancer Biology, Department of Pathology and Laboratory Medicine, Dr. John T. Macdonald Foundation Biomedical Nanotechnology Institute, University of Miami Miller School of Medicine, Miami, FL, USA

Per Augustsson Department of Biomedical Engineering, Lund University, Lund, Sweden

Kiran Bhattacharyya Department of Biomedical Engineering, Northwestern University, Evanston, IL, USA

Farideh Z. Bischoff Silicon Biosystems, Inc., San Diego, CA, USA

Benjamin P. Casavant Department of Biomedical Engineering, University of Wisconsin, Madison, WI, USA

Cesar M. Castro Center for Systems Biology, Massachusetts General Hospital, Harvard Medical School, Boston, MA, USA

Jeffrey J. Chalmers William G. Lowrie Department of Chemical and Biomolecular Engineering, The Ohio State University, Columbus, OH, USA

Chun-Liang Chen Molecular Medicine, University of Texas Health Science Center at San Antonio, San Antonio, TX, USA

Jie-fu Chen Urologic Oncology Research Program, Samuel Oschin Comprehensive Cancer Institute, Cedars-Sinai Medical Center, Los Angeles, CA, USA

Kangfu Chen Department of Mechanical and Aerospace Engineering, University of Florida, Gainesville, FL, USA

Shang-Fu Chen Department of Life Science, National Taiwan University, Taipei City, Taiwan (ROC)

Sharon Chen Department of Molecular and Medical Pharmacology, Crump Institute for Molecular Imaging (CIMI), California NanoSystems Institute (CNSI), University of California, Los Angeles, CA, USA

Daniel T. Chiu Department of Chemistry, University of Washington, Seattle, WA, USA

Mark C. Connelly Oncology Biomarkers, Janssen R&D, Huntingdon Valley, PA, USA

Richard J. Cote Department of Pathology and Laboratory Medicine, Department of Biochemistry and Molecular Biology, Department of Pathology, Jackson Memorial Hospital, Dr. John T. Macdonald Foundation Biomedical Nanotechnology Institute, University of Miami Miller School of Medicine, Miami, FL, USA

Ram H. Datar Department of Pathology and Laboratory Medicine, Department of Biochemistry and Molecular Biology, Dr. John T. Macdonald Foundation Biomedical Nanotechnology Institute, University of Miami Miller School of Medicine, Miami, FL, USA

Clayton Deighan William G. Lowrie Department of Chemical and Biomolecular Engineering, The Ohio State University, Columbus, OH, USA

Paulina Do Department of Molecular and Medical Pharmacology, Crump Institute for Molecular Imaging (CIMI), California NanoSystems Institute (CNSI), University of California, Los Angeles, CA, USA

Simon P. Duffy Department of Mechanical Engineering, University of British Columbia, Vancouver, British Columbia, Canada

Z. Hugh Fan Department of Mechanical and Aerospace Engineering, University of Florida, Gainesville, FL, USA; J. Crayton Pruitt Family Department of Biomedical Engineering, University of Florida, Gainesville, FL, USA; Department of Chemistry, University of Florida, Gainesville, FL, USA

Anna Fong Department of Molecular and Medical Pharmacology, Crump Institute for Molecular Imaging (CIMI), California NanoSystems Institute (CNSI), University of California, Los Angeles, CA, USA

Morgan Fong Department of Molecular and Medical Pharmacology, Crump Institute for Molecular Imaging (CIMI), California NanoSystems Institute (CNSI), University of California, Los Angeles, CA, USA

Brad W. Foulk Oncology Biomarkers, Janssen R&D, Oncology Biomarkers, Spring House, PA, USA

Maria E. Gaczynska Molecular Medicine, University of Texas Health Science Center at San Antonio, San Antonio, TX, USA

Benjamin S. Goldschmidt Department of Bioengineering, University of Missouri, Columbia, MO, USA

Aaron M. Horning Integrated Biomedical Science Graduate Program, University of Texas Health Science Center, San Antonio, TX, USA

Shuang Hou Department of Molecular and Medical Pharmacology, Crump Institute for Molecular Imaging (CIMI), California NanoSystems Institute (CNSI), University of California, Los Angeles, CA, USA

Joshua M. Jackson Department of Chemistry, University of North Carolina Chapel Hill, Chapel Hill, NC, USA

Stefanie S. Jeffrey Department of Surgery, Stanford University School of Medicine, Stanford, CA, USA

Sangmoo Jeong Center for Systems Biology, Massachusetts General Hospital, Harvard Medical School, Boston, MA, USA

Michael R. King Nancy E. and Peter C. Meinig School of Biomedical Engineering, Cornell University, Ithaca, NY, USA

David Kosoff Department of Medicine, University of Wisconsin, Madison, WI, USA

Joshua M. Lang Department of Medicine, University of Wisconsin Carbone Cancer Center, Madison, WI, USA

Thomas Laurell Department of Biomedical Engineering, Lund University, Lund, Sweden

Hakho Lee Center for Systems Biology, Massachusetts General Hospital, Harvard Medical School, Boston, MA, USA

Qingyu Li Department of Molecular and Medical Pharmacology, Crump Institute for Molecular Imaging (CIMI), California NanoSystems Institute (CNSI), University of California, Los Angeles, CA, USA

An-Jou Liang Department of Life Science, National Taiwan University, Taipei City, Taiwan (ROC)

Hans Lilja Department of Surgery, Memorial Sloan Kettering Cancer Center, New York, NY, USA

Chun-Lin Lin Molecular Medicine, University of Texas Health Science Center at San Antonio, San Antonio, TX, USA

Millicent Lin Department of Molecular and Medical Pharmacology, Crump Institute for Molecular Imaging (CIMI), California NanoSystems Institute (CNSI), University of California, Los Angeles, CA, USA

Anna D. Louie School of Medicine, University of Nevada, Reno, NV, USA

Maryam B. Lustberg Stefanie Spielman Comprehensive Breast Center, Wexner Medical Center, The Ohio State University, Columbus, OH, USA

Hongshen Ma Department of Mechanical Engineering, University of British Columbia, Vancouver, British Columbia, Canada; Vancouver Prostate Centre, Vancouver General Hospital, Vancouver, British Columbia, Canada; Department of Urologic Science, University of British Columbia, Vancouver, British Columbia, Canada

Cecilia Magnusson Department of Translational Medicine, Lund University, Lund, Sweden

Devalingam Mahalingam Department of Medicine, University of Texas Health Science Center, San Antonio, TX, USA

Nicolò Manaresi Silicon Biosystems, SpA, Bologna, Italy

Jocelyn R. Marshall Nancy E. and Peter C. Meinig School of Biomedical Engineering, Cornell University, Ithaca, NY, USA

Robert T. McCormack Oncology Biomarkers, Janssen R&D, Raritan, NJ, USA

Gianni Medoro Silicon Biosystems, SpA, Bologna, Italy

Changwook Min Center for Systems Biology, Massachusetts General Hospital, Harvard Medical School, Boston, MA, USA

Nivedita Nivedita Department of Electrical Engineering and Computing Systems, University of Cincinnati, Cincinnati, OH, USA

Pawel Osmulski Molecular Medicine, University of Texas Health Science Center at San Antonio, San Antonio, TX, USA

Klaus Pantel Department of Tumour Biology, Center of Experimental Medicine, University Cancer Center Hamburg, University Medical Centre Hamburg-Eppendorf, Hamburg, Germany

Ian Papautsky Department of Electrical Engineering and Computing Systems, University of Cincinnati, Cincinnati, OH, USA

Kyoung-Joo J. Park William G. Lowrie Department of Chemical and Biomolecular Engineering, The Ohio State University, Columbus, OH, USA

Kyle Rood Biodesign Program, One Hospital Drive, Columbia, MO, USA

Perry G. Schiro MiCareo Inc., Taipei, Taiwan

Huilin Shao Center for Systems Biology, Massachusetts General Hospital, Harvard Medical School, Boston, MA, USA

Evelyn K. Sigal Department of Surgery, Stanford University School of Medicine, Stanford, CA, USA

Denis A. Smirnov Oncology Biomarkers, Janssen R&D, Oncology Biomarkers, Spring House, PA, USA

Min Song Department of Molecular and Medical Pharmacology, Crump Institute for Molecular Imaging (CIMI), California NanoSystems Institute (CNSI), University of California, Los Angeles, CA, USA

Steven A. Soper Department of Chemistry, University of North Carolina Chapel Hill, Chapel Hill, NC, USA; Department of Biomedical Engineering, University of North Carolina Chapel Hill, Chapel Hill, NC, USA; Department of Chemistry, Louisiana State University, Baton Rouge, LA, USA; Department of Mechanical Engineering, Louisiana State University, Baton Rouge, LA, USA

Hsian-Rong Tseng Department of Molecular and Medical Pharmacology, Crump Institute for Molecular Imaging (CIMI), California NanoSystems Institute (CNSI), University of California, Los Angeles, CA, USA

Jonathan W. Uhr Department of Immunology, University of Texas Southwestern Medical Center, Dallas, TX, USA

John A. Viator Biomedical Engineering Program, Duquesne University, Pittsburgh, PA, USA

Jose Varillas J. Crayton Pruitt Family Department of Biomedical Engineering, University of Florida, Gainesville, FL, USA

Chiou-Miin Wang Molecular Medicine, University of Texas Health Science Center at San Antonio, San Antonio, TX, USA

Ralph Weissleder Center for Systems Biology, Massachusetts General Hospital, Harvard Medical School, Boston, MA, USA

Anthony Williams Department of Surgery, Section of Urology, University of Chicago - Pritzker School of Medicine, Chicago, IL, USA and
Department of Pathology and Laboratory Medicine, University of Miami Miller School of Medicine, Miami, FL, USA

Małgorzata A. Witek Department of Biomedical Engineering, University of North Carolina Chapel Hill, Chapel Hill, NC, USA

Yongqi Wu William G. Lowrie Department of Chemical and Biomolecular Engineering, The Ohio State University, Columbus, OH, USA

Pauline Yang Department of Molecular and Medical Pharmacology, Crump Institute for Molecular Imaging (CIMI), California NanoSystems Institute (CNSI), University of California, Los Angeles, CA, USA

Jinling Zhang Department of Mechanical and Aerospace Engineering, University of Florida, Gainesville, FL, USA

Yang Zhang Biomedical Engineering Program, The University of Texas at El Paso, El Paso, TX, USA

Mengxia Zhao Department of Chemistry, University of Washington, Seattle, WA, USA

FOREWORD

Primary tumor and/or metastasized sites release a small number of tumor cells into the blood circulation. As a result, the detection of circulating tumor cells (CTCs) requires specialized technologies for the enrichment and detection of single tumor cells [1]. Since most CTC assays rely on epithelial biomarkers, they have a tendency to miss CTCs undergoing an epithelial-to-mesenchymal transition (EMT). Newly discovered biomarkers such as the actin-bundling protein plastin-3 [2] could address this concern because they are not downregulated during EMT and they are not expressed in normal blood cells. As demonstrated by several research groups, CTC isolation, enumeration, and analysis offer reliable information for cancer prognosis, potentially serving as a method of liquid biopsy [3]. Functional characterization using specialized in vitro and in vivo test systems has started [4–6]. Furthermore, monitoring the change in the CTC number before, during, and after anticancer therapy (e.g., chemotherapy or hormonal therapy in breast cancer [7]) and determination of therapeutic targets (e.g., HER2 [8] or PD-L1 [9]) might serve as a surrogate marker for response to therapy. These individualized therapy responses can be employed as a tool of companion diagnostics to help the stratification of anticancer therapies and to gain insights into therapy-induced selection of tumor cells [10–12].

This book is the first one to focus on the current platforms of CTC isolation. CTCs can be enriched positively or negatively based on biological properties (e.g., expression of protein markers). For example, the positive enrichment involves using antiepithelial antibody (Ab), antimesenchymal Ab, or a combination of antiepithelial and antimesenchymal Ab, whereas the negative depletion employs anti-CD45 Ab to remove the unwanted leukocytes (see Chapter 14 for the details). Moreover, physical properties (e.g., size, density, deformability, electric charges) can be applied to capture CTCs. For example, CTCs can be isolated by using a membrane filter system based on the CTC size, microposts in a microchip based on the CTC size plus deformability, centrifugation on a Ficoll density gradient based on the CTC density, dielectrophoresis (DEP) based on CTC dipole moment, or spiral CTC chips based on the CTC size. Some of them may combine with protein-expression-based CTC strategies (e.g., CTCs are first selected based on their presumably larger size, while the smaller leukocytes are removed; then, CTCs are conjugated with beads-tagged antiepithelial Abs and captured in a magnetic field [13]). This book gives an excellent overview of current technologies and discusses the potential and challenges for their future development as diagnostic tools in oncology.

Better understanding of CTC biology will help CTC assay development. As a result, tumor-specific CTC assays could be advantageous than a CTC technology developed for all cancer types [1]. Most importantly, clinical validity must be achieved for newly developed CTC assays through clinical trials, and clinical utility must be identified in the appropriate context of use [14]. This may take considerable efforts, time, and budget, which might be one important reason why after 10 years of intense development work in the CTC field only one assay (CellSearch®, see Chapter 19 for the details) has received FDA clearance. One of the future goals of CTC characterization is to identify metastasis-initiation cells and drug-resistant clones. The genetic analysis of mutations relevant to cancer therapies in these subpopulations by concurrent monitoring of CTCs might lead to the development of new companion diagnostics in cancer therapy. The validation of liquid biopsy assays is an important task of the new European consortium CANCER-ID that comprises more than 30 institutions from academia and industry (www.cancer-id.eu). This important book will stimulate further developments in new technologies for the detection and characterization of CTCs as liquid biopsy.

Klaus Pantel, M.D., Ph.D.
University Medical Centre Hamburg-Eppendorf, Germany

REFERENCES

1. Alix-Panabieres, C. and Pantel, K. (2014) Challenges in circulating tumour cell research. *Nature Reviews Cancer*, **14**, 623–631.
2. Yokobori, T., Iinuma, H., Shimamura, T. *et al.* (2013) Plastin3 is a novel marker for circulating tumor cells undergoing the epithelial-mesenchymal transition and is associated with colorectal cancer prognosis. *Cancer Research*, **73**, 2059–2069.
3. Pantel, K. and Alix-Panabieres, C. (2013) Real-time liquid biopsy in cancer patients: fact or fiction? *Cancer Research*, **73**, 6384–6388.
4. Baccelli, I., Schneeweiss, A., Riethdorf, S. *et al.* (2013) Identification of a population of blood circulating tumor cells from breast cancer patients that initiates metastasis in a xenograft assay. *Nature Biotechnology*, **31**, 539–544.
5. Hodgkinson, C.L., Morrow, C.J., Li, Y. *et al.* (2014) Tumorigenicity and genetic profiling of circulating tumor cells in small-cell lung cancer. *Nature Medicine*, **20**, 897–903.

6. Cayrefourcq, L., Mazard, T., Joosse, S. *et al.* (2015) Establishment and characterization of a cell line from human circulating colon cancer cells. *Cancer Research*, **75**, 892–901.
7. Bidard, F.C., Peeters, D.J., Fehm, T. *et al.* (2014) Clinical validity of circulating tumour cells in patients with metastatic breast cancer: a pooled analysis of individual patient data. *Lancet Oncology*, **15**, 406–414.
8. Riethdorf, S., Muller, V., Zhang, L. *et al.* (2010) Detection and HER2 expression of circulating tumor cells: prospective monitoring in breast cancer patients treated in the neoadjuvant GeparQuattro trial. *Clinical Cancer Research*, **16**, 2634–2645.
9. Mazel, M., Jacot, W., Pantel, K. *et al.* (2015) Frequent expression of PD-L1 on circulating breast cancer cells. *Molecular Oncology*, **9**, 1773–1782.
10. Wan, L., Pantel, K., and Kang, Y. (2013) Tumor metastasis: moving new biological insights into the clinic. *Nature Medicine*, **19**, 1450–1464.
11. Antonarakis, E.S., Lu, C., Wang, H. *et al.* (2014) AR-V7 and resistance to enzalutamide and abiraterone in prostate cancer. *New England Journal of Medicine*, **371**, 1028–1038.
12. Miyamoto, D.T., Zheng, Y., Wittner, B.S. *et al.* (2015) RNA-Seq of single prostate CTCs implicates noncanonical Wnt signaling in antiandrogen resistance. *Science*, **349**, 1351–6.
13. Ozkumur, E., Shah, A.M., Ciciliano, J.C. *et al.* (2013) Inertial focusing for tumor antigen-dependent and -independent sorting of rare circulating tumor cells. *Science Translational Medicine*, **5**, 179ra47.
14. Danila, D.C., Fleisher, M., and Scher, H.I. (2011) Circulating tumor cells as biomarkers in prostate cancer. *Clinical Cancer Research*, **17**, 3903–3912.

PREFACE

Cancer is a crucial public health problem. According to the World Health Organization (WHO), 8.2 million people die each year from cancer, equivalent to >22,000 per day. Since more than 90% of cancer deaths result from metastasis, fully understanding how cancer spreads from the primary tumor to the secondary sites, and subsequently identifying ways to prevent metastasis, would have significant societal impacts.

It is generally agreed that circulating tumor cells (CTCs) in the peripheral blood play a key role in cancer metastasis because cancer cells must transport through the circulatory system before colonizing the secondary sites. CTC enumeration is less invasive than biopsy while providing quantifiable information. It has been hailed as one of the potential "liquid biopsy" methods. It can also be used to monitor the response of an individual patient to a therapy and then tailor the treatment. Hence, CTCs have been advocated as potential biomarkers for cancer diagnosis, prognosis, and theragnosis or precision medicine. As a result, CTC isolation and analysis are an important topic in research, medical, and clinical communities.

The number of CTCs in the peripheral blood is extremely low. At large, there are a few CTCs in 1 mL of blood, which contains billions of red blood cells, millions of white blood cells, and hundreds of millions of platelets. Therefore, CTC isolation is truly a needle-in-a-haystack challenge.

The challenge makes CTC research exciting to many scientists and engineers who are developing various methods for efficient isolation and accurate enumeration of CTCs. Biomedical researchers have been studying CTCs for understanding the metastasis mechanisms, identifying cancer stem cells, and investigating their genomic and proteomic profiles for helping develop therapeutic drugs. The significance of CTCs and potential market values make companies and investors interested in the field. Many researchers, as well as students, are jumping into the field to make contributions.

Because CTC isolation and analysis are important topics that have attracted much interest recently from academics, government agencies, and industry, this field has been the subject of many international symposia, calls-for-proposals from funding agencies, and articles in high-impact journals. The strong interest of CTC-related research is partially indicated by the exponential increase in publications in the past decade. The figure below plots the number of publications in PubMed as a function of the year from 1995 to 2015, using "circulating tumor cells" in the title/abstract as the search term (the number for 2015 – indicated by a star at the top – is up to September of the year).

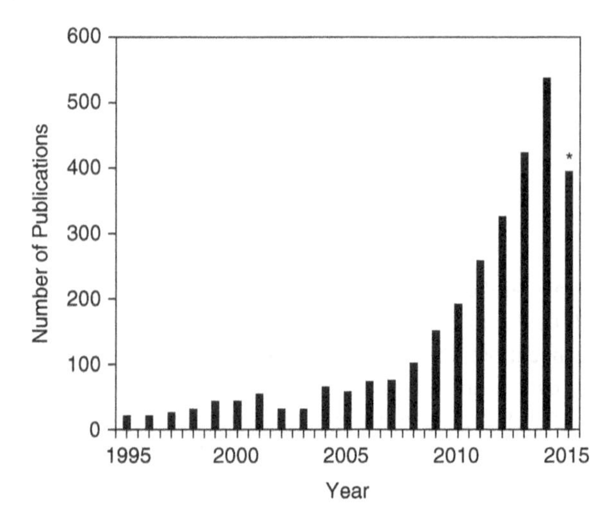

This book is aimed to those who are in the field, those entering the field, and those who just want to learn about the field. Readers who are new to this field will benefit from the introduction to CTCs and the historical perspectives offered in early chapters. Subsequent chapters explore a variety of state-of-the-art isolation methods, capture reagents used for CTC isolation and enumeration, the analysis of isolated CTCs, clinical translation, and commercial platforms. All of these chapters are written by experts who have developed remarkable techniques and made considerable contributions to the field.

The content in this book is organized into five parts.

- Part I is the introduction, which presents CTCs from historic perspectives. Chapter 1 is written by Professor Jonathan Uhr, whose pioneering research with his colleagues on CTCs in the 1990s led to the birth of CellSearch®. In Chapter 2, microfluidics is introduced, as it is a technology used for the work described in several chapters.
- Part II consists of 12 chapters, each of which describes different CTC isolation methods. The methods range from the macro- to the microscale, from positive isolation to negative depletion, and from biological-property-enabled to

physical-property-based approaches. These chapters are written by the inventors who developed the methods.

- Part III is comprised of three chapters, covering post-isolation analysis and clinical translation. Topics include tumor heterogeneity, single-cell analysis, regulatory policy, and clinical practice, all of which are important in realizing the full potential of CTCs and having impacts on clinical outcomes. The authors of these chapters are from the school of medicine, and they are pioneers in translating technologies from the bench to the bedside.
- Part IV focuses on commercialization. One commercialized platform is CellSearch, the only CTC platform currently approved by the US Food and Drug Administration (FDA). The majority of clinical studies on CTCs have been performed using this apparatus, and it has played a significant role in shaping the field. The other platform is DEPArray™, which is an instrument that can identify, quantify, and recover individual CTCs, and it has been placed in many laboratories in the world.
- Part V is glossary, consisting of the definition of scientific terms related to the field and those used in this book.

This is the first book focusing on CTC isolation and analysis. It grows out of and is a continuation of a CTC-themed issue of *Lab on a Chip* in 2014 (Vol. 14, Issue 1) that I edited with Professor David Beebe of University of Wisconsin and two invited symposia I organized at The Pittsburgh Conference in 2013 (Philadelphia, USA) and 2015 (New Orleans, USA). Some of the authors in this book, as well as other CTC investigators, contributed to the journal issues and presented their advancement at the conferences.

There are many people who deserve special thanks and recognition for their support of this book. Dr. Mark Vitha, the editor of the Chemical Analysis Series, initiated the conversation about this book after noticing my CTC session at Pittcon 2013. I am grateful for his help in the format guideline and content review and, more importantly, for nudging me to complete this book on time. I would like to thank all authors for their time and expert contribution. My appreciation also goes to Dr. Jinling Zhang for helping in the editorial work, to Ms. Lauren Miller for creating the cover art, and to all reviewers for their comments and suggestions. I am grateful to Dr. Klaus Pantel, who is an authority on CTCs with one book and dozens of review articles, for writing the forward. I am also thankful to Bob Esposito, Michael Leventhal, and their team at John Wiley & Sons, Inc. for their efforts in transforming the initial idea and the manuscripts into a viable book. Lastly, but not least, I appreciate my wife for her love and support by allowing me to spend time on this book.

Thanks for reading.

Z. Hugh Fan, Ph.D.
Gainesville, FL
September 18, 2015

PART I

INTRODUCTION

1

CIRCULATING TUMOR CELLS AND HISTORIC PERSPECTIVES

JONATHAN W. UHR

Department of Immunology, University of Texas Southwestern Medical Center, Dallas, TX, USA

This introduction reviews the history of research on circulating tumor cells (CTCs) with brief reviews of the present opportunities and challenges using CTCs for diagnostic and treatment decisions.

1.1 EARLY STUDIES ON CANCER DORMANCY LED TO THE DEVELOPMENT OF A SENSITIVE ASSAY FOR CTCs (1970–1998)

Prior to my involvement in the development of the first Food and Drug Administration (FDA)-approved capture device for CTCs, our laboratory's research was involved with investigations of the signaling pathways in B lymphocytes. We were studying the mechanisms underlying induction of replication, differentiation, apoptosis, and cell cycle arrest when an important event took place at Stanford University that affected our future plans. In 1980, Slavin and Strober [1] isolated the first murine B-cell lymphoma that spontaneously arose in a BALB/c mouse and allowed us to propagate the tumor. We found that an antibody to the tumor immunoglobulin [anti-idiotype (Id)] injected into the tumor-bearing mice could induce a state of cancer dormancy [2] by its ability to induce cell cycle arrest and apoptosis (antibodies are immunoglobulins; each antibody has a unique antigenicity, so it can stimulate production of an antibody to it, that is, an antibody to an antibody called an anti-idiotypic antibody). Dormancy could last up to 2 years. The population of dormant lymphoma cells in the spleen was

Circulating Tumor Cells: Isolation and Analysis, First Edition. Edited by Z. Hugh Fan.

stable for the 210 days of observation, although a subpopulation of tumor cells were replicating; loss of dormancy occurred at a steady rate during the 2 years of observation; and, about 90% of the lymphoma cells that were now replicating remained Id+ [3–6]. However, the majority of such antibodies had undergone minor changes that made them Id+ variants with decreased or no susceptibility to anti-Id-mediated induction of dormancy. With time, these lymphoma cells regained their full malignant potential; as few as three of these cells transferred progressive tumor growth to syngeneic recipients [6]. Thus, anti-Id suppresses the malignant phenotype, observed in control mice that do not receive anti-Id, by signal transduction mechanisms that override the genetic lesions that cause neoplasia. In searching for potential signaling molecules, alterations in Syk, Lyn, and HS1 were suggested by either the loss of an epitope recognized by a monoclonal antibody or the loss of functional kinase activity [5]. Significant advances in characterizing cancer dormancy followed from studies by Meltzer [7], Stewart [8], and Demichelli [9].

At the same time, the Stevensons reported impressive results of anti-Id therapy of murine B-cell tumors [10, 11], and Levy obtained groundbreaking results in treating patients with B-cell lymphomas with anti-Id serum [12, 13]. Prior conventional treatment could induce long-term remission, but dormant lymphoma cells were not eliminated and virtually all patients eventually died of the disease. Levy treated such patients with anti-Id and the vast majority went into long-term remissions even though tumor cells remained [14]. As with the B-cell tumor lymphoma 1 (BCL1) mouse tumor model, the clinical studies also indicated that the antitumor effect of anti-Id and several other B-cell reactive antibodies related to their ability to act as agonists rather than conventional effector antibodies. Levy's results initiated a burst of research into immunological methods for cancer treatment, which continue to the present day. There are currently over 30 immunologically based drugs in clinical trials to treat patients with various cancers.

These observations in human B-cell lymphoma and in the BCL1 mouse tumor model – that long-lasting tumor dormancy can be induced by antibodies to the tumor immunoglobulin in the face of persisting tumor cells, a portion of which are replicating in the BCL1 tumor – gave us the impetus to study clinical dormancy. We were interested in analyzing those tumor types in which metastases can develop many years after primary tumor removal even though the patient has appeared clinically disease-free. Breast cancer was an excellent example because of the prevalence of the disease and because recurrences occur at a steady rate from 7 to 20 years after mastectomy in patients who appear well [9, 15]. To study the tumor cell dynamics in clinically healthy humans, we needed a relatively noninvasive procedure, namely, a very sensitive blood test to determine whether CTCs were present in a proportion of these patients. However, the methods under development were insensitive, could not be quantified, frequently gave false positives, and were impractical for the challenge.

The history of CTCs began in 1869, when Ashworth [16] described cells in the blood that appeared similar to those observed in the tumor at autopsy. In the mid-twentieth century, there were many claims that CTCs as determined by cytology were commonly seen in cancer patients. However, further studies indicated that hematopoietic cells, particularly megakaryocytes, were responsible for almost all of these results [17] and such studies were then abandoned. However, there were positive exceptions. Drye *et al.* [18] had convincing evidence that the presence of cancer cells in peripheral blood of 17 patients related to their clinical progress. Engell [19] claimed to have found cancer cells in venous blood draining the tumor as well as in peripheral blood.

Beginning in the 1970s, experimental models were developed to study the events that led to metastases [20, 21]. Metastases had been generally regarded as a late event in the development of epithelial tumors. However, the poor prognosis of patients with clinically localized lung cancer suggested that micrometastases may have taken place before the primary tumor was diagnosed. Pantel *et al.* [22, 23] searched for tumor cells in the bone marrow by immunohistochemistry in patients with breast, gastrointestinal tract, or non-small-cell lung carcinomas with and without evidence of metastases. In the majority of lung cancer patients without metastases, cytokeratin-positive cells were detected at significant concentrations, whereas they were rarely found in controls. The authors concluded that early dissemination of isolated tumor cells is a frequent occurrence in non-small-cell lung carcinomas; in breast and gastrointestinal carcinomas, the majority of these disseminated cells in the bone marrow were in a dormant state. In a series of pioneering experiments, Folkman *et al.* [24, 25] defined a critical role for angiogenesis in the metastatic cascade. Liotta *et al.* [26] studied mechanisms of metastases and focused on the role of angiogenesis. The primary tumor must first develop an adequate vascular supply. This is achieved by balancing angiogenesis-promoting and -inhibiting factors released by tumor cells, inflammatory cells, and extracellular matrices. Although there were problems with the methods for quantification of angiogenesis, the results indicated that increased angiogenesis within the primary tumor resulted in a worse prognosis.

Using a fibrosarcoma model, Liotta also quantified some of the major processes that occur following transplantation of tumor into the leg muscle of a mouse and the subsequent rapid development of pulmonary metastases [27]. He showed that by about Day 4, a vascular network first appeared in the peripheral regions of the tumor and grew throughout the tumor mass by Day 10. Following intratumoral perfusion (using a solution of human hemoglobin, calf serum, Eagle's medium, amino acids, glucose, and insulin), fibrosarcoma tumor cells were detectable in the draining venous blood vessels by Day 5. The number of these cells in the blood stream, both as single cells and tumor clumps (2–30 tumor cells, comprising about 10% of tumor cells), increased rapidly until Day 10–12, after which it diminished. There was a linear relationship between the density of perfused vessels and the concentration of detectable tumor cells in the effluent. Metastases, which were first detectable on Day 10, increased with time and were directly related to the concentration of tumor cell clumps (four or more cells) in the effluent. The aforementioned studies suggested that tumor vascularization and the resultant entry of tumor cell clumps into the circulation were critical events in the initiation of metastases.

These results propagated new types of experiments to look for CTCs [28]. The first procedure was to enrich the small number of tumor cells in the blood. The different techniques that were tried included lysis of RBCs or enrichment of non-RBCs, including positive [29–33] and negative [34] selection by an immunomagnetic procedure.

Antibodies to either an epithelial or a hematopoietic marker were attached to metal beads, mixed with blood, and then removed from the mixture by placement in a magnetic field. Molecular analysis following enrichments was then performed using either immunocytochemical staining or polymerization chain reaction (PCR) or its variants. Immunocytochemical assays, which relied on time-consuming searches for rare cells, were unable to measure tumor burden or to phenotype tumor cells in depth. PCR assays for detecting CTCs involved amplification of specific target DNA sequences and were very sensitive, for example, one tumor cell in 10^6 normal lymphocytes could be detected. Another CTC detection strategy was to amplify tissue-specific mRNA by reverse transcriptase PCR (RT-PCR) for markers characteristic of that tumor's tissue of origin (e.g., to detect circulating prostate or melanoma cells). However, knowledge of that tissue's gene sequence and intron–exon junctions was required for selection of appropriate oligonucleotide primers for the RT-PCR. Fusion genes – due to translocations, interstitial deletions, or chromosomal inversions – may also be detected using RT-PCR, although diagnostic tumor tissue is generally needed to confirm the presence of tumor cells containing the rearranged sequence. However, PCR assays were difficult to quantify and may show false-positive results (due to PCR's high sensitivity, presence of pseudogenes, or introduction of normal tissue-specific cells during biopsy or surgery) and false-negative results (due to low level signals, CTC heterogeneity, or sampling issues). The interpretation and limitations of the assay and its variations are elegantly described by Ghossein *et al.* [28]. Nevertheless, these approaches demonstrated that patients with cancer frequently had evidence of CTCs, even though it was not possible to quantify their number.

1.2 MODERN ERA FOR COUNTING CTCs: 1998–2007

This began with our publication of a very sensitive method to detect, quantify, and immunophenotype CTCs in patients with breast and prostate cancer [35]. Circulating epithelial cells (CECs) were collected by a positive immunomagnetic approach that involved coating metal particles with an antibody to an epithelial cell surface antigen, epithelial cell adhesion molecule (EpCAM), a transmembrane glycoprotein expressed exclusively in epithelia and epithelial-derived neoplasms. (Note that EpCAM is distinct from cytokeratin proteins, which are part of the cytoskeleton and are exclusively intracytoplasmic.) The conjugated particles were mixed with the blood sample and tumor cells were captured in a magnetic field, while the remaining nonadherent blood cells were washed away. The captured epithelial cells were then fluorescently stained with an anti-cytokeratin antibody (CAM 5.2), a nuclear stain, and an anti-CD45 antibody that detects hematopoietic cells. One aliquot of the captured cells was analyzed by flow cytometry and a second by microscopy of the stained slides (in this case, stained with antibodies against cytokeratins 5, 6, 8, 18, Muc-1 glycoprotein, or prostate-specific membrane antigen. The development of this assay was a collaboration between my laboratory and Immunicon Corp. The latter was headed by Liberti [36, 37], an immunologist and an expert in development of magnetic particles; the senior scientist was Terstappen [38], an excellent immunologist and skilled in flow cytometry. The utility of our method was due to multiple factors: a very effective ferrofluid that maintained its colloidal properties without reacting with blood components and yet could still be magnetically separated; an avid and specific anti-EpCAM antibody; absence of prior enrichment steps, thus eliminating cell loss and preserving cytomorphology; exclusion of circulating cells that stained with a hematopoietic marker; initial analysis using flow cytometry, and, when positive, purification of another simultaneously obtained aliquot of blood for direct staining on slides. Hence, the assay was highly specific and very sensitive. Applying threshold values as a cut-off, excess CECs were detected in 12 of 14 patients with localized breast cancer and in all 3 patients with early prostate cancer, with no false positives in healthy controls.

This prototype is probably the most sensitive detection instrument to the present day. The evidence for calling these CECs, CTCs, was strong. The cells were epithelial, only detected in patients with cancer but not in normal age-matched controls, and there was an excellent correlation in individual patients between their clinical status before and after chemotherapy and the number of CTCs detected.

1.3 PROOF OF MALIGNANCY OF CTCs

The critical nature of this issue made it important to use genetic techniques to support the conclusion that the cells in question were malignant. Therefore, fluorescent *in situ* hybridization (FISH) was used to determine the patterns for chromosomes 1, 8, and 17, which were indicative of malignancy in touch preparations from 74 primary breast cancer patients [39]. Use of the three probes detected aneusomy in 92% of the samples. The genetic abnormalities in the CTCs were compared with those in the primary tumor. CECs from 15 patients with organ-confined breast, kidney, prostate, or colon cancer were analyzed by dual or tricolor FISH using the numerator DNA probes for chromosomes 1, 3, 4, 7, 8, 11, and 17 [40]. In 10 of 13 patients in which touch preparations of the primary tumor tissue were available, the patterns of aneusomy in their CECs matched one or more clones in the primary tumor. We concluded that the CECs were CTCs and that the three negative results were readily explainable by the small number of CECs in several patients (one to two CECs) and the known technical difficulties in counting chromosomes.

1.4 NEW EXPERIMENTS INVOLVING CTCs

There proceeded to be many new experiments extending the immunomagnetic approach for quantifying tumor burden [41–45]. Terstappen [46] led experiments to determine the accuracy and linearity of the CellSearch system in enumerating CTCs in healthy subjects, patients with nonmalignant diseases, and patients with a variety of metastatic carcinomas. The results culminated in the development by Immunicon of the CellSearch instrument (see Chapter 19 for details). It was based on the patent of the 1998 prototype owned by Immunicon and the University of

Texas Southwestern Medical Center. It remains the only FDA-approved instrument for counting CTCs. Johnson and Johnson, Inc. obtained the patent and supported additional clinical trials to refine the ability of CellSearch to help with care of patients with recurrent cancer. The results of clinical trials led by Cristofanilli *et al.* [47] and Hayes *et al.* [48] indicated that the number of CTCs/7.5 ml blood could differentiate between two groups of patients with metastatic breast cancer (MBC) and strikingly different outcomes. Five or more CTCs indicated a very aggressive outcome, whereas less CTCs indicated a better response to treatment with a longer progression-free and overall survival (OS). The prognostic value of CTC enumeration was independent of disease subtype and line of therapy and was superior to or augmented standard anatomical and functional imaging assessment [49]. Similar results have been observed in colorectal, prostate, and lung cancer [50–54]. Also, the longitudinal monitoring of CTCs indicated patients that would fail to benefit from systemic therapies whose disease is associated with extremely rapid progression [47, 48]. These data were confirmed in a recently reported prospective randomized clinical trial conducted by SWOG (formerly Southwest Oncology Group) [55], showing that failure to clear CTCs to below 5/7.5 ml whole blood in patients with MBC after receiving one cycle of first-line chemotherapy was associated with a very poor prognosis. The median OS for this group of patients was 13 months, compared to 23 months for those who presented with elevated CTCs but "cleared" them after one cycle of chemotherapy, and 35 months for those patients who did not have elevated CTC at baseline. However, in this randomized trial, switching from whatever the clinician chose as first-line chemotherapy to an alternative chemotherapeutic regimen did not improve the dismal prognosis for those whose CTCs failed to decline to less than 5/7.5 ml whole blood, most likely because the second-line alternate therapy was no more effective than the first-line therapy. This dampens enthusiasm for only performing enumeration of CTCs, which has previously been the gold standard and major guiding force for prognostication and help with treatment decisions. Additional evidence to be discussed indicates the need for extensive immunophenotypic and genotypic analyses of CTCs and for development of a more sensitive instrument.

1.5 CLINICAL CANCER DORMANCY

The sensitivity of our assay for CTCs allowed us to look for them in patients with breast cancer dormancy. In breast cancer, 15–20% of clinically disease-free patients relapse 7–25 years after mastectomy and, from 10–20 years, the rate of relapse is relatively steady at about 1.5% [9, 15]. Clinical cancer dormancy is also frequently observed in thyroid, renal, and prostate carcinomas and in B-cell lymphoma and melanoma. The current concept is that the persisting cells are nondividing cells. However, in our study involving 36 patients whose mastectomy for cancer occurred 7–22 years previously and who were clinically disease-free, 13 had CTCs [56]. These patients continued to have sustained low levels of CTCs during the next year. Since CTCs develop an apoptotic program after entering the circulation [57–59] and their half-life is 1–2.4 h in patients whose primary breast cancer was just removed [56], there must coexist a replicating population of tumor cells in these patients with a population that dies at precisely the same rate for many years, perhaps decades. This is one way for keeping a population of persisting cancer cells "dormant." It does not exclude that there is another tumor cell population in these patients that is not replicating. Also, it appears that some patients with early-stage breast cancer and disseminated tumor cells (DTCs) in their bone marrow may not relapse for long periods of time [60–62]. Their tumor cells may represent a dormant population or the same replicating population described earlier or both.

Several possible mechanisms underlying the dormant state have been extensively investigated. These include an antitumor immune response, as first proposed by Thomas [63] and Burnett [64], and lack of vascularization, as proposed by Folkman and colleagues [24, 65]. Aguirre-Ghiso [66] has accumulated considerable evidence for a role for stress signaling in inducing dormancy in models of human cancer.

We hypothesize that the size of the persisting cancer cell population may be stabilized by some of the same mechanisms that control the size of organs and subsets of cells [67]. Our attraction to this hypothesis is that the organ-control mechanisms are precise as is the balance between replication and cell death of the persisting tumor cells in breast cancer dormancy. This does not exclude contributions from other mechanisms less likely to be so precise. Evolution has given mammals and other species the regulatory equipment to keep organs at a precise size and to restore that size if the organ is altered in size [68, 69]. Thus, each cell type in the blood is kept at a relatively constant level barring disease. The same is true for solid organs. For example, if half the liver is removed, liver cells that are quiescent suddenly begin to divide and stop dividing only when the liver has reached normal size [70]. The stem cells that give rise to the differentiated cells in each organ control the size of that organ. Organisms from Drosophila to mammals have evolved complex mechanisms to coordinate cell proliferation with cell death to prevent inappropriate proliferation of somatic cells. Asymmetric replication in which one cell differentiates and dies and the other is a renewal cell is one strategy.

Many molecules and signaling pathways have been implicated in organ size control in Drosophila, and there is considerable conservation of the signaling molecules between Drosophila and humans [71]. These include, but are not limited to, IGF-1 and IGF-2 [72], IRS1-4 [73], TSC1, and TSC2 [74].

Another important example is the Hippo signaling pathway. It plays a major role in controlling eye and wing growth in Drosophila and size control of some organs in mice [75]. Again, there is significant conservation in the components of this pathway between humans and Drosophila and there is mounting evidence that this pathway can act as a tumor suppressor as well as controlling organ size in humans [75, 76]. Very recent studies have revealed some of its downstream transcriptional targets and their potential role as a suppressor of cancer. In breast cancer cells, the mevalonate metabolic pathway, or its inhibitor simvastatin, exerts regulation upon the Hippo signaling pathway and appears to be involved in tumor suppression [77]. Statins, which block the mevalonate pathway, are reported to have multiple anticancer effects, for example, antiproliferation and anti-invasive properties against mammary carcinoma [78, 79] The mevalonate pathway also regulates the transcription of *RHAMM* (receptor for hyaluronan-mediated motility or CD168), a breast cancer susceptibility gene [80, 81]. Its hyperexpression is associated with tumor development and progression to metastases [82], and its expression is upregulated in a variety of human cancers, including breast [83]. These results emphasize the possibility that size-controlling pathways may affect the persisting tumor cell population in clinical breast cancer dormancy.

It is possible that all "survivors" of breast cancer have a small number of tumor cells somewhere in their body, the growth of which is controlled usually for the lifetime of the host. If the mechanisms underlying clinical tumor dormancy and relapse were understood, it is possible that appropriate targeting drugs could be developed, which could eliminate or control these persistent tumor cells and prevent their transformation into growing metastases. Thus, cancers with late recurrences could be treated after removal of the primary tumor as chronic diseases to be controlled by relatively nontoxic therapy.

1.6 HUMAN EPIDERMAL GROWTH FACTOR RECEPTOR 2 (HER2) GENE AMPLIFICATION CAN BE ACQUIRED AS BREAST CANCER PROGRESSES

Use of trastuzumab (Herceptin), a humanized monoclonal antibody that targets HER2 on breast cancer cells, to treat patients with HER2-overexpressing breast cancers provided impressive clinical results in both the metastatic and adjuvant settings [84–86]. However, the accepted dogma that HER-negative tumors could never become HER2 amplified with recurrence was puzzling. Patients whose primary tumors were HER2-negative and who developed breast cancer recurrence were ineligible for treatment with Herceptin under the assumption that they would not have tumor cells that would be blocked by this antibody. However, the plausibility of this assumption is under question, considering the ability of HER2 amplification to confer aggressiveness on the tumor and the known evolution of neoplastic cells to constantly mutate to more aggressive phenotypes. Studies of HER2 status of primary and discontinuous recurrences of breast cancer tumors indicated that there could be discordance among them, with HER2-positive tumors developing HER2-negative metastases and HER2-negative tumors developing HER2-positive metastases [87, 88]. Wülfing *et al.* [89] demonstrated that HER2-positive CTCs were associated with a poor prognosis and identified HER2-positive CTCs in patients with primary HER2-negative breast cancer, thus further challenging this concept.

We, therefore, embarked on a study to determine the relationship between the HER2 status of CTCs and the primary tumor tissue and to determine the HER2 status of CTCs in recurrent breast cancer [90]. CTCs were captured and evaluated for HER2 gene status by fluorescence *in situ* hybridization. In 31 patients with primary breast cancer who had not received Herceptin, there was 97% concordance in the HER2 status of the primary tumor and corresponding CTCs, with no false positives. Amplification of HER2 in CTCs was shown to be a reliable surrogate marker for amplification in the primary tumor, and HER2 protein expression in CTCs was sufficient to make a definitive diagnosis of HER2 gene status. However, of 24 patients with recurrent breast cancer whose primary tumor was HER2-negative, nine (38%) acquired HER2 gene amplification in their CTCs during cancer progression. Four of these were treated with therapy that included Herceptin. One patient who was terminal from liver and kidney failure had a complete response and lived for another 2 ½ years, and two had partial responses.

Subsequently, Fehm *et al.* [91] showed that between 32% and 49% of patients with HER2-negative breast cancers (78 patients) had HER2-positive CTCs after recurrence depending upon the CTC test used. In a study by Riethdorf *et al.* [92], examining blood samples from patients with large operable or locally advanced nonmetastatic breast cancer eligible to undergo neoadjuvant chemotherapy, among 26 patients with primary tumors classified as HER2-negative, approximately 20% had CTCs with strong HER2 expression. However, some of the patients had only one CTC. Others also confirmed discordance between HER2-negative primary tumors and HER2-positive CTCs and DTCs in breast and gastrointestinal cancers [93–95]. Paik *et al.* [96] followed 1787 patients in a study comparing standard adjuvant chemotherapy with and without trastuzumab and concluded that some patients, who were later classified as having HER2-negative primary tumors by immunohistochemistry and FISH testing, appeared to benefit from trastuzumab treatment. It is important to develop more sensitive CTC capture techniques to increase the number of CTCs that can be examined in nonmetastatic patients and to understand the biology of cells surviving trastuzumab treatment that are likely responsible for metastatic spread. Further, amplification may not be the only mechanism of HER2 overexpression in micrometastases. For example, Ithimakin *et al.* [97] have shown that MCF-7 breast cancer cells, which normally only express low levels of HER2, upregulate HER2 expression without amplification when placed into estrogen-free murine bone marrow. Moreover, bone metastases in 12 of 14 (87%) patients with hormone-receptor-positive HER2-negative breast cancer showed significantly higher HER2 protein expression than present in matched primary tumors. These data suggest that HER2 regulation may be plastic in some types of breast cancer cells (specifically, the luminal subtype) and highlight the importance of using the proper assay to measure protein levels. A current clinical trial, National Surgical Adjuvant Breast and Bowel Project B-47, is evaluating whether the addition of trastuzumab to chemotherapy improves invasive disease-free survival in patients with low HER2 expression.

1.7 uPAR AND HER2 CO-AMPLIFICATION

Activation of the urokinase plasminogen activator (UPA) system is associated with a poor prognosis in breast cancer, with the greatest amount of evidence for UPA overexpression. However, it is well documented that UPA receptor (uPAR) overexpression is also associated with increased tumor aggressiveness and worse disease-free survival and OS in breast and other cancers [95–97]. The uPAR gene on chromosome 19, encodes a 35 kDa protease, with a glycosyl phosphatidyl anchor linking it to the outer cell membrane. Interaction of uPAR and UPA causes cleavage of plasminogen and activation of plasmin (a serine protease), which degrades several extracellular matrix (ECM) components and also activates matrix metalloproteinases (MMPs). The result is proteolytic degradation of ECM, thereby allowing tissue penetration, thereby facilitating metastasis. Additional signaling pathways are activated resulting in cellular replication, motility, and remodeling of the ECM. Therefore, overexpression of uPAR in breast and other cancers is associated with a poor prognosis [98–100].

Since overexpression of HER2 also predicts a poor prognosis, we thought it is important to analyze both HER2 and uPAR protein and gene expression in CTCs and touch preps of primary breast carcinomas and CTCs of patients with advanced breast carcinomas. Our results showed amplification of the uPAR gene in 20–25% of breast cancer patients and a marked bias for amplification of both oncogenes in the same tumor cell [101]. The amplification was responsible for overexpression of both genes. Coamplification suggests cross talk and cooperation between the HER2 and uPAR signaling pathways. This represents an intriguing finding for further studies that could lead to additional antitumor therapy.

Recent studies of Meng and coworkers [102] showed that tumor cell growth could be suppressed and apoptosis induced by using RNA interference (RNAi) to deplete either HER2 or uPAR or both in cell lines. With depletion of both genes, MAPK signaling pathways were suppressed and there was decreased ERK activity and a high P38/ERK activity ratio. Growth suppression and induction of apoptosis were further augmented when uPAR downregulation was used in conjunction with trastuzumab. It was concluded that targeting HER2 and uPAR has a synergistic inhibitory effect on breast cancer cells.

LeBeau *et al.* [103] investigated a possible role for recombinant human anti-uPAR antibodies in treating breast cancer. *In vitro*, these antibodies exhibited strong binding to the surface of cancer cells expressing uPAR. Using the antibodies for *in vivo* imaging, they detected uPAR expression in triple-negative breast cancer (TNBC) tumor xenografts and micrometastases. When the antibodies were used for monotherapy and radioimmunoassay studies, there was significant decrease in tumor growth and tumor burden in the TNBC xenograft model. These results suggest use of such uPAR targeting for therapy of highly aggressive breast cancers. Moreover, the simultaneous targeting of both uPAR and HER2 might also be considered.

1.8 EPITHELIAL–MESENCHYMAL TRANSITION (EMT)

EMT is a conversion of an epithelial to mesenchymal phenotype that occurs during embryonic development in all vertebrates. It was first recognized as a distinct process in 1982 by Greenburg and Hay [104] and currently has been well reviewed by Bednarz-Knoll *et al.* [105] and others [106, 107]. Epithelial markers include cytokeratins 8, 18, and 19, E-cadherins, Mucin-1, occludins, and others; mesenchymal markers include vimentin, N-cadherin, F-actin, nuclear beta-catenin, and others. The mesenchymal morphology causes increased motility and thereby facilitates migration into the circulation. Conversely, circulating mesenchymal cells can participate in the formation of epithelial organs through mesenchymal–epithelial transition (MET). This process is not well understood, for example, whether it may start in the blood or only after extravasation into the site of the future organ.

EMT may be stimulated by growth factors such as EGF or TGF-beta or transcription factors such as Twist, Snail, or Slug. Thiery [108] implicated EMT in tumor metastasis based on studies with human cancer cell lines and mouse models. This idea was extended in the stem cell hypothesis that states that the EMT-transformed stem cells are the tumor cells responsible for the development and propagation of cancer [109]. This has profound implications for the diagnosis and treatment of cancer. CTCs with stem cell markers such as CD44+, CD24-/low and ALDH1+ are frequently seen in epithelial cancers [110]. We will not discuss this important hypothesis, but there are extensive reviews evaluating the evidence [111–115].

EMT is usually demonstrated by downregulation or relocalization of E-cadherin and appearance of mesenchymal markers such as vimentin [105–107]. CTCs are detected when the size of the vimentin-positive areas within the primary tumor has increased [116]. Decreased E-cadherin expression in primary tumors also correlates with appearance of CTCs [117]. In several tumor types, the majority of CTCs were negative for E-cadherin expression and stained positive or overexpressed vimentin [117–119]. Notably, when analyzed as single cells or clusters, CTCs show heterogeneous expression of EMT-associated markers, including TGF-beta, FOXC1, ZEB1, ZEB2, Fibronectin, Twist, and others [118–122].

There is accumulating evidence that CTCs showing EMT or a cancer stem cell phenotype may have enhanced capacity for metastatic growth [123, 124]. Therefore, it is particularly important to both count and analyze such CTCs in depth. There are still several issues with counting CTCs with a mesenchymal phenotype. Some of the EMT-related markers are present on nonneoplastic cells – for example, vimentin is expressed in monocytes and activated macrophages [125]. Hence, using an epithelial marker for CTC capture is necessary, but it eliminates the possibility of detection of a CTC that may have lost some of its epithelial phenotype and displays only mesenchymal markers [126, 127]. Because there are a variety of CTCs expressing different proportions of epithelial versus mesenchymal markers [121], more studies are needed on the different combinations of mesenchymal and epithelial markers to determine the contribution of each type to the development of metastases.

This is an extremely important field with a great deal of "catching-up-to-do." There is every reason to assume that adding mesenchymal markers to the various capture assays for CTCs under development will improve sensitivity. However, there is ample evidence that patients with TNBC, which lack expression of estrogen receptor, progesterone receptor, and HER2 and which comprise the majority of the basal-like molecular subtype, are equally likely to have elevated CTCs when measured by CellSearch as do patients with other intrinsic subtypes [128, 129]. Moreover, CTCs, when enumerated by CellSearch, are highly prognostic, suggesting that cells with metastatic potential are being counted. Nonetheless, there is sufficient evidence for redoing many of the clinical trials that used CellSearch when an instrument employing both antiepithelial and antimesenchymal antibodies is developed that can also meet the criteria for performing multicenter trials. Indeed, the studies of Chalmers and coworkers [130] using negative selection raise new questions concerning which biomarkers of CTCs are important for identification of those that can cause metastases.

1.9 NEW INSTRUMENTS TO CAPTURE CTCs

In 2007, another major advance took place when a microfluidic platform (the CTC chip) capable of capturing CTCs was described by the Toner and Haber group [131]. In contrast to immunomagnetic purification, CTCs were captured by EpCAM-coated microposts under controlled laminar flow conditions without any preprocessing of samples. Prior theoretical analyses determined the geometric arrangement, distance between microposts, and volumetric flow rate to maximize cell attachment. The resultant CTC chip contained 78,000 microposts within a 9.7 cm^2 surface area. The advantages of this instrument included flow velocity that optimized duration of cell–micropost interaction; low shear stress to minimize damage to the CTCs; the ability to do on-chip immunostain counting and characterization of live cells; and its very small size. The instrument was able to capture CTCs in patients with different kinds of metastatic disease. Although there were major concerns about specificity that affected interest in this particular instrument, it stimulated a major effort to develop microfluidic-based instruments to capture CTCs. Improvements were then made, and a new high-throughput microfluidic mixing device, the herringbone chip (HB-Chip), providing greater specificity and sensitivity was introduced by the same group [132]. The new design used microvortices to maximize the number of interactions between CTCs and the antibody-coated chip surface. CTCs were detected in 14 of 15 (93%) patients with metastatic disease. Captured CTCs could be visualized using both immunofluorescence conjugated antibodies or standard immunocytochemical stains. More recently, this group has developed another microfluidic cell sorting device, the CTC-iChip, which uses size-based lateral displacement, inertial focusing, and magnetophoresis to achieve both antigen-dependent and antigen-independent CTC capture from whole blood. It is capable of isolating EpCAM-positive (with the ^{pos}CTC-iChip) and EpCAM-negative (with the ^{neg}CTC-iChip) CTCs [133].

Many new instruments, as well as multiple filter devices, have been produced based on differences between cancer and hematopoietic cells. The pros and cons of these are elegantly discussed by Alix-Panabières and Pantel [134] and in the following chapters of this book. It is not yet clear which approach will provide the best capture, which also will depend on specific application needs.

Another major issue is that current methods for CTC detection result in capture of tumor cells of which the majority are apoptotic [57–59]. It is important to have an alternative procedure of isolating viable CTCs that are responsible for metastases and characterizing this population. Yao *et al.* [135] placed individual CTCs in nanowells and measured short-term viability, invasiveness, and secretory profiles. Only a rare subset showed anoikis resistance or invasive capacity; most viable cells were nonproliferative (quiescent). Paris *et al.* [136] used a novel functional cell separation method to isolate viable CTCs from patients with prostate cancer. This method relies on the ability of CTCs to invade a collagen adhesion matrix (CAM) and, hence, collects viable cells for subsequent cellular and genomic analyses. Another functional approach for counting viable CTCs is the epithelial immunospot (EPISPOT) assay, which detects target proteins (e.g., cytokeratin-19) secreted by live cells remaining in leukocyte-depleted blood [137]. In this prospective multicenter study of OS in 254 patients with MBC, OS correlated with live CTC status as determined by the EPISPOT assay; when EPISPOT and CellSearch system results were combined, there was better prediction of OS than either method alone.

Also, of particular interest is the development of a new immunomagnetic cell separator, the MagSweeper, which gently enriches live target cells from whole blood and is highly effective at eliminating any nonspecifically bound cells without significantly affecting cell viability or gene expression [138]. The MagSweeper thus enables capture of viable CTCs or DTCs whose RNA or DNA can be profiled at the single cell level [120, 139–141] or propagated in culture [139, 142]. There are multiple other recent microfluidic approaches that isolate live unlabeled CTCs, including those developed by Lim and coworkers [143], Gascoyne and coworkers [144], and Di Carlo and coworkers [145]. Also, under evaluation is an internally based EpCAM-functionalized capture wire that is inserted into a peripheral arm vein for 30 min, exposing the capture wire to 1.5–3 l of blood. Of note is that median numbers of CTCs captured from patients with breast cancer and those with nonsmall-cell lung cancer were 5.5 and 16, respectively [146].

1.10 GENOTYPIC ANALYSES

An important future goal is the development of personalized therapy for patients with primary malignant tumors or recurrent cancer. As previously emphasized, it is not known which cells from the primary tumor will cause metastases, if and when they are in the circulation at sufficient concentrations for this to occur, nor the appropriate treatment for each patient to either prevent metastases or treat metastatic disease. A critical step to answer these questions will be the genotypic, transcriptional, and/or protein signatures of the particular cells in question. CTCs are the obvious population to be so analyzed, for at the single cell level, they can show both genomic and transcriptional heterogeneity and changes over time. However, results of CTC analysis can be discordant from those in the primary tumor and distant metastases [139]. Hence, analysis of both CTCs and DTCs may provide independent clinical information relevant to treatment decisions [94, 139, 147, 148]. For example, persistence of bone marrow metastases after diagnosis of breast cancer was associated with increased risk for subsequent relapses and death [60–62]. Also, isolated DTCs can remain viable and some can be propagated in culture while maintaining the original mutational status [139, 149, 150]. They may serve as a resource for investigating new drug therapies. Recently, improved techniques for propagating CTCs in culture represent a major step forward for facilitating both *in vitro* and *in vivo* personalized drug testing [151–153].

There are already several examples of genotypic analyses that could eventually impact drug selection on individual patients. In colorectal cancer, *KRAS*, *BRAF*, *NRAS*, and *PIK3CA* exon 20 (but not *PIK3CA* exon 9) mutations have been established as negative predictors for treatment with EGFR inhibitors [154–156]. The first comprehensive genomic profiling of CTCs using array comparative genomic hybridization (arrayCGH) and next-generation sequencing was performed by Speicher's group [157] in primary tumors, metastases, and single CTCs from stage IV colorectal carcinoma patients; the sequencing was performed using a panel of 68 colorectal cancer-associated genes. In individual patients, similar copy number changes or driver gene mutations in the primary tumor and metastases were also identified in corresponding CTCs. However, additional mutations were detected exclusively in some CTCs (including "private mutations" defined as being found in only one CTC). Ultradeep sequencing of the primary tumor and metastases was then performed, which identified most of the private mutations as present at the subclonal level in the tumors from these same patients. These findings emphasize the importance of CTC analysis and monitoring tumor genomes that are likely to change during progression, treatment, and relapse.

A single-cell arrayCGH assay from Klein was optimized to enable reliable detection of structural copy number changes as small as 0.1 Mb in single CTCs, allowing identification of CTC heterogeneity and observation of chromosomal changes that occur over the course of treatment; in particular, this method identified the emergence of aberrant clones likely selected by therapy that may contribute to chemotherapy resistance [158].

Whole-exome sequencing of single CTCs from patients with metastatic castration-resistant prostate cancer has been recently performed using a census-based approach in a proof of principal study by Love's group that showed impressively high fidelity [141]: 70% of CTC mutations were present in multiple cores of matched tissue, and 90% of "early trunk mutations" identified in tumor tissue were detected in the CTC exomes.

Lianidou's group measured multiplex gene expression of enriched CTCs using a scalable liquid bead array assay capable of handling minute samples [159]. Another study by Sieuwerts *et al.* [160] measured the expression of highly expressed genes in pooled CTCs and demonstrated that transcriptional profiling of low numbers of CTCs within a high background of leukocytes was feasible.

High-dimensional single-cell transcriptional profiling was first performed by Jeffrey's group on individual CTCs from patients with breast cancer [120] and later by Huang and coworkers [118] in prostate cancer, both studies demonstrating CTC heterogeneity within even a single blood draw, a finding that would have been missed in pooled samples of CTCs, and leading to speculation regarding whether different populations of CTCs may require different therapies (similar to the discussion about mixed populations of HER2-positive and -negative CTCs). Moreover, marked gene expression differences between CTCs and single cells from proliferating cell lines brought into question the "fit for purpose" of standard cancer cell lines in new drug discovery [120].

The employment of leukapheresis enabled CTC screening from large volumes of blood and identified larger numbers of CTCs, a technique that could be potentially helpful in molecularly evaluating CTCs in early disease and in defining which cells contribute to the metastatic process for determining treatment options [161].

CTCs can also be interrogated for metastasis suppressors that inhibit metastasis without blocking primary tumor growth. This is accomplished by regulating signaling pathways that inhibit proliferation, cell migration, and growth at the secondary site. One example is breast cancer metastasis suppressor-1 (*BRMS1*) [162], which differentially regulates the expression of multiple genes leading to suppression of metastases without blocking orthotropic tumor growth. Chimonidou *et al.* [163] showed that the *BRMS1* promoter is methylated in CTCs isolated from 1/3 of patients with early breast cancer and in 44% of those with metastatic disease, compared to 8.7% in healthy individuals. *BRMS1* promoter methylation results in the transcriptional repression of this metastasis suppressor gene and, when found in the primary tumor, predicts poorer disease-free survival [164]. These results indicate the potential clinical relevance of identifying this methylation event in CTCs from patients with operable breast cancer.

Another recent step forward is the identification and characterization of breast cancer CTCs that can cause brain metastasis by Zhang *et al.* [151]. By successfully culturing CTCs from patients with MBC, they developed CTC cell lines including three from patients who had brain metastases. Selecting markers HER2+/EGFR+/HPSC+/Notch1+, a high proportion metastasized to the brain and lungs after injection into immunodeficient animals, whereas parental CTCs metastasized only to the lung. Hence, there may be therapeutic interventions that can prevent CTCs from colonizing distant organs such as the brain.

Is there a role for DNA analysis in addition to the aforementioned cellular analyses? Cell-free DNA (cfDNA) or circulating tumor DNA (ctDNA) may derive from lysed tumor cells, from secretion by macrophages that have ingested necrotic cells, or potentially from CTCs themselves. Mutations associated with therapeutic resistance can be detected in ctDNA up to 10 months prior to detection by imaging; this may thereby allow clinicians to change therapy sooner [165–167]. ctDNA has also been used for early detection of cancer [168–170]. Such information is likely to supplement that obtained from CTC analysis.

However, there are issues for the future. Is ctDNA derived from dying tumor cells sensitive to a given therapy and/or derived from resistant tumor cells that persist? Another concern is determining those patient- and/or tumor-specific mutations or epigenetic alterations that should be analyzed. Finally, prior to acceptance as a technique for clinical biomarker detection, both ctDNA and CTCs will require proof of analytical validation with standardized, reproducible assays and proven clinical utility in particular patient populations, as aptly discussed by Danila *et al.* [171].

1.11 CONCLUSIONS

The potential importance of CTCs is generally accepted. The buzzword "liquid biopsy" confirms the relatively noninvasive nature of this means for monitoring or investigating the biology of both primary malignant tumors and recurrent disease. However, perhaps even more importantly, CTCs represent a critical tumor cell population for understanding metastatic potential and, eventually, determining choice of treatment for individual patients. It is likely that they would also aid in the development of novel therapies. However, the degree to which this helps the oncologist treat his or her patient at present has come into question. CTC enumeration is moderately prognostic in early-stage breast cancer but probably does not outweigh what is already known from tumor size/characteristics and evaluation of lymph node involvement. It can monitor patients for recurrence, but so can tumor markers such as CA-15.3 [172, 173] although such markers may show tumor subtype dependence [174]. There is ample evidence for the relation between CTC count and prognosis in metastatic breast, prostate, and colon cancer, with prediction of disease progression long before symptoms or imaging changes. But what is the evidence that it changes treatment or that earlier treatment is helpful? Early change in therapy based on persistently elevated CTC counts after starting a drug has not yet changed patient outcome [55]. Of course, this could be caused by ineffective treatment. Regardless, measurement of biomarkers and genotype will be needed in addition to CTC enumeration. Also, the increasing evidence that a portion of CTCs that have undergone EMT may have escaped detection by the only FDA-approved instrument for counting CTCs is a major issue. This concern is heightened by the probability that CTCs with a mesenchymal phenotype may play a major role in development of metastases. Thus, mesenchymal as well as epithelial markers must now be used to capture CTCs, which must in turn be genotypically characterized in clinical trials. Indeed, negative selection methodology for CTCs represents another important pathway for determining the biomarkers and immunophenotype of those CTCs responsible for metastatic disease.

However, even the most extensive characterization of CTCs does not necessarily solve the problem of the complexity of cancer. When will a particular cancer metastasize? The primary tumor cell population and the CTCs are extremely heterogeneous and constantly change over time. This is true also for metastatic seeding, which can be dormant for the lifetime of the patient; there may be dormant cells that then regrow, or such cells may develop metastases from the onset. Even when a treatment strategy has been developed for a particular cancer population, it may not apply for a specific patient's tumor. Jeffrey [139, 165] has stressed that additional information, such as analyses of DTCs from the bone marrow and/or ctDNA, may be used to augment in-depth genotypic analyses of CTCs to maximize the patient's specific information needed for monitoring and optimizing individual therapy in real time.

What is the potential role for a CTC assay to become a general blood test to detect cancers at an earlier stage? Progress in treating epithelial cancers over the last 60 years has been relatively modest. Patients with clinical recurrences still succumb to disease, although their lifespan can be prolonged with chemotherapy. There are a large number of new immunologically based therapies in clinical trials, but it is too early to make efficacy predictions. It seems reasonable, therefore, to develop a diagnostic blood test as a high priority. This is not a generally accepted view for several reasons: it is argued that earlier diagnosis might not be beneficial; there can be false positives from mucosal inflammation and other factors; and localizing the organ source might be a problem. These are valid concerns but not of sufficient weight to stop the effort. Earlier diagnosis will certainly save some lives and perhaps a significant proportion; the other issues have possible solutions. The most important concern and the one less articulated is the presence of CTCs in a patient who should not be treated. The finding in breast cancer dormancy that a substantial proportion of patients many years after mastectomy who are disease-free, most of whom will live a normal lifespan, have CTCs and the evidence that the presence of breast cancer cells in the bone marrow does not necessarily lead to growing metastases over a period as long as one decade both raise this issue [56, 60–62]. Of course, the dormancy patients have had their primary tumor removed and patients with bone marrow metastases have received treatment. Nevertheless, there may be a significant proportion of patients, particularly in the elderly age group, who develop CTCs and would be best left untreated. However, this is an issue for the future when a diagnostic blood test has been developed.

The prospects are favorable for developing such a blood test. There is a myriad of new CTC assays under development. However, a recurrent problem reflected in the literature is the unstated assumption that a cell from a tumor line that has the same approximate phenotype as a CTC can be used as an accurate guide to the level of sensitivity of a particular capture procedure. CTCs are far more fragile. Many of the current instruments are overrated with regard to sensitivity. That is solely determined by CTC capture in very early primary cancers. It is too early to make a choice on a particular approach, for example, immunomagnetic, microfluidic, or the many other approaches. In the next 5 years or so, there may be sufficient clinical data to move ahead with a government-sponsored multicenter trial using an optimal instrument. In this regard, it is critical to also make full use of an optimal immunological "cocktail" [175]. The capacity to immunologically engineer antibodies with low picomolar (pM) affinity constants makes it possible to use univalent Fv antibody fragments of molecular weight 35,000. Because the affinity of divalent antibodies can be 1000–10,000-fold higher than conventional secondary antibodies from immunized animals, such univalent fragments still have strong avidity. This is a critical point because they allow combinations of antibodies to different epitopes to be used without the problems of steric hindrance that would develop if native antibodies of 150,000 molecular weight were employed. Hence, specificities to more than one epithelial, mesenchymal, and organ-specific marker can be tried on the same specimen in various combinations and in different time sequences. Deng *et al.* [176] have shown that antibody to intracellular cytokeratin (CK) together with anti-EpCAM antibodies increases assay sensitivity. It is also possible to attach a short polypeptide to the polypeptide holding the Fv light (L)- and heavy (H)-chain antigen-combining regions together, thereby increasing the Fv's potential range of attachment to the corresponding epitopes on the CTCs [175]. Such an arrangement could be likened to individual arms of an "immunological octopus."

Nonimmunoglobulin molecules with a high degree of specificity and picomolar affinities can be engineered from either polypeptides (protein scaffolds) [177] or nucleic acids (aptamers) [178]. The latter can be as small as 19 nucleotides with 55 pM affinity to its specific antigen [179]. Protein scaffolds with pM affinities have been produced to EpCAM [180], EGFR2/HER2 [181], and HER2, the latter being a 6 kDa Affibody [182]. Combinations between antibodies and scaffolds have also been produced, for example, a llama H chain variable domain fragment and a non-Ig scaffold protein resulted in a protein with an affinity of 0.54 pM to its specific antigen [183]. The aforementioned changes to the immunological portion of the assays should help to increase their sensitivity.

This is a particularly exciting time for scientists studying CTCs. There are opportunities for obtaining large amounts of new data concerning the genomes and potential behavior of these cells and for utilizing that information to better understand the metastatic cascade. At the same time, it is necessary to redefine the criteria and instrumentation for capturing and enumerating CTCs and to determine how this new knowledge should be funneled into particular clinical trials designed to improve diagnosis and therapy. The latter will include a host of recently developed immunologic- and inflammatory-specific cancer drugs that together with earlier diagnosis could result in a major advance in treatment.

REFERENCES

1. Slavin, S. and Strober, S. (1978) Spontaneous murine B-cell leukaemia. *Nature*, **272**, 624–626.
2. Krolick, K.A., Isakson, P.C., Uhr, J.W., and Vitetta, E.S. (1979) BCL1, a murine model for chronic lymphocytic leukemia: use of the surface immunoglobulin idiotype for the detection and treatment of tumor. *Immunological Reviews*, **48**, 81–106.
3. Knapp, M.R., Jones, P.P., Black, S.J. *et al.* (1979) Characterization of a spontaneous murine B cell leukemia (BCL1). I. Cell surface expression of IgM, IgD, Ia, and FcR. *Journal of Immunology*, **123**, 992–999.
4. Uhr, J.W., Tucker, T., May, R.D. *et al.* (1991) Cancer dormancy: studies of the murine BCL1 lymphoma. *Cancer Research*, **51**, 5045s–5053s.
5. Vitetta, E.S., Tucker, T.F., Racila, E. *et al.* (1997) Tumor dormancy and cell signaling. V. Regrowth of BCL1 tumor after dormancy is established. *Blood*, **89**, 4425–4436.
6. Scheuermann, R.H., Racila, E., Tucker, T. *et al.* (1994) Lyn tyrosine kinase signals cell cycle arrest but not apoptosis in B-lineage lymphoma cells. *Proceedings of the National Academy of Sciences of the United States of America*, **91**, 4048–4052.
7. Meltzer, A. (1990) Dormancy and breast cancer. *Journal of Surgical Oncology*, **43**, 181–188.
8. Stewart, T.H., Hollinshead, A.C., and Raman, S. (1991) Tumour dormancy: initiation, maintenance and termination in animals and humans. *Canadian Journal of Surgery*, **34**, 321–325.

9. Demichelli, R., Abbattista, A., Miceli, R. *et al.* (1996) Time distribution of the recurrence risk for breast cancer patients undergoing mastectomy: further support about the concept of tumor dormancy. *Breast Cancer Research and Treatment*, **41**, 177–185.
10. George, A.J. and Stevenson, F.K. (1989) Prospects for the treatment of B cell tumors using idiotypic vaccination. *International Reviews of Immunology*, **4**, 271–310.
11. Stevenson, F.K., George, A.J., and Glennie, M.J. (1990) Anti-idiotypic therapy of leukemias and lymphomas. *Chemical Immunology*, **48**, 126–166.
12. Meeker, T., Lowder, J., Cleary, M.L. *et al.* (1985) Emergence of idiotype variants during treatment of B-cell lymphoma with anti-idiotype antibodies. *New England Journal of Medicine*, **312**, 1658–1665.
13. Levy, R. and Miller, R.A. (1990) Therapy of lymphoma directed at idiotypes. *Journal of the National Cancer Institute Monographs*, **10**, 61–68.
14. Miller, R.A., Maloney, D.G., Warnke, R., and Levy, R. (1982) Treatment of B-cell lymphoma with monoclonal anti-idiotype antibody. *New England Journal of Medicine*, **306**, 517–522.
15. Karrison, T.G., Ferguson, D.J., and Meier, P. (1999) Dormancy of mammary carcinoma after mastectomy. *Journal of the National Cancer Institute*, **91**, 80–85.
16. Ashworth, T. (1869) A case of cancer in which cells similar to those in the tumours were seen in the blood after death. *Australian Medical Journal*, **14**, 146–147.
17. Christopherson, W.M. (1965) Cancer cells in the peripheral blood: a second look. *Acta Cytologica*, **9**, 169–174.
18. Drye, J.C., Rumage, W.T. Jr., and Anderson, D. (1962) Prognostic import of circulating cancer cells after curative surgery: a long time follow up study. *Annals of Surgery*, **155**, 733–740.
19. Engell, H.C. (1955) Cancer cells in the circulating blood; a clinical study on the occurrence of cancer cells in the peripheral blood and in venous blood draining the tumour area at operation. *Acta Chirurgica Scandinavica. Supplementum*, **201**, 1–70.
20. Fidler, I.J. (1973) The relationship of embolic homogeneity, number, size and viability to the incidence of experimental metastasis. *European Journal of Cancer*, **9**, 223–227.
21. Fidler, I.J. (1978) Tumor heterogeneity and the biology of cancer invasion and metastasis. *Cancer Research*, **38**, 2651–2660.
22. Pantel, K., Schlimok, G., Braun, S. *et al.* (1993) Differential expression of proliferation-associated molecules in individual micrometastatic carcinoma cells. *Journal of the National Cancer Institute*, **85**, 1419–1424.
23. Pantel, K., Izbicki, J., Passlick, B. *et al.* (1996) Frequency and prognostic significance of isolated tumour cells in bone marrow of patients with non-small-cell lung cancer without overt metastases. *Lancet*, **347**, 649–653.
24. Folkman, J. (1971) Tumor angiogenesis: therapeutic implications. *New England Journal of Medicine*, **285**, 1182–1186.
25. Folkman, J., Watson, K., Ingber, D., and Hanahan, D. (1989) Induction of angiogenesis during the transition from hyperplasia to neoplasia. *Nature*, **339**, 58–61.
26. Liotta, L.A., Steeg, P.S., and Stetler-Stevenson, W.G. (1991) Cancer metastasis and angiogenesis: an imbalance of positive and negative regulation. *Cell*, **64**, 327–336.
27. Liotta, L.A., Kleinerman, J., and Saidel, G.M. (1974) Quantitative relationships of intravascular tumor cells, tumor vessels, and pulmonary metastases following tumor implantation. *Cancer Research*, **34**, 997–1004.

28. Ghossein, R.A., Bhattacharya, S., and Rosai, J. (1999) Molecular detection of micrometastases and circulating tumor cells in solid tumors. *Clinical Cancer Research*, **5**, 1950–1960.
29. Hardingham, J.E., Kotasek, D., Farmer, B. *et al.* (1993) Immunobead-PCR: a technique for the detection of circulating tumor cells using immunomagnetic beads and the polymerase chain reaction. *Cancer Research*, **53**, 3455–3458.
30. Hardingham, J.E., Kotasek, D., Sage, R.E. *et al.* (1995) Detection of circulating tumor cells in colorectal cancer by immunobead-PCR is a sensitive prognostic marker for relapse of disease. *Molecular Medicine*, **1**, 789–794.
31. Wong, L.S., Bateman, W.J., Morris, A.G., and Fraser, I.A. (1995) Detection of circulating tumor cells with the magnetic activated cell sorter. *British Journal of Surgery*, **82**, 1333–1337.
32. Makarovskiy, A.N., Ackerley, W. 3rd, Wojcik, L. *et al.* (1997) Application of immunomagnetic beads in combination with RT-PCR for the detection of circulating prostate cancer cells. *Journal of Clinical Laboratory Analysis*, **11**, 346–350.
33. Martin, V.M., Siewert, C., Scharl, A. *et al.* (1998) Immunomagnetic enrichment of disseminated epithelial tumor cells from peripheral blood by MACS. *Experimental Hematology*, **26**, 252–264.
34. Shibata, K., Mori, M., Kitano, S., and Akiyoshi, T. (1998) Detection of ras gene mutations in peripheral blood of carcinoma patients using CD45 immunomagnetic separation and nested mutant allele specific amplification. *International Journal of Oncology*, **12**, 1333–1338.
35. Racila, E., Euhus, D., Weiss, A.J. *et al.* (1998) Detection and characterization of carcinoma cells in the blood. *Proceedings of the National Academy of Sciences of the United States of America*, **95**, 4589–4594.
36. Liberti, P.A. and Feeley, B. (1991) Analytical- and process-scale cell separation with bioreceptor ferrofluids and high-gradient magnetic separation, in *Cell Separation Science and Technology* (eds D.S. Kompala and P. Todd), American Chemical Society, Washington, DC, pp. 268–288.
37. Kemshead, J.T., Hancock, J.P., and Liberti, P.A. (1994) A model system for the enrichment of tumor cells from peripheral blood and bone marrow using immunomagnetic ferrofluids. *Progress in Clinical and Biological Research*, **389**, 593–600.
38. Terstappen, L.W., Meiners, H., and Loken, M.R. (1989) A rapid sample preparation technique for flow cytometric analysis of immunofluorescence allowing absolute enumeration of cell subpopulations. *Journal of Immunological Methods*, **123**, 103–112.
39. Fehm, T., Morrison, L., Saboorian, H. *et al.* (2002) Patterns of aneusomy for three chromosomes in individual cells from breast cancer. *Breast Cancer Research and Treatment*, **75**, 227–239.
40. Fehm, T., Sagalowsky, A., Clifford, E. *et al.* (2002) Cytogenetic evidence that circulating epithelial cells in patients with carcinoma are malignant. *Clinical Cancer Research*, **8**, 2073–2084.
41. Witzig, T.E., Bossy, B., Kimlinger, T. *et al.* (2002) Detection of circulating cytokeratin-positive cells in the blood of breast cancer patients using immunomagnetic enrichment and digital microscopy. *Clinical Cancer Research*, **8**, 1085–1091.
42. Ulmer, A., Schmidt-Kittler, O., Fischer, J. *et al.* (2004) Immunomagnetic enrichment, genomic characterization, and prognostic impact of circulating melanoma cells. *Clinical Cancer Research*, **10**, 531–537.

43. Mangan, K.F., Leonardo, J., Mullaney, M.T. *et al.* (1999) A rapid two-step method for elimination of bcl-2/IgH positive non-Hodgkin's lymphoma cells from human blood or marrow stem cells, employing immunomagnetic purging with streptavidin-coated ferrofluids. *Cytotherapy*, **1**, 287–293.
44. Liberti, P.A., Rao, C.G., and Terstappen, L.W.M.M. (2001) Optimization of ferrofluids and protocols for the enrichment of breast tumor cells in blood. *J. Magnetism Magnetic Materials*, **225**, 301–307.
45. Tibbe, A.G., de Grooth, B.G., Greve, J. *et al.* (2001) Cell analysis system based on immunomagnetic cell selection and alignment followed by immunofluorescent analysis using compact disk technologies. *Cytometry*, **43**, 31–37.
46. Allard, W.J., Matera, J., Miller, M.C. *et al.* (2004) Tumor cells circulate in the peripheral blood of all major carcinomas but not in healthy subjects or patients with nonmalignant diseases. *Clinical Cancer Research*, **10**, 6897–6904.
47. Cristofanilli, M., Hayes, D.F., Budd, G.T. *et al.* (2005) Circulating tumor cells: a novel prognostic factor for newly diagnosed metastatic breast cancer. *Journal of Clinical Oncology*, **23**, 1420–1430.
48. Hayes, D.F., Cristofanilli, M., Budd, G.T. *et al.* (2006) Circulating tumor cells at each follow-up time point during therapy of metastatic breast cancer patients predict progression-free and overall survival. *Clinical Cancer Research*, **12**, 4218–4224.
49. Budd, G.T., Cristofanilli, M., Ellis, M.J. *et al.* (2006) Circulating tumor cells versus imaging – predicting overall survival in metastatic breast cancer. *Clinical Cancer Research*, **12**, 6403–6409.
50. Cohen, S.J., Punt, C.J., Iannotti, N. *et al.* (2009) Prognostic significance of circulating tumor cells in patients with metastatic colorectal cancer. *Annals of Oncology*, **20**, 1223–1229.
51. de Bono, J.S., Scher, H.I., Montgomery, R.B. *et al.* (2008) Circulating tumor cells predict survival benefit from treatment in metastatic castration-resistant prostate cancer. *Clinical Cancer Research*, **14**, 6302–6309.
52. Hodgkinson, C.L., Morrow, C.J., Li, Y. *et al.* (2014) Tumorigenicity and genetic profiling of circulating tumor cells in small-cell lung cancer. *Nature Medicine*, **20**, 897–903.
53. Hou, J.M., Krebs, M.G., Lancashire, L. *et al.* (2012) Clinical significance and molecular characteristics of circulating tumor cells and circulating tumor microemboli in patients with small-cell lung cancer. *Journal of Clinical Oncology*, **30**, 525–532.
54. Krebs, M.G., Sloane, R., Priest, L. *et al.* (2011) Evaluation and prognostic significance of circulating tumor cells in patients with non-small-cell lung cancer. *Journal of Clinical Oncology*, **29**, 1556–1563.
55. Smerage, J.B., Barlow, W.E., Hortobagyi, G.N. *et al.* (2014) Circulating tumor cells and response to chemotherapy in metastatic breast cancer: SWOG S0500. *Journal of Clinical Oncology*, **32**, 3483–3489.
56. Meng, S., Tripathy, D., Frenkel, E.P. *et al.* (2004) Circulating tumor cells in patients with breast cancer dormancy. *Clinical Cancer Research*, **10**, 8152–8162.
57. Méhes, G., Witt, A., Kubista, E., and Ambrose, P.F. (2001) Circulating breast cancer cells are frequently apoptotic. *American Journal of Pathology*, **159**, 17–20.
58. Smerage, J.B., Budd, G.T., Doyle, G.V. *et al.* (2013) Monitoring apoptosis and Bcl-2 on circulating tumor cells in patients with metastatic breast cancer. *Molecular Oncology*, **7**, 680–692.

59. Kallergi, G., Konstantinidis, G., Markomanolaki, H. *et al.* (2013) Apoptotic circulating tumor cells in early and metastatic breast cancer patients. *Molecular Cancer Therapeutics*, **12**, 1886–1895.
60. Pantel, K., Müller, V., Auer, M. *et al.* (2003) Detection and clinical implications of early systemic tumor cell dissemination in breast cancer. *Clinical Cancer Research*, **9**, 6326–6334.
61. Wiedswang, G., Borgen, E., Kåresen, R. *et al.* (2004) Isolated tumor cells in bone marrow three years after diagnosis in disease-free breast cancer patients predict unfavorable clinical outcome. *Clinical Cancer Research*, **10**, 5342–5348.
62. Klein, C.A. (2009) Parallel progression of primary tumors and metastases. *Nature Reviews Cancer*, **9**, 302–312.
63. Thomas, L. (1959) Reactions to homologous tissue antigens in relations to hypersensitivity, in *Cellular and Humoral Aspects of the Hypersensitive States* (ed H.S. Lawrence), Hoeber-Harper, New York, p. 529.
64. Burnet, F.M. (1970) The concept of immune surveillance. *Progress in Experimental Tumor Research*, **13**, 1–27.
65. Gimbrone, M.A., Leapman, S.B., Cotran, R.S., and Folkman, J. (1972) Tumor dormancy *in vivo* by prevention of neovascularization. *Journal of Experimental Medicine*, **136**, 261–276.
66. Aguirre-Ghiso, J.A. (2007) Models, mechanisms and clinical evidence for cancer dormancy. *Nature Reviews Cancer*, **7**, 834–846.
67. Uhr, J.W. and Pantel, K. (2011) Controversies in clinical cancer dormancy. *Proceedings of the National Academy of Sciences of the United States of America*, **108**, 12396–12400.
68. Conlon, I. and Raff, M. (1999) Size control in animal development. *Cell*, **96**, 235–244.
69. Day, S.J. and Lawrence, P.A. (2000) Measuring dimensions: the regulation of size and shape. *Development*, **127**, 2977–2987.
70. Michalopoulous, G.K. and DeFrances, M.C. (1997) Liver regeneration. *Science*, **276**, 60–66.
71. Verdu, J., Buratovich, M.A., Wilder, E.L., and Birnbaum, M.J. (1999) Cell-autonomous regulation of cell and organ growth in Drosophila by Akt/PKB. *Nature Cell Biology*, **1**, 500–506.
72. Zaina, S. and Squire, S. (1998) The soluble type 2 insulin-like growth factor (IGF-II) receptor reduces organ size by IFG-II-mediated and IGF-II-independent mechanisms. *Journal of Biological Chemistry*, **273**, 28610–28616.
73. Böhni, R., Riesgo-Escovar, J., Oldham, S. *et al.* (1999) Autonomous control of cell and organ size by CHICO, a Drosophila homolog of vertebrate IRS1-4. *Cell*, **97**, 865–875.
74. Gao, X., Zhang, Y., Arrazola, P. *et al.* (2002) Tsc tumour suppressor proteins antagonize amino-acid-TOR signalling. *Nature Cell Biology*, **4**, 699–704.
75. Pan, D. (2007) Hippo signaling in organ size control. *Genes and Development*, **21**, 886–897.
76. Harvey, K. and Tapon, N. (2007) The Salvador-Warts-Hippo pathway - an emerging tumour-suppressor network. *Nature Reviews Cancer*, **7**, 182–191.
77. Wang, Z., Wu, Y., Wang, H. *et al.* (2014) Interplay of mevalonate and Hippo pathways regulate RHAMM transcription via YAP to modulate breast cancer cell motility. *Proceedings of the National Academy of Sciences of the United States of America*, **111**, E89–98.

78. Demierre, M.F., Higgins, P.D., Gruber, S.B. *et al.* (2005) Statins and cancer prevention. *Nature Reviews Cancer*, **5**, 930–942.

79. Clendening, J.W. and Penn, L.Z. (2012) Targeting tumor cell metabolism with statins. *Oncogene*, **31**, 4967–4978.

80. Pujana, M.A., Han, J.D., Starita, L.M. *et al.* (2007) Network modeling links breast cancer susceptibility and centrosome dysfunction. *Nature Genetics*, **39**, 1338–1349.

81. Maxwell, C.A., Benítez, J., Gómez-Baldó, L. *et al.* (2011) Interplay between BRCA1 and RHAMM regulates epithelial apicobasal polarization and may influence risk of breast cancer. *PLoS Biology*, **9**, e1001199.

82. Wang, C., Thor, A.D., Moore, D.H. 2nd *et al.* (1998) The overexpression of RHAMM, a hyaluronan-binding protein that regulates ras signaling, correlates with overexpression of mitogen-activated protein kinase and is a significant parameter in breast cancer progression. *Clinical Cancer Research*, **4**, 567–576.

83. Velseh, M. and Turley, E.A. (2011) Hyaluronan metabolism in remodeling extracellular matrix: probes for imaging and therapy of breast cancer. *Integrative Biology*, **3**, 304–315.

84. Slamon, D.J., Clark, G.M., Wong, S.G. *et al.* (1987) Human breast cancer: correlation of relapse and survival with amplification of the HER-2/neu oncogene. *Science*, **235**, 177–182.

85. Romond, E.H., Perez, E.A., Bryant, J. *et al.* (2005) Trastuzumab plus adjuvant chemotherapy for operable HER2-positive breast cancer. *New England Journal of Medicine*, **353**, 1673–1684.

86. Piccart-Gebhart, M.J., Procter, M., Leyland-Jones, B. *et al.* (2005) Herceptin Adjuvant (HERA) Trial Study Team. Trastuzumab after adjuvant chemotherapy in HER2-positive breast cancer. *New England Journal of Medicine*, **353** (16), 1659–1672.

87. Sekido, Y., Umemura, S., Takekoshi, S. *et al.* (2003) Heterogeneous gene alterations in primary breast cancer contribute to discordance between primary and asynchronous metastatic/recurrent sites: HER2 gene amplification and p53 mutation. *International Journal of Oncology*, **22**, 1225–1232.

88. Houssami, N., Macaskill, P., Balleine, R.L. *et al.* (2011) HER2 discordance between primary breast cancer and its paired metastasis: tumor biology or test artefact? Insights through meta-analysis. *Breast Cancer Research and Treatment*, **129**, 659–674.

89. Wülfing, P., Borchard, J., Buerger, H. *et al.* (2006) HER2-positive circulating tumor cells indicate poor clinical outcome in stage I to III breast cancer patients. *Clinical Cancer Research*, **12**, 1715–1720.

90. Meng, S., Tripathy, D., Shete, S. *et al.* (2004) HER-2 gene amplification can be acquired as breast cancer progresses. *Proceedings of the National Academy of Sciences of the United States of America*, **101**, 9393–9398.

91. Fehm, T., Müller, V., Aktas, B. *et al.* (2010) HER2 status of circulating tumor cells in patients with metastatic breast cancer: a prospective multicenter trial. *Breast Cancer Research and Treatment*, **124**, 403–412.

92. Riethdorf, S., Müller, V., Zhang, L. *et al.* (2010) Detection and HER2 expression of circulating tumor cells: prospective monitoring in breast cancer patients treated in the neoadjuvant GeparQuattro trial. *Clinical Cancer Research*, **16**, 2634–2645.

93. Pestrin, M., Bessi, S., Galardi, F. *et al.* (2009) Correlation of HER2 status between primary tumors and corresponding circulating tumor cells in advanced breast cancer patients. *Breast Cancer Research and Treatment*, **118**, 523–530.

94. Krishnamurthy, S., Bischoff, F., Ann Mayer, J. *et al.* (2013) Discordance in HER2 gene amplification in circulating and disseminated tumor cells in patients with operable breast cancer. *Cancer Medicine*, **2**, 226–233.
95. Iwatsuki, M., Toyoshima, K., Watanabe, M. *et al.* (2013) Frequency of HER2 expression of circulating tumour cells in patients with metastatic or recurrent gastrointestinal cancer. *British Journal of Cancer*, **109**, 2829–2832.
96. Paik, S., Kim, C., and Wolmark, N. (2008) Her2 status and benefit from adjuvant trastuzumab in breast cancer. *New England Journal of Medicine*, **358**, 1409–1411.
97. Ithimakin, S., Day, K.C., Malik, F. *et al.* (2013) HER2 drives luminal breast cancer stem cells in the absence of HER2 amplification: implications for efficacy of adjuvant trastuzumab. *Cancer Research*, **73**, 1635–1646.
98. Sidenius, N. and Blasi, F. (2003) The urokinase plasminogen activator system in cancer: recent advances and implication for prognosis and therapy. *Cancer Metastasis Reviews*, **22**, 205–222.
99. Hemsen, A., Riethdorf, L., Brünner, N. *et al.* (2003) Comparative evaluation of urokinase-type plasminogen activator receptor expression in primary breast carcinomas and on metastatic tumor cells. *International Journal of Cancer*, **107**, 903–909.
100. Riisbro, R., Christensen, I.J., Piironen, T. *et al.* (2002) Prognostic significance of soluble urokinase plasminogen activator receptor in serum and cytosol of tumor tissue from patients with primary breast cancer. *Clinical Cancer Research*, **5**, 1132–1141.
101. Meng, S., Tripathy, D., Shete, S. *et al.* (2006) uPar and HER-2 gene status in individual breast cancer cells from blood and tissues. *Proceedings of the National Academy of Sciences of the United States of America*, **103**, 17361–17365.
102. Li, C., Cao, S., Liu, Z. *et al.* (2010) RNAi-mediated downregulation of uPAR synergizes with targeting of HER2 through the ERK pathway in breast cancer cells. *International Journal of Cancer*, **127**, 1507–1516.
103. LeBeau, A.M., Duriseti, S., Murphy, S.T. *et al.* (2013) Targeting uPAR with antagonistic recombinant human antibodies in aggressive breast cancer. *Cancer Research*, **73**, 2070–2081.
104. Greenburg, G. and Hay, E.D. (1982) Epithelia suspended in collagen gels can lose polarity and express characteristics of migrating mesenchymal cells. *Journal of Cell Biology*, **95**, 333–339.
105. Bednarz-Knoll, N., Alix-Panabières, C., and Pantel, K. (2012) Plasticity of disseminating cancer cells in patients with epithelial malignancies. *Cancer and Metastasis Reviews*, **31**, 673–687.
106. Thompson, E.W., Newgreen, D.F., and Tarin, D. (2005) Carcinoma invasion and metastasis: a role for epithelial-mesenchymal transition? *Cancer Research*, **65**, 5991–5995.
107. Polyak, K. and Weinberg, R.A. (2009) Transitions between epithelial and mesenchymal states: acquisition of malignant and stem cell traits. *Nature Reviews Cancer*, **9**, 265–273.
108. Thiery, J.P. (2002) Epithelial-mesenchymal transitions in tumour progression. *Nature Reviews Cancer*, **2**, 442–454.
109. Wicha, M.S., Liu, S., and Dontu, G. (2006) Cancer stem cells: an old idea – a paradigm shift. *Cancer Research*, **66** (4), 1883–1890.
110. Kasimir-Bauer, S., Hoffmann, O., Wallwiener, D. *et al.* (2012) Expression of stem cell and epithelial-mesenchymal transition markers in primary breast cancer patients with circulating tumor cells. *Breast Cancer Research*, **14**, R15.

111. Krawczyk, N., Meier-Stiegen, F., Banys, M. *et al.* (2014) Expression of stem cell and epithelial-mesenchymal transition markers in circulating tumor cells of breast cancer patients. *Biomed Research International*, **2014**, 415721.
112. Gupta, P.B., Chaffer, C.L., and Weinberg, R.A. (2009) Cancer stem cells: mirage or reality? *Nature Medicine*, **15**, 1010–1012.
113. Farnie, G. and Clarke, R.B. (2007) Mammary stem cells and breast cancer – role of Notch signalling. *Stem Cell Reviews*, **3**, 169–175.
114. Lathia, J.D., Venere, M., Rao, M.S., and Rich, J.N. (2011) Seeing is believing: are cancer stem cells the Loch Ness monster of tumor biology? *Stem Cell Reviews*, **7**, 227–237.
115. Patel, P. and Chen, E.I. (2012) Cancer stem cells, tumor dormancy, and metastasis. *Front Endocrinology*, **3**, 125.
116. Bonnomet, A., Syne, L., Brysse, A. *et al.* (2012) A dynamic *in vivo* model of epithelial-to-mesenchymal transitions in circulating tumor cells and metastases of breast cancer. *Oncogene*, **31**, 3741–3753.
117. Loric, S., Paradis, V., Gala, J.L. *et al.* (2001) Abnormal E-cadherin expression and prostate cell blood dissemination as markers of biological recurrences in cancer. *European Journal of Cancer*, **37**, 1475–1481.
118. Chen, C.L., Mahalingam, D., Osmulski, P. *et al.* (2013) Single-cell analysis of circulating tumor cells identifies cumulative expression patterns of EMT-related genes in metastatic prostate cancer. *Prostate*, **73**, 813–826.
119. Hou, J.M., Krebs, M., Ward, T. *et al.* (2011) Circulating tumor cells as a window on metastasis biology in lung cancer. *American Journal of Pathology*, **178**, 989–996.
120. Powell, A.A., Talasaz, A.H., Zhang, H. *et al.* (2012) Single cell profiling of circulating tumor cells: transcriptional heterogeneity and diversity from breast cancer cell lines. *PloS One*, **7**, e33788.
121. Yu, M., Bardia, A., Wittner, B.S. *et al.* (2013) Circulating breast tumor cells exhibit dynamic changes in epithelial and mesenchymal composition. *Science*, **339**, 580–584.
122. Kallergi, G., Papadaki, M.A., Politaki, E. *et al.* (2011) Epithelial to mesenchymal transition markers expressed in circulating tumour cells of early and metastatic breast cancer patients. *Breast Cancer Research*, **13**, R59.
123. Raimondi, C., Gradilone, A., Naso, G. *et al.* (2011) Epithelial-mesenchymal transition and stemness features in circulating tumor cells from breast cancer patients. *Breast Cancer Research and Treatment*, **130**, 449–455.
124. Gradilone, A., Naso, G., Raimondi, C. *et al.* (2011) Circulating tumor cells (CTCs) in metastatic breast cancer (MBC): prognosis, drug resistance and phenotypic characterization. *Annals of Oncology*, **22**, 86–92.
125. Mor-Vaknin, N., Punturieri, A., Sitwala, K., and Markovitz, D.M. (2003) Vimentin is secreted by activated macrophages. *Nature Cell Biology*, **5**, 59–63.
126. Königsberg, R., Obermayr, E., Bises, G. *et al.* (2011) Detection of EpCAM positive and negative circulating tumor cells in metastatic breast cancer patients. *Acta Oncologica*, **50**, 700–710.
127. Gorges, T.M., Tinhofer, I., Drosch, M. *et al.* (2012) Circulating tumour cells escape from EpCAM-based detection due to epithelial-to-mesenchymal transition. *BMC Cancer*, **12**, 178.
128. Peeters, D.J., van Dam, P.J., Van den Eynden, G.G. *et al.* (2014) Detection and prognostic significance of circulating tumour cells in patients with metastatic breast cancer according to immunohistochemical subtypes. *British Journal of Cancer*, **110**, 375–383.

129. Wallwiener, M., Hartkopf, A.D., Baccelli, I. *et al.* (2013) The prognostic impact of circulating tumor cells in subtypes of metastatic breast cancer. *Breast Cancer Research and Treatment*, **137**, 503–510.
130. Yang, L., Lang, J.C., Balasubramanian, P. *et al.* (2009) Optimization of an enrichment process for circulating tumor cells from the blood of head and neck cancer patients through depletion of normal cells. *Biotechnology and Bioengineering*, **102**, 521–534.
131. Nagrath, S., Sequist, L.V., Maheswaran, S. *et al.* (2007) Isolation of rare circulating tumour cells in cancer patients by microchip technology. *Nature*, **450**, 1235–1239.
132. Stott, S.L., Hsu, C.H., Tsukrov, D.I. *et al.* (2010) Isolation of circulating tumor cells using a microvortex-generating herringbone-chip. *Proceedings of the National Academy of Sciences of the United States of America*, **107**, 18392–18397.
133. Karabacak, N.M., Spuhler, P.S., Fachin, F. *et al.* (2014) Microfluidic, marker-free isolation of circulating tumor cells from blood samples. *Nature Protocols*, **9**, 694–710.
134. Alix-Panabières, C. and Pantel, K. (2014) Technologies for detection of circulating tumor cells: facts and vision. *Lab on a Chip*, **14**, 57–62.
135. Yao, X., Choudhury, A.D., Yamanaka, Y.J. *et al.* (2014) Functional analysis of single cells identifies a rare subset of circulating tumor cells with malignant traits. *Integrative Biology*, **6**, 388–398.
136. Paris, P.L., Kobayashi, Y., Zhao, Q. *et al.* (2009) Functional phenotyping and genotyping of circulating tumor cells from patients with castration resistant prostate cancer. *Cancer Letters*, **277**, 164–173.
137. Ramirez, J.M., Fehm, T., Orsini, M. *et al.* (2014) Prognostic relevance of viable circulating tumor cells detected by EPISPOT in metastatic breast cancer patients. *Clinical Chemistry*, **60**, 214–221.
138. Talasaz, A.H., Powell, A.A., Huber, D.E. *et al.* (2009) Isolating highly enriched populations of circulating epithelial cells and other rare cells from blood using a magnetic sweeper device. *Proceedings of the National Academy of Sciences of the United States of America*, **106**, 3970–3975.
139. Deng, G., Krishnakumar, S., Powell, A.A. *et al.* (2014) Single cell mutational analysis of PIK3CA in circulating tumor cells and metastases in breast cancer reveals heterogeneity, discordance, and mutation persistence in cultured disseminated tumor cells from bone marrow. *BMC Cancer*, **14**, 456.
140. Cann, G.M., Gulzar, Z.G., Cooper, S. *et al.* (2012) mRNA-Seq of single prostate cancer circulating tumor cells reveals recapitulation of gene expression and pathways found in prostate cancer. *PloS One*, **7**, e49144.
141. Lohr, J.G., Adalsteinsson, V.A., Cibulskis, K. *et al.* (2014) Whole-exome sequencing of circulating tumor cells provides a window into metastatic prostate cancer. *Nature Biotechnology*, **32**, 479–484.
142. Ameri, K., Luong, R., Zhang, H. *et al.* (2010) Circulating tumour cells demonstrate an altered response to hypoxia and an aggressive phenotype. *British Journal of Cancer*, **102**, 561–569.
143. Khoo, B.L., Warkiani, M.E., Tan, D.S. *et al.* (2014) Clinical validation of an ultra high-throughput spiral microfluidics for the detection and enrichment of viable circulating tumor cells. *PloS One*, **9**, e99409.
144. Shim, S., Stemke-Hale, K., Tsimberidou, A.M. *et al.* (2013) Antibody-independent isolation of circulating tumor cells by continuous-flow dielectrophoresis. *Biomicrofluidics*, **7**, 11807.

145. Sollier, E., Go, D.E., Che, J. *et al.* (2014) Size-selective collection of circulating tumor cells using Vortex technology. *Lab on a Chip*, **14**, 63–77.

146. Saucedo-Zeni, N., Mewes, S., Niestroj, R. *et al.* (2012) A novel method for the in vivo isolation of circulating tumor cells from peripheral blood of cancer patients using a functionalized and structured medical wire. *International Journal of Oncology*, **41**, 1241–1250.

147. Wiedswang, G., Borgen, E., Schirmer, C. *et al.* (2006) Comparison of the clinical significance of occult tumor cells in blood and bone marrow in breast cancer. *International Journal of Cancer*, **118**, 2013–2019.

148. Fehm, T., Hoffmann, O., Aktas, B. *et al.* (2009) Detection and characterization of circulating tumor cells in blood of primary breast cancer patients by RT-PCR and comparison to status of bone marrow disseminated cells. *Breast Cancer Research*, **11**, R59.

149. Ross, A.A., Cooper, B.W., Lazarus, H.M. *et al.* (1993) Detection and viability of tumor cells in peripheral blood stem cell collections from breast cancer patients using immunocytochemical and clonogenic assay techniques. *Blood*, **82**, 2605–2610.

150. Solakoglu, O., Maierhofer, C., Lahr, G. *et al.* (2002) Heterogeneous proliferative potential of occult metastatic cells in bone marrow of patients with solid epithelial tumors. *Proceedings of the National Academy of Sciences of the United States of America*, **99**, 2246–2251.

151. Zhang, L., Ridgway, L.D., Wetzel, M.D. *et al.* (2013) The identification and characterization of breast cancer CTCs competent for brain metastasis. *Science Translational Medicine*, **5**, 180ra48.

152. Pizon, M., Zimon, D., Carl, S. *et al.* (2013) Heterogeneity of circulating epithelial tumour cells from individual patients with respect to expression profiles and clonal growth (sphere formation) in breast cancer. *Ecancermedicalscience*, **7**, 343.

153. Yu, M., Bardia, A., Aceto, N. *et al.* (2014) Cancer therapy. Ex vivo culture of circulating breast tumor cells for individualized testing of drug susceptibility. *Science*, **345**, 216–220.

154. Misale, S., Yaeger, R., Hobor, S. *et al.* (2012) Emergence of KRAS mutations and acquired resistance to anti-EGFR therapy in colorectal cancer. *Nature*, **486**, 532–536.

155. Diaz, L.A. Jr., Williams, R.T., Wu, J. *et al.* (2012) The molecular evolution of acquired resistance to targeted EGFR blockade in colorectal cancers. *Nature*, **486**, 537–540.

156. De Roock, W., Claes, B., Bernasconi, D. *et al. et al.* (2010) Effects of KRAS, BRAF, NRAS, and PIK3CA mutations on the efficacy of cetuximab plus chemotherapy in chemotherapy-refractory metastatic colorectal cancer: a retrospective consortium analysis. *Lancet Oncology*, **11**, 753–762.

157. Heitzer, E., Auer, M., Gasch, C. *et al.* (2013) Complex tumor genomes inferred from single circulating tumor cells by array-CGH and next-generation sequencing. *Cancer Research*, **73**, 2965–2975.

158. Czyż, Z.T., Hoffmann, M., Schlimok, G. *et al.* (2014) Reliable single cell array CGH for clinical samples. *PloS One*, **9**, e85907.

159. Markou, A., Strati, A., Malamos, N. *et al.* (2011) Molecular characterization of circulating tumor cells in breast cancer by a liquid bead array hybridization assay. *Clinical Chemistry*, **57**, 421–430.

160. Sieuwerts, A.M., Mostert, B., Bolt-de Vries, J. *et al.* (2011) mRNA and microRNA expression profiles in circulating tumor cells and primary tumors of metastatic breast cancer patients. *Clinical Cancer Research*, **17**, 3600–3618.

161. Fischer, J.C., Niederacher, D., Topp, S.A. *et al.* (2013) Diagnostic leukapheresis enables reliable detection of circulating tumor cells of nonmetastatic cancer patients. *Proceedings of the National Academy of Sciences of the United States of America*, **110**, 16580–16585.
162. Seraj, M.J., Samant, R.S., Verderame, M.F., and Welch, D.R. (2000) Functional evidence for a novel human breast carcinoma metastasis suppressor, BRMS1, encoded at chromosome 11q13. *Cancer Research*, **60**, 2764–2769.
163. Chimonidou, M., Strati, A., Tzitzira, A. *et al.* (2011) DNA methylation of tumor suppressor and metastasis suppressor genes in circulating tumor cells. *Clinical Chemistry*, **57**, 1169–1177.
164. Chimonidou, M., Kallergi, G., Georgoulias, V. *et al.* (2013) Breast cancer metastasis suppressor-1 promoter methylation in primary breast tumors and corresponding circulating tumor cells. *Molecular Cancer Research*, **11**, 1248–1257.
165. Kidess, E. and Jeffrey, S.S. (2013) Circulating tumor cells versus tumor-derived cell-free DNA: rivals or partners in cancer care in the era of single-cell analysis? *Genome Medicine*, **5**, 70.
166. Murtaza, M., Dawson, S.J., Tsui, D.W. *et al.* (2013) Non-invasive analysis of acquired resistance to cancer therapy by sequencing of plasma DNA. *Nature*, **497**, 108–112.
167. Dawson, S.J., Tsui, D.W., Murtaza, M. *et al.* (2013) Analysis of circulating tumor DNA to monitor metastatic breast cancer. *New England Journal of Medicine*, **368**, 1199–1209.
168. Schwarzenbach, H., Hoon, D.S., and Pantel, K. (2011) Cell-free nucleic acids as biomarkers in cancer patients. *Nature Reviews Cancer*, **11**, 426–437.
169. Bettegowda, C., Sausen, M., Leary, R.J. *et al.* (2014) Detection of circulating tumor DNA in early- and late-stage human malignancies. *Science Translational Medicine*, **6**, 224ra24.
170. Newman, A.M., Bratman, S.V., To, J. *et al.* (2014) An ultrasensitive method for quantitating circulating tumor DNA with broad patient coverage. *Nature Medicine*, **20**, 548–554.
171. Danila, D.C., Pantel, K., Fleisher, M., and Scher, H.I. (2011) Circulating tumors cells as biomarkers: progress toward biomarker qualification. *Cancer Journal*, **17**, 438–450.
172. Barak, V., Carlin, D., Sulkes, A. *et al.* (1988) CA15-3 serum levels in breast cancer and other malignancies--correlation with clinical course. *Israel Journal of Medical Sciences*, **24**, 623–627.
173. Hashim, Z.M. (2014) The significance of CA15-3 in breast cancer patients and its relationship to HER-2 receptor status. *International Journal of Immunopathology and Pharmacology*, **27**, 45–51.
174. Yerushalmi, R., Tyldesley, S., Kennecke, H. *et al.* (2012) Tumor markers in metastatic breast cancer subtypes: frequency of elevation and correlation with outcome. *Annals of Oncology*, **23**, 338–345.
175. Uhr, J.W. (2013) The clinical potential of circulating tumor cells; the need to incorporate a modern "immunological cocktail" in the assay. *Cancers*, **5**, 1739–1747.
176. Deng, G., Herrier, M., Burgess, D. *et al.* (2008) Enrichment with anti-cytokeratin alone or combined with anti-EpCAM antibodies significantly increases the sensitivity for circulating tumor cell detection in metastatic breast cancer patients. *Breast Cancer Research*, **10**, R69.
177. Skerra, A. (2000) Engineered protein scaffolds for molecular recognition. *Journal of Molecular Recognition*, **13**, 167–187.
178. McKeague, M. and Derosa, M.C. (2012) Challenges and opportunities for small molecule aptamer development. *Journal of Nucleic Acids*, **2012**, 748913.

179. Shigdar, S., Lin, J., Yu, Y. *et al.* (2011) RNA aptamer against a cancer stem cell marker epithelial cell adhesion molecule. *Cancer Science*, **102**, 991–998.

180. Stefan, N., Martin-Killias, P., Wyss-Stoeckle, S. *et al.* (2011) DARPins recognizing the tumor-associated antigen EpCAM selected by phage and ribosome display and engineered for multivalency. *Journal of Molecular Biology*, **413**, 826–843.

181. Theurillat, J.P., Dreier, B., Nagy-Davidescu, G. *et al.* (2010) Designed ankyrin repeat proteins: a novel tool for testing epidermal growth factor receptor 2 expression in breast cancer. *Modern Pathology*, **23**, 1289–1297.

182. Orlova, A., Magnusson, M., Eriksson, T.L. *et al.* (2006) Tumor imaging using a picomolar affinity HER2 binding affibody molecule. *Cancer Research*, **66**, 4339–4348.

183. Järviluoma, A., Strandin, T., Lülf, S. *et al.* (2012) High-affinity target binding engineered via fusion of a single-domain antibody fragment with a ligand-tailored SH3 domain. *PloS One*, **7**, e40331.

2

INTRODUCTION TO MICROFLUIDICS

KANGFU CHEN

Department of Mechanical and Aerospace Engineering, University of Florida, Gainesville, FL, USA

Z. HUGH FAN

Department of Mechanical and Aerospace Engineering, University of Florida, Gainesville, FL, USA; J. Crayton Pruitt Family Department of Biomedical Engineering, University of Florida, Gainesville, FL, USA; Department of Chemistry, University of Florida, Gainesville, FL, USA

2.1 INTRODUCTION

2.1.1 Brief History

In 1959, Feynman [1] gave a talk entitled "There's Plenty of Room at the Bottom" at the Annual Meeting of American Physical Society, opening the door for the field of miniaturization. Since then, microelectronics has changed the world through its impacts on computers and communication with its adherence to Moore's Law [2, 3]. Miniaturized nonelectronic technologies have shown great potential for a variety of applications, even though they have not developed as fast as microelectronics. Microfluidics is an important part of microtechnology, and it has seen significant growth in the last couple of decades with applications in different fields, especially in life sciences and chemistry [4].

The history of microfluidics dates back to 1979 when a miniature gas chromatography device was fabricated on a silicon wafer by Terry *et al.* [5, 6]. Silicon-based microflow meters, micropumps, and microvalves were developed in the 1980s [4]. The concept of Miniaturized Total chemical Analysis System (μTAS) was formulated

Circulating Tumor Cells: Isolation and Analysis, First Edition. Edited by Z. Hugh Fan.

by Andreas Manz and his colleagues in 1990, which truly initiated the field [7]. μTAS is a device that integrates all necessary components for analysis of a sample, and it was envisioned as an alternative to chemical sensors. The device is often called as Lab-on-a-Chip [8, 9] because it contains all functional elements for a complete analytical operation. Explosive growth in the field was observed in the 1990s, and the rapid development of microfluidics afterward was partially helped by using new materials such as polydimethylsiloxane (PDMS) and thermoplastics for device fabrication. Microfluidics has been developed for a large range of applications, including DNA sequencing [10, 11], PCR [12, 13], protein analysis [14–16], cell studies, and others.

2.1.2 Fluids

Before introducing microfluidics, we discuss what fluids are. A fluid is a substance that deforms continuously under the application of a tangential force, no matter how small the force is. In other words, a fluid cannot withstand shear force – this is the basic difference between fluids and solids. Fluids include both liquids and gases. Compressibility is the main difference between liquid and gas: liquid is usually considered to be incompressible, whereas gas is compressible.

For fluid mechanics, the following fluid properties are often studied: (i) energy transport: heat conduction or convection within a fluid field or between a fluid system and an outer environment. Fluids are used to carry power from a power source to certain destinations or for cooling mechanical systems. On the microscale, fluids have been used for cooling integrated circuits [17–19]; (ii) momentum transport: fluids are used to transmit or create forces. For instance, fluids are used for microthrusters [20, 21]; (iii) material transport: fluids are widely used as a carrier of materials and even as a medium for chemical reactions. For example, a solution is generally employed for cell separations, chemical syntheses, and as a medium for chemical sensors [22–24].

2.1.3 Microfluidics

There are several definitions for microfluidics in the literature. Whitesides defined microfluidics as "the science and technology of systems that process or manipulate small (10^{-9} to 10^{-18} L) amounts of fluids, using channels with dimensions of tens to hundreds of micrometers [2]." Stone *et al.* [25] referred to microfluidics as "devices and methods for controlling and manipulating fluid flow with length scales less than a millimeter." Other literature considered microfluidics as a subdiscipline of microelectromechanical system (MEMS) or a discipline that originated from MEMS [4, 5, 26]. The key difference of microfluidics from the traditional MEMS is that microfluidics is characterized by the small size of fluidic channels.

We define microfluidics as a field to study fluids on the micron scale involving three aspects: microfabrication, microflows, and applications. Microfabrication refers to a range of methods to fabricate and manufacture microfluidic devices, including photolithography, molding, casting, bonding, and packaging, some of which are detailed in Section 2.3. Microflows involve the studies of fluids on the microscale as they may behave differently in comparison with the macroscale (e.g., they tend to be laminar). It also involves microscale fluid manipulation such as micropumps, microvalves, and micromixers. Applications are one of key drivers of the field, ranging from the analysis of small molecules, DNA, proteins, cells, tissue, and organs, to diagnostics, drugs, and energy.

The key characteristic of microfluidics is the small scale of fluid flows. However, it does not necessarily mean a microfluidic device itself is small (though miniaturization of devices is, to some extent, an important subject). Instead, it contains microchannels, through which fluids flow [4]. Then what is the size to distinguish macro- and microscale fluidics? Microfluidics is generally classified on the scale of submillimeter or smaller. However, in practice, specific application is of more importance. It is the fluid behavior on the microscale that is attractive, rather than just the small size of fluids.

2.1.4 Advantages of Microfluidics

As a fluid is shrunk down to the microscale, there are several advantages. First, the surface-area-to-volume ratio is inversely proportional to the characteristic length of a fluid. As a result, the surface-area-to-volume ratio increases significantly as the fluid dimension decreases from the macroscale to the microscale. A higher surface-area-to-volume ratio creates more opportunities for interactions among molecules within a confined space, leading to more efficient chemical reactions. In addition, the diffusion time for a molecule to migrate from the bulk of a solution to the channel surface on the microscale is much shorter than that on the macroscale, resulting in faster diffusion-based mixing within the solution and faster interaction between the solutes in the solution and those immobilized on the channel surfaces.

Another major advantage is less sample and reagent consumption on the microscale. Low sample volume requirement is beneficial in the situations where the amount of samples is limited, such as blood samples of newborn babies. Small volumes in flow manipulation also lead to enhanced thermal transfer and more effective reactions [26, 27].

In addition, integration of different procedures – that traditionally were performed in multiple steps in a lab – into one single device dramatically simplifies the process of chemical analysis, decreases the analysis time, and reduces cost. Miniaturization also leads to portability, which makes it friendly to users. Additionally, many devices offer high sample throughput due to highly parallel integration, allowing multiplexed detections and multiple analyses in one single device [28, 29].

It should be noted that each advantage comes with caveats. For instance, the small sample volume requirement is an advantage as discussed earlier. However, a certain volume of a sample is required for low abundant analytes. Too low a volume (say 1 pL) of a sample means it statistically contains no analytes of interest (e.g., at 10^{-12} M or lower).

As will be discussed in the subsequent chapters, microfluidic devices are used to isolate circulating tumor cells (CTCs). Advantages of microfluidic devices for CTC isolation include high CTC detection sensitivity and spatial resolution, possible integration with a reference system with little human intervention, low cost, and potential for portability [30–32]. Among several microfluidic devices designed for CTC isolation, one with micropost arrays or herringbone-based mixers showed increased interaction between CTCs and antibodies immobilized on the device [31, 32].

2.2 SCALING LAW

To analyze and study microflows, classical fluid mechanics theories are often employed except for special conditions. The foundation of fluid mechanics theories is the continuum assumption, which implies two main points: (i) physical properties such as density, temperature, and velocity should be distributed throughout the control volume of interest with finite values and without singular points; and (ii) properties of one point in the control volume may be interpreted as an average of a great number of microscopic particles [33]. To justify the validity of classical fluid mechanics applicable in most cases of microfluidics, an approximate argument can be made as follows. The volume of a water molecule is 3×10^{-29} m^3. Thus, for a volume of 0.001 μm^3 (i.e., a cube with a length of 0.1 μm for each side), there are 3.3×10^7 water molecules, which is big enough for averaging properties of the volume. Most microchannels are around 10 μm or larger in each dimension.

2.2.1 Laminar Flow

According to classical fluid mechanics theory, the Reynolds number (*Re*) is the key factor that describes the flow condition of a fluid. The Reynolds number is defined as

$$Re = \frac{uL}{\nu} \tag{2.1}$$

where u is the flow velocity, L is a length scale characteristic of the flow, and ν is the kinematic viscosity of the flow. The physical significance implied in *Re* is that it measures the ratio of inertial force and viscous force. From the expression of Reynolds number, we can conclude that when $Re \gg 1$, the inertial force dominates the flow field; such a flow is called an inviscid flow. When this condition is not satisfied, viscous force cannot be neglected and the flow is called viscous flow. For viscous fluids, the no-slip condition is often valid, which states that the flow velocity at a solid boundary is zero.

For a fluid flow in a pipe, when *Re* is smaller than 1500, the flow is laminar. When *Re* is greater than 2300, the flow becomes turbulent. Between being laminar and turbulent, there is a transitional region [4]. For microfluidics, *Re* values are usually low. For example, the Reynolds number is 0.01 for a microchannel with a hydraulic diameter of 10 μm, with a flow velocity of 1 mm/s and kinematic viscosity of 1×10^{-6} m^2/s. This is much smaller than 1500. Therefore, a microscale flow is generally laminar.

However, the continuum assumption is not necessarily valid in all cases for microfluidics applications. When gas, instead of liquid, is used in a microchannel, several issues have been reported [34–36]. Noncontinuum effects can be judged by the Knudsen number (Kn), which is defined as

$$Kn = \frac{\lambda}{d} \tag{2.2}$$

where λ is the mean free path of the gas molecules and d is the channel diameter. Noncontinuum effects increase as Kn increases. For gas flows in a microchannel, when $Kn \gg 1$, it is more probable for a molecule to collide with the channel than with another molecule [37]. In 1879, Maxwell [38] predicted that the no-slip condition was violated as noncontinuum effects increase. It is widely agreed that gas is in the slip condition at the boundaries as $Kn \gg 1$, where the continuum assumption breaks [34, 36, 37].

2.2.2 Flow Rate

The flow rate in microchannels or microchambers is a critical parameter for CTC isolation and some other applications. The velocity distribution field in a microchannel is important when analyzing the interactions between cells in the flow and antibodies on the channel surfaces. To calculate flow velocity, V, in a circular channel with radius of R when a fluid is introduced with a pressure P, the Navier–Stokes equation for an incompressible flow can be used as follows:

$$\frac{\partial \boldsymbol{V}}{\partial t} + \boldsymbol{V} \cdot \nabla \boldsymbol{V} = -\frac{1}{\rho}\nabla p + \nu \nabla^2 \boldsymbol{V} \tag{2.3}$$

where $\nu=\mu/\rho$, μ is the dynamic viscosity of the fluid, ρ is the density of the fluid, and t is the time.

For a steady, fully developed, and axisymmetric flow, the axial velocity profile along the channel can be derived from the Navier–Stokes equation:

$$V_z = \frac{R^2}{4\mu}\left(-\frac{dP}{dz}\right)\left(1 - \frac{r^2}{R^2}\right) \tag{2.4}$$

where r is the distance from the center of the channel toward the wall, while dP/dz is the pressure gradient along the channel. The resulting parabolic velocity profile is shown in Figure 2.1, just as in a macroscale pipe. Note that the velocity profile in a microchannel could be different in some cases. For example, it is a flat profile in an electroosmotically pumped flow.

Often in practice, a constant pressure difference is applied to the fluid domain. As a result, $-dP/dz = \Delta P/L$, where ΔP is the pressure drop along the channel and L is the length of the channel. Therefore, the volume flow rate, Q, can be calculated by integrating V_z along the cross-sectional area:

$$Q = \frac{\pi R^4 \Delta P}{8\mu L} \tag{2.5}$$

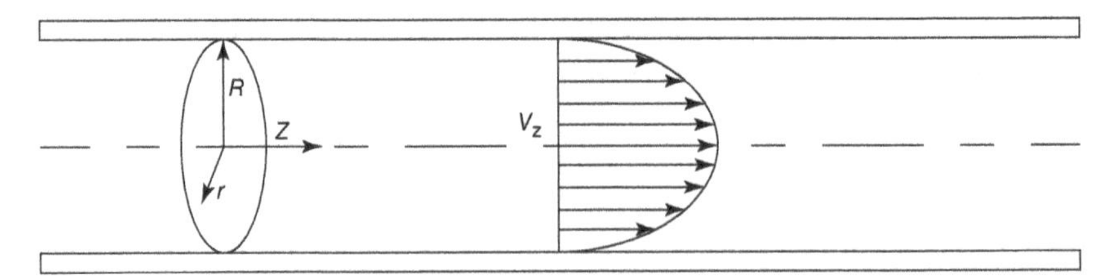

Figure 2.1 Fluid flow velocity profile in a microchannel under a pressure gradient.

Note that the flow rate has fourth power relationship with the channel diameter. When the dimension is reduced from a macroscale pipe (say 10 mm) to a microscale channel (say 10 μm), the volume flow rate is decreased by 10^{12} times if other conditions are the same. Due to this scaling relationship, the reagent consumption can be significantly reduced.

2.2.3 Diffusion

For most microfluidic systems, laminar flow dominates the flow field as discussed earlier. Therefore, mass transport among fluid streams depends on the diffusion and advection of molecules [4]. Diffusion is defined as the stochastic process by which molecules drift from one region to another in response to the concentration gradient [5]. Fick's first law gives a mathematical description of diffusion:

$$-J(x,t) = D\frac{\partial C(x,t)}{\partial x} \tag{2.6}$$

where J is the mass flux (which means net mass transfer in units of amount per unit time per unit area), C is the sample concentration, and D is the diffusion coefficient of a solute in the sample.

Fick's first law shows mass transport across a stationary fluid interface. As molecules transfer within a fluid, the sample concentration in a control volume will change over time. Fick's second law describes the time rate of change in concentration in a stationary fluid domain.

$$\frac{\partial C(x,t)}{\partial t} = D\frac{\partial^2 C(x,t)}{\partial x^2} \tag{2.7}$$

The scale analysis of this equation leads to

$$x \sim \sqrt{Dt} \tag{2.8}$$

Therefore, the diffusion length x of a molecule is proportional to $\sqrt{Dt}$. In a microfluidic channel, x could be the width or depth of the channel. Through the aforementioned relationship, the time for molecules to diffuse from the center of the channel to the channel surface can be estimated for the scale comparison.

Alternatively, the Einstein equation for Brownian motion can be used to calculate the diffusion time:

$$x = \sqrt{2Dt} \tag{2.9}$$

These equations suggest that the diffusion time (t) will be reduced significantly (10^6 times for the said example) when the dimension (x) is reduced from a macroscale pipe (say 10 mm) to a microscale channel (say 10 μm). The time required to mix reagents in microchannels will be reduced accordingly.

2.3 DEVICE FABRICATION

2.3.1 Materials

The properties of the material used for fabricating microfluidic devices are of fundamental importance; they include machinability, surface charges, water adsorption, surface properties, optical properties, ability to modify surfaces, and easiness of bonding/assembly [39]. Many materials have been used for fabricating microfluidic devices, including silicon, glass, and a variety of polymers. Next is a brief introduction to each material.

Silicon is the material used for early microfluidic device fabrication because it is used in microelectronics. Although it is no longer popular, silicon is still used in some efforts. Silicon with extreme purity can be controllably produced with low cost. Thanks to the development of microelectronics, silicon wafer processing such as thin-film deposition is amenable to miniaturization. Pattern generation techniques in silicon such as photolithography are capable of high precision. The downsides of silicon include being not transparent and its electrical conductivity (which is not desirable in applications involving electroosmotic pumping).

Glass, a cheaper material than silicon, is attractive for micromachining. Compared with polymers, most of which are not compatible with organic solvents, glass has much better chemical resistance. Because glass is transparent in the visible and a part of UV region of the spectrum, it has an advantage over silicon in terms of optical detection. Moreover, glass is an insulator, which is another advantage over silicon. These differences lead to a lot of devices being fabricated in glass rather than in silicon.

PDMS is one of the most popular materials for microfluidic device fabrication. PDMS is a thermosetting polymer, and it is one kind of silicone. Its easy and fast fabrication by simply mixing and curing makes it well liked. Its low Young's modulus makes PDMS an ideal material for soft lithography. Furthermore, PDMS is transparent, similar to glass. It can also be either permanently or reversibly bonded to a glass plate. PDMS has high permeability to gases, which is beneficial for applications such as cell culture. One drawback of PDMS is that it swells when exposed to organic solvents and expands with increasing temperature.

Thermoplastics are also polymer materials that are popular for microfluidic device fabrication, especially with a goal of a commercial end product. The most common thermoplastic materials used for microfluidic devices include poly(methylmethacrylate) (PMMA), cyclic olefin copolymers (COCs), polycarbonate, and polyester. Advantages of these polymer materials include established manufacturing methods, low cost to produce a large quantity of devices, disposability, chemical and biological compatibility, and a large range of materials from which to choose (e.g., for the transparency property).

2.3.2 Fabrication Methods

2.3.2.1 Photolithography Photolithography is the most common technique for fabricating microscale structures [4]. Its typical process consists of the following steps.

Step 1 is the design of a photomask. Computer-aided design (CAD) software (e.g., AutoCAD) is available for designing a photomask [40]. Patterns, such as microchannels or other microfeatures, can be designed with high accuracy. Transparency mask and chrome (or emulsion) patterned glass mask are two types of photomasks that are widely used in microfluidic device fabrication [41]. Photomasks could be made in two types: clear mask and dark mask. A clear field mask refers to a mask wherein the features are light-proof while the background is clear; for a dark field mask, it is the opposite. After the photomask design is finalized, a chrome mask is often sent out to a photomask manufacturer, who will make it and ship it back to the designer. A transparency mask can be printed in-house or by a professional printing shop that has higher resolution printers.

Step 2 is to prepare surface preparation. For a silicon wafer, this step is to coat hexamethyldisilazane (HMDS) for promoting the adhesion of photoresist (from the next step) onto the wafer. For a glass substrate, this step may include surface cleaning and coating of a thin layer of chromium (as an adhesion layer) and a layer of gold as the etching mask.

Step 3 is to apply photoresist to the wafer (or glass substrate) through spin coating, which is an important technique in device fabrication. To implement the process, the wafer is first held onto the vacuum chuck of the spinner. Then, the photoresist is dispensed on the center of the wafer. As the spinner spins, the photoresist is distributed evenly on the wafer. The thickness of the photoresist depends on the spin speed, solution concentration, and molecular weight of the photoresist. The amount of the photoresist used depends on the expected thickness of the photoresist.

Photoresist is a photosensitive polymer, and it consists of a base resin, a sensitizer, and a casting solvent. There are two types of photoresist: positive photoresist and negative photoresist. For positive photoresist, the base resin will go through chemical reactions when exposed to UV light, and it becomes soluble in its developer solution. Conversely, negative photoresist forms an insoluble polymer after UV exposure, whereas the nonexposed area is soluble in its developer solution.

Step 4 is soft bake. The purpose of soft bake is to improve the adhesion of the photoresist onto the wafer, uniformity, etch resistance, line width control, and light absorbance characteristics of photoresist. The soft bake is often done by placing the wafer under vacuum for a few minutes or under the ambient atmosphere at 75–100 °C for 10–30 min.

Step 5 is to align the wafer to the photomask and then expose it to light. This step transfers the photomask image to the photoresist-coated wafer. Specific equipment called a mask aligner is often used for alignment and UV light exposure.

After exposure, step 6 is to soak the wafer in a developer solution to remove the exposed photoresist (if it is a positive one) so that a pattern appears on the wafer. The development time depends on the thickness of the photoresist layer. Too short a time causes underdevelopment when the exposed photoresist is not thoroughly removed. Too long a time may cause unexpected undercut of the photoresist. In a worse case, the designed feature can be damaged.

Step 7 is hard bake. This step is to evaporate solvents in the remaining photoresist as well as to improve the adhesion of the photoresist onto the wafer. The hard bake procedure requires a higher temperature and longer time than the soft bake step.

After inspection under a microscope, a properly processed wafer will go through the etching step.

2.3.2.2 Etching Etching is another important technique in microfluidic device fabrication. There are two types of etching methods, namely, wet etching and dry etching.

Wet etching refers to a process of etching a solid material in a chemical solution called etchant. The major advantages of wet etching include high selectivity, smooth channel surface, repeatability, and control of the etching depth by controlling the etchant concentration and etching time [4]. Most etching methods result in either isotropic etching or anisotropic etching. For isotropic etching, the etching rate is equal in all directions regardless of the crystalline orientation of the silicon wafer. Undercutting is a significant characteristic for isotropic etching, which refers to unintentional etching of the substrate material underneath the etching mask. As a result, microchannels formed by isotropic etching are usually D-shaped. Because glass has amorphous structures, it always produces isotropic etching profiles. An example of etchants used for isotropic etching is a mixture of hydrofluoric acid, nitric acid, and acetic acid for silicon or a mixture of hydrofluoric acid and nitric acid for glass. For anisotropic etching, the etching rate is different according to the crystalline orientation of the silicon wafer. An example of etchants used for anisotropic etching is a potassium hydroxide solution for silicon.

Dry etching techniques are often carried out using commercial equipment (in comparison to the wet etching, which may be performed in a beaker or other containers). They generally use plasmas to drive chemical reactions or employ energetic ion beams to remove material. One of these techniques is called reactive ion etching (RIE). After a silicon wafer is placed in a dry etching reaction chamber, reactive ions are accelerated toward the wafer. These reactive ions are in a plasma, which is generated under vacuum by an electromagnetic field that dissociates a gas (e.g., sulfur hexafluoride) to produce high-energy ions, electrons, photons, and reactive radicals. Bombardment of these high-energy particles onto a surface produces outgoing fluxes of ions and other particles from the substrate, resulting in etching.

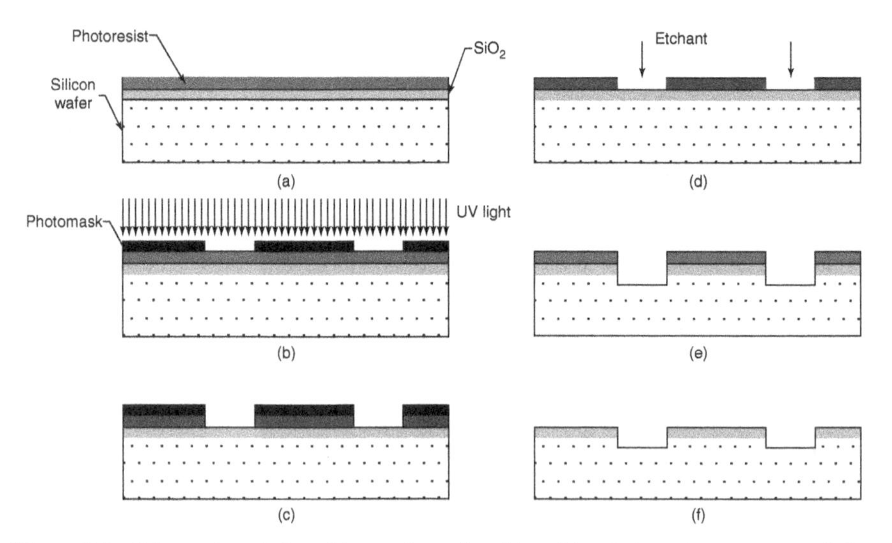

Figure 2.2 Schematics of photolithography and etching. (a)–(c) show the processes of photolithography using a positive photoresist. (d)–(f) show the processes of etching and the removal of photoresist.

After etching, the next step is to remove the photoresist using a solvent or the developer. Microgrooves or other microstructures are properly formed in the substrate. The steps of photolithography and etching steps are illustrated in Figure 2.2.

2.3.2.3 ***Bonding*** After forming microgrooves or other microstructures, the silicon or glass substrate must be bonded with a cover layer to form microchannels or other features such as reaction chambers. Similarly, traditional microelectronics or microfabricated devices must be packaged properly before use even thought they might not need a cover to seal them.

Thermal bonding is one of the techniques often used for glass devices as discussed in the literature [42]. It basically allows two pieces to melt together at an elevated temperature, although it is very critical to have an appropriate temperature. Too high a temperature distorts microchannels due to melting, whereas too low a temperature will not bond the device. Anodic bonding is a method to address the critical temperature issue since it enables to bond a silicon device (or a glass device) with a glass cover at a temperature much lower than the softening temperature [43]. Thermal lamination is one of the popular methods to bond thermoplastic devices, although there are several other methods as reviewed in the literature [44].

2.3.2.4 ***Molding*** While PDMS devices are made by curing as mentioned earlier, thermoplastic devices are often fabricated using one of several commercially proven manufacturing methods. For thermoplastic microfluidic devices, the fabrication methods include hot or cold embossing, injection molding, compression molding, and thermoforming. Injection molding is often used for industrial-scale production such as for compact discs, which contain microgrooves on the backside. Compression molding is generally used on the laboratory scale [45]. The detailed protocols for making microfluidic devices using these fabrication methods have been reviewed in the literature [46, 47].

2.4 FUNCTIONAL COMPONENTS IN MICROFLUIDIC DEVICES

There are many steps required to analyze real-world samples, including sample collection, sample preparation (e.g., dilution), separation (e.g., chromatography), detection, and result interpretation. To carry out some of these steps, a microfluidic device often integrates a variety of components. In this section, we discuss three commonly used components: micropumps, microvalves, and micromixers.

2.4.1 Micropumps

A micropump is necessary for sample/reagent delivery in microfluidic devices. Therefore, integrating pumping components into a microfluidic system is of great importance. While a high flow rate may not be easily achieved in a microchannel due to the high backpressure induced, a reasonable flow rate is often achievable. In addition to pumping, precise metering sometimes is also important [48, 49]. Related considerations include reliability, power consumption, cost, and biocompatibility [50, 51]. In general, micropumps can be classified as mechanic pumps and nonmechanic pumps [4, 52], according to the principle of energy conversion.

2.4.1.1 Mechanical Pumps An actuator is the key component of a mechanical pump [53, 54]. Externally actuated mechanisms include electromagnetic, piezoelectric, pneumatic, and those using shape memory materials. Some actuators can be integrated into the device, including electrostatic, thermopneumatic, electromagnetic, and thermomechanic actuators [55]. External actuators are able to produce relatively large driving forces and sufficient pumping, but they tend to be too big to be integrated into a microfluidic device [53, 55]. On the other hand, integrated actuators usually work well within a microfluidic device with good reliability, but they tend to produce small pumping forces [55]. Therefore, choosing an actuator depends on the requirements of specific applications.

Figure 2.3 shows an example of a micropump with an external piezoelectric actuator. The diaphragm of the pump travels up or down as it is actuated by an external piezoelectric actuator. When the diaphragm moves up, the inlet valve opens and fluid is infused to the flow chamber. When the diaphragm moves down, the output valve opens and fluid is driven out of the chamber. The mechanism of these valves is similar to that of the ball check valves used in liquid chromatography instruments and other systems. The vibration of the diaphragm achieves the overall fluid pumping from the inlet to the outlet.

Other mechanical pumps include those using pneumatic pressure, syringe drivers, bubble generation, and thermal expansion.

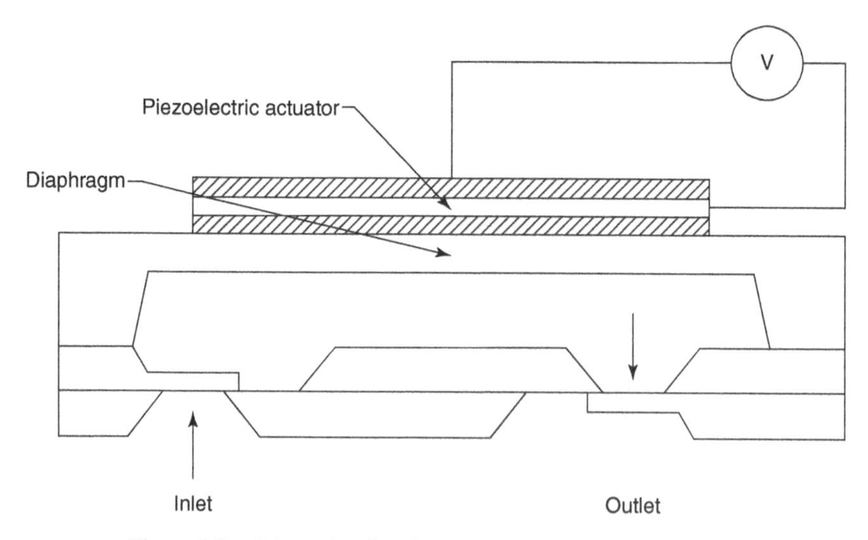

Figure 2.3 Schematic of a micropump with an external actuator.

*2.4.1.2 **Nonmechanical Pumps*** Pumps that convert nonmechanical energy to kinetic energy of fluids are nonmechanical pumps. Electroosmotic pumping is one example, in which electrokinetic phenomena are used as a means to pump fluids. Electroosmosis also takes place in a separation technique called capillary electrophoresis, in which a fused silica capillary is often used.

The fundamentals of electrokinetic pumping are briefly explained here. When a microchannel of glass or silica is filled with a buffer solution with $pH > 3$, the channel walls are negatively charged due to the ionization of silanol groups. The negative charges on the wall attract positive ions from the buffer solution, forming an ionic double-layer structure. Each mobile positive ion in the diffuse layer carries several solvent molecules, and these solvated ions are attracted to the cathode under an electric field. The movement of this sheath of solvated ions drags along the solution in the rest of the channel, resulting in electroosmotic pumping.

In addition to electrokinetic pumping, magnetokinetic pumps drive the bulk fluid through Lorentz force [56]. Electrohydrodynamic pumping occurs because an electric field acts on induced electric charges in a fluid [52]. Dielectrophoretic pumping can be achieved by using an electrical voltage traveling along an array of electrodes.

2.4.2 Microvalves

Microvalves are another important component for microfluidic devices. In many applications, microvalves are used to control and regulate the flow of fluids. They are also used to contain reagents and/or isolate one region of the device from the other. A variety of microvalves have been designed and applied in microfluidic devices. Some examples of these microvalves are discussed next.

Quake and his coworkers fabricated a pneumatically controlled microvalve based on multilayer PDMS [57]. The valve has a crossed channel architecture, where the upper channels are designed for pressure manipulation and the lower channels are for fluids such as samples or reagents. Between these two channel layers, there is a thin layer of PDMS that functions as an elastomer. When sufficient pressure is applied to the upper channel, the elastomer deflects downward and closes the channel. It should be noted that this valve concept has been extensively used by a large number of research groups. Some variations and improvement have been reported, including our efforts in using thermopneumatic pressure to control elastomer-based valve in a thermoplastic device [58, 59].

Another important aspect of Quake's microvalves is their large-scale integration; they integrated 3574 valves in a single device [60]. A variety of applications with massively parallel microvalves have been demonstrated by his group as well as others. Moreover, Fluidigm has commercialized these devices for genomic and single-cell analysis.

Smart-gel-based valves have also been explored in microfluidic devices. Smart gel refers to those gel polymers that respond to stimuli, which include temperature, light, pH, and electric forces. In one example, microposts are fabricated in a microchannel, followed by coating pH-sensitive hydrogel jackets around the posts [61]. Hydrogels expand or shrink according to the pH of a fluid introduced, leading to closing or opening the microchannel [61].

A variety of other microvalves have been reported in the literature. In addition to the flap-based check valves at the inlet and outlet of Figure 2.3 and pneumatically actuated elastomer valves discussed earlier, different microvalves are actuated using magnetic, electric, bubble, thermal, and other mechanisms. Interested readers are directed to review articles on this topic [62].

2.4.3 Micromixers

Mixing is crucial to chemical reactions, biological assays, and other functions of microfluidic devices [63, 64]. Because laminar flow dominates the fluid region as discussed earlier, mixing in microfluidic devices largely depends on molecular diffusion. However, diffusion-based mixing takes a long time. Therefore, methods for rapidly enhanced mixing are desirable for many microfluidic applications.

Mixing methods in microfluidic devices are classified into active mixing and passive mixing [65]. During active mixing, an external energy is introduced into the fluid channels to enhance mixing. For instance, acoustic energy is introduced into the system and bubbles induce vibration in the flow channel [66, 67]. Passive mixing refers to enhanced mixing through rectifying the structures of flow channels without adding external energy.

An example of passive mixers is the incorporation of herringbone structures into a microchannel as demonstrated by Stroock *et al.* [68]. They demonstrated chaotic mixing in the device. Theoretically, twisting of streamlines can significantly increase the chaotic effect of fluids. To generate transverse flows in microchannels, ridges were placed on the floor of the channel at an oblique angle with respect to the long axis of the channel. When a fluid was introduced, transverse velocity of the flow was produced because of the different directions between the bulk fluid velocity and the shear force on the ridge-coated surfaces. Simulation showed that herringbone structures can produce large vortex areas in the microchannel, increasing mixing effects significantly [32].

In addition to the bubble mixer and herringbone mixer discussed earlier, other mixing enhancement methods in microfluidic devices include stream splitting and recombination, microstirrers, fluid pulsation, and electrokinetic mixing, as we have recently reviewed [69].

2.4.4 Other Components

In addition to micropumps, microvalves, and micromixers, there are many other components that are important to microfluidics. They include sample injectors, separation channels, microheaters, detectors, and other functional elements. A typical microfluidic device contains one or a few of these components depending on its particular application. Interested readers may look into monographs for references [70].

Some of these components are very beneficial for CTC isolation discussed in this book. For example, herringbone-based micromixers have been used for CTC isolations [32] as discussed in Chapter 13. It was shown that microfluidic devices with micromixers increased the interactions between CTCs and antibodies immobilized on the channel surfaces, significantly enhancing the cell capture efficiency.

2.5 CONCLUDING REMARKS

While a lot of progress has been made in the past couple of decades, there is still plenty of room for advances in microfluidics in terms of fabrication methods, micropumps, microvalves, flow control, micromixers, sample preparation, surface modification, analyte separation and detection, reproducible manufacturing, and device integration. More importantly, innovative and sometimes unexpected applications enabled by microfluidics are continuously evolving, and these applications could be on the molecular, cellular, tissue, and organ level.

In this chapter, we introduced the fundamentals of microfluidics, which involves microfabrication, microflows, and applications. Microfluidic devices have been fabricated using different materials, including silicon, glass, PDMS, and various thermoplastics. Microfluidics mainly deals with laminar flows characterized by low Reynolds number. A variety of components are needed to control and regulate microflows, including micropumps, microvalves, and micromixers discussed in this chapter.

Microfluidic devices have been explored for a broad range of applications, including DNA sequencing and genomics, protein analysis and proteomics, cellular studies, medical diagnosis, energy research, and many others. Several microfluidic devices have been commercialized for DNA, protein, and cell analysis. As the technology develops, fully integrated and multifunctional microfluidic devices will be available, fulfilling the original goal of lab-on-a-chip.

In addition to the aforementioned applications, microfluidics could play an important role in the isolation and analysis of CTCs. As discussed in Chapter 1, the number of CTCs is extremely low, typically a few CTC cells in billions of healthy blood cells. Microfluidics is in a unique position to address this challenge because of its fundamental attributes discussed in this chapter. First, cells in a sample must diffuse from the bulk of the solution to the solid surface before interacting with the immobilized capture reagents. Microchannels have unique advantages due to much shorter diffusion distances than in a macroscale container. As a result, the capture efficiency is significantly increased and the reaction time can be reduced. Second, microchannels offer significantly higher surface-to-volume ratio than a macroscale container. As a result, the amount of the capture reagents on the surfaces could be sufficiently high for the interactions with the cells. Additionally, microposts and other geometries may be created to enhance the surface areas. These attributes make microfluidic devices attractive for CTC isolation as demonstrated in Chapters 3–6 and other sections of this book.

REFERENCES

1. Feynman, R.P. (1959) There's plenty of room at the bottom. *Journal of Micromechanical Systems*, **1** (1), 60–66.
2. Whitesides, G.M. (2003) The 'right' size in nanobiotechnology. *Nature Biotechnology*, **21** (10), 1161–1165.
3. Moore, G.E. (2000) Cramming more components onto integrated circuits. *Readings in Computer Architecture*, **56**, 82–85.
4. Nguyen, N.-T. and Wereley, S.T. (2002) *Fundamentals and Applications of Microfluidics*, Artech House, pp. 1–110.
5. Greg, J., Sommer, D.S.C., Jain, A. *et al.* (2008) Introduction to Microfluidics, in *Microfluidics for Biological Applications* (ed E.F. Wei-cheng Tian), Springer-Verlag, pp. 1–29.
6. Terry, S.C., Jerman, J.H., and Angell, J.B. (1979) A gas chromatographic air analyzer fabricated on a silicon wafer. *Electron Devices, IEEE Transactions on*, **26** (12), 1880–1886.
7. Manz, A., Graber, N., and Widmer, H.M. (1990) Miniaturized total chemical analysis systems: a novel concept for chemical sensing. *Sensors and Actuators A: Physical*, **B 1**, 244–248.
8. Reyes, D.R. *et al.* (2002) Micro total analysis systems. 1. Introduction, theory, and technology. *Analytical Chemistry*, **74** (12), 2623–2636.
9. Lee, S.J. and Lee, S.Y. (2004) Micro total analysis system (μ-TAS) in biotechnology. *Applied Microbiology and Biotechnology*, **64** (3), 289–299.
10. Woolley, A.T. and Mathies, R.A. (1995) Ultra-high-speed DNA sequencing using capillary electrophoresis chips. *Analytical Chemistry*, **67** (20), 3676–3680.
11. Woolley, A.T. and Mathies, R.A. (1994) Ultra-high-speed DNA fragment separations using microfabricated capillary array electrophoresis chips. *Proceedings of the National Academy of Sciences of the United States of America*, **91** (24), 11348–11352.
12. Kopp, M.U., Mello, A.J.d., and Manz, A. (1998) Chemical Amplification: Continuous-Flow PCR on a Chip. *Science*, **280** (5366), 1046–1048.

13. Krishnan, M., Ugaz, V.M., and Burns, M.A. (2002) PCR in a Rayleigh-Bénard convection cell. *Science*, **298** (5594), 793.
14. Sanders, G.H.W. and Manz, A. (2000) Chip-based microsystems for genomic and proteomic analysis. *TrAC Trends in Analytical Chemistry*, **19** (6), 364–378.
15. Hadd, A.G., Jacobson, S.C., and Ramsey, J.M. (1999) Microfluidic assays of acetylcholinesterase inhibitors. *Analytical Chemistry*, **71** (22), 5206–5212.
16. Hatch, A. *et al.* (2001) A rapid diffusion immunoassay in a T-sensor. *Nature Biotechnology*, **19** (5), 461–465.
17. Linan, J. *et al.* (2002) Closed-loop electroosmotic microchannel cooling system for VLSI circuits. *Components and Packaging Technologies, IEEE Transactions on*, **25** (3), 347–355.
18. Darabi, J. and Ekula, K. (2003) Development of a chip-integrated micro cooling device. *Microelectronics Journal*, **34** (11), 1067–1074.
19. Peterson, G.P. and Ma, H.B. (1996) Theoretical analysis of the maximum heat transport in triangular grooves: a study of idealized micro heat pipes. *Journal of Heat Transfer*, **118** (3), 731–739.
20. Ye, X.Y. *et al.* (2001) Study of a vaporizing water micro-thruster. *Sensors and Actuators A: Physical*, **89** (1–2), 159–165.
21. Juergen, M., *et al.* (1997) *Design, Analysis and Fabrication of a Vaporizing Liquid Micro-Thruster.* 33rd Joint Propulsion Conference and Exhibit, American Institute of Aeronautics and Astronautics.
22. Inglis, D.W. *et al.* (2004) Continuous microfluidic immunomagnetic cell separation. *Applied Physics Letters*, **85** (21), 5093–5095.
23. Lee, C.C. *et al.* (2005) Multistep synthesis of a radiolabeled imaging probe using integrated microfluidics. *Science*, **310** (5755), 1793–1796.
24. Ahn, C.H. *et al.* (2004) Disposable Smart lab on a chip for point-of-care clinical diagnostics. *Proceedings of the IEEE*, **92** (1), 154–173.
25. Stone, H.A., Stroock, A.D., and Ajdari, A. (2004) Engineering flows in small devices: microfluidics toward a lab-on-a-chip. *Annual Review of Fluid Mechanics*, **36**, 381–411.
26. Tesar, V. (2007) *Introduction and Basic Concepts, in* Pressure-Driven Microfluidics, Artech House, Boston, pp. 1–28.
27. Bayraktar, T. and Pidugu, S.B. (2006) Characterization of liquid flows in microfluidic systems. *International Journal of Heat and Mass Transfer*, **49** (5–6), 815–824.
28. Melin, J. and Quake, S.R. (2007) Microfluidic large-scale integration: the evolution of design rules for biological automation. *Annual Review of Biophysics and Biomolecular Structure*, **36**, 213–31.
29. Fair, R.B. (2007) Digital microfluidics: is a true lab-on-a-chip possible? *Microfluidics and Nanofluidics*, **3** (3), 245–281.
30. Bhagat, A. *et al.* (2010) Microfluidics for cell separation. *Medical & Biological Engineering & Computing*, **48** (10), 999–1014.
31. Nagrath, S. *et al.* (2007) Isolation of rare circulating tumour cells in cancer patients by microchip technology. *Nature*, **450** (7173), 1235–1239.
32. Stott, S.L. *et al.* (2010) Isolation of circulating tumor cells using a microvortex-generating herringbone-chip. *Proceedings of the National Academy of Sciences of the United States of America*, **107** (43), 18392–18397.

33. Panton, R.L. (2013) *Incompressible Flow*, John Wiley & Sons.
34. Squires, T.M. and Quake, S.R. (2005) Microfluidics: fluid physics at the nanoliter scale. *Reviews of Modern Physics*, **77** (3), 977–1026.
35. Ho, C.M. and Tai, Y.C. (1998) Micro-electro-mechanical-systems (MEMS) and fluid flows. *Annual Review of Fluid Mechanics*, **30**, 579–612.
36. Gad-el-Hak, M. (1999) The fluid mechanics of microdevices – The Freeman scholar lecture. *Journal of Fluids Engineering*, **121** (1), 5–33.
37. Hardt, S. (2007) Microfluidics: fundamentals and engineering, in *Microfluidic Technologies for Miniaturized Analysis Systems* (ed F.S.S. Hardt), Springer-Verlag, pp. 8–14.
38. Maxwell, J.C. (1879) On stresses in rarified gases arising from inequalities of temperature. *Philosophical Transactions of the Royal Society of London*, **170**, 231–256.
39. Becker, H. and Locascio, L.E. (2002) Polymer microfluidic devices. *Talanta*, **56**, 267–287.
40. Qin, D., Xia, Y., and Whitesides, G.M. (1996) Rapid prototyping of complex structures with feature sizes larger than 20 μm. *Advanced Materials*, **8** (11), 917–919.
41. Qin, D., Xia, Y., and Whitesides, G.M. (2010) Soft lithography for micro- and nanoscale patterning. *Nature Protocols*, **5** (3), 491–502.
42. Fan, Z.H. and Harrison, D.J. (1994) Micromachining of capillary electrophoresis injectors and separators on glass chips and evaluation of flow at capillary intersections. *Analytical Chemistry*, **66** (1), 177–184.
43. Pomerantz, D.I. (1968) Anodic bonding, U.S.A. Patent US3397278A.
44. Tsao, C.-W. and DeVoe, D.L. (2008) Bonding of thermoplastic polymer microfluidics. *Microfluidics and Nanofluidics*, **6** (1), 1–16.
45. Fredrickson, C.K. *et al.* (2006) Effects of fabrication process parameters on the properties of cyclic olefin copolymer microfluidic devices. *Journal of Microelectromechanical System*, **15** (5), 1060–1068.
46. Boone, T.D. *et al.* (2002) Plastic advances microfluidic devices. *Analytical Chemistry*, **74** (3), 78A–86A.
47. Heckele, M. and Schomburg, W.K. (2004) Review on micro molding of thermoplastic polymers. *Journal of Micromechanics and Microengineering*, **14** (3), R1.
48. Dash, A. and Cudworth Ii, G. (1998) Therapeutic applications of implantable drug delivery systems. *Journal of Pharmacological and Toxicological Methods*, **40** (1), 1–12.
49. Laser, D.J. and Santiago, J.G. (2004) A review of micropumps. *Journal of Micromechanics and Microengineering*, **14** (6), R35–R64.
50. Burns, M.A. *et al.* (1998) An integrated nanoliter DNA analysis device. *Science*, **282** (5388), 484–487.
51. Hanaire-Broutin, H. *et al.* (1995) Feasibility of intraperitoneal insulin therapy with programmable implantable pumps in IDDM: a multicenter study. *Diabetes Care*, **18** (3), 388–392.
52. Francis, E.H. and Tay, W.O.C. (2002) in *Microfluidics and BioMEMS Applications* (ed F.E.H. Tay), Kluwer Academic Publisher, Boston, pp. 1–50.
53. Amirouche, F., Zhou, Y., and Johnson, T. (2009) Current micropump technologies and their biomedical applications. *Microsystem Technologies*, **15** (5), 647–666.
54. Shoji, S. and Esashi, M. (1994) Microflow devices and systems. *Journal of Micromechanics and Microengineering*, **4** (4), 157–171.

55. Nguyen, N.-T., Huang, X., and Chuan, T.K. (2002) MEMS-Micropumps: a review. *Journal of Fluids Engineering*, **124** (2), 384.
56. Iverson, B.D. and Garimella, S.V. (2008) Recent advances in microscale pumping technologies: a review and evaluation. *Microfluidics and Nanofluidics*, **5** (2), 145–174.
57. Unger, M.A. *et al.* (2000) Monolithic microfabricated valves and pumps by multilayer soft lithography. *Science*, **288** (5463), 113–116.
58. Pitchaimani, K. *et al.* (2009) Manufacturable plastic microfluidic valves using thermal actuation. *Lab on a Chip*, **9** (21), 3082–3087.
59. Gu, P. *et al.* (2011) Chemical-assisted bonding of thermoplastics/elastomer for fabricating microfluidic valves. *Analytical Chemistry*, **83**, 7.
60. Thorsen, T., Maerkl, S.J., and Quake, S.R. (2002) Microfluidic large-scale integration. *Science*, **298** (5593), 580–584.
61. Beebe, D.J. *et al.* (2000) Functional hydrogel structures for autonomous flow control inside microfluidic channels. *Nature*, **404** (6778), 588–590.
62. Kwang, W.O. and Chong, H.A. (2006) A review of microvalves. *Journal of Micromechanics and Microengineering*, **16** (5), R13.
63. Ottino, J.M. and Wiggins, S. (2004) Introduction: mixing in microfluidics. *Philosophical Transactions of the Royal Society of London Series A-Mathematical Physical and Engineering Sciences*, **362** (1818), 923–935.
64. Knight, J. (2002) Microfluidics: Honey, I shrunk the lab. *Nature*, **418** (6897), 474–475.
65. Hessel, V., Löwe, H., and Schönfeld, F. (2005) Micromixers – a review on passive and active mixing principles. *Chemical Engineering Science*, **60** (8–9), 2479–2501.
66. Liu, R.H. *et al.* (2002) Bubble-induced acoustic micromixing. *Lab on a Chip*, **2** (3), 151–157.
67. Liu, R.H. *et al.* (2003) Hybridization enhancement using cavitation microstreaming. *Analytical Chemistry*, **75** (8), 1911–1917.
68. Stroock, A.D. *et al.* (2002) Chaotic mixer for microchannels. *Science*, **295** (5555), 647–651.
69. Ward, K. and Fan, Z.H. (2015) Mixing in microfluidic devices and enhancement methods. *Journal of Micromechanics and Microengineering*, **25** (9), 094001.
70. Folch, A. (2012) *Introduction to BioMEMS*, CRC Press.

PART II

ISOLATION METHODS

3

ENSEMBLE-DECISION ALIQUOT RANKING (eDAR) FOR CTC ISOLATION AND ANALYSIS

Mengxia Zhao
Department of Chemistry, University of Washington, Seattle, WA, USA
Perry G. Schiro
MiCareo Inc., Taipei, Taiwan
Daniel T. Chiu
Department of Chemistry, University of Washington, Seattle, WA, USA

3.1 OVERVIEW OF eDAR

An ideal method for isolating rare cells, such as circulating tumor cells (CTCs), should be (1) capable of high throughput, so that milliliters of whole blood can be processed within tens of minutes; (2) highly efficient and sensitive, with a recovery efficiency higher than 90% and a limit of detection (LOD) close to 1 CTC per blood sample; (3) capable of easily recovering live CTCs with high purity, which would facilitate downstream molecular analyses or cell culture; and (4) robust and economical to operate [1–6]. To meet these challenges, we developed an approach that is different from what has been explored so far in this area of research and which we believe offers significant advantages.

Our method is based on positive selection, where blood samples are labeled with specific antibodies conjugated with fluorophores, after which we divide the blood sample into nanoliter aliquots (typically 2 nl) that are optically interrogated and

Circulating Tumor Cells: Isolation and Analysis, First Edition. Edited by Z. Hugh Fan.

ranked for the presence or absence of CTCs. This method is called ensemble-decision aliquot ranking (eDAR) because an ensemble of cells within each aliquot is ranked in a single scan [7]. It helps us to select the aliquots of cells that are worth a closer investigation without sacrificing the throughput, which is a key bottleneck in applying flow cytometry for CTC isolation. In traditional flow cytometry, the throughput is limited by the sequential analyses of individual cells, which may take over 24 h for 1 ml of whole blood containing more than 5 billion cells [8]. This throughput constraint is one reason why most CTC detection methods based on flow cytometry can only analyze fractionated blood, either by lysing red blood cells (RBCs) or by centrifuging the blood with density gradient buffer to extract nucleated cells.

eDAR is operationally similar to flow cytometry but with significant differences. To improve the throughput, eDAR probes for rare cells in nanoliter aliquots of blood in a way that each aliquot contains thousands of cells. Due to the high sensitivity of the detection scheme, we are able to detect one labeled CTC per a few nanoliters of blood in less than a millisecond. Aliquots containing CTCs are automatically sorted and collected into a small region on the microfluidic chip, where a further purification method can be applied. The isolated CTCs can be imaged on the microchip or restained with more biomarkers for further analyses.

In comparison with other CTC methods, there are some key advantages and disadvantages of eDAR. A main advantage of eDAR is the three to four orders of magnitude improvement in throughput compared to flow cytometry. For example, we were able to analyze 1 ml of whole-blood sample in less than 20 min with a zero false-positive rate and a recovery efficiency of about 95%. Unlike most existing approaches, where the isolated cells are spread over a large area or multiple slides, in eDAR, the cells are trapped in a much smaller area allowing rapid subsequent imaging. eDAR is extremely efficient at enriching rare cells, discarding 99.999% of the blood volume if there are only five CTCs present per 1 ml of whole blood.

However, the ensemble ranking process limits eDAR to the isolation of rare cells and prevents it from working well for target cells that are abundant. In exchange for the vast improvement in throughput of isolating rare cells, the selection in eDAR is limited to using only fluorescence information. Information from light scattering data, such as cell size and morphology, cannot be obtained because there are many cells within the detection volume that may cause scattering. In contrast, flow cytometry detects in a single-profile format, making it possible to obtain both fluorescence and light-scattering information to select a subset of cells. For the isolation of rare cells, such as CTCs, the lack of light-scattering information is not important because the targeted cells are captured downstream for imaging and individual analysis. Another advantage of eDAR is the reduced hydrodynamic-induced cell stress. In eDAR, a volume containing thousands of cells flows past the interrogation region, resulting in a linear flow rate up to 50 times lower than in flow cytometry, which minimizes shear stress on the cells. Preserving the viability of each isolated cell is particularly important for these rare cells.

3.2 INDIVIDUAL COMPONENTS AND ANALYTICAL PERFORMANCE OF eDAR

Figure 3.1 summarizes the general workflow of eDAR, which can be divided into four steps, aliquot ranking, aliquot sorting, on-chip purification, and secondary labeling and imaging analysis. As an "all-in-one" platform for CTC analysis, eDAR integrates all these four steps into a single microfluidic chip. Most of the steps, such as aliquot ranking, aliquot sorting, and on-chip purification, are highly automated. The step of secondary labeling and imaging is also semiautomated.

3.2.1 Aliquot Ranking

To detect CTCs in aliquots, whole-blood samples need to be stained with fluorescent antibodies specific to CTCs. For most of the clinical samples, we applied epithelial cell adhesion molecules (EpCAMs) as the biomarker to select CTCs, which is consistent with most other CTC methods [9]. This process of sample preparation has been optimized to ensure the best labeling efficiency, and the procedures have been simplified and minimized to almost entirely eliminate fragile cell loss and fragmentation. The sample preparation can also be performed in parallel, which significantly increases the throughput of the assay. Figure 3.2 shows our cell labeling procedure, in which the labeled blood is diluted with buffer and then centrifuged to remove free antibodies. The supernatant is then removed, leaving a final volume that is the same as the initial volume of the sample. Other labeling schemes could also be applied based on the application and the criteria to select CTCs.

Figure 3.3a depicts our eDAR system. The first step in eDAR is to generate or define an aliquot. We initially focused on aliquoting blood into droplets; Figure 3.3b shows a continuous stream of aliquots defined by droplets surrounded by an immiscible phase. We had presumed that encapsulation of the cells in a droplet

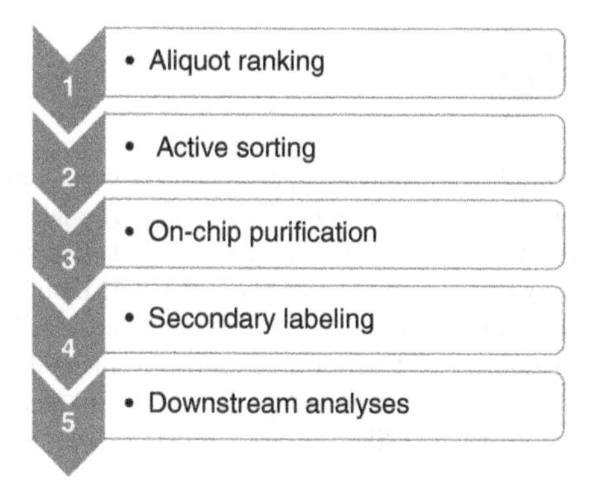

Figure 3.1 General workflow of eDAR having four individual steps.

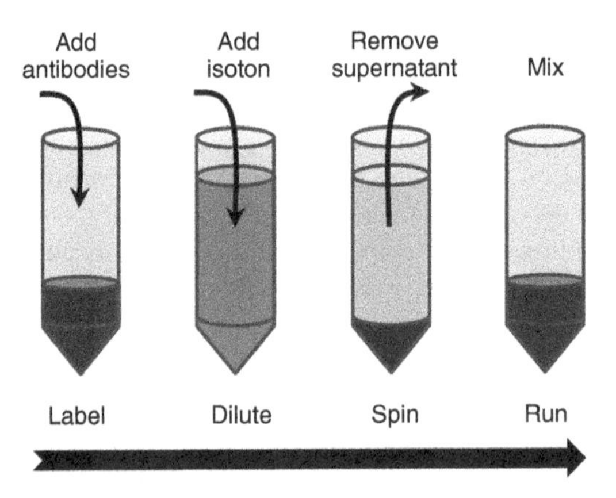

Figure 3.2 Sample preparation of eDAR [7]. Typically, 2 ml of whole blood was incubated on a rocker at room temperature for 30 min with the chosen antibodies. After incubation, the sample was diluted to 14 ml with Isoton buffer and centrifuged for 10 min to remove the supernatant containing free antibodies. Schiro *et al.* [7], p. 10, Figure S5. Reproduced with permission of John Wiley and Sons, Inc.

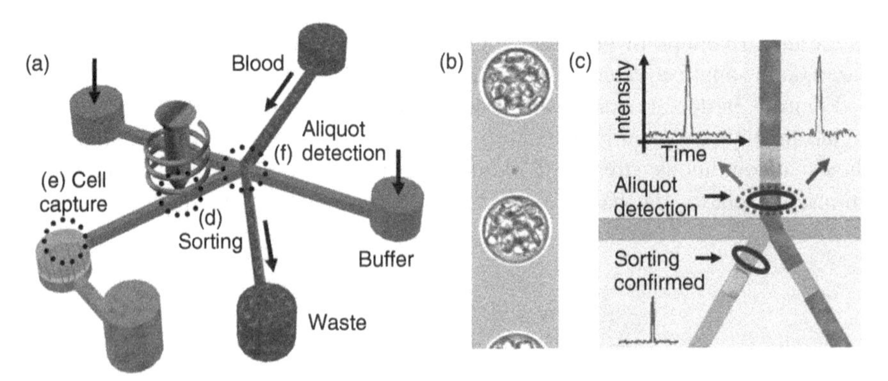

Figure 3.3 Schematic and images showing the general concept of eDAR and the notion of virtual aliquots [7]. (a) Overview of the microfluidic chip. The labeled blood sample flows through the main channel of the microfluidic chip with desired aliquots actively sorted into the cell capture chamber. (b) A high-speed camera image of whole blood aliquoted into a continuous stream of droplets surrounded by silicone oil. (c) Laser-induced fluorescence triggers the sorting of an aliquot containing a CTC (shown in yellow) to the collection channel, which is confirmed by the second detection window. Adapted from Schiro *et al.* [7], p. 4619, Figure 1. Reproduced with permission of John Wiley and Sons, Inc. (*See color plate section for the color representation of this figure.*)

may reduce the associated stresses on those cells. We have since found that this assumption might be too simplistic. Additionally, because droplets were spaced by immiscible phase, the throughput would be reduced by at least half at a given flow rate through the detection window. More importantly, the in-line generation and stability of the droplets were highly sensitive to fluidic conditions. And finally, droplets were not compatible with the on-chip purification method that we used to isolate the targeted CTCs from other blood cells. Because of these issues, we pursued the notion of virtual aliquots instead of creating physically separated aliquots with droplets. In this method, each aliquot was defined by a combination of laser illumination volume, the response time of the detectors, the volumetric flow rate of the blood sample, and the active sorting speed. This approach would offer the highest throughput and operational robustness. Typically, 2 nl aliquots were found to be optimal at a flow rate of 50 μl/min.

Aliquots were ranked based on the presence or absence of the target CTCs by sequentially interrogating aliquots in a flow-through format (Figure 3.3c). If the aliquot had a CTC labeled with the fluorescent antibodies, the fluorescence would be excited by the laser beams and then detected by the fiber-coupled avalanche photodiodes (APDs). As a result, that aliquot would be ranked as "positive," compared to those "negative" aliquots that did not present any fluorescent signal.

To detect a single CTC in a background of thousands of blood cells, we developed a line-confocal detection scheme [10, 11] with a probe volume that spanned the width (150 μm) and height (50 μm) of the microfluidic channel. In our system [7], the two laser sources (488 and 633 nm) were shaped by cylindrical lenses, overlapped and aligned using a dichroic mirror and a series of lens and mirrors, and finally focused into a 20× objective to form a 200 μm by 5 μm two-color laser beam. Fluorescence from this region was collected through a rectangular confocal aperture (4 mm long by 0.8 mm wide) and a series of dichroics and filters to fiber-coupled APDs operating in single-photon counting mode. Figure 3.4a depicts the optical layout of the eDAR system. In our current setup, APD1 is used to detect the yellow fluorescence (560–590 nm) typically from anti-EpCAM-phycoerythrin (PE); APD2 detects the signals in the green wavelength range (500–550 nm), which often serves as a negative control to eliminate false-positive events from broadly emitting fluorescent contaminates. APD3 detects the fluorescence from other antibody, such as human epidermal growth factor receptor 2 (Her2) conjugated with Alexa647 or allophycocyanin (APC) in the red wavelength band (640–690 nm). A second detection window using only the 488 nm laser is located immediately after the sorting junction (Figure 3.3c); yellow fluorescence (560–590 nm) is collected in this spatially distinct detection window through a separate confocal slit to confirm in real time the sorted aliquot with the targeted cell.

Figure 3.4b shows a small portion of the data trace from a stage IV breast cancer patient sample [7]. For every APD signal in the EpCAM channel higher than the sorting threshold, the corresponding aliquot was sorted to the collection channel, which was further confirmed by the APD traces in the confirmation channel (note the scale in Figure 3.4b has been compressed in the HER2 and EpCAM sorting channels relative to the EpCAM channel). The average signal-to-noise (S/N) ratio was 32 for the

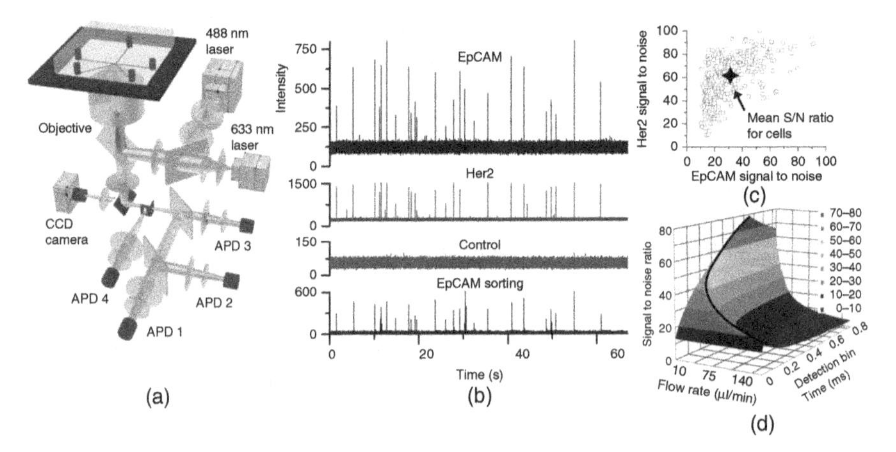

Figure 3.4 Aliquot ranking system [7]. (a) The setup for aliquot ranking is comprised of two excitation lasers and four APDs. APDs 1, 2, and 3 collect fluorescence at three wavelength regions (560–590, 500–550, 640–690 nm) from the aliquot to determine the presence or absence of a single CTC. APD 4 confirms the sorting of the desired aliquot. (b) A segment of the APD traces at the three different colors from a breast cancer sample, showing aliquots positive for EpCAM (top trace) and Her2 (second trace) that was correctly sorted (bottom trance). (c) The S/N ratio for each single CTC detected in an aliquot of blood for both antibody markers from a breast cancer sample. (d) The S/N ratio for EpCAM-labeled single CTCs as a function of flow rate and signal bin time. Adapted from Schiro *et al.* [7], p. 4619, Figure 2. Reproduced with permission of John Wiley and Sons, Inc. (*See color plate section for the color representation of this figure.*)

EpCAM-PE marker and 64 for the Her2-Alexa 647 marker (Figure 3.4c). The S/N ratio was highly dependent on the flow rate of the cells, as well as the detection bin time, because faster flow rates resulted in fewer photons detected from the CTCs and a relatively higher background from the blood cells in the aliquot (Figure 3.4d). In our current eDAR system, the maximum flow rate that still allows for an average S/N higher than 20 is 133 μl/min; the optimal bin time is 0.1 ms, which is comparable to the transit time of a CTC through the detection region.

For this version of eDAR, we simply ranked the aliquots as "negative" (no CTCs) or "positive" (with CTCs). Additionally, we have tested more sophisticated multiparameter rankings, which will be beneficial in the future and is a unique feature of the eDAR technology. For example, in comparison to the affinity-based [12, 13] or immunomagnetic [14] methods, eDAR offers more flexible and complicated logical operations. We could collect aliquots based on the presence of EpCAM and Her2 and absence of fluorescence in the green channel, while in other methods, cells that express either EpCAM or Her2 will be retained. This would be extremely important in isolating other important types of rare cells, such as circulating cancer stem cells (CSCs), because negative control markers, for example, CD24, are frequently used with other positive control markers such as CD44 to define this population [15]. Affinity-based or immunomagnetic methods could only employ the "OR" logical operation, while eDAR can make more complex logic decisions. It is also difficult for those methods to capture rare cells that do not express a particular marker without the depletion process prior to the experiment.

3.2.2 Aliquot Sorting

Once an aliquot is ranked by the line-confocal detection scheme, we actively sort this aliquot to a specific collection channel on the microfluidic chip. This active sorting step needs to be fast enough to capture the target cell and minimize the volume of each aliquot. Moreover, this active sorting scheme had to be robust enough to handle the heterogeneity of the human blood samples. To meet these criteria, we designed the hydrodynamic switching scheme for the active sorting step in eDAR, which is faster and more robust than other fluidic switching methods, such as the electroosmotic flow [16] or sol–gel transition [17]. In this scheme, the flow of the blood sample is controlled by the solenoid, which has a response time on the order of milliseconds or less. We have developed two different hydrodynamic switching schemes based on two different types of solenoids and related microfluidic designs, on-chip solenoid or off-chip solenoid.

3.2.2.1 Active Sorting Scheme Based on an On-Chip Solenoid To sort and collect the aliquots with target CTCs, we used an externally actuated solenoid piston valve set above the elastomeric polydimethylsiloxane (PDMS) microchannel (Figure 3.5a) [7]. When an aliquot containing the target cell was detected, the solenoid piston was released to direct the sample flow through the collection microchannel (Figure 3.3b). The flow rate in the two side buffer channels was balanced by air-pressure regulators, so that when the collection channel was open,

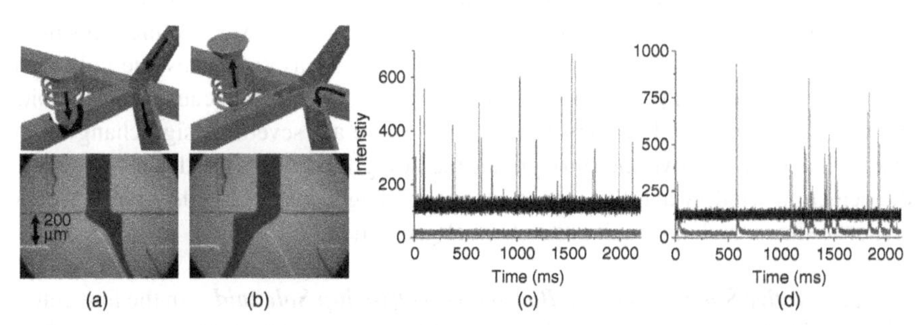

Figure 3.5 General scheme and effectiveness of the aliquot sorting [7]. (a) The solenoid piston is pushed down to stop the flow through the collection channel. (b) The solenoid piston is released to allow flow through the collection channel. (c) When the sorting scheme was configured as "off," APD traces from the first detection window (black) and the second detection area (gray) show that no cells were collected. (d) When the sorting scheme was applied, APD traces from the first (black) and the second (gray) detection area show that all the CTCs were collected. Adapted from Schiro *et al.* [7], p. 8, Figure S3. Reproduced with permission of John Wiley and Sons, Inc.

the entire aliquot of blood with the target cell would be directed toward the collection channel. To control the automated sorting process, the raw photon counts from the APDs were first sent to the sorting electronics. At a rate of 10,000 Hz, the field-programmable gate array (FPGA) processor compared the photon count at each time point to a threshold value defined by the blood's background level. If there was a signal higher than the background threshold value, the solenoid piston was released by switching the voltage from high (22 V) to low (0 V). The elastomeric collection channel reverted to its native open state, forcing the solenoid piston up. Using a high-speed camera, we found that this hydrodynamic switching process took less than 3 ms at the flow rate of 50 μl/min [7]. After the aliquot was sorted, the 22 V was reapplied to the solenoid to close the collection channel.

To test the effectiveness of the solenoid piston at preventing unwanted aliquots from being collected, we prepared a sample using healthy donors' blood to which we spiked SKBr-3 cells stained with anti-EpCAM-PE. We used the confirmation APD (Figures 3.3c and 3.4b) to monitor the performance of the eDAR system (gray bottom APD trace in Figure 3.5c and d). As expected, when the sorting processor was configured to not release the solenoid, we found that none of the aliquots that contained SKBr-3 cells (top black trace in Figure 3.5c) were detected in the collection channel. When the sorting processor was operating in the normal eDAR mode (Figure 3.5d), each aliquot containing a labeled SKBr-3 cell that was detected by the decision APD (black trace) was detected by the collection APD (gray trace) again, confirming successful sorting.

As a modeling system to characterize the performance of the active sorting step, we spiked cultured breast cancer cells (SKBr-3 or MCF-7) into whole blood and then labeled the samples with anti-EpCAM-PE. The sorting efficiency was defined as the number of aliquots triggered to be sorted compared to the number of cells detected as they entered the collection channel [7]. At a slow flow rate (<25 μl/min), we got 100% sorting efficiency as the entire volume of each aliquot was collected. At the fastest flow rate of 80 μl/min, only a portion of each aliquot was captured, resulting in a reduced sorting efficiency of 85% [7]. The loss was caused by the collection channel not being completely open before the aliquot arrived, because the solenoids do not have reproducible timing below 2 ms. There are several design changes that can be made to improve the throughput, including increasing the distance between the detection window and the sorting junction, using a solenoid with submillisecond response times, and optimizing the channel geometries.

3.2.2.2 Active Sorting Scheme Based on an Off-Chip Solenoid In the first generation of eDAR, we used an on-chip solenoid to control the hydrodynamic switching step, which has been successfully used for analyzing blood samples from breast cancer patients [7]. Although promising, some design factors of this active sorting scheme may constrain the potential applications of eDAR. To form the mechanical valve on the chip, three individual structural layers were required: the solenoid, its PDMS thread, and the microchannels on a spin-coated 150 μm PDMS film, which made the chip preparation complicated and time consuming. Another shortcoming was the direct contact between the isolated blood aliquots and the mechanical valve, which increased the risk of damage on CTCs. To overcome these drawbacks, we replaced the on-chip solenoid with an off-chip model, which is normally closed but can be opened in 2–3 ms when a 5 V DC voltage is applied [18]. The preparation of the microchip was significantly simplified because this in-line solenoid was not a part of the chip.

These off-chip solenoids could be easily connected with any microchannels, so we could test many possible hydrodynamic switching schemes to screen for the optimal one. After characterization and optimization, the new eDAR platform and the corresponding active sorting scheme (Figure 3.6) were designed and fabricated [18]. Blood sample labeled with fluorescent antibodies was injected into the top channel of the chip using a syringe pump (Figure 3.6a). Two side channels, where buffer flowed through, were used to control the active sorting step; three pressurized buffer sources were connected to the side buffer channels. The normally closed solenoid was connected to the port near the sorting junction to control the hydrodynamic switch. The channel on the bottom left was used to collect selected aliquots and transfer them to the filtration and collection area for further purification; the last channel was the waste collection channel where all the negative aliquots flowed through.

When negative aliquots flowed through the sorting junction, there was no voltage applied on the solenoid, so it was closed (Figure 3.6b). A pressure drop was preset between the no. 1 and no. 3 buffer sources shown in Figure 3.6a, so the blood flow was focused into the channel that collected the waste, which is also shown in the bright field image in Figure 3.6b. When a positive aliquot was detected in the first detection window, a 5 V DC voltage was immediately applied to the solenoid to open the buffer flow from the no. 2 buffer reservoir. As a result, the blood flow was pushed from the right side to the left, so this positive aliquot could be sorted (Figure 3.6c). After the second detection window confirmed this aliquot, the solenoid was then closed to switch the blood flow back to the waste collection channel (Figure 3.6d). The time required for the switchover and back was 2–3 ms for each, and the process was stable enough for eDAR even after more than 10^5 on–off cycles [18]. Compared to the previous design, the in-line solenoid was placed on the buffer line, so blood could not come into contact with the solenoid, which eliminated the possibility of cross-contamination and blood coagulation. Moreover, in this scheme, there was a constant flow of buffer in the CTC collection channel during the eDAR process, which improved the efficiency of the subsequent purification and prevented the formation of cell aggregates.

The efficiency of this active sorting step could be monitored in real time. Figure 3.6e shows a small portion of the APD data from a pancreatic cancer patient sample. The signals in black were from the first detection window that ranked the aliquots and controlled the sorting. The two peaks at 978 and 1298 ms were two CTCs labeled with anti-EpCAM-PE that triggered the aliquot sorting; the two peaks in gray confirmed the two cancer cells flowing through the second detection window. It is worthwhile to point out that the background change from the second detector (Figure 3.6e) also confirmed that only a small portion of blood was collected by eDAR, contributing to the ultrahigh enrichment ratio of CTCs (up to a million fold for a typical clinical sample) [18]. We could observe a time difference between the

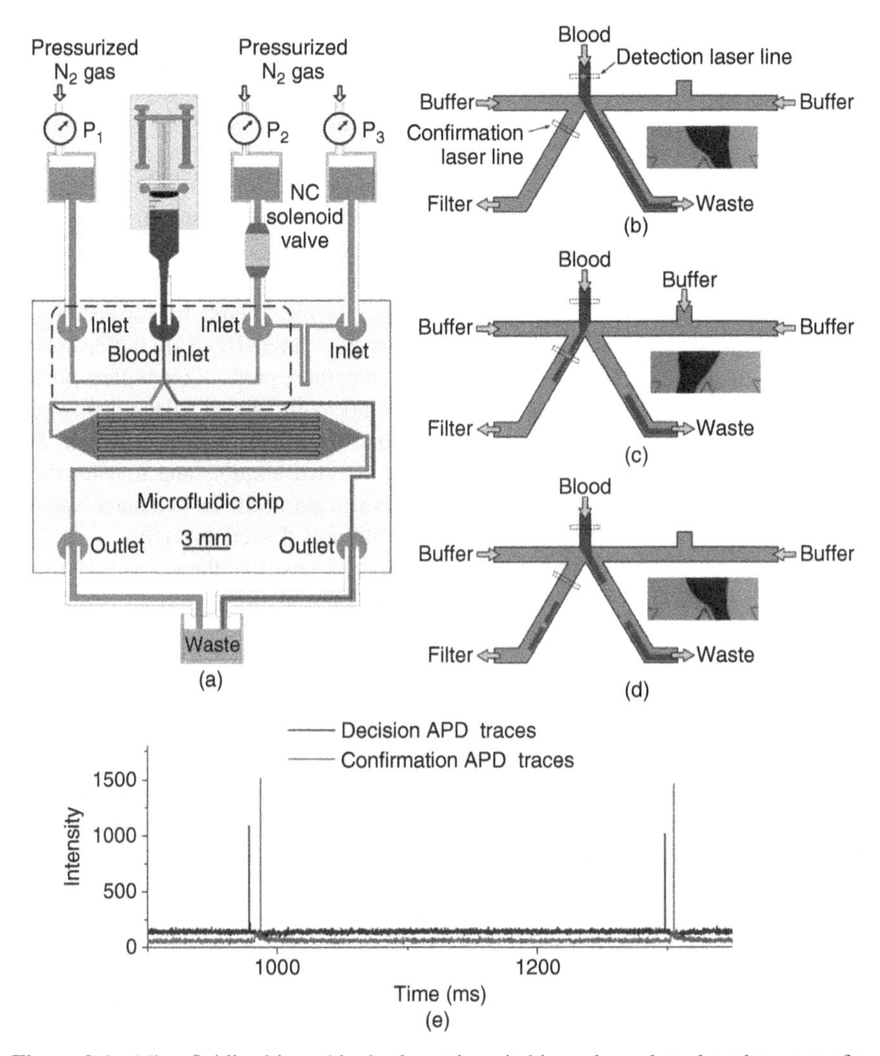

Figure 3.6 Microfluidic chip and hydrodynamic switching scheme based on the usage of an off-chip solenoid [18]. (a) General structure of the microfluidic chip and the configuration of the eDAR platform. The normally closed (NC) solenoid was connected to the center port on the left side channel. The bottom left channel was to collect sorted aliquots and transfer them to the subsequent purification area, which had 20,000 microslits. The area marked with a dashed box is further explained in (b)–(d). (b) The fluidic condition when no positive aliquot was ranked. (c) The blood flow was switched to the CTC collection channel by opening the solenoid, and the sorted aliquot was confirmed by the second APD. (d) The blood flow was switched back after the aliquot was sorted. (e) The segment of APD data from a pancreatic cancer sample that shows two events triggered the sorting, which then were confirmed by the second detection window. (f) The distribution of transit time at flow rates of 40 and 80 μl/min, respectively. (g) A plot shows the fastest average transit time was about 4 ms when the flow rate was 90 μl/min. Adapted from Zhao *et al.* [18], pp. 9673–9675.

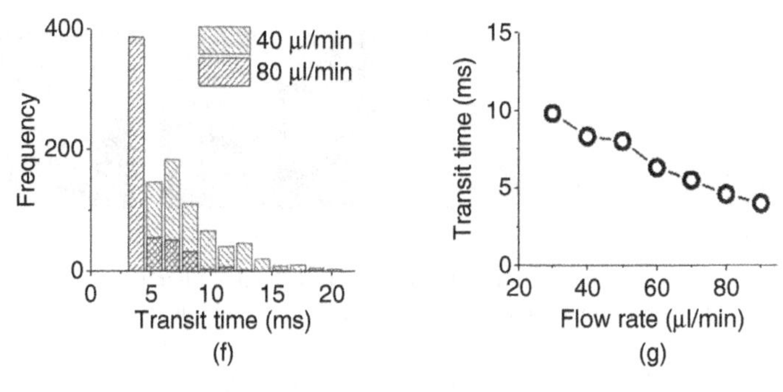

Figure 3.6 *(continued)*

decision APD peak and its confirmation signal, which was from the transit time of the sorted CTCs. This value might vary because the CTCs could have different linear flow rates due to the nature of laminar flow in the microchannel. Figure 3.6f shows the distribution histogram of the transit time at flow rates of 40 and 80 μl/min, respectively. Generally, a higher volumetric flow rate of the sample resulted in a shortened transit time of the sorted CTCs (Figure 3.6g). When the sample flow rate was 90 μl/min, the mean transit time was lowered to 4 ms, very close to the switching time of the sorting scheme (2–3 ms), which implies that this might be the upper limit of the throughput for this format of eDAR [18].

3.2.3 On-Chip Purification

Sorted aliquots flowed to the cell capture chamber where the target rare cells were retained by a filter integrated in the same microchip. The majority of the blood cells, including all RBCs and many small white blood cells (WBCs), passed through the filter. After this further purification step, we could stain, image, and enumerate the target cells with high accuracy and efficiency. We have developed two types of on-chip purification schemes based on different microfluidic designs and fabrication techniques. These are discussed in the next section.

3.2.3.1 Purification via Integrated Planar Filter In this scheme, the filter was a transparent, track-etched, polycarbonate membrane with 5 μm pores that spanned the 1-mm-diameter chamber (Figure 3.7a). With such a small volume of blood coming into the collection chamber, the capacity of about 1000 pores was more than adequate. Isolated cells, which clogged the pores once trapped, did not significantly increase the back pressure of the system, unlike bulk blood filtration methods where capacity issues are a primary concern [19–22]. The small size of the filtration chamber allowed for high-magnification imaging with just a few fields of view. Alternatively, the entire chamber could be imaged in a single frame using a 10× objective and a camera with a large charge-coupled device (CCD) sensor.

Figure 3.7a shows that the chamber was open on the top and easily accessible from there. Additional reagents, such as washing buffer, antibodies, or nuclear stains, could be transferred into the chamber and perfused over the targeted cells retained

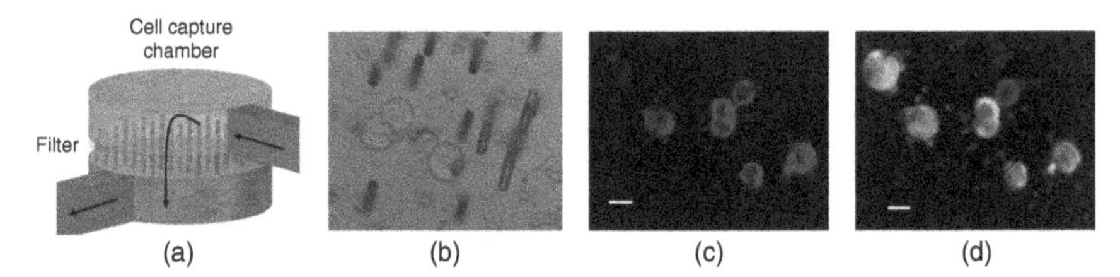

Figure 3.7 The on-chip purification scheme based on planar filter [7]. (a) The schematic drawing shows the structure of the cell capture chamber, which can trap the CTCs and further purify it by removing all the RBCs and most of the WBCs. (b) The bright field image shows the four MCF-7 cells isolated by eDAR and trapped on the polycarbonate filter. (c) Fluorescence images of the captured CTCs on the filtration membrane labeled with anti-EpCAM-PE. (d) Fluorescence images of the captured CTCs on the filtration membrane labeled with anti-Her2-Alexa647. Adapted from Schiro *et al.* [7], p. 4620, Figure 4. Reproduced with permission of John Wiley and Sons, Inc. (*See color plate section for the color representation of this figure.*)

by the filter. Because polycarbonate filter is somewhat transparent (Figure 3.7b), micropipettes could also be positioned above the filter to remove individual CTCs for analysis or culture. With a 5-μm pore size, some WBCs were also retained on the polycarbonate filter. But with highly enriched CTCs, these WBCs were present in comparable numbers to CTCs and were easily distinguishable from CTCs with anti-CD45 specific for WBCs (Figure 3.7c and d).

3.2.3.2 Purification via Microfabricated Slits In the first generation of eDAR, we used a piece of track-etched polycarbonate filter to trap and purify the isolated CTCs. This structure required two additional layers in the microchip and the corresponding procedures to bond the filter with PDMS. Moreover, although the throughput of fluorescence imaging and cell enumeration was greatly improved because of the small area where CTCs were trapped, the imaging quality could be adversely affected by the track-etched polycarbonate filter, generating a nonuniform background [7]. To overcome these issues, we developed a new scheme based on an on-chip purification method, which relied on microfabricated structures. Those PDMS microslits were a monolithic structure and did not require any additional layers.

Figure 3.8a shows the 2D schematic structure of these microslits, which trapped the CTCs without retaining any RBCs. A 5 μm gap was chosen to match the original planar filter design [7]. With this size of microslits, we minimized the risk of losing small CTCs, while still allowing many WBCs to deform and flow through the filter. The purity of isolated CTCs was similar to the design based on polycarbonate filters. Because the microfilter was made of PDMS and bonded with a piece of coverslip, the imaging quality was improved significantly (Figure 3.8c and d) compared to the track-etched filter (Figure 3.7), which is only semitransparent and generates scattering and aberration. Moreover, because the cells could only be trapped along the array of microslits, they could be easily referenced and tracked, allowing faster and more accurate enumeration. In many other CTC isolation methods, the cells are distributed randomly on a surface and may have different positions on the

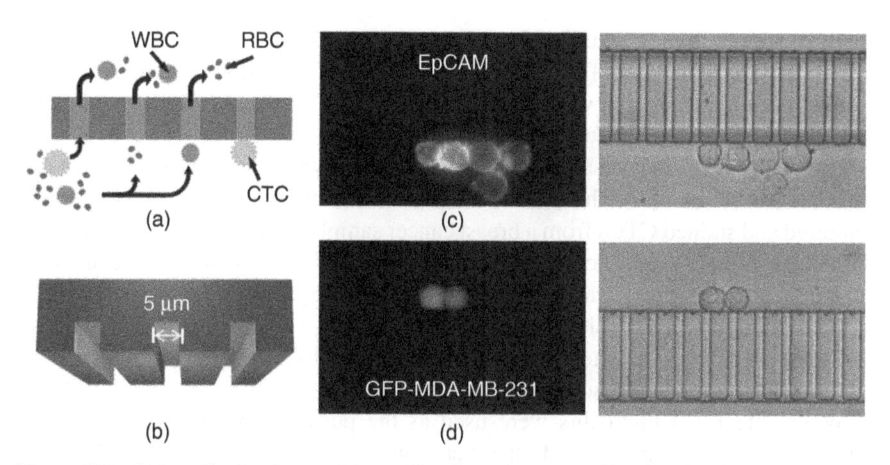

Figure 3.8 Microslits for the on-chip purification scheme used in eDAR [18]. (a) The sorted aliquots were further purified through the array of microslits. Objects in yellow represent CTCs; red and gray objects represent RBCs and WBCs, respectively. The curved arrows show the flow paths across the microslits. (b) The 3D model of the microslits with a 5 μm width. (c) Fluorescence (left) and bright field (right) images of five MCF-7 cells captured via eDAR. (d) Fluorescence (left) and bright field (right) images of two MDA-MB-231 cells captured via eDAR. Adapted from Zhao *et al.* [18], p. 9674. (*See color plate section for the color representation of this figure.*)

Z dimension, making enumeration and analysis more difficult due to the limited depth of field of microscope.

To optimize the performance, we prepared and tested microchips with 1000, 5000, 20,000, and 30,000 microslits. These microfluidic chips helped us to determine the flow resistance across the purification area, which could affect the hydrodynamic switching and the stress on the trapped cells [18]. The eDAR chips with 20,000 or more microslits required a low pressure (less than 4 psi) on the two side buffer channels to balance the fluidic switching. The pressure drop across the microfilter is also lower, which minimizes the deformation and stress of fragile cells.

3.2.4 Secondary Labeling and the Enumeration of CTCs

After the on-chip purification step, CTCs were trapped on a small area for imaging and subsequent analyses. Those isolated CTCs could be directly enumerated based on their morphology and the fluorescence from the primary immunostain. However, to ensure accuracy of the enumeration results, we labeled the cells with more biomarkers to confirm the identity of the cells. Here, the most accepted and common working definition of CTCs is positive to cytokeratin, positive to a nuclear stain, such as DAPI or Hoechst, and negative to CD45, which is typically only expressed on WBCs. We generally adapted these standard criteria in the secondary labeling step after eDAR isolation. In this step, the cells were first fixed with 2% paraformaldehyde (PFA) and permeabilized with a surfactant to allow binding of antibodies to cytoskeletal or cytoplasmic proteins, such as cytokeratin or vimentin. After washing the purification chamber with buffer, a cocktail of fluorescent antibodies and a nuclear stain was introduced onto the same area. Figure 3.9a shows three cells isolated using eDAR from a breast cancer sample that were positively labeled with EpCAM, Her2, cytokeratin, and DRAQ 5 but negative for CD45.

eDAR affords us the versatility of trapping the CTCs enriched on a small open area and retained by the filter where we could pipette the labeling and washing reagents. Captured and stained CTCs from a breast cancer sample were first photobleached and then stained with the stem cell marker set, $CD24^-$ and $CD44^+$, which has been widely used in identifying breast cancer stem cells [15, 23]. Figure 3.9b shows a stem CTC in this sample that was positive to EpCAM and CD44 but negative to CD24. Using eDAR, we were able to detect the presence of this subset of cells within the sorted CTCs from breast cancer patients.

When integrated microslits were used as the purification method, the isolated CTCs were trapped in a closed chamber, which has an internal volume less than 500 nl. Although lacking the benefits of the open-access design, it allows for the automation of the secondary labeling process using a programmable peristaltic pump. The small internal volume minimizes the amount of antibodies and other reagents required for the labeling process, which is usually expensive and time-consuming in many other methods. Again, the premium optical property of PDMS and glass

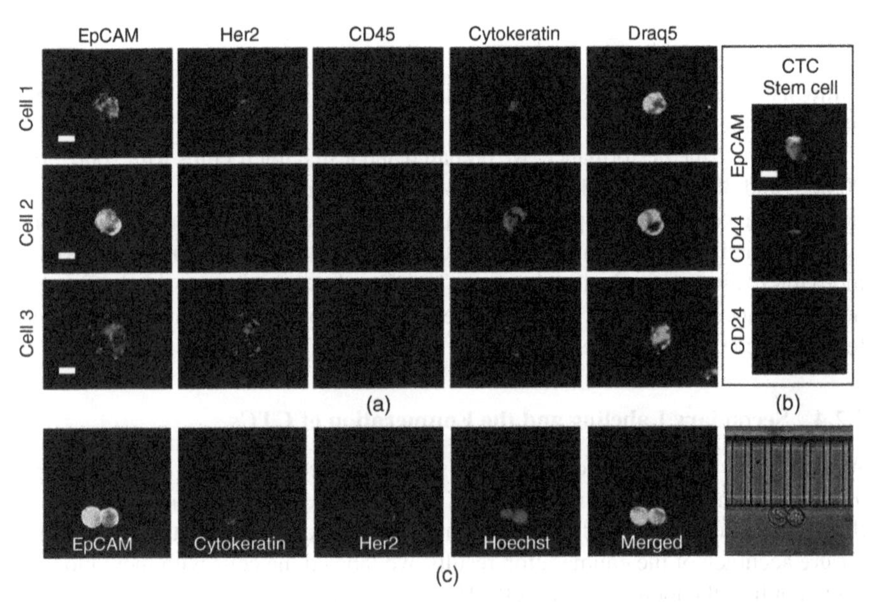

Figure 3.9 Multicolor fluorescence images for the CTCs trapped on eDAR microchip after the secondary labeling step [7, 18]. (a) Images of three CTCs isolated from a breast cancer sample; scale bar is 20 μm. (b) Images from a CTC labeled with fluorescent antibodies against the breast cancer stem cell marker (CD44+/CD24–). (c) Two SKBr-3 cells were captured by eDAR and further stained with additional markers. Adapted from Schiro *et al.* [7] and Zhao *et al.* [18]. (*See color plate section for the color representation of this figure.*)

gave us a much improved imaging quality using this scheme. Figure 3.9c shows two SKBr-3 cells labeled with anti-EpCAM-PE were retained on the microslits. They were further fixed, permeabilized, and then labeled with anti-Cytokeratin-Alexa488, anti-Her2-Alexa647, and Hoechst. Fluorescence images in Figure 3.9c showed the expression of these markers on these two cells clearly, and the bright field image also confirmed their morphology. We labeled these two cells with anti-CD45-Alexa700 as a negative control marker and did not find any signal from the fluorescence channel that corresponded to this tag [18].

3.2.5 Characterization and Analytical Performance of eDAR

Among the many reasons why recovery efficiency is typically low for CTCs is the loss or damage to cells during the preparation process. Depending on the method, these losses can be caused by mechanical or chemical lysis or cell adhesion to the tubing and vials during multiple transfer steps. eDAR was optimized to include minimum sample preparation that did not use cell lysing. The blood processing consisted of labeling with antibodies, dilution with buffer, followed by centrifugation and removal of the supernatant that contained the free antibodies.

In the first generation of eDAR, which had an on-chip solenoid and a piece of planar filter integrated on the same microchip, recovery experiments were performed with two breast cancer cell lines, MCF-7 and SKBr-3. A known number of cells ranging from 5 to 311 were spiked into a tube containing 1–2 ml of healthy donors' blood. Typically, the cell recovery experiments are conducted by calculating the concentration of cells using a hemocytometer before the serial dilution step. For the low cell numbers that are most relevant in CTC isolation experiments, the statistics for serial dilution break down and are inaccurate. For our studies, the quantity of cells was counted individually using a capillary cell-spiking method developed in our group [24]. This technique allowed us to investigate the recovery rates conveniently for as few as five cells rather than the much higher range that is typically reported elsewhere. The isolated cells were enumerated in the filtration area after secondary labeling. From nine separate experiments at a flow rate of 50 μl/min, the average recovery rate of eDAR was higher than 93% [7].

In the second generation of eDAR, we combined an off-chip solenoid to drive the hydrodynamic switch and the microfabricated slits to further purify the sorted aliquots. As discussed previously, if the transit time for a sorted CTC was shorter than the fluidic switching time, the target CTC could not be collected on this platform. The sorting efficiency was thus defined as the number of confirmed APD events versus the overall number of events that triggered the eDAR sorting. Figure 3.10a shows the values of sorting efficiency at flow rates ranging from 30 to 100 μl/min. When the flow rate was at the low end, the sorting efficiency was nearly 100%, because the mean transit time at that flow rate was about 10 ms (Figure 3.6g), which was much longer than the fluidic switching time (2–3 ms). The sorting efficiency dropped slightly to 90% at the flow rate of 80 μl/min and then dropped significantly to 49% at a flow rate of 90 μl/min. Figure 3.10a also shows the overall recovery rate of eDAR, which is similar in the trend compared to the sorting efficiency. However, the recovery

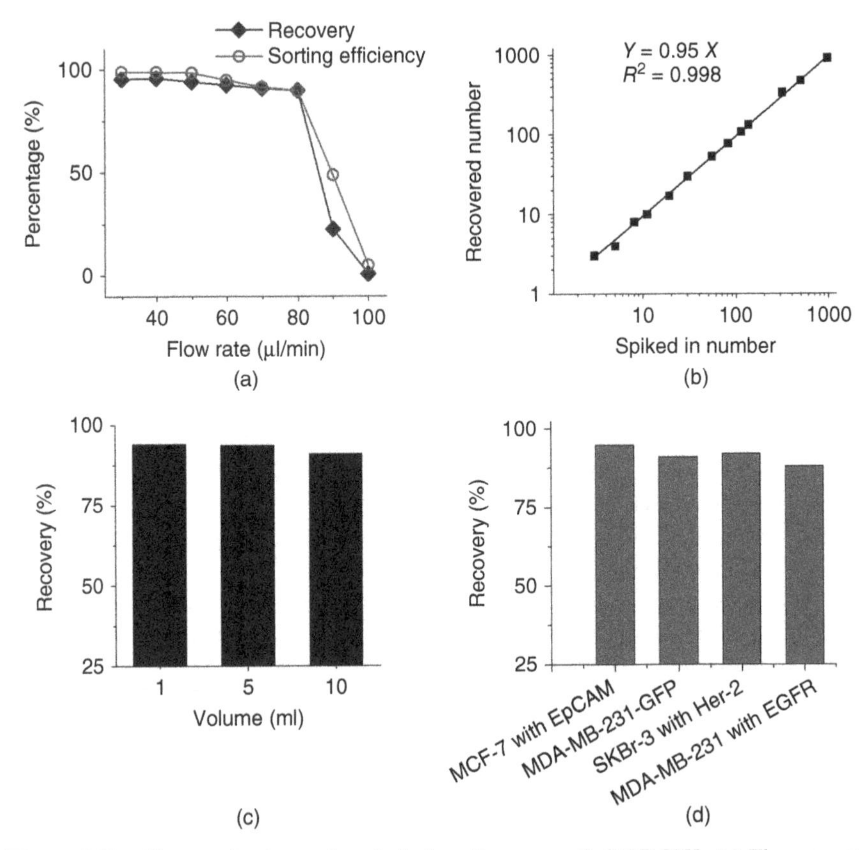

Figure 3.10 Characterization and analytical performance of eDAR [18]. (a) The recovery and sorting efficiency value versus different flow rates. (b) The recovery ratio of MCF-7 cells spiked into whole blood. (c) The recovery ratio of 300 MCF-7 cells spiked into 1, 5, and 10 ml of whole-blood aliquots. (d) The recovery ratio of our selection schemes of four breast cancer cell lines spiked into whole blood. Adapted from Zhao *et al.* [18], p. 9675.

rate was defined as the quantity of spiked-in cells versus the number of recovered cells enumerated using multicolor fluorescence microscopy. It is a combination of several factors, including the immunostaining efficiency, the line-confocal detection efficiency, the sorting efficiency, and the filtration efficiency. This could explain the difference between the sorting and recovery efficiency at the same flow rate. At the maximum throughput for this design (80 μl/min, 4.8 ml of blood per hour), the overall recovery rate was 88%. Although this throughput is higher than most methods for the analysis of CTCs from whole blood, it can be improved by optimizing the optical and fluidic designs of eDAR.

To test the recovery efficiency of the second generation of eDAR, from 3 to 975 MCF-7 cells were spiked into 1 ml of healthy donor blood, which was then stained with anti-EpCAM-PE. We analyzed the labeled samples using eDAR at a flow rate of 50 μl/min. The average recovery ratio was 95% with an R^2 value of 0.998 (Figure 3.10b), which is slightly higher than the first generation of eDAR (93%). Because CTCs are typically rare in blood samples, the enumeration results can be affected by Poisson distribution. In this case, the capacity to analyze whole-blood samples with a larger volume becomes more and more important. When we used the track-etched polycarbonate filter in the first generation of eDAR, this capacity was limited by the number of micropores on the filtration area, typically around 1000. However, when we used the second generation of eDAR with 20,000 built-in microslits, the capacity was increased significantly to over 5,000 CTCs. We spiked the same number of MCF-7 cells into 1, 5, and 10 ml of healthy donors' blood and then analyzed them at the flow rate of 50 μl/min. Figure 3.10c shows that the recovery rates for these three samples did not change significantly, which means that our method is capable of analyzing a large amount of whole-blood sample with a high efficiency and throughput. Equally important is the false-positive ratio. We performed negative control experiments using healthy blood. A total of 1 ml of whole blood was labeled with anti-EpCAM-PE and analyzed using eDAR following the same protocol as used for clinical samples. We did not find any target cell in all the eight samples, yielding a false-positive ratio of 0 ($n = 8$).

As discussed previously, EpCAM has historically been used in most CTC studies to select tumor cells; however, increasingly more studies have reported that CTCs with a low expression of EpCAM have more mesenchymal characteristics and are more aggressive [25–27]. Our eDAR platform is sufficiently flexible to use any fluorescent biomarker to select rare cells, so we are not constrained by the expression of EpCAM. To prove this point, we designed three schemes to select different cultured breast cancer cell lines spiked in healthy donors' blood (Figure 3.10d). MCF-7 was selected by the expression of EpCAM, SKBr3 was selected by Her2, and MDA-MB-231 was selected by the expression of epidermal growth factor receptor (EGFR). All three of these schemes isolated and trapped the target cells on the eDAR microchip with a recovery rate higher than 88%. Another unique and important feature of this method is the compatibility of using intracellular markers to select CTCs. Other technologies, such as the immunoaffinity-based methods or immunomagnetic methods, can only capture the antigens expressed on the cell surface. Yet our method can select rare cells using intracellular markers, such as green fluorescence protein (GFP) (Figure 3.8d). The recovery rate of the MDA-MB-231-GFP cells spiked into healthy blood was 91% (Figure 3.10d). Because fluorescent proteins have been widely used in animal models to study the mechanisms and progression of metastasis [28], eDAR could be easily used to select CTCs in these animal models without any labeling steps.

3.3 APPLICATION AND DOWNSTREAM ANALYSES OF eDAR

3.3.1 Enumeration of CTCs from Cancer Patients using eDAR

The enumeration of CTCs from cancer patients is still the most important application of CTC analysis as multiple independent clinical trials have confirmed the prognostic value of the CTC counts [29–31]. The number of CTCs correlates well with the prognosis of patients with various types of metastatic cancer, such as breast, lung, prostate, and ovarian cancers [32, 33]. CTCs could also be counted from early-stage breast or lung cancer patients and correlate with their prognosis as well [30, 34, 35]. We have used eDAR to analyze CTCs from patients with breast, lung, pancreatic, and prostate cancers.

In one study, we performed a side-by-side comparison between eDAR and CellSearch, the only CTC method approved by U.S. Food and Drug Administration (FDA) at present (see Chapter 19 for details), using 20 stage IV breast cancer samples. Multiple tubes of venipunctured blood were collected in each draw, which was performed in an outpatient cancer clinic as a part of their office visit. The first tube was not used for CTC analysis because there might be potential contamination from the epithelial skin cells during the venipuncture process. The remaining blood was divided into two parts. One was drawn into a CellSave tube for CTC enumeration using CellSearch. A second sample from the same draw was collected in a Vacutainer tube with EDTA for the analysis with eDAR. The side-by-side comparison was performed independently: CellSearch experiments were performed by a certified clinical technologist; eDAR was run by our lab. The results of the CellSearch analysis were unknown to us until after we had completed the study using eDAR. This arrangement allowed for a direct side-by-side comparison between the two platforms [7].

For the eDAR experiments, each sample was analyzed using a new disposable microchip made of PDMS integrated with a piece of polycarbonate filter to retain captured cells. CTCs were selected by the presence of the fluorescence from anti-EpCAM-PE. EpCAM was used because it is the primary marker used in the immunomagnetic capture in the CellSearch method. We could analyze 1 ml of whole-blood sample in 20 min at a flow rate of 50 μl/min. The captured cells at the end of each run were stained with more fluorescent antibodies against cell markers and then imaged using multicolor epifluorescence, which is described in detail in the second section of this chapter. To make a consistent comparison, we applied the same criteria used in CellSearch to define a CTC, which is Cytokeratin-positive, nuclear-stain-positive, and CD45-negative (Figure 3.9a).

Figure 3.11 summarizes the results of this side-by-side comparison of the two systems. The number of CTCs counted using eDAR ranged from 11 to 105 per 7.5 ml of blood, with an average of 45 CTC/7.5 ml. The CellSearch counts of CTCs from the same set of samples ranged from 0 to 111 CTCs/7.5 ml, with an average of 10 CTCs/7.5 ml. In those 20 patient samples, CellSearch did not find any CTC in 12 samples, with only 4 samples containing more than 2 CTCs/ml, while the clinical threshold determined using CellSearch was 5 CTCs/7.5 ml for breast cancer patients. In comparison, eDAR analysis identified the presence of CTCs in all clinical samples in this study. The average of 10 CTCs/7.5 ml was highly affected by sample No 14. If we omit that sample from the data set, CellSearch found 0 to 64 CTCs per 7.5 ml of blood with an average of 5 CTCs/7.5 ml, while eDAR recovered 11–105 CTCs with an average of 44. This study clearly illustrates that eDAR is more sensitive than CellSearch in analyzing metastatic breast cancer samples.

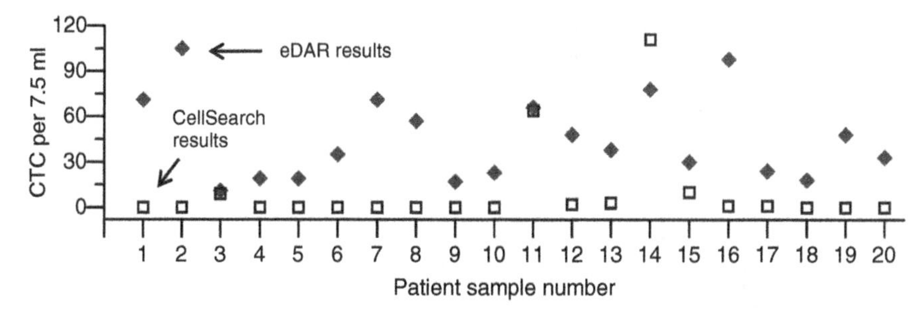

Figure 3.11 Clinical results for CTCs isolated in blood samples from patients with metastatic breast cancer [7]. A side-by-side study was performed to compare the clinical results from 20 breast cancer samples analyzed by CellSearch and eDAR. Adapted from Schiro *et al.* [7], p. 4621, Figure 5. Reproduced with permission of John Wiley and Sons, Inc.

In another study, we analyzed the blood samples from patients with metastatic pancreatic cancer and compared the clinical performance of the two generations of eDAR [18]. A total of 26 clinical samples were analyzed in this study; 16 of them were analyzed using the first generation of eDAR; the other 10 samples were analyzed using the second generation of eDAR. We also analyzed blood samples from 15 healthy donors to evaluate the false-positive rate of the second generation of eDAR, and we did not find CTCs in any of those samples. Figure 3.12 summarizes the distribution of these three data sets: the control blood analyzed by the second generation of eDAR, pancreatic cancer samples analyzed by the first generation of eDAR, and the pancreatic cancer samples analyzed by the second generation of eDAR. We detected CTCs in 80% (8 of 10) of the clinical samples ranging from 2 to 872 cells/ml using the second generation of eDAR. In comparison, CTCs were found in 88% (14 of 16) of the pancreatic cancer samples ranging from 2 to 183 cells/ml with the first-generation eDAR. CTC clusters, reported by other groups, were also isolated from the pancreatic cancer samples. We compared the clinical performance of the two generations of eDAR based on the analysis of variance (ANOVA) method, which showed that the two data sets were not significantly different ($p = 0.30$, $\alpha = 0.05$) [18]. Using eDAR, we also isolated CTCs from patients with other types of cancer, such as lung or prostate cancer.

3.3.2 Downstream Analysis of CTCs Isolated by eDAR

Many studies have shown that the molecular and cellular analyses of CTCs may provide more information than simple enumeration [32]. These molecular studies of CTCs can potentially verify important biological hypotheses by analyzing the expression pattern of biomarkers on CTCs and comparing them with the corresponding expression profile of the primary tumor. Downstream analysis postenumeration can also improve our understanding of the mechanism of metastasis and help with the management of the clinical treatment. As an "all-in-one" platform, eDAR is capable of running some downstream molecular assays directly on the same

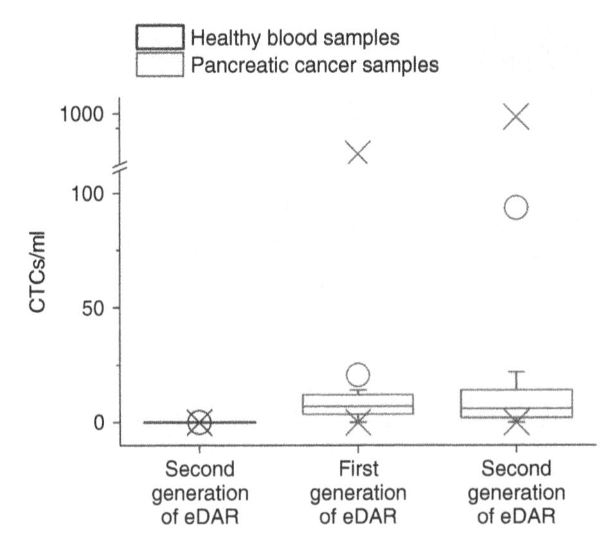

Figure 3.12 The distribution of 15 control samples and 10 pancreatic cancer samples analyzed by the second generation of eDAR, as well as the distribution of 16 pancreatic cancer samples analyzed by the first generation of eDAR [18]. O shows the average values for each data set; X shows the minimum and maximum values we found in each data set. Zhao *et al.* [18]. Reproduced with permission of the American Chemical Society.

microchip and also compatible with many other assays that need to be performed off the chip.

Among various potential downstream analyses, analyzing the expression of protein markers is an easy and important starting point for three reasons: (1) Some protein biomarkers are direct targets of anticancer drugs or immunotherapy, such as Her2, EGFR, or MUC1. By studying the expression profile of these protein markers, more molecular details can be revealed, which can benefit the management of the disease. (2) Some protein markers can be used to define the subpopulation of CTCs, such as epithelial CTCs, epithelial–mesenchymal transition (EMT) CTCs, apoptotic CTCs, or CTCs with stem-cell characteristics, which will help us understand the heterogeneity of CTCs and the nature of metastasis. (3) Most of the current CTC technologies rely on immunostaining and fluorescence imaging methods to count and verify CTCs. These techniques are fully compatible with the downstream protein marker analysis after CTC enumeration. Therefore, the immunostaining and fluorescence analyses of CTCs can be readily performed once the cells are isolated with minimum additional preparation or transfer processes.

eDAR is an ideal platform to isolate and perform the subsequent analysis of the protein markers. Cells are enriched onto a small area ($\sim$1 mm^2) on the microchip, so the imaging process can be done much faster than most other methods. The small area also minimizes the use of staining reagents. Additionally, eDAR has an open-access design, which facilitates the manipulation of single CTCs, such as picking up a CTC of interest or delivering certain reagents to single CTCs. Based on these aspects, we developed a simple and semiautomated method for the analysis of protein markers expressed on CTCs isolated on the platform of eDAR [36]. We designed an in-line immunostaining and photobleaching system, which allowed us to perform labeling and fluorescence imaging on selected CTCs with a panel of antibodies conjugated with different fluorophores, followed by the photobleaching and relabeling with another set of fluorescent antibodies against protein markers. This process can be repeated multiple times to study a large set of biomarkers [36].

We developed an in-line washing and staining system coupled with the eDAR platform in order to minimize the dead volume, lower the amount of antibodies used, avoid air bubbles, and automate the process. Figure 3.13a shows that after the eDAR isolation was done, two ports on the eDAR chip were left open to perform the perfusion and washing procedures, while all the other three ports were completely closed. We used a peristaltic pump and a six-way valve to deliver the washing and labeling reagents to the eDAR microchip. When we ran the eDAR experiment, the six-way valve was switched to the pressurized buffer side to provide a driving force for the active sorting step. After eDAR sorting, the valve was turned to the peristaltic pump side, so that we could deliver accurate amounts of reagents to the microchip without introducing any air bubbles. Typically, a few nanograms of antibodies were injected to the trapped CTCs in less than 5 min; a typical incubation step took less than 20 min (Figure 3.13b). If we needed to perform intracellular marker labeling, the captured cells were fixed and permeabilized on the chip prior to the test. Multiple rounds of immunostaining, washing, imaging, and photobleaching could be performed sequentially. In our current method, we monitor four colors of fluorescence – yellow (PE), red (Alexa 647 or APC), green (FITC), and blue (nuclear stain) – in each round.

We have carefully characterized and optimized the two critical factors that determine the efficiency of the photobleaching step – exposure power and time. Based on

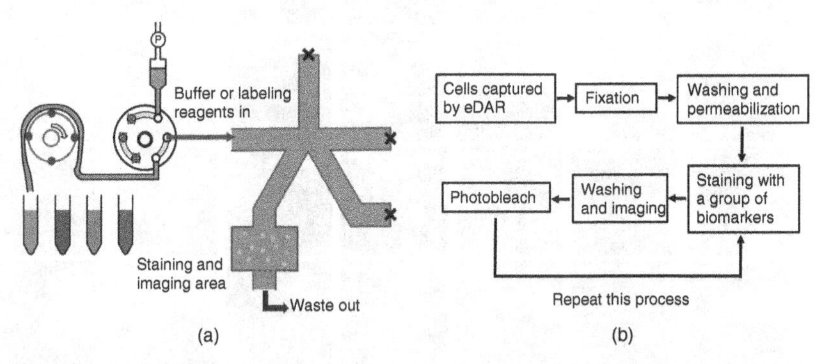

Figure 3.13 General scheme and procedure of the sequential immunostaining and photobleaching tests performed on the eDAR microchip [36]. (a) The in-line labeling system coupled to the current eDAR system. A peristaltic pump delivered the labeling reagents and washing buffer. The crossbars in this figure mean that the corresponding ports were closed during the experiments. (b) The general process flow of the sequential immunostaining and photobleaching experiment. Adapted from Zhao *et al.* [36], p. 111, Figure 2. Reproduced with permission of Elsevier.

their photobleaching curves, the fluorescence of PE, FITC, and Alexa488 could be bleached to less than 10% of their maximum intensity in less than 5 min; the photobleaching time for Alexa 647 took longer, partly because the power of the red excitation source was lower and partly because of the stability of the fluorophore. As a result, we set the photobleaching time to be 15 min to achieve a high bleaching efficiency with an acceptable throughput.

As a "proof-of-concept" study, we spiked a mixture of MCF-7, SKBr-3, and MDA-MB-231 cells into healthy donors' blood to generate a population of breast cancer CTCs with a heterogeneous expression of protein markers. We performed four rounds of the immunostaining and photobleaching process for those CTCs trapped on the eDAR microchip (Figure 3.14). In each round, we monitored four individual biomarkers through four different channels using epifluorescence microscopy. Each group of markers had a nuclear stain (Hoechst) serving as a positive control marker, CD45-FITC as a negative control marker, as well as two other protein markers conjugated with a yellow fluorophore, such as PE, or a red fluorophore, such as Alexa 647 or APC, respectively. As shown in Figure 3.14, the first set of protein markers we used had EpCAM and cytokeratin; the second set had MUC1 and Her2;

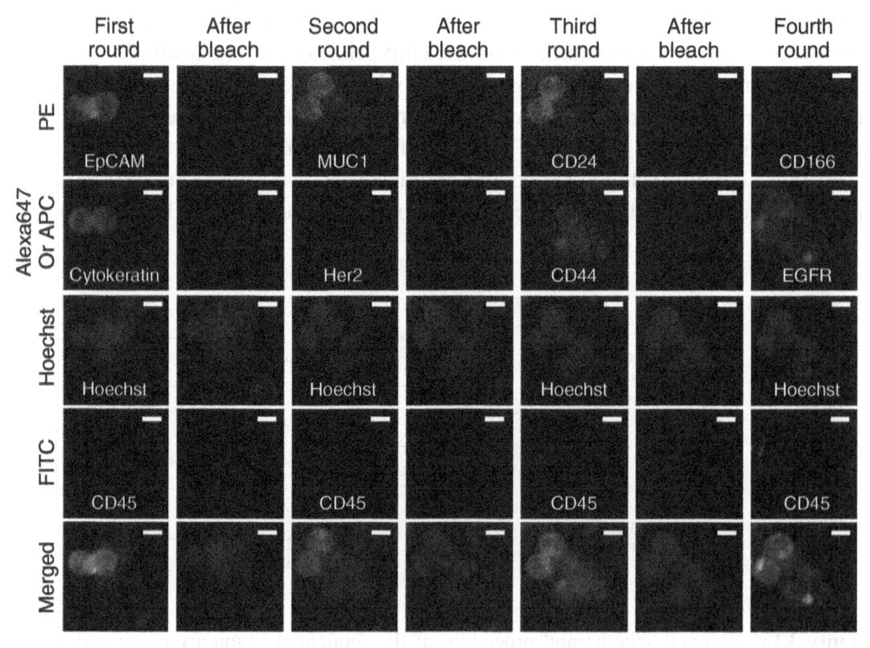

Figure 3.14 Sequential immunostaining and photobleaching results for eight breast cancer cells trapped on an eDAR chip [36]. The Hoechst nuclear stain was used as positive control marker and CD45 was used to exclude the potential interference from WBCs. Eight protein markers were studied, including EpCAM/cytokeratin, MUC1/Her2, CDD44/CD24, and CD166/EGFR. Scale bar represents 20 μm. Zhao *et al.* [36], p. 112, Figure 4. Reproduced with permission of Elsevier. (*See color plate section for the color representation of this figure.*)

the third set had CD24 and CD44; the last set had CD166 and EGFR. Downstream assays other than immunostaining were also compatible with the perfusion and washing methods performed on the same eDAR chip. For example, fluorescence *in situ* hybridization (FISH) is compatible with the perfusion scheme on PDMS microchip [37], so we can also study cytogenetic markers expressed on the CTCs trapped on the eDAR chip using this technique.

Although successful, the on-chip analysis of CTCs constrained the application of many other molecular or cellular assays, which may provide more information on CTCs and metastasis. One important advantage of eDAR is its ability to easily access the isolated CTCs, which are enriched and retained within a small area on the microchip. Using a micropipette controlled by a micromanipulator, we could selectively pick up single CTCs from the eDAR chip and deliver them into centrifuge tubes for other bioassays. We have also successfully cultured those CTCs isolated by eDAR by transferring them to cell culture plates. Many additional assays could be performed on the amplified population of single CTCs.

3.3.3 Automated High-Throughput Counting of CTCs – A "Simplified" Version of eDAR

eDAR has been successfully applied to the analysis of CTCs from clinical samples with a high detection efficiency and throughput. However, the whole process of eDAR is still not fully automated because CTCs trapped on the microchip need to be imaged and identified manually. Additionally, it requires micropipetting by a skilled operator to harvest CTCs trapped on the planar filter integrated on the eDAR microchip. These requirements imply that eDAR might not be ideally efficient for some special applications, such as fast "coarse screening" of CTCs that may not require very detailed cellular and molecular information. This application might be especially important if CTCs' diagnostic value is proved.

To address this need, we developed a "simplified" version of eDAR for high-throughput and automated enumeration of CTCs [38]. This technology was based on the multicolor line-confocal detection scheme used in eDAR. The sample preparation and experimental procedure is similar to that used in eDAR without any enrichment step. Moreover, because immunostained blood samples were simply flowed through the microchannel and were not immobilized, destroyed, fractionated, or physically retained on any substrate, this method is highly compatible with any other assays. Indeed, the same sample can be reanalyzed using eDAR or other CTC techniques after the initial "coarse screening" for the presence of CTCs.

To detect CTCs on this platform, blood samples need to be labeled with fluorescent antibodies, injected into the microchip pneumatically, and analyzed by the line-confocal detection method (Figure 3.15a). When the labeled sample flowed through the detection window, fluorescence was excited by the combined two-color laser beam (488 and 633 nm) and detected by APDs. The information collected by APDs was used to identify and enumerate CTCs.

As discussed previously, simply relying on one biomarker, such as EpCAM, is not adequate to define the population of CTCs. As a result, a set of criteria is

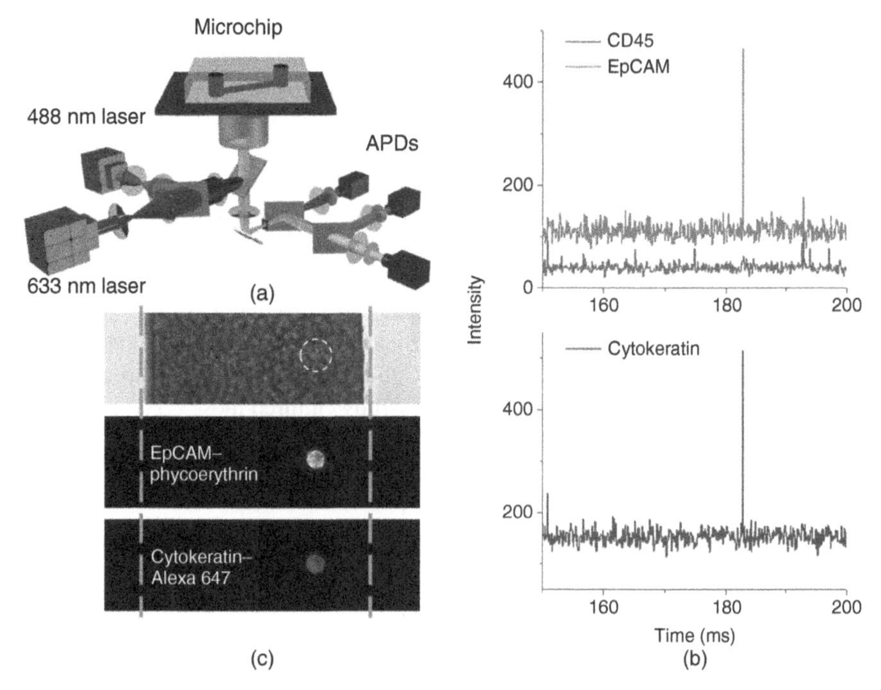

Figure 3.15 Schematic and data illustrating the flow detection platform. (a) Depiction of the microfluidics and optics [38]. (b) CTC detection and identification scheme using APD signals. A typical CTC event at 183 ms is positive for EpCAM (yellow signal) and cytokeratin (red signal), but negative for CD45 (green signal). (c) A MCF-7 cell imaged in the microfluidic channel filled with whole blood. The top panel shows a bright field image of the blood in the microchannel. The dashed circle shows the location of the MCF-7 cell that is not visible beneath the many blood cells. The gray dashed lines show the location of the microchannel walls. Fluorescence images of the same location show the MCF-7 labeled with both anti-EpCAM-PE and anti-Cytoekratin-Alexa647. Zhao *et al.* [38]. Reproduced with permission of the American Chemical Society.

necessary to identify CTCs. We partially followed the most widely used criteria to enumerate CTCs, EpCAM$^+$/Cytokeratin$^+$/CD45$^-$/Nuclear$^+$. The only exception was the lack of nuclear stain because we would need another APD detector for that channel of signal, which we did not have when we performed those experiments. We labeled the whole-blood sample with anti-EpCAM-PE, anti-Cykokeratin-Alexa647, and anti-CD45-FITC. Any CTC counted by this method should have fluorescence representing the expression of both EpCAM and Cytokeratin, but no signal associated with CD45 (Figure 3.15b). Imaging results in Figure 3.15C show a MCF-7 cell stained with anti-EpCAM-PE and anti-Cytokeratin-Alexa647 in the microfluidic channel filled with whole blood. This MCF-7 cell could not be identified in the bright field image, because it was surrounded by thousands of blood cells. However, using fluorescence imaging, we were able to clearly distinguish this cell from the blood cells in the microchannel, using both yellow (PE channel) and red (Alexa647 channel) emission.

To improve the signal-to-noise (S/N) ratio of the fluorescence detection, we have optimized the labeling condition as well as the data processing techniques. After this optimization, we measured S/N values for both spiked and clinical samples using the same labeling scheme. Over 95% of the data points have a S/N ratio higher than 20, which is high enough to identify CTCs. Taking a breast cancer sample as an example, the average and median values of S/N in EpCAM-PE channel are 56 and 44, respectively; the average and median values of S/N in the Cytokeratin-Alexa647 channel are 29 and 24, respectively [38].

Blood is a highly heterogeneous biological fluid; so, it is important to evaluate the background levels for any CTC methods. We analyzed 10 healthy donors' blood using the same experimental parameters used for clinical samples. Similar to many other CTC methods, which have reported a nonzero background value [39, 40], this method also has an average background of CTCs at 1.2 counts/ml. These background counts may result from several different sources. For example, similar to flow cytometry, antibodies can aggregate due to random colocalization, which leads to false-positive events. To address this issue, we can use more biomarkers with more fluorescent colors, which will lower the probability of detecting a colocalization event. Alternatively, background counts may come from the fluorescent dust particles in blood samples that have relatively wide emission spectra compared to the fluorophores conjugated with antibodies. Finally, it is also possible that the background counts were caused by the presence of circulating epithelial cells in healthy individuals, although this is unlikely based on our previous work [7, 18]. In this method, discarding the first tube of blood drawn from the patients could also discard the potential contamination of epithelial cells during the blood collection process. Regardless of the cause, as long as the background counts are much lower than the true counts in cancer patients, it should not affect the accuracy of the results.

To determine the recovery rate of this method, we spiked 25–1000 SKBr-3 cells into eight healthy blood samples and enumerated CTCs using the same protocol used for patient samples. The average recovery was about 94% with an R^2 value of 0.983, which is consistent with the recovery value of eDAR and is acceptable for screening CTCs from clinical samples.

Over a 2-year study, we collected 90 whole-blood samples from 24 patients with stage IV breast cancer. Using these samples, we performed a side-by-side study comparing the performance of this high-throughput counting method with the only FDA-approved CellSearch method (Figure 3.16). In this study, CTCs were identified as $EpCAM^+/Cytokeratin^+/CD45^-$, similarly to the criteria used in CellSearch method, which has an additional nuclear stain marker and does not image the EpCAM expression of the captured cells.

CTC-positive events were found using the "simplified" eDAR method in 91% of the samples (82 of 91), ranging from 15 to 3375 counts/7.5 ml, with an average CTC level of 305 counts/7.5 ml. Using the CellSearch method, we found CTCs in 44% of the samples, ranging from 1 to 846 CTCs/7.5 ml; only 22% of the samples had the number of CTCs higher than the clinical threshold (5 CTCs/7.5 ml) determined by

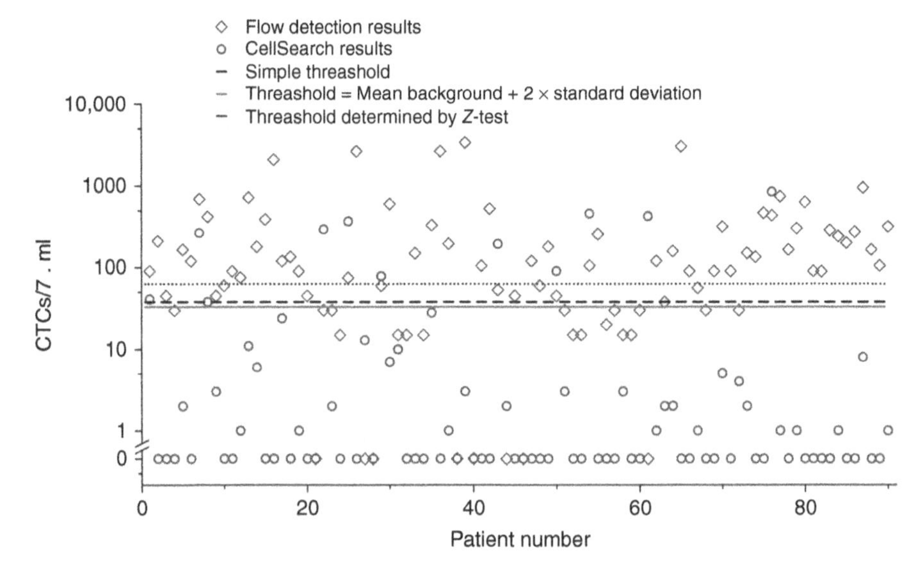

Figure 3.16 Clinical results from the "simplified" eDAR system and CellSearch [38]. Our method found a median of 90 CTCs per 7.5 ml of blood compared to a median of 0 for the CellSearch system. The dashed line is a simple threshold (38 counts/7.5 ml) set based on the range of detected CTCs in healthy donors' blood. The solid line is the threshold (33 counts/7.5 ml) set using the mean background level plus two times its standard deviation. The dotted line is the threshold (63 counts/7.5 ml) determined by Z-test with a 95% confidence level. Zhao *et al.* [38]. Reproduced with permission of the American Chemical Society.

the CellSearch method in previous studies [14]. Considering our background level of 1.2 counts/ml, it is necessary to set a threshold with a certain confidence level, so we can determine the data points that might be affected by the presence of false-positive events. Figure 3.16 summarizes the three possible methods that can determine the threshold level for the analysis of clinical samples. The most stringent threshold was determined by using the one-dimensional Z-test, assuming that any error in the measurements of the clinical samples was Poisson distributed, and the threshold was for a one-sided test. For this method, the threshold was 63 counts/7.5 ml for a 95% confidence level. With this threshold, this "simplified" eDAR found that 60% of the patient samples had CTCs with a 95% confidence level.

Similar to eDAR, this high-throughput counting method is also flexible in using other labeling and detecting scheme to detect other populations of CTCs. Of the 90 breast cancer samples, 30 were additionally analyzed using EpCAM and CD44 as positive markers, and CD24 as negative markers to enumerate CTCs with stem cell characteristics. Figure 3.17 shows a side-by-side comparison of three assays, the stem CTC and regular CTC numbers enumerated by the flow detection method, as well as the CTCs counted by the CellSearch method. $EpCAM^{+}/CD44^{+}/CD24^{+}$ events were found in 90% of the samples with average and median values of 150 and 53 counts/7.5 ml, respectively.

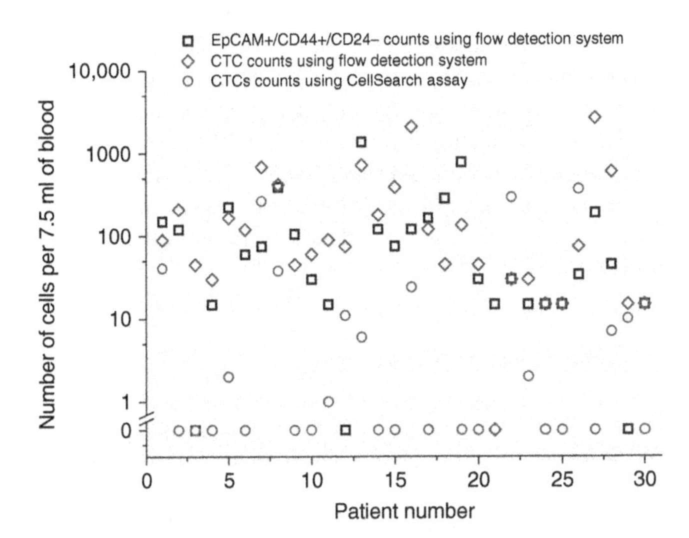

Figure 3.17 Side-by-side clinical results for regular CTCs and circulating cells with $EpCAM^+/CD44^+/CD24^-$ expression from the CTC flow detection system and CellSearch method [38]. Normal CTCs were determined by both flow detection and CellSearch method. The average number of the $EpCAM^+/CD44^+/CD24^-$ cells for these 30 breast cancer samples is 150 cells/7.5 ml. Zhao *et al.* [38]. Reproduced with permission of the American Chemical Society.

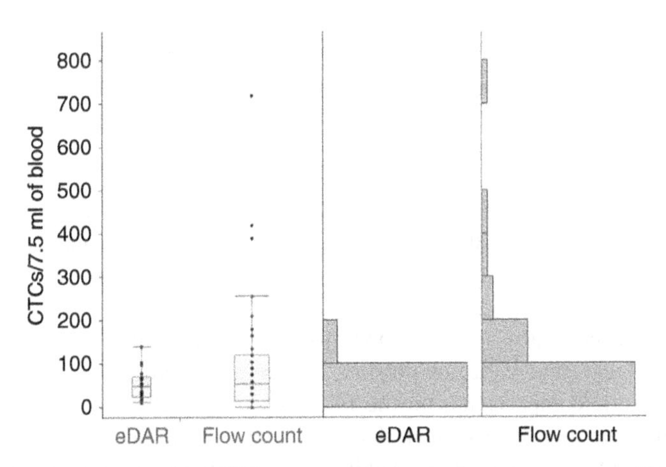

Figure 3.18 Comparison of the CTC enumeration results from the same set of breast cancer patients using eDAR and flow detection system [38]. The left part is the box plots that show the smallest observation, lower quartile, median, upper quartile, and the largest observation of the two data sets (eDAR vs flow detection), respectively. The right part shows the histogram of the two data sets. Zhao *et al.* [38]. Reproduced with permission of the American Chemical Society.

It is also interesting to compare this "simplified" version of eDAR with the regular eDAR method. In this comparison, CTC counts were analyzed in 22 blood samples from 9 breast cancer patients, using the regular eDAR approach; CTCs were also enumerated from another 40 blood samples collected from the same nine patients using the high-throughput flow detection method. Figure 3.18 shows the box plots and histograms for each data set. The data set from flow detection averaged 102 CTCs/7.5 ml, whereas that from eDAR averaged 52 CTCs/7.5 ml. However, given the large variance in the flow detection data, the two data sets are not different significantly in statistics ($p = 0.14$, $\alpha = 0.05$, ANOVA test).

3.4 CONCLUSION AND PERSPECTIVE

eDAR is an "all-in-one" platform designed for the analysis of CTCs from peripheral blood. eDAR integrates ultrasensitive detection, high-throughput active sorting, on-chip purification, and secondary immunostaining on a single microfluidic chip. Using eDAR, we were able to isolate CTCs in 1 ml of whole-blood sample in 12.5 min with a recovery efficiency of about 95% and with a zero false-positive rate. CTCs were detected and selected from billions of blood cells based on the expression of biomarkers. We have successfully applied different markers for the isolation of CTCs, such as EpCAM, Her2, EGFR, or GFP expressed intracellularly, with a recovery rate higher than 88% for each scheme. Besides these flexible selection schemes, we also can apply more complicated selection logics in eDAR, such as "AND" or "OR" logics. Those would be extremely useful in exploring CTCs with unusual expression profiles or the ones that lack EpCAM or any other biomarkers.

Understanding the molecular and cellular characteristics of CTCs is an important goal for any CTC technologies in the future. eDAR has great compatibility with many downstream assays. We can perform directly the subsequent analyses of protein markers expressed on the CTCs trapped on the same eDAR microchip. By applying a sequential immunostaining and photobleaching scheme, we can extend the number of protein markers that can be investigated for the CTCs to a large set. Because of the open-access design of the cell trapping area, we also can select and pick up individual CTCs for any downstream single-cell analysis that needs to be performed out of the eDAR chip.

We have successfully applied eDAR for the analysis of CTCs from patients with breast, lung, pancreatic, or prostate cancer with a high detection efficiency. In a side-by-side comparison with the FDA-approved method, CellSearch, eDAR was shown to be more sensitive in detecting CTCs from metastatic breast cancer patients. We are performing more studies on other types of metastatic cancer or early-stage cancer samples to examine the performance of eDAR with a broader range of clinical samples. To have an automated method for the "coarse screening" of CTCs in clinical samples, we have developed a simplified version of eDAR that can count CTCs from peripheral blood without enrichment of CTCs. This method has been successfully applied for analyzing stage IV breast cancer samples with a higher sensitivity compared to CellSearch. We believe the different versions of eDAR that we have developed will effectively address the needs of the research and clinical community in the enumeration, isolation, and study of CTCs.

REFERENCES

1. Harris, J.L., Stocum, M., Roberts, L. *et al.* (2013) Quest for the ideal cancer biomarker: an update on progress in capture and characterization of circulating tumor cells. *Drug Development Research*, **74**, 138–147.
2. Alix-Panabieres, C. and Pantel, K. (2013) Circulating Tumor Cells: Liquid Biopsy of Cancer. *Clinical Chemistry*, **59**, 110–118.
3. Maheswaran, S. and Haber, D.A. (2010) Circulating tumor cells: a window into cancer biology and metastasis. *Current Opinion in Genetics and Development*, **20**, 96–99.
4. Li, P., Stratton, Z.S., Dao, M. *et al.* (2013) Probing circulating tumor cells in microfluidics. *Lab on a Chip*, **13**, 602–609.
5. Dong, Y., Skelley, A.M., Merdek, K.D. *et al.* (2013) Microfluidics and Circulating Tumor Cells. *Journal of Molecular Diagnostics*, **15**, 149–157.
6. Paterlini-Brechot, P. and Benali, N.L. (2007) Circulating tumor cells (CTC) detection: Clinical impact and future directions. *Cancer Letters*, **253**, 180–204.
7. Schiro, P.G., Zhao, M., Kuo, J.S. *et al.* (2012) Sensitive and high-throughput isolation of rare cells from peripheral blood with ensemble-decision aliquot ranking. *Angewandte Chemie International Edition*, **51**, 4618–4622.
8. Gross, H.J., Verwer, B., Houck, D. *et al.* (1995) Model Study Detecting Breast-Cancer Cells in Peripheral-Blood Mononuclear-Cells at Frequencies as Low as 10^{-7}. *Proceedings of the National Academy of Sciences of the United States of America*, **92**, 537–541.
9. Patriarca, C., Macchi, R.M., Marschner, A.K., and Mellstedt, H. (2012) Epithelial cell adhesion molecule expression (CD326) in cancer: A short review. *Cancer Treatment Reviews*, **38**, 68–75.
10. Jeffries, G.D.M., Lorenz, R.M., and Chiu, D.T. (2010) Ultrasensitive and High-Throughput Fluorescence Analysis of Droplet Contents with Orthogonal Line Confocal Excitation. *Analytical Chemistry*, **82**, 9948–9954.
11. Schiro, P.G., Kuyper, C.L., and Chiu, D.T. (2007) Continuous-flow single-molecule CE with high detection efficiency. *Electrophoresis*, **28**, 2430–2438.
12. Stott, S.L., Hsu, C.H., Tsukrov, D.I. *et al.* (2010) Isolation of circulating tumor cells using a microvortex-generating herringbone-chip. *Proceedings of the National Academy of Sciences of the United States of America*, **107**, 18392–18397.
13. Wang, S.T., Liu, K., Liu, J.A. *et al.* (2011) Highly efficient capture of circulating tumor cells by using nanostructured silicon substrates with integrated chaotic micromixers. *Angewandte Chemie International Edition*, **50**, 3084–3088.
14. Riethdorf, S., Fritsche, H., Mueller, V. *et al.* (2007) Detection of circulating tumor cells in peripheral blood of patients with metastatic breast cancer: a validation study of the CellSearch system. *Clinical Cancer Research*, **13**, 920–928.
15. Jaggupilli, A. and Elkord, E. (2012) Significance of CD44 and CD24 as cancer stem cell markers: an enduring ambiguity. *Clinical and Developmental Immunology*, **2012**, 708036.
16. Schiro, P.G., Gadd, J.C., Yen, G.S., and Chiu, D.T. (2012) High-Throughput Fluorescence-Activated Nanoscale Subcellular Sorter with Single-Molecule Sensitivity. *Journal of Physical Chemistry B*, **116**, 10490–10495.

17. Shirasaki, Y., Tanaka, J., Makazu, H. *et al.* (2006) On-chip cell sorting system using laser-induced heating of a thermoreversible gelation polymer to control flow. *Analytical Chemistry*, **78**, 695–701.
18. Zhao, M., Nelson, W.C., Wei, B. *et al.* (2013) New Generation of Ensemble-Decision Aliquot Ranking Based on Simplified Microfluidic Components for Large-Capacity Trapping of Circulating Tumor Cells. *Analytical Chemistry*, **85**, 9671–9677.
19. Kuo, J.S., Zhao, Y.X., Schiro, P.G. *et al.* (2010) Deformability considerations in filtration of biological cells. *Lab on a Chip*, **10**, 837–842.
20. Lara, O., Tong, X.D., Zborowski, M., and Chalmers, J.J. (2004) Enrichment of rare cancer cells through depletion of normal cells using density and flow-through, immunomagnetic cell separation. *Experimental Hematology*, **32**, 891–904.
21. Vona, G., Sabile, A., Louha, M. *et al.* (2000) Isolation by size of epithelial tumor cells - A new method for the immunomorphological and molecular characterization of circulating tumor cells. *American Journal of Pathology*, **156**, 57–63.
22. Zabaglo, L., Ormerod, M.G., Parton, M. *et al.* (2003) Cell filtration-laser scanning cytometry for the characterisation of circulating breast cancer cells. *Cytometry Part A*, **55A**, 102–108.
23. Theodoropoulos, P.A., Polioudaki, H., Agelaki, S. *et al.* (2010) Circulating tumor cells with a putative stem cell phenotype in peripheral blood of patients with breast cancer. *Cancer Letters*, **288**, 99–106.
24. Zhao, Y., Schiro, P.G., Kuo, J. *et al.* (2009) Method for the Accurate Preparation of Cell-Spiking Standards. *Analytical Chemistry*, **81**, 1285–1290.
25. Farace, F., Massard, C., Vimond, N. *et al.* (2011) A direct comparison of CellSearch and ISET for circulating tumour-cell detection in patients with metastatic carcinomas. *British Journal of Cancer*, **105**, 847–853.
26. Gorges, T.M., Tinhofer, I., Drosch, M. *et al.* (2012) Circulating tumour cells escape from EpCAM-based detection due to epithelial-to-mesenchymal transition. *BMC Cancer*, **12**, 178.
27. Sieuwerts, A.M., Kraan, J., Bolt, J. *et al.* (2009) Anti-Epithelial Cell Adhesion Molecule Antibodies and the Detection of Circulating Normal-Like Breast Tumor Cells. *Journal of the National Cancer Institute*, **101**, 61–66.
28. Yagublu, V., Ahmadova, Z., Hafner, M., and Keese, M. (2012) Fluorescent Protein-based Tumor Models. *In Vivo*, **26**, 599–607.
29. Andreopoulou, E. and Cristofanilli, M. (2010) Circulating tumor cells as prognostic marker in metastatic breast cancer. *Expert Review of Anticancer Therapy*, **10**, 171–177.
30. Aurilio, G., Sciandivasci, A., Munzone, E. *et al.* (2012) Prognostic value of circulating tumor cells in primary and metastatic breast cancer. *Expert Review of Anticancer Therapy*, **12**, 203–214.
31. Cohen, S.J., Punt, C.J.A., Iannotti, N. *et al.* (2008) Relationship of circulating tumor cells to tumor response, progression-free survival, and overall survival in patients with metastatic colorectal cancer. *Journal of Clinical Oncology*, **26**, 3213–3221.
32. Lianidou, E.S., Markou, A., and Strati, A. (2012) Molecular characterization of circulating tumor cells in breast cancer: challenges and promises for individualized cancer treatment. *Cancer Metastasis Reviews*, **31**, 663–671.
33. Mocellin, S., Keilholz, U., Rossi, C.R., and Nitti, D. (2006) Circulating tumor cells: the 'leukemic phase' of solid cancers. *Trends in Molecular Medicine*, **12**, 130–139.

34. Hou, J.M., Krebs, M., Ward, T. *et al.* (2011) Circulating tumor cells as a window on metastasis biology in lung cancer. *American Journal of Pathology*, **178**, 989–996.
35. Krebs, M.G., Sloane, R., Priest, L. *et al.* (2011) Evaluation and prognostic significance of circulating tumor cells in patients with non-small-cell lung cancer. *Journal of Clinical Oncology*, **29**, 1556–1563.
36. Zhao, M., Wei, B., and Chiu, D.T. (2013) Imaging multiple biomarkers in captured rare cells by sequential immunostaining and photobleaching. *Methods*, **64**, 108–113.
37. Choudhury, D., van Noort, D., Iliescu, C. *et al.* (2012) Fish and chips: a microfluidic perfusion platform for monitoring zebrafish development. *Lab on a Chip*, **12**, 892–900.
38. Zhao, M., Schiro, P.G., Kuo, J.S. *et al.* (2013) An automated high-throughput counting method for screening circulating tumor cells in peripheral blood. *Analytical Chemistry*, **85**, 2465–2471.
39. Stott, S.L., Lee, R.J., Nagrath, S. *et al.* (2010) Isolation and characterization of circulating tumor cells from patients with localized and metastatic prostate cancer. *Science Translational Medicine*, **2**, 25ra23.
40. Goda, K., Ayazi, A., Gossett, D.R. *et al.* (2012) High-throughput single-microparticle imaging flow analyzer. *Proceedings of the National Academy of Sciences of the United States of America*, **109**, 11630–11635.

4

SINUSOIDAL MICROCHANNELS WITH HIGH ASPECT RATIOS FOR CTC SELECTION AND ANALYSIS

JOSHUA M. JACKSON

Department of Chemistry, University of North Carolina Chapel Hill, Chapel Hill, NC, USA; Center for Biomodular Multi-scale Systems for Precision Medicine, University of North Carolina at Chapel Hill, Chapel Hill, NC, USA

MAŁGORZATA A. WITEK

Department of Biomedical Engineering, University of North Carolina Chapel Hill, Chapel Hill, NC, USA; Center for Biomodular Multi-scale Systems for Precision Medicine, University of North Carolina at Chapel Hill, Chapel Hill, NC, USA

STEVEN A. SOPER

Department of Chemistry, University of North Carolina Chapel Hill, Chapel Hill, NC, USA; Department of Biomedical Engineering, University of North Carolina Chapel Hill, Chapel Hill, NC, USA; Department of Chemistry, Louisiana State University, Baton Rouge, LA, USA; Center for Biomodular Multi-scale Systems for Precision Medicine, University of North Carolina at Chapel Hill, Chapel Hill, NC, USA

4.1 INTRODUCTION

4.1.1 Brief Perspective

The only Food and Drug Administration (FDA)-certified circulating tumor cell (CTC) assay currently available is Johnson & Johnson's Veridex CellSearch™ system (see Chapter 19 for details). The CellSearch system uses functionalized immunomagnetic nanoparticles to target the tumor-specific antigen, epithelial cell adhesion molecule (EpCAM) that is often associated with CTCs. But the assay does have limitations such as: (i) the test must be conducted at a dedicated laboratory due to its semiautomated format requiring lengthy sample preprocessing steps,

Circulating Tumor Cells: Isolation and Analysis, First Edition. Edited by Z. Hugh Fan.

generating long analysis times (4–6 h); (ii) it uses fixatives, which are necessary to stabilize the sample for transport to the dedicated laboratory, that have been shown to compromise EpCAM binding and thereby limit the assay's clinical sensitivity [1]; (iii) it targets only EpCAM, but many reports indicate that only certain CTC subpopulations express EpCAM while other subpopulations, such as CTCs that possess EMT (epithelial to mesenchymal transition) or CSC (cancer stem cell) characteristics, express little if any EpCAM [2–6] – recent data has shown that 7/9 normal-type breast cancer cell lines could not be recovered using the CellSearch system [7]; and (iv) molecular analyses are currently difficult because CTCs are isolated with extremely low purities (0.01–0.1%) [8].

The utility of microfluidics for the selection and analysis of CTCs when invoking either biochemical and/or physical selection processes has been well documented. Over the last 5 years, a plethora of new technologies have been reported that can isolate CTCs directly from clinical samples with high recovery [9–46]. The principal advantages associated with the use of microfluidics for CTC analyses stem from the fact that point-of-care (PoC) instruments can be realized, which can offer the ability to fully automate the CTC processing pipeline with shorter assay turnaround times and smaller footprint instruments. In addition, microfluidic platforms can be operated in a fully automated manner, minimizing the loss of rare cells due to sample handling or contamination upon exposing the sample to the surroundings. Besides, most microfluidic devices can be operated at low shear, which reduces cell damage due to high shear conditions. Finally, the CTC selection process can be integrated with downstream processing to secure additional information besides just enumeration data, such as molecular analyses of the selected CTCs.

Arguably, research in the CTC microfluidics community is currently in a critical phase of translating laboratory-based research (evaluating small cohorts of tens to hundreds of patients) to clinical implementation (case studies on the order of thousands to tens of thousands of patients). Clinical CTC testing requires that devices are manufactured as disposable units, at low cost, with tight compliance to avoid sample carryover and cross-contamination artifacts and with high assay reproducibility. Thus, microfluidic devices used in the laboratory need to be produced in high volumes and at low unit cost to be successfully implemented [47]. The work reported in this chapter has been guided by these criteria, involving the design of devices that use thermoplastic substrates appropriate for microreplication, the same technology used to produce consumables such as CDs and DVDs [12, 46, 48].

Throughout this chapter, focus is placed on a microfluidic design for CTC selection that employs microchannels possessing high aspect ratios (aspect ratio = channel depth/channel width) with a sinusoidal shape, which were implemented to increase sampling throughput and improve recovery, respectively (see overview in Section 4.2). Information on high-throughput manufacturing and functionalization of thermoplastic microfluidic devices for CTC analyses is presented, and we highlight specific barriers in translating laboratory-based technologies to thermoplastic devices specifically dedicated to the selection of CTCs directly from clinical samples. We also present details on the physical dynamics of positive affinity CTC selection in terms of encouraging CTC interactions with surface-confined recognition elements, discouraging nonspecific interactions of highly abundant cells, and methods of scaling throughput for rapid processing of large sample volumes (i.e., 7.5 ml blood draw). Finally, clinical results obtained using microfluidic devices employing these channel architectures are outlined, as are molecular analyses of selected CTCs from clinical samples.

4.1.2 Overview of CTC Selection Modalities and Assay Metrics

CTC selection methods can be broadly differentiated into two modes: selection by targeting *biological or physical* CTC properties that differ from normal hematopoietic cells (Table 4.1) [55]. Biological properties range from surface and cytoplasmic protein expression, viability, and invasion capacity. Physical properties include size, density, cell membrane electric potential, and deformability.

CTC expression of the surface protein EpCAM is a ubiquitous biological property exploited for isolating CTCs directly by positive affinity selection. The primary argument against this selection modality is EMT, meaning that some important CTC subpopulations may not express EpCAM and may therefore be lost by selecting only EpCAM positive cells [2–6]. There is currently no single antigenic marker that is uniformly expressed on all CTCs that can be used to identify the various subpopulations that may be present in circulation and be key effectors in the metastatic cascade [56]. These arguments have driven extensive research in physical selection methods or negative selection strategies such as CD45 selection to clear hematopoietic cells and obtain CTCs indirectly [57]. However, multiple positive affinity selection devices employing different selection aptamers or antibodies (Abs, targeting both epithelial and mesenchymal markers) can be used to broaden the search and isolate a range of clinically relevant CTC subpopulations.

Five important figures of merit are relevant to all CTC selection technologies [52, 53]:

1. Throughput (maximum volumetric processing rate) enables processing milliliter blood volumes to isolate the extremely rare CTCs (only 1–100 per milliliter blood) in a practical amount of time and enrich them into volumes typically encountered in microfluidics.
2. Recovery (the number of target cells selected from the total number of target cells input) establishes assay sensitivity.
3. Purity (the ratio of CTCs selected to the total number of cells enriched) complicates or simplifies CTC identification and permits or precludes molecular analyses.
4. Clinical sensitivity (the ability of the assay to correctly identify patients with the disease) prevents false negatives.
5. Clinical specificity (the ability of the assay to correctly identify patients without the disease) prevents false positives.

Generally, positive affinity selection strategies achieve higher purities compared to physical or negative selection methods but at the expense of throughput; Section 4.2.4

TABLE 4.1 Operational Metrics for Many CTC Selection Technologies

Method Name	Principle	Throughput	Recovery	Purity	Clinical Demonstration	References
Micropillar chip	**Biological** Ab EpCAM	1 ml/h	65% from PBS (NCI-H1650)	50% 9%	99.1% clinical sensitivity (5–1281 CTCs/ml for different cancers)	[13, 66]
Micropillar chip (GEDI)	**Biological** Ab EpCAM, Ab PSMA	1 ml/h	97 ± 3% from PBS 85 ± 5% from blood	68 ± 8%	Prostate cancer; 90% clinical sensitivity (30 CTCs/ml)	[87]
Sinusoidal chip	**Biological** Ab EpCAM, Ab/Aptamer PSMA	15 ml/h	98% from whole blood for MCF-7 82% for MDA-MB231 from whole blood	>80%	Prostate, pancreatic, colorectal >95% clinical sensitivity for metastatic disease (53 CTCs/ml for PDAC)	[12, 45, 46, 48, 55]
Herringbone chip	**Biological** Ab EpCAM	1.2 ml/h	PC3: 91.8 ± 5.2% (5×10^4 Ag/cell) NB508: 35%	14% 0.1–6%	Prostate 93% clinical sensitivity (63 CTCs/ml) Pancreatic mouse models (31 cells/0.1 ml)	[13, 121]
Si nanopillars	**Biological** Ab EpCAM	1 ml/h	>95% for cell lines	ND	Prostate (localized 12 CTCs/ml and metastatic 51 CTCs/ml)	[11]
DEP	**Physical** Dielectric	1.5 ml/h	92%	~7.5%	Colorectal (~1.3 CTCs/ml)	[122–124]

Microfilters (parylene slot filters)	**Physical** Size	90 ml/h	90%	<0.1%	Breast, colorectal, prostate, and bladder (0–12.5 CTCs/ml)	[13]
Flow cytometry	**Biological** Antigen expression	3.6 ml/h	85%	<0.1%	SCCHN 33–43% clinical sensitivity 3 CTCs/7.5 ml) Breast (16 CTCs/20 ml)	[55, 58, 59]
Invasive assay	**Biological** CAM	NA	77.3%	0.5–35%	Breast (18–256 CTCs/ml)	[125]
Negative depletion	**Biological** Ab CD45	1.3 ml/h	83%	0.26%	Squamous cell head and neck (0–281 CTCs/ml)	[57]
IMACS	**Biological** Ab EpCAM	2 ml/h	85%	0.01–0.1%	Most solid tumors but poor results for pancreatic and ovarian	[126]
iChip	**Biological** Ab EpCAM **Physical** Size	8 ml/h	SKBR3: 98% PC3-9: 89% MBA-MD231: 9%	1–10%	Prostate (50.3 CTCs/ml average)	[127]
Halloysite nanotubes	**Biological** E-selectin + Ab EpCAM	4.8 ml/h	ND	18–80%	Breast, prostate, lung, ovarian (5–188 CTCs/ml)	[128]
Ephesia	**Biological** Ab EpCAM	9 μl/h	MCF-7: 80 ± 20%	89.5%	Epithelial cancers ND	[129]
μHall	**Biological** Ab EpCAM	3.25 ml/h	MDA-MB 468: 99%	100%	Ovarian (7.6 CTCs/ml)	[130]

demonstrates that careful microfluidic engineering can be used to expand assay throughput without compromising other figures of merit [21, 45]. More specifically, developing a positive affinity CTC selection technology that is highly effective in terms of all five figures of merit requires a firm handle of material properties, Ab densities and activities, and fluid dynamics of CTC–surface interactions.

Table 4.1 shows results for different microfluidic platforms that have been used for CTC analyses and some of their operational figures of merit [55]. While flow cytometry has been used extensively for characterizing different biological cells based on antigen expression, extensive sample preprocessing causes significant loss of CTCs and low clinical sensitivity even at large blood volumes, which is especially problematic as processing throughput is only 3.6 ml/h [54, 58, 59]. For comparison, microfluidics can be considered as a low-throughput technology due to scaling issues; however, when the microfluidic is engineered to accommodate the need for sampling large input volumes (~7.5 ml of whole blood), these platforms can produce throughputs that rival even macroscale sampling systems. For example, the sinusoidal chip can produce throughputs approaching 15 ml/h when employing multiple sampling channels (>50 microcapillaries) [60].

4.2 PARALLEL ARRAYS OF HIGH-ASPECT-RATIO, SINUSOIDAL MICROCHANNELS FOR CTC SELECTION

In the subsequent sections, we detail microfluidic CTC selection technologies that employ an array of high-aspect-ratio microchannels configured in a sinusoidal architecture.

Before proceeding, we provide a brief overview of the sinusoid technology. CTCs are isolated from unprocessed and unfractionated blood using devices that are comprised of parallel arrays of microchannels. In each high aspect ratio, sinusoidal microchannel (Figure 4.1g), CTCs are selected by Abs, for example, anti-EpCAM, or aptamers coated on the microchannel surfaces (for brevity, we will refer to the selection moieties as Abs). Critical to selection, CTC–Ab interactions are promoted by several aspects of the channel design: (i) the narrow channel width (~25 μm), which approaches CTC diameter (12–20 μm) and increases the probability of CTC–wall interactions while avoiding leukocyte (7–15 μm) surface interactions; (ii) 150 μm channel depth, which increases throughput while reducing pressure drop, especially when an isolated CTC could potentially "clog" the channel; (iii) high-aspect-ratio channels generate an order of magnitude greater fluidic shear stress than in comparable devices (Figure 4.1j) to interrupt nonspecific interactions; (iv) sinusoid's tight radius of curvature (125 μm) induces centrifugal forces that propel CTCs out of laminar streamlines and toward channel walls; and (v) the channel's continuous design (as opposed to discrete microposts) enables uninterrupted CTC rolling along the affinity surface, improving the probability of CTC-Ab binding. All of these factors contribute to efficient CTC recovery (>90%) with high purity (>80%) [21, 45, 46, 48, 60]. Furthermore, while the volumetric throughput of each cell isolation channel is small (~30 μl/h), arraying channels in parallel (Figure 4.1f)

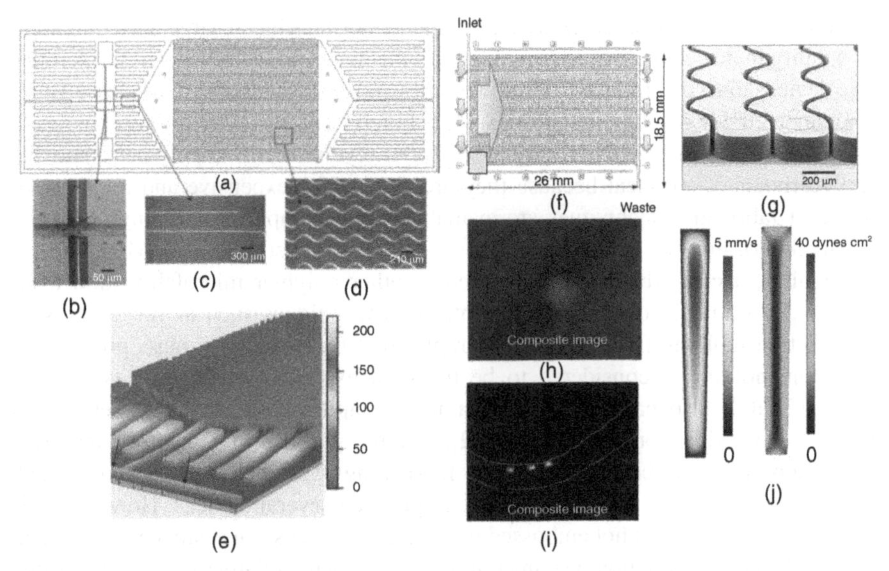

Figure 4.1 (Left) Schematics of the first-generation, high-throughput CTC selection device showing: (a) A scaled AutoCAD diagram of the sinusoid capture channels with bright-field optical micrographs; (b) the integrated conductivity sensor consisting of cylindrical Pt electrodes that were 75 µm in diameter with a 50 µm gap; (c) single port exit where the device tapers from 100 µm wide to 50 µm while the depth tapers from 150 to 80 µm over a 2.5 mm region that ends 2.5 mm from the Pt electrodes; (d) micrograph taken at 5× magnification showing the sinusoidal cell capture channels; and (e) 3D projection of the topology of the device obtained at 2.5 µm resolution using noncontact optical profilometry (arrows indicate the Pt electrode conduits). Adapted from Adams *et al.* [48]. Copyright (2008) American Chemical Society. (Right) Second-generation sinusoidal device: (f) Schematic operation of the CTC selection module with 50 parallel, sinusoidal microchannels and inlet/outlet channels arranged in the Z-configuration. The large arrow indicates the sample flow direction through the selection channels. (g) SEM of the selection bed showing high-aspect-ratio (~25 × 150 µm, $w \times d$), sinusoidal microchannels and the output channel. (h) Composite fluorescence images of a CTC stained with DAPI, CK8/19, and CD45. (i) Three WBCs staining positively for DAPI and CD45 and negatively for CD8/19. (j) Fluid dynamics simulation results showing the distribution of flow velocities and shear stress in microfluidic selection channels. Jackson *et al.* [45]. Reproduced with permission of the Royal Society of Chemistry. (*See color plate section for the color representation of this figure.*)

increases throughput and enables tuning devices for processing a range of sample volumes (from 7.5 ml to 1 ml or less) in a practical amount of time (<30 min) [21, 61].

4.2.1 Production of CTC *In Vitro* Diagnostic Devices in Thermoplastics

Clinical CTC testing requires that devices be manufactured as disposable units in order to avoid sample carryover and/or cross-contamination artifacts during analysis. Moreover, disposable devices avoid the need to reactivate the devices between samples, which in many cases is impossible or not feasible at large scale, creating variability in device performance and compromising reproducibility. Thus, microfluidic devices must be produced in high volumes and at low unit cost to be successfully implemented for *in vitro* diagnostics [47]. For this purpose, thermoplastics, such as PMMA (poly(methyl methacrylate)), COC (cyclic olefin copolymer), and PC (polycarbonate), are ideal because they are generally inexpensive and a variety of low-cost fabrication techniques are available for both rapid prototyping and mass production of finished devices, even those with high aspect ratios [62–64].

Significant cost advantages compared to other polymer microfabrication techniques can be realized when using replication techniques such as hot embossing or injection molding [64]. Among the replication techniques for mass production, injection molding is considered to be the least expensive and fastest (<10 s per device), but it also carries the highest initial capital costs, machine setup, and molding condition optimization [49, 62]. For the purpose of laboratory research, hot embossing can be employed for fabricating CTC selection devices. Each embossing cycle takes 10–15 min and can produce several devices. However, it is important to realize that hot embossed microfluidic devices can easily be transitioned into high-scale production via injection molding with relative ease as the same thermoplastic polymers are used and similar design rules are followed during device development. This approach avoids or at least minimizes well-recognized problems associated with transferring academic ideas into manufacturable products due to incompatibility of the materials and microfabrication methods between early development and high-volume production (see Section 4.2.2) [47].

The first step in fabricating microfluidic devices by hot embossing or injection molding is the fabrication of the molding master (Figure 4.2a), which can be produced nonlithographically using high-precision micromilling (HPMM) of a brass substrate

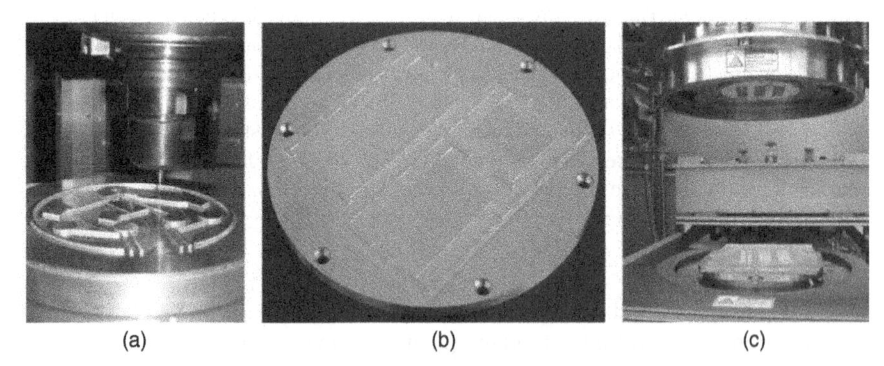

Figure 4.2 (a) High-precision micromilling of a brass substrate using solid carbide bits on the order of 500 to 50 μm or smaller in diameter. (b) A completed brass master mold that can be used for. (c) repeated microreplication by hot embossing the mold master into a thermoplastic substrate. Reproduced with permission of Springer, from Hupert *et al.* [60]. (*See color plate section for the color representation of this figure.*)

via a commercial milling machine and micrometer-sized solid carbide tools [65]. HPMM offers significant advantages over other mold master fabrication methods such as X-ray-LiGA (Lithographie, Galvanoformung, Abformung in German for Lithography, Electroplating, and Molding, respectively) or UV-LiGA as it allows for shorter turnaround times, low cost, and single-step fabrication of multilevel structures [63–65]. Although HPMM is not capable of achieving the fine resolution or minimum feature size of lithography-based techniques (i.e., submicrometer for X-ray LiGA) due to milling bit size limitations, it is well suited for many microfluidic devices such as those shown in Figure 4.1, which usually employ structures in the range of 10–500 μm with aspect ratios smaller than 20 and interstructure spacing greater than 30 μm that are easily obtainable using micromilling [60, 65].

Hot embossing involves compressing a thermoplastic that is heated to above its glass transition temperature, T_g (or softening temperature), into the microstructured mold master with precise control over both temperature and force [49]. When the polymer is cooled and released, the inverted microstructures are imprinted into the thermoplastic with high precision; nanometer-sized micromilling marks remain on the sidewalls of the brass mold after transfer into the thermoplastic (Figure 4.3b inset). Note that the replicated channels (Figure 4.3c and d) have a trapezoidal cross section with slightly tapered walls with an average draft angle of 1.4°, which is not intentionally imposed by the mold fabrication but rather the result of selected milling conditions and flexing of the straight (not tapered) micromilling tools during brass removal. Draft angles of this magnitude are very advantageous for the replication process as they reduce friction between the molded part and mold master during demolding, thus reducing replication errors due to polymer pullout and prolonging the molding master usable lifetime [60].

One can perform more than 1000 replication cycles using a single brass mold master without noticeable reduction in the quality of the embossed structures or deterioration of the mold master. For the device shown in Figure 4.1g, the average microchannel dimensions measured after the first 30 embossing cycles were $20.0 \pm 1.0\,\mu m \times 27.4 \pm 0.8\,\mu m \times 152 \pm 1\,\mu m$ (bottom width × top width × depth), and after 300 embossing cycles, these dimensions were $19.7 \pm 0.5\,\mu m \times 27.8 \pm 1.0\,\mu m \times 152 \pm 1\,\mu m$, well within measurement error, indicating high-dimensional stability of the mold master and long-term reproducibility of the hot embossing process [60].

Following hot embossing, each microstructured sheet is diced into individual devices, and the devices are sealed using a cover plate made from the same material as the substrate. Typically, thermal fusion bonding is employed in which the cover plate and substrate are brought into conformal contact under a fixed pressure (~1 N/cm^2) and slowly heated to a temperature slightly below the T_g of the thermoplastic so that the microstructured substrate and cover plate thermally fuse. Bonding conditions must be carefully selected to preserve structural integrity of the high-aspect-ratio microchannels as much as possible while still yielding high bond strengths that easily withstand the fluidic pressure of pumping whole blood through the device [45, 60].

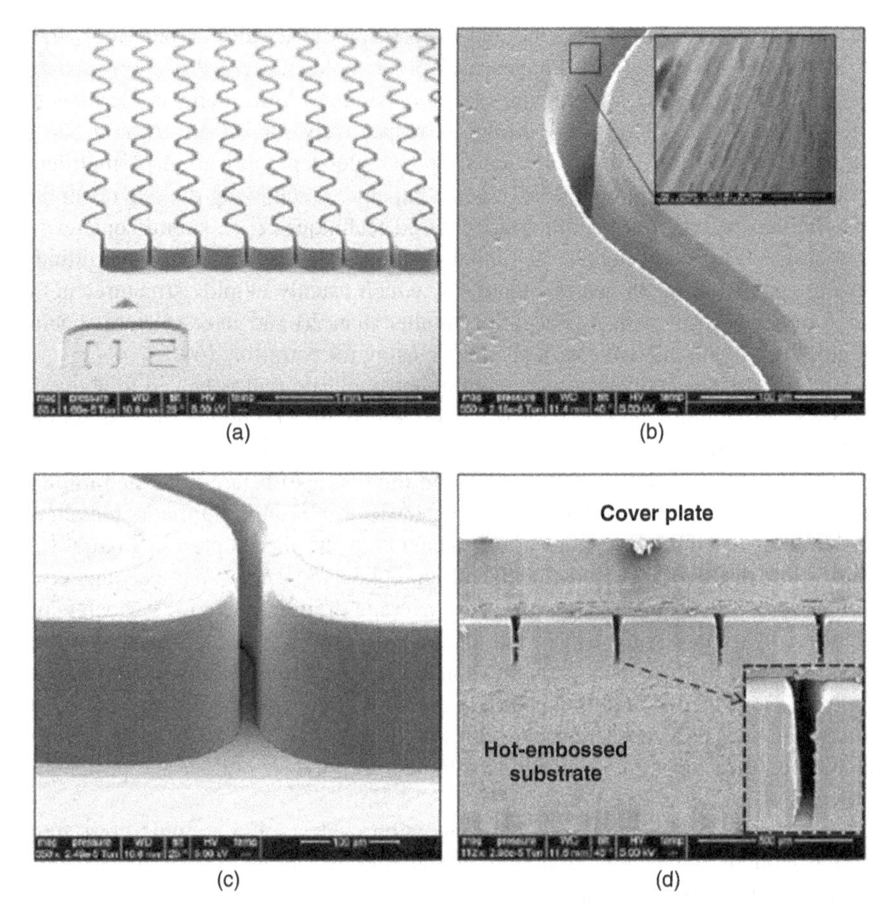

Figure 4.3 SEMs of a CTC selection device shown in Figure 4.1f hot embossed into COC thermoplastic. (a) Low-resolution SEM showing a series of sinusoidal, high-aspect-ratio channels. (b) High-resolution SEM of one channel with the *inset* showing surface roughness due to milling. For reference, the marks left by the milling bit have an average roughness of 115 nm and mean peak height of 290 nm when measured vertically along the channel wall and 55 and 200 nm when measured horizontally, respectively, while the typical average roughness of the polymer used for embossing is <20 nm [49]. (c) SEM of the channel prior to thermal fusion bonding of the cover plate. (d) SEM of a channel following thermal fusion bonding of the cover plate. The cover plate is also made from COC. Hupert *et al.* [60]. Reproduced with permission of Springer.

4.2.2 Activation of High-Aspect-Ratio Microchannels for Efficient Ab Immobilization

To enable CTC isolation in the high-aspect-ratio, sinusoidal microchannels, Abs must be covalently attached to the microchannel surfaces. When a CTC passes over the selection channel surface, Ab binding to the target antigen generates an adhesion force between the selection Abs and the CTC. The dynamics of this selection process are discussed in detail in Section 4.2.3. Briefly, this adhesion force (F_A) can be calculated from

$$F_A = \left(\frac{2A_c k_B \theta N_r}{l_b}\right) \ln\left(1 + \frac{N_L}{\eta K_D}\right) \tag{4.1}$$

where A_c is the cell contact area with the surface, k_B is Boltzmann's constant, θ represents the absolute temperature, N_r is the receptor density (antigen expression level of the CTC), l_b is the extent of stretch to reach the critical force before breaking the antigen/Ab association, N_L is the surface density of active mAbs, K_D is the antigen/mAb dissociation constant, and η is an adjustable fitting parameter [48].

Because a CTC can be lifted from the surface when the hydrodynamic shear force (F_s) exceeds F_A, selected CTCs can be more easily removed when N_r or N_L is small (i.e., when the selection Ab load is low or when CTCs have low target antigen expression levels). One could operate at extremely low shear to improve CTC recovery in these cases [66] but at the expense of both throughput and purity because F_s is necessary to break high K_D and low F_A nonspecific interactions [46]. Therefore, to improve CTC clinical yields, which is defined as the number of CTCs selected from a cancer patient, it is imperative to maximize N_L by maximizing Ab loading to the selection bed's surfaces.

For many nonfunctional surfaces, passive adsorption of the biologic to the surface may be used but can result in high loss of the adsorbate's activity; greater than 90% Ab denaturation and deactivation have been reported [67, 68]. Alternatively, activity may be retained using covalent coupling chemistry, which requires surface functional groups on the substrate. A well-established strategy (Figure 4.4) uses a substrate containing surface-confined carboxylic acids that can be reacted with EDC/NHS reagents to form an ester intermediate that subsequently reacts with primary-amine-bearing biologics, such as lysine and arginine residues found on an Ab surface or pendant amino groups on oligonucleotides or aptamers [46, 48, 68].

Many thermoplastics do not contain surface functional groups, and, therefore, activation protocols are employed to create the appropriate surface scaffolds [64, 69–72]. UV irradiation prior to device assembly (i.e., thermal fusion bonding of the cover

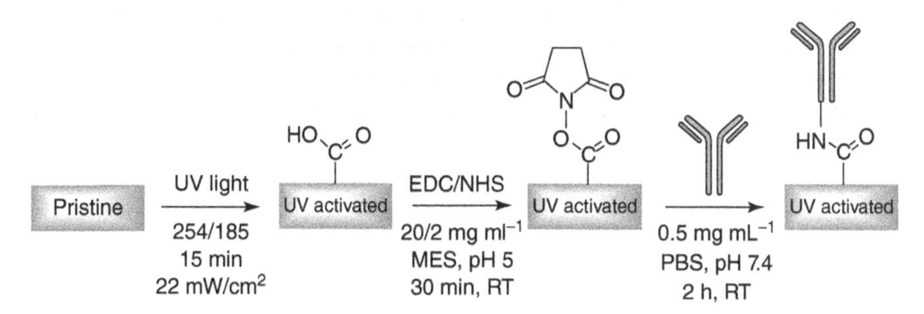

Figure 4.4 Pristine polymer is activated via UV light to generate carboxylic acid moieties to which Abs are covalently linked via EDC/NHS chemistry.

plate to the substrate) is an activation modality that is amenable to high-aspect-ratio microstructures and is arguably the most efficient method for mass production, requiring only quartz Hg lamps operated at ambient pressure.

The UV activation process is more accurately described as UV/O_3 treatment because the UV light continually generates and destroys O_3 by 185 nm O_2 absorption and 254 nm O_3 absorption, respectively, yielding a steady-state concentration of strongly oxidizing atomic O [45]. At sufficiently high energy, both UV exposure and oxidative stress can generate radicals within the polymer, which may break or scission polymer chains into smaller fragments, cross-link polymer chains, cause intramolecular rearrangements, and/or react with water or oxidative species to form carboxylic acid scaffolds and/or other O-containing species [73–83]. In brief, polymer surfaces are exposed to both intense UV light and highly reactive oxidizing species that heavily oxidize the polymer's surface, thereby generating carboxylic acid scaffolds for covalent biologic attachment [84].

The extensive literature available on PMMA activation uses planar substrates or thin films as models, but it was unclear if these studies translated to microstructured surfaces embedded beneath the bulk polymer surface, especially high-aspect-ratio microstructures [50, 73–81, 85]. Moreover, it is well known that fragmentation (scissioning of the polymer backbone into smaller, more thermally mobile polymer fragments) can be a by-product of UV activation, but it was not known how the carboxylated polymer surface would then respond to near T_g heating during the thermal fusion bonding process necessary to seal the activated microfluidic device. Also, there was not a literature precedent for generating carboxylic acid functionalities on COC surfaces by UV activation. COC is a relatively new polymer for lab-on-a-chip devices (Figure 4.5) and has some attractive properties, especially for CTC assays, including its exquisite optical transparency and low levels of autofluorescence [86].

After fabricating, UV activating, and thermally bonding the high-aspect-ratio CTC device shown in Figure 4.3d using either PMMA or COC substrates, carboxylic acid formation and biologic immobilization throughout the microchannel surfaces were evaluated [45]. Cy3-labeled oligonucleotides were immobilized to activate surfaces to serve as fluorescent reporters. Then, the microchannels were cut open for visualization of Cy3-oligonucleotide immobilization along the channel's cross section by fluorescence microscopy (Figure 4.6). Cy3-oligonucleotides were only apparent in the top third of the PMMA microchannels, indicating a strong inverse correlation between UV activation efficiency and microstructure aspect ratio that was validated using a colorimetric assay for carboxylic acid quantification. To contrast, COC microchannels were uniformly modified with Cy3-oligonucleotides. This effect was

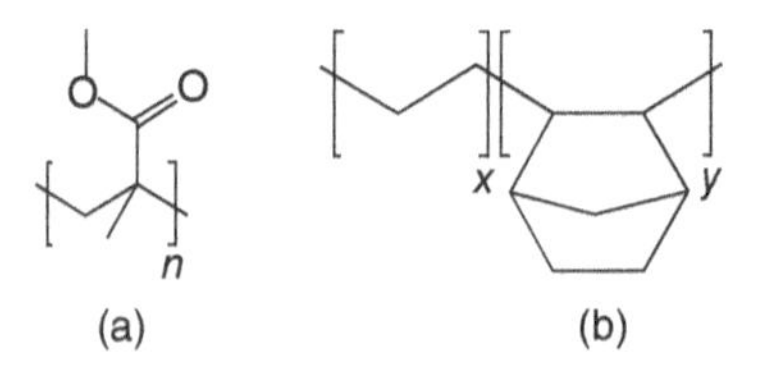

Figure 4.5 Chemical structures of (a) PMMA and (b) COC.

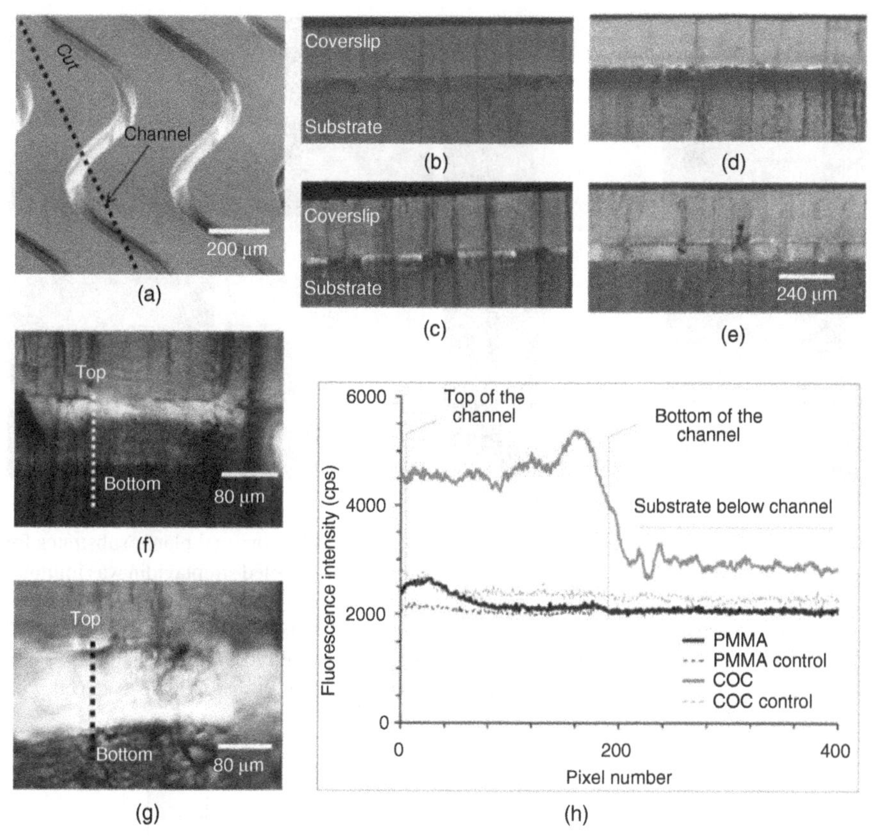

Figure 4.6 Sinusoidal channels fabricated in PMMA or COC were UV activated, thermal fusion bonded, and labeled with Cy3-oligonucleotides covalently attached via EDC chemistry to serve as fluorescent reporters of successful activation. (a) Channels were cut along their length to expose Cy3-oligonucleotides immobilized along sidewalls. 5× fluorescent images of controls and Cy3-oligonucleotides are shown for (b,c) PMMA and (d,e) COC, respectively. 20× fluorescence images for (f) PMMA and (g) COC are presented along with (h) line plots as indicated by the thick, dotted lines. Controls were Cy3-labeled oligonucleotides immobilized without the EDC coupling agent. Only control images are scaled to the same intensity as their counterparts. Jackson *et al.* [45]. Reproduced with permission of the Royal Society of Chemistry.

a consequence of the bulk polymer damping the flux of modifying UV radiation due to absorption. Pristine PMMA's 1.5% transmittance at 254 nm was reduced to 0.5% after UV activation, whereas native COC showed a transmittance of 53.8% that remained at 36.8% after irradiation. Thus, absorption of UV radiation can prevent surfaces that are entrenched in the bulk polymer from being activated, and this effect scales with aspect ratio [45].

The effect of thermal fusion bonding on carboxylic acid surface densities was determined using planar substrate models to enable a variety of surface analyses

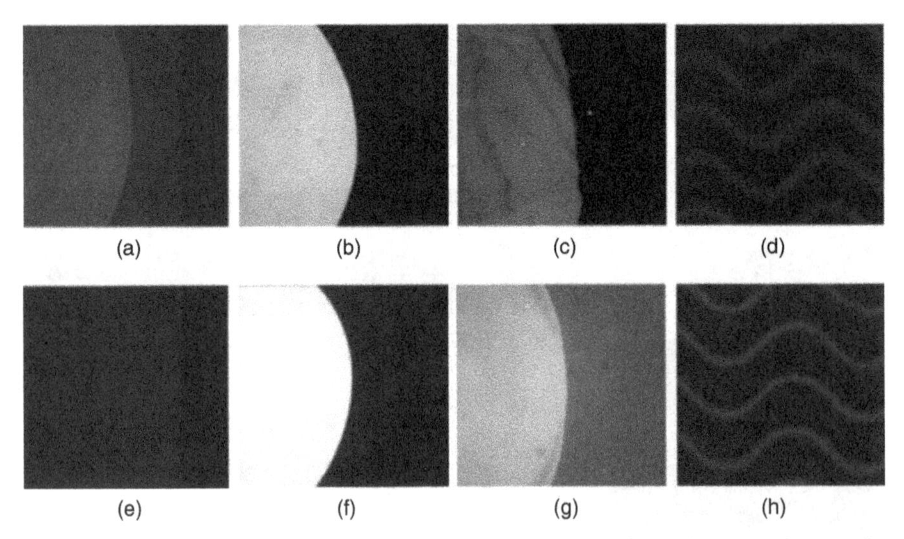

Figure 4.7 Cy3-oligonucleotides immobilized on UV and UV/thermal planar substrates for (a,b) PMMA and (c,d) COC, respectively. (e,f) Fluorescently labeled streptavidin was immobilized on UV and UV/thermal COC, respectively. Images of Cy3-oligonucleotides immobilized within UV-modified and thermal-fusion-bonded (g) PMMA and (h) COC microchannels. All fluorescence images of Cy3-labeled oligonucleotides are scaled to the same intensity. Fluorescent streptavidin images are scaled independently. Jackson *et al.* [45]. Reproduced with permission of the Royal Society of Chemistry.

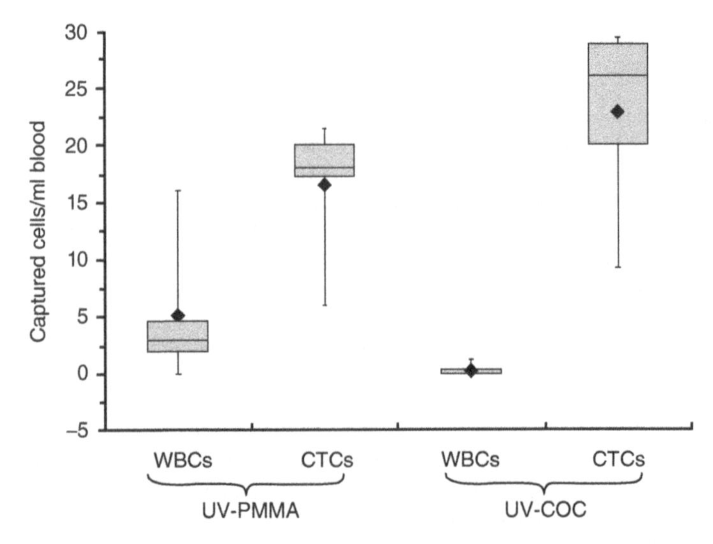

Figure 4.8 Box plots presenting CTCs and WBCs selected in UV-PMMA ($N=5$) and UV-COC ($N=4$) chips from blood samples secured from patient-derived xenograft mouse models for pancreatic ductal adenocarcinoma (PDAC). Data are normalized to 1 ml. Lower and upper edges of box show 25th and 75th percentiles, respectively. Solid line in box represents median, and solid diamond represents mean. Bars show maximum and minimum values (range). Jackson *et al.* [45]. Reproduced with permission of the Royal Society of Chemistry.

such as water contact angle measurements, atomic force microscopy (AFM), attenuated total reflectance-Fourier transform infrared (ATR-FTIR) spectroscopy, Raman spectroscopy, and X-ray photoelectron spectroscopy (XPS). We found that the UV activation of COC surfaces generated carboxylic acid scaffolds far more efficiently than on PMMA, largely due to competitive processes occurring within the PMMA polymer, especially polymer fragmentation. Immobilization of the same Cy3-oligonucleotides reporters used in Figure 4.6 generated fluorescence signals of 1978 ± 229 cps on UV-activated COC surfaces but only 282 ± 98 cps on UV-activated PMMA (Figure 4.7). When heated near the pristine material's T_g, rearrangement of the fragmented PMMA polymer significantly reduced carboxylic acid surface densities and oligonucleotide loads while COC remain unaffected by the heating process, at least to the extent that excess carboxylic acid moieties were available for oligonucleotide or protein immobilization [45].

Both UV absorption and heating artifacts hindered activation of PMMA microfluidic devices, and these phenomena directly affected CTC selection because high Ab loading is critical to efficient CTC recovery (Eq. 4.1) and to mitigate nonspecific normal blood cell interactions with the affinity surface (Figure 4.8). *Microfluidic devices fabricated in COC exceeded PMMA devices in both CTC recovery and purity* [45]. These results further highlight the importance of using substrates from the onset of laboratory-based research that are conducive to high-scale production modalities; exchanging substrate materials to enable mass-production commercialization is not a trivial matter and requires extensive knowledge of polymer properties and activation modalities.

4.2.3 CTC Selection in Sinusoidal Microchannels from a Fluid Dynamics Perspective

Microfluidic devices for positive affinity selection employ various microchannel geometries to improve CTC recovery by increasing the probability and duration of CTC–antigen Ab interactions. Because CTCs do not typically cross laminar flow streamlines and approach Ab-coated surfaces through diffusion processes, geometries are designed to induce some cross-stream forces on CTCs such as centrifugal forces or cause laminar flow streamlines to interact with surfaces, such as arrayed micropillars that obstruct flow [13, 66, 87] and herringbone grooves that induce flow convection [13]. A more subtle aspect is that fluid shear stress must be generated on all surfaces to disrupt nonspecific interactions throughout the entire device [45], meaning that wide channels, where substantial shear forces are not generated, and low-velocity regions or "dead zones" should be avoided.

We provide here a framework of the physical dynamics of CTC selection in sinusoidal channels. In Section 4.2.3.1, we detail how CTCs are delivered to Ab-coated channel walls, and in Section 4.2.3.2, we discuss the dynamics of Ab–CTC binding as the CTC rolls along the channel wall. Both of these sections are supported through the following experimental observations (Figure 4.9):

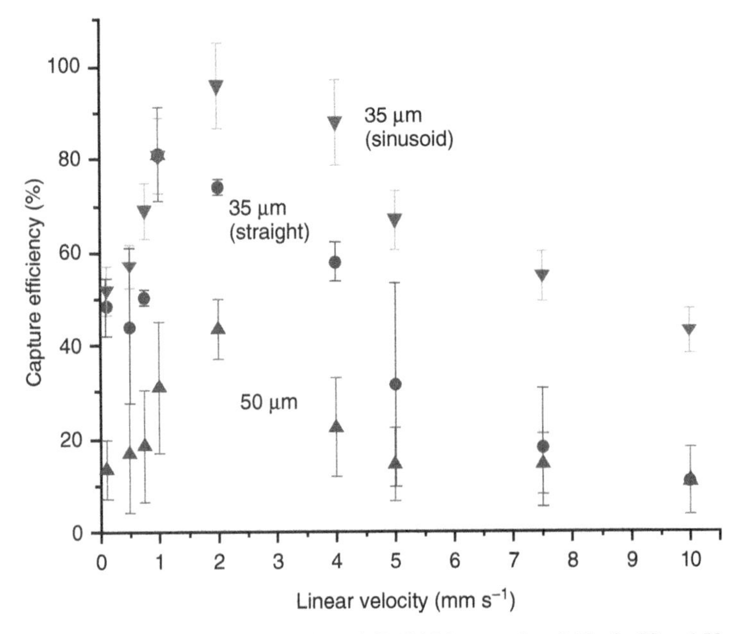

Figure 4.9 Recovery of MCF-7 cells via anti-EpCAM monoclonal Abs in 35 and 50 μm wide channels that are either straight or sinusoid. In all configurations, optimal recovery occurs at 2 mm/s linear velocity. Recovery increased by ∼30% due to decreasing channel width and increased by ∼20% due to channel curvature. Adams *et al.* [48]. Reproduced with permission of the American Chemical Society.

(i) Relative to straight channels, curved channels increased recovery by ∼20% [48].

(ii) Narrowing the channel width from 50 to 35 μm improved recovery by ∼30% [48]. In our most recent devices (Figure 4.1f–j), channel width has been reduced to ∼25 μm, which approaches the CTC diameter (12–20 μm) while remaining larger than leukocytes (7–15 μm) [21, 45].

(iii) The recovery of MCF-7 CTC surrogates was optimal at an average linear velocity of 2 mm/s [46, 48].

4.2.3.1 Centrifugal Forces in Curved Microchannels The delivery of CTCs to selection surfaces is critical to microfluidic devices that employ solid-phase affinity selection. In sinusoidal-based devices, as CTCs traverse about the channel's curvature, they experience a centrifugal force that propels them out of laminar streamlines and toward the Ab-coated channel walls. It is important to note that channel width is critical to the centrifugal force's effectiveness. For example, in narrow channels (25 μm wide), only a 4.5 μm shift in position is needed for a 16 μm CTC to interact with the Ab-coated wall. Because a CTC traveling through a single channel experiences 102 alternating curvatures (Figure 4.10a), even small centrifugal forces can accumulate into a significant shift in CTC position. Computational fluid dynamics

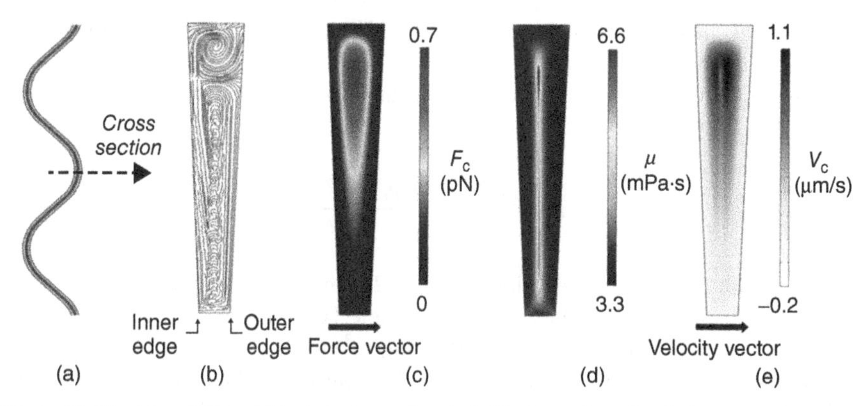

Figure 4.10 Three-dimensional computational fluid dynamics simulations conducted on blood flow through the sinusoid cell isolation channel utilizing the Carreau model for blood's non-Newtonian viscosity. Shown are: (a) Longitudinal velocity profile of blood flow; (b) cross-sectional velocity streamlines of weak Dean flow on the order of 0.1 μm/s; (c) cross-sectional centrifugal forces acting on a 16 μm CTC due to the channel's radius of curvature, where positive forces act from left to right as shown by the force vector; (d) cross section of non-Newtonian viscosity profile generating fluidic drag that opposes the centrifugal forces; and (e) cross-sectional centrifugal velocities obtained by balancing centrifugal and drag forces, where positive velocity is a left-to-right motion as shown by the velocity vector. Note that negative centrifugal velocities in (e) are due to the velocity streamlines in (b). (*See color plate section for the color representation of this figure.*)

(CFD) simulations were used to assess the magnitude of the centrifugal forces likely experienced by CTCs traveling in the curved microchannels and how these forces translate to shifts in the CTC's axial positions.

Centrifugal forces are experienced by both CTCs and other blood components. It is well known that incompressible fluids such as blood recirculate in response to centrifugal forces to conserve mass and energy. The magnitude of the secondary, recirculating flow profiles in the channel's cross section (known as Dean flow) is gauged by the Dean number;

$$De = Re\sqrt{\frac{D_\mathrm{h}}{2r_\mathrm{c}}} \tag{4.2}$$

where r_c is the radius of the channel's curvature, D_h is the channel's hydraulic diameter and in rectangular channels is given by the channel width W and height H by $D_\mathrm{h} = 2WH/(W + H)$, and the Reynolds number $Re = \rho D_\mathrm{h} V/\mu$; where ρ, μ, and V are the fluid's density, viscosity, and average velocity, respectively. When De exceeds 35.92, flow instability occurs [88], but below this, stable Dean flow has been utilized in inertial microfluidics to separate particles or cells based on size [89, 90]. However, in the sinusoidal channels of the devices shown in Figures 1 and 3, Re and De are limited to 0.02 and 0.01, respectively, due to the relatively slow linear velocity (2 mm/s) and the small channel dimensions. Thus, the secondary rotational flows in the sinusoidal channel's cross section (Figure 4.10b) are very small in magnitude with velocities on the order of 0.1 μm/s.

As blood slowly recirculates through the channel's cross section, CTCs also experience a centrifugal force (F_c) given by

$$F_c = \frac{((4/3)\pi a^3 \rho_{CTC})V^2}{r_c} \tag{4.3}$$

where V is the CTC's forward velocity, $r_c = 125\,\mu m$ is the channel's radius of curvature, and the term in parentheses rephrases the CTC's mass in terms of the CTC's radius ($a = 8\,\mu m$) and density ($\rho_{CTC} = 1056\,kg/m^3$) [91].

In Figure 4.10c, we plot F_c throughout the sinusoidal channel's cross section. A CTC experiences the highest centrifugal forces (up to 0.7 pN) in the center of the curving channel; elsewhere, the channel walls reduce V and F_c by viscous drag (Figure 4.1j). In a sense, a centrifugal force of 0.7 pN is somewhat arbitrary because the number does not convey how fast the CTC moves in response to this force.

The CTC's axial velocity due to F_c (V_c, which is positive toward the outer channel wall) is given by balancing F_c with the blood's reciprocal drag force (F_d);

$$F_d = 6\pi\mu a(V_c - V_f) \tag{4.4}$$

F_d increases with the blood's viscosity (μ), the CTC's radius a, and as the CTC moves faster relative to the Dean flow's axial velocity (V_f). Because F_d must equal and oppose F_c, the CTC's V_c can be determined by balancing Eqs 4.3 and 4.4;

$$V_c = \frac{2\rho_{CTC}}{9\mu r_c} a^2 V^2 + V_f \tag{4.5}$$

For an average 16 μm CTC traveling at 2 mm/s (Figure 4.10e), the magnitude of V_c ranges from 1.1 μm/s toward the outer wall to –0.2 μm/s (negative where F_c is too small to counter the minor Dean flow), which is significant considering the average 15 s required for a CTC to traverse the 30 mm long channel at 2 mm/s and the 4.5 μm shift needed to deliver the CTC to the microchannel wall. Note that in Eq. 4.5, V_c scales with the square of the CTC diameter and forward linear velocity (Figure 4.11). Thus, centrifugal forces operating on smaller cells such as 8 μm leukocytes propel the cells with a velocity four times slower than the CTCs with 16 μm diameter, and in general, centrifugal forces are very sensitive to linear velocity, becoming negligible at lower flow rates. This latter point likely causes reduced recovery at lower linear velocities (Figure 4.9), whereas at higher velocities, recovery drops due to the dynamics discussed in Section 4.2.3.2.

Note that a major limitation for this model's viscosity (Figure 4.10d) is that the non-Newtonian nature of blood arises from red blood cells. The cell membranes of red blood cells rotate around the cytoplasm under shear, spinning the cytosolic fluid with the surrounding plasma and transitioning the cell from a particle-like object to a less viscous flowing liquid. Red blood cells are biconcave discs that are 8 μm in

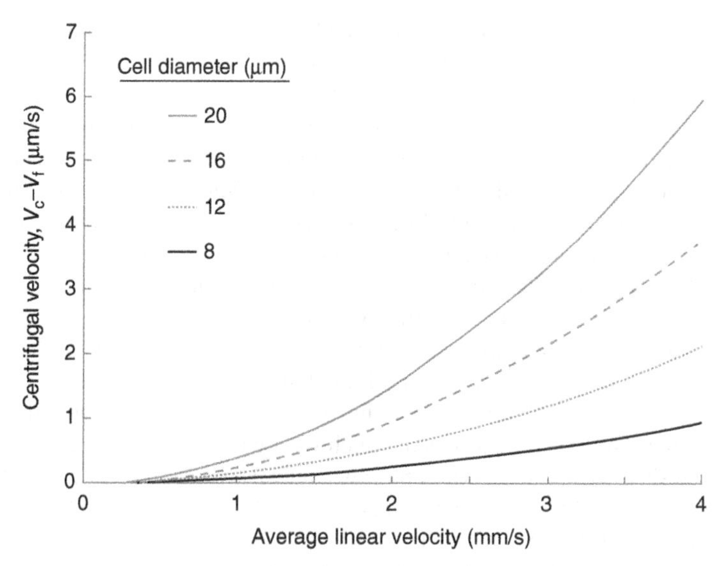

Figure 4.11 The centrifugal velocity V_c minus the axial velocity V_f (Eq. 4.5) experienced by a CTC traversing through sinusoidal microfluidic channels. V_c is critical to force CTCs out of laminar streamlines and toward microchannel surfaces and scales with the CTC's size and the fluid's forward linear velocity squared.

diameter and 2 μm thick; [A,B] assuming a red blood cell cross section of 16 μ^2, the channel cross section of 3750 μ^2, and 40% hematocrit, on the order of 100 red blood cells are likely to occupy a given cross section of the sinusoidal microchannel. Especially when a WBC or CTC simultaneously occupies the microchannel, the continuum-based Carreau model is likely to break down and become inaccurate. [C] The dynamics in the actual device are likely to be dominated by cell-cell interactions. However, conceptually speaking, the layer of high viscosity blood, which was hypothesized to bias CTC position away from the channel midline, could be envisioned to be red blood cells that occupy the remaining fluid space and resist axial movement of the cell. This argument is surely only conceptual, and truly accurate fluid dynamics simulations would require extensive modeling of deformable particles. Rather, particle imaging velocimetry would be a more viable strategy for capturing enough single-cell dynamics to statistically assembled CTC trajectories.

The influence of the fluid's viscosity, μ, shown in Figure 4.10e is especially interesting. Blood is a non-Newtonian fluid; in response to shearing, blood viscosity decreases [92]. In these narrow channels, drag along the walls rapidly slows fluid flow and induces a large shear effect. The same shear that produces large fluidic shear stresses for disrupting nonspecific interactions in turn roughly halves blood viscosity near the channel walls (Figure 4.10d). Effectively, if a CTC is at a central position (midway between the walls) and is "pushed" toward the wall, the CTC experiences less drag as it approaches the wall. However, at the next alternating curvature, where centrifugal forces act in the opposite direction, a CTC encounters an increase in fluidic drag as it is "pushed" back toward the channel's midline, resulting in a net centrifugal velocity toward the channel walls.

Finally, we must comment that particle lift forces are purposefully disregarded in the aforementioned analysis. Briefly, two primary lift forces are imposed on cells flowing in a channel: (i) cells lag behind the fluid's velocity and are forced toward lower velocity regions near the channel wall; and (ii) fluid flows faster over the cell's surface that faces the channel midline, producing a pressure lift toward the channel midline (as well as the Magnus effect due to the cell's shear-induced rotation) [89, 90, 93]. These two opposing forces result in an equilibrium position nearer the wall (*not* the channel's midline), which was first observed by Segre and Silberberg [94]. Despite being laminar flow, the lift forces acting on a cell in a single shear profile are surprisingly complex (see Chapter 5 for details) [89, 90, 93].

Lift forces were not incorporated in this analysis because traditional models are not applicable. The existing physics of which we are aware assumes that a cell experiences one shear profile [93], but in narrow channels where $a > W/2$, a CTC experiences the shear profiles *along both walls*. Thus, cell rotation and the forces causing CTCs to lift off of the channel walls are likely to be dampened but to an unknown extent. Nevertheless, the centrifugal effects described are independent and additive to these lift forces.

4.2.3.2 Transient Dynamics of CTC-Ab Binding With increasing linear velocity, centrifugal forces increase the delivery of CTCs to the selection surfaces, but as clearly shown in Figure 4.9, there was an optimal linear velocity of 2 mm/s for CTC recovery [46, 48]. Above this linear velocity, slow binding kinetics of the CTC-Ab reaction supersede centrifugal delivery of CTCs to the surface. The Chang–Hammer model [95] thoroughly describes the CTC–Ab binding process taking into account Ab–antigen binding kinetics, the transient motion of the antigen and its associated residence time in proximity to the surface-confined Ab, and the distance over which the cell rolls along the surface (Figure 4.12). In the high Péclet number regime, where convective mass transport of the antigen to the Ab dominates over diffusion, which is clearly applicable as the CTC rolls over the selection surface, the Chang–Hammer

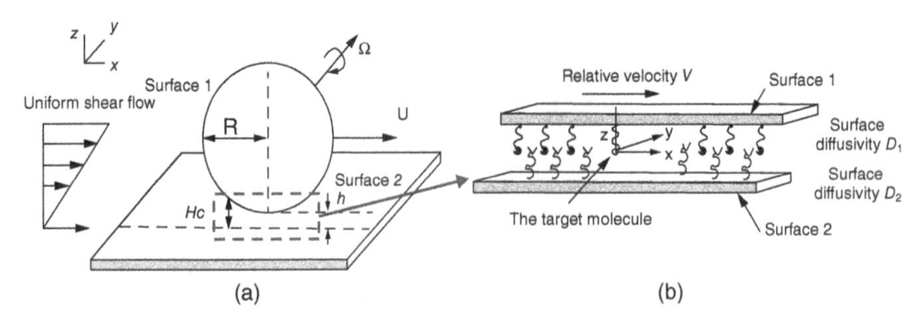

Figure 4.12 Schematic of the Chang–Hammer model describing (a) a cell rolling along a surface coated with recognition elements such as Abs. (b) The probability of cell adherence is governed by the relative motion of the cell's antigens with the surface elements. Sourced from Chang and Hammer [95].

model [95] can be simplified to a few key equations. Firstly, as the transient cell rolls, the forward rate constant k_o for the encounter of antigens with a surface-confined Ab is given as

$$k_o = 2a_i V_{eff} \tag{4.6}$$

In Eq. 4.6, a_i is the Ab–antigen interaction radius (2 nm), and V_{eff} is the velocity of the antigen relative to the surface, which is roughly half (0.47 times) the rolling cell's velocity due to the opposing rotational motion. Furthermore, as the antigen encounters the Ab, the probability that they complex (P) is a function of both the Ab's binding kinetics, k_{in}, and the encounter duration $\tau = 8a_i/3\pi V_{eff}$;

$$P = \frac{k_{in}}{k_{in} + 1/\tau} \tag{4.7}$$

As the cell's linear velocity increases, τ decreases, meaning that less time is available for the Ab and antigen to form a complex, and P decreases as well. Both the encounter rate, k_o, and the binding probability, P, are weighted against one another to yield an effective forward rate constant, k_f;

$$k_f = k_o P \tag{4.8}$$

Finally, the overall rate of cell adhesion k_{ad} combines k_f with the cell's antigen surface density C_∞;

$$k_{ad} = k_f C_\infty \tag{4.9}$$

To review, k_{ad} takes into account the cell's antigen expression and the velocity of the cell's antigens, both in terms of how often the antigens encounter Abs and how probable a binding event is given the balance of antigen–Ab interaction time and the Ab's binding kinetics. To relate k_{ad} to experimental systems, consider a cell rolling along an Ab-coated surface at a linear velocity (V) for only a limited distance (L). The percent of cells that will adhere is

$$\%_{bound} = 1 - 1/e^{k_{ad}L/V} \tag{4.10}$$

Two aspects of Eq. 4.10 that improve CTC recovery are immediately apparent: (i) decreased linear velocity and (ii) maximized interaction length between the cell and the surface.

Note that improving recovery by decreasing linear velocity is limited by throughput and purity because high shear forces are required to break nonspecific interactions (Eq. 4.1). For example, due to the narrow width of the sinusoidal channels (Figure 4.13), the average fluid shear stress observed in the devices shown in Figures 1 and 3 is 3.4 dyne/cm^2 for buffer flow and 13.3 dyne/cm^2 for blood flow, which is approximately an order of magnitude larger than comparable devices and leads to purities >80% (Table 4.1) [45].

Figure 4.13 provides an example of CTCs rolling along different surfaces. In one case, CTCs are isolated by 100-μm-diameter micropillars used in many microfluidic devices (at a linear velocity of ~0.65 mm/s [87, 96]), and in the other case, CTCs are isolated in sinusoidal channels (at ~2 mm/s [46, 48]). Despite the lower linear

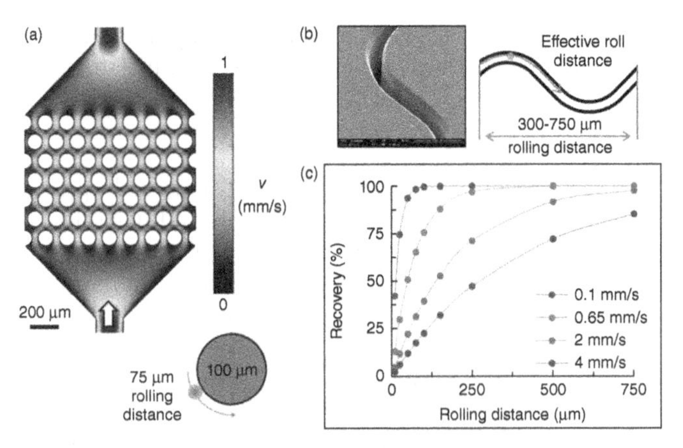

Figure 4.13 (a) Schematic of flow fields generated in a micropillar device (0.65 mm/s), and a close-up of a pillar with a roll distance of approximately 75 μm. From Battle *et al.* [96]. (b) An SEM of the high-aspect ratio sinusoid channel (2 mm/s) with a period of 750 μm, 125 μm radius of curvature, and an effective roll distance of >300 μm. (c) The recovery of CTCs (EpCAM expression = 49,700 molecules/cell) for different roll lengths and translational velocities according to the Chang-Hammer kinetic model. (*See color plate section for the color representation of this figure.*)

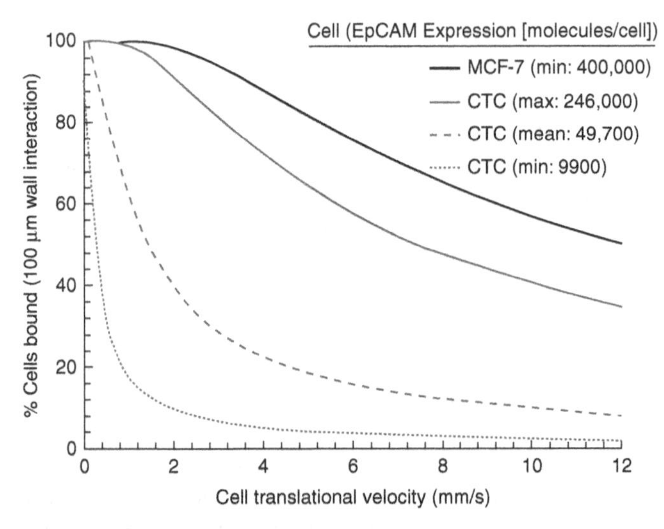

Figure 4.14 Recovery of the MCF-7 CTC surrogates (cell line), which expresses high levels of EpCAM and is commonly used to assess recovery, and clinical CTCs, which vary significantly in EpCAM expression. Cell recoveries were determined by the Chang–Hammer model assuming a 100 μm rolling distance and various translational velocities.

velocity used in the micropillar designs, which enhances cell binding, the limited rolling distance (~75 μm) yields a lower probability of cell recovery per surface interaction (~65%) than in the sinusoidal channels (~78% for a 300 μm rolling distance; ~98% for a 750 μm roll), but the sinusoidal device operates at nearly triple the linear velocity and thereby offers increased processing throughput and increased shear stress for disrupting non-specific interactions. For a brief comparison, the CellSearch technology offers an effective rolling distance of 0 and depends solely on random collisions of the CTCs with the anti-EpCAM magnetic nanoparticles. In this case, encounter rates can be increased by increasing nanoparticle concentrations. However, the probability of nonspecific interactions increases as well, and without any fluidic shear stress to break these nonspecific interactions (Eq. 4.1), poor purity results in the CellSearch system.

A less immediate observation of Eq. 4.10 is that CTC recovery depends on C_{∞}, the CTC's selection antigen expression. In Figure 4.13, EpCAM expression was fixed to 49,700 EpCAM molecules (mean) distributed over the surface area of a 16 μm CTC, but EpCAM expression levels are known to be clinically heterogenous and have been reported to range from 9,900 to 246,000 EpCAM molecules/CTC. As a reference, MCF-7 cells overexpress EpCAM relative to clinical samples (400,000 EpCAM molecules/cell) [97]. Figure 4.14 shows that as EpCAM expression decreases, the probability of CTC recovery drops significantly (over a 100 μm rolling distance), and long rolling distances become even more critical to clinical CTC recovery.

4.2.4 Parallel Arrays for High-Throughput Sample Processing

Considering the rarity of CTCs (1–100 CTCs/ml blood), relatively large blood volumes (ideally an entire 7.5 ml blood draw) must be processed through the microfluidic CTC selection device. Due to the Chang–Hammer kinetics discussed in Section 4.2.2, throughput is limited by linear velocity; thus, balancing recovery and analysis time is a common issue for positive CTC selection devices [60]. For example, the device shown in Figure 4.1f has an internal volume of only ~10 μl and a throughput of 25 μl/min [45]. Clearly, processing 7.5 ml blood using this device over the course of 5 h is not realistic for PoC applications.

A relatively straightforward method to increase throughput without affecting the fluid dynamics of CTC recovery (namely optimal channel width and linear velocity) is to either deepen the channels or increase the number of parallel channels (Figure 4.15). By increasing the channel depth to 250 μm (a practical fabrication limit) and increasing the number of channels, 7.5 ml of blood can be processed in less than 30 min with 300 channels or less than 17 min using 500 channels. Parallel arrays with more than 300 sinusoid channels are advantageous for rapidly isolating CTCs [60]. However, there is a practical question to consider; namely, can more than 300 channels be addressed while ensuring that flow through the entire parallel array remains uniform at 2 mm/s?

The first-generation sinusoidal devices (Figure 4.1a) utilized large triangular inlet/outlet regions to address a CTC isolation bed comprised of 51 parallel channels (Figure 4.16b) [12, 46, 48]. CFD simulations confirmed uniform flow between the parallel channels in this triangular configuration with the exception that the flow was slightly reduced in the outer channels due to viscous drag along the inlet/outlet walls by the no-slip condition (Figure 4.16a). However, slow linear velocities in the large, triangular regions (~0.2 mm/s) did not generate high enough fluidic shear stress to remove all blood cells nonspecifically adhering in the triangular regions, resulting in

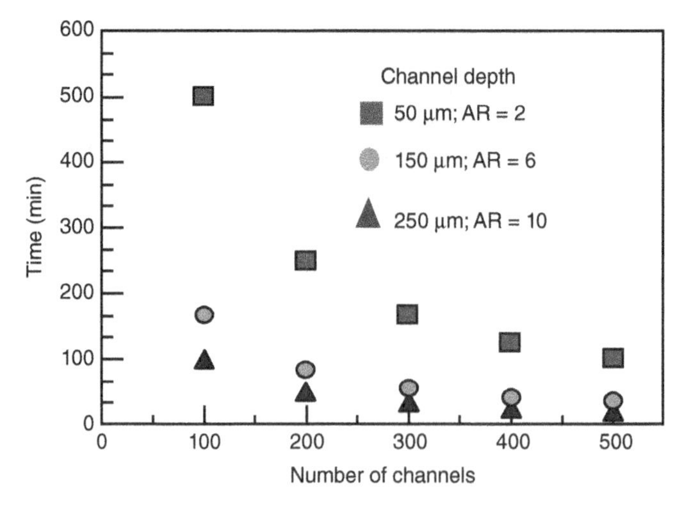

Figure 4.15 Time to process 7.5 ml of blood as a function of microchannel depth and channel number for a channel width of 25 μm and a linear flow velocity of 2 mm/s. AR = aspect ratio. Hupert *et al.* [60]. Reproduced with permission of Springer.

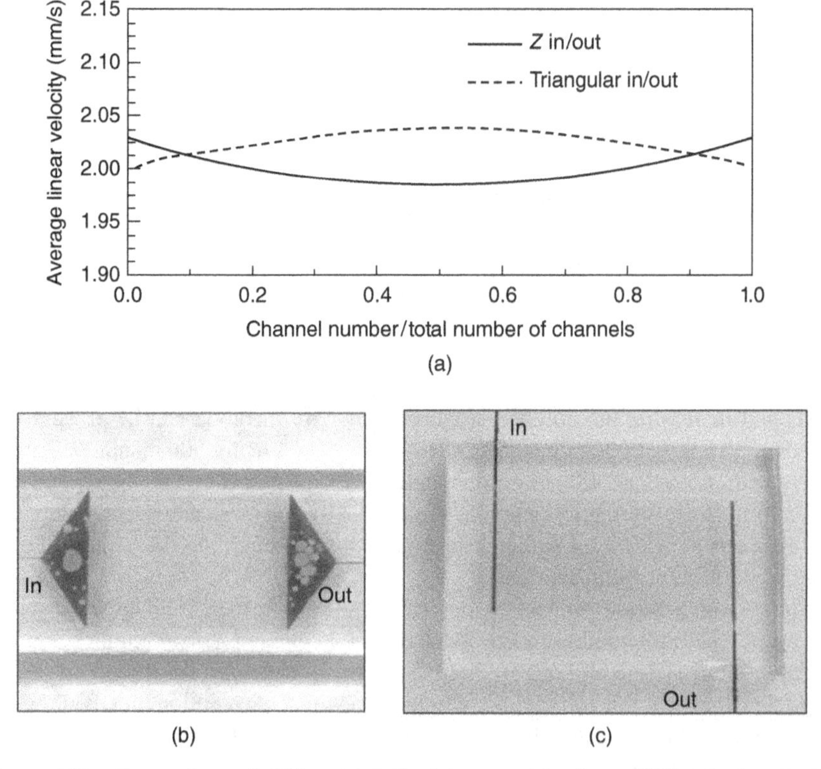

Figure 4.16 Comparison of different inlet/outlet geometries for a CTC selection device. (a) Results of computer simulations for the distribution of flow velocities within the CTC isolation bed with 50 microchannels arranged in the Z-configuration or 51 microchannels with triangular inlets and outlets. Pictures of devices filled with blood for (b) the triangular configuration and (c) the Z-configuration. Hupert *et al.* [60]. Reproduced with permission of Springer.

lowered purity (data not shown). Also, pressure differences in the triangular regions were not large enough to reliably displace air bubbles inadvertently introduced into the system during operation (Figure 4.16b). Disruption of the flow dynamics by compressible air bubbles is well known in microfluidics, where air bubbles can unpredictably perturb the flow profile as shown in Figure 4.16b. Both CTCs and contaminating blood cells can be trapped about the air–liquid interface at the bubble surface, which are extremely difficult to control during washing and staining procedures [60].

To alleviate issues associated with triangular inlets and outlets, a different fluidic architecture was designed (Figure 4.1f) that offered improved scalability and consisted of the so-called *Z*-configuration (Figure 4.16c) that is well known in the field of fuel cells [98]. Fluid enters the device through a single inlet channel that is poised perpendicular to the parallel channels and exits through a single outlet channel that is also perpendicular to the sinusoidal microchannels. The *Z*-configuration offers: (i) high pressure drops throughout the inlet and outlet channels, which efficiently remove air bubbles; (ii) easily scaled to any number of parallel channels by simply lengthening the inlet and outlet channels; and (iii) the smallest possible footprint relative to other designs [60]. For example, a bifurcation network was used to address parallel channels in a recently proposed device for neutrophil isolation [99]. The device's footprint nearly doubled to allocate space for the fluid distribution network, and of course, this problem scales with the number of parallel channels.

The *Z*-configuration is a highly parallelized fluid network, and uniform flow distribution is uniquely sensitive to the number of parallel channels and their length, or more accurately, the fluidic resistance of the parallel channels. To conceptualize this, the network is analogous to parallel, electrical resistors. It is well known that if one connects a set of resistors in parallel using wires with negligible resistance, the current (or flow rate) will be identical if all the resistors have the same resistance (and all CTC isolation channels have the same fluidic resistance). What happens if the wire connecting the parallel resistors has nonnegligible resistance? If the parallel resistors are not very resistive (e.g., short parallel channels for cell isolation) or if the inlet and outlet channels have significant resistance (narrow channels), then the current or flow distribution is no longer uniform. For this exact reason, numerically tractable CFD simulations with short, narrow parallel channels cannot be used to assess flow uniformity in *Z*-configuration networks, and an alternative numerical model must be used. The most tractable model available in the field of fuel cells (where *Z*-configuration networks are popular) actually uses the analogy just posed and treats the entire network as a set of fluidic resistors to construct a so-called network analysis model (a small model with only three parallel channels is shown in Figure 4.17 for visualization) [61]. This model is briefly described here because it is useful when CFD simulations are numerically intractable.

Every fluidic channel or segment of the inlet and outlet channels in a *Z*-configuration network can be modeled as a simple fluidic resistor with a resistance (R) to fluid flow predicted by its geometry and the fluid's viscosity (μ):

$$R = \frac{1}{2}\frac{(Re\,f)\mu \cdot P \cdot L}{D_{\mathrm{h}} \cdot A} \tag{4.11}$$

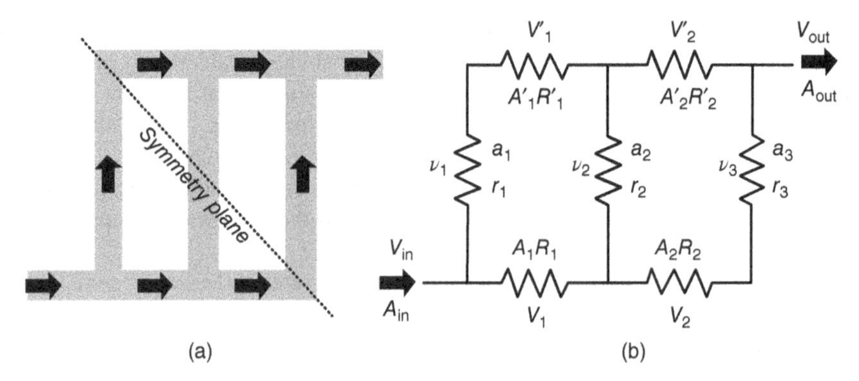

Figure 4.17 (a) Schematic diagram and (b) electrical representation of a Z-configuration network with three parallel channels. In (a), gray regions represent fluidic channels and black arrows represent the direction of flow. Jackson *et al.* [61].

In Eq. 4.11, the channel's geometry is defined by its cross-sectional area (A), perimeter (P), length (L), width (W), height (H), and hydraulic diameter (D_h). The product of the Reynolds number and friction factor ($Re\,f$) is approximated by Kays and Crawford [100] as $13.84 + 10.38 \cdot \exp(-3.4/\alpha)$, where α is the channel's aspect ratio (≥ 1) [61].

The velocity field throughout the three-channel Z-configuration network shown in Figure 4.17 is comprised of seven unknown velocities, the velocities through the three parallel channels and the four velocities in the inlet and outlet channel segments. Thus, we need a system of seven equations to solve for the velocity field. Pressure balance equations provide two of these equations (each term in these equations is a pressure drop):

$$v_1 r_1 + V_1' R_1' = V_1 R_1 + v_2 r_2 \tag{4.12a}$$

$$v_2 r_2 + V_2' R_2' = V_2 R_2 + v_3 r_3 \tag{4.12b}$$

The remaining five equations are mass balance equations (each term is a volumetric flow rate):

$$V_{in} A_{in} = V_1 A_1 + v_1 a_1 \tag{4.12c}$$

$$V_1 A_1 = V_2 A_2 + v_2 a_2 \tag{4.12d}$$

$$V_2 A_2 = v_3 a_3 \tag{4.12e}$$

$$V_{in} A_{in} = V_1 A_1 + V_1' A_1' \tag{4.12f}$$

$$V_{in} A_{in} = V_2 A_2 + V_2' A_2' \tag{4.12g}$$

In the given equations, v_i, r_i, and a_i are the average linear velocity, resistance, and cross-sectional area, respectively, of the parallel channels. Similarly, V_i, R_i, and A_i

describe the inlet channel segments, and V_i', R_i', and A_i' correspond to the outlet channel segments [61]. These definitions are illustrated in Figure 4.17b as well.

After minor rearrangements to Eqs 4.12a–g and rewriting the system of equations in matrix form that is easily expanded to any number of parallel channels, only a few linear algebraic operations need to be performed to solve for the velocities through the parallel channels. The flow distribution in a Z-configuration CTC selection device with 500 channels can be solved in a matter of seconds, whereas with modern computing, an accurate CFD simulation of the 50 channel device as shown in Figure 4.16c is difficult [60]. Moreover, the network analysis model agrees extremely well with CFD simulations [61] as well as experimental results [21, 101] (Figure 4.18a and c).

Numerical analysis of high-throughput cell isolation devices similar to those shown in Figure 4.16c shows that a parabolic flow distribution can be generated in certain Z-configuration networks, where fluid velocities increase in the outer channels while the fluid velocities in the central channels decrease (Figure 4.18b and c). This is reflected in the profiles observed when filling the device as shown in Figure 4.18a. Qualitatively, if each channel of the array was filling under a constant linear velocity, the filling profile of the array should show a linear gradient [21].

Uniformity of the fluid flow in Z-configuration networks is highly dependent on the ratio of fluidic resistance of the inlet/outlet channels to the fluidic resistance of the parallel channels. In general, wider (lower fluidic resistance) inlet channels and longer and/or narrower (higher fluidic resistance) parallel channels show higher flow uniformity throughout the array. Inhomogeneity of the flow rate is exacerbated by the number of parallel channels in the array (Figure 4.18b) [21, 61]. As can be seen from this data, the 320-channel device with 8.4 mm selection channel lengths generates maximum and minimum linear velocities that range from roughly 1 to 4 mm/s, which have been shown to provide above 80% recovery of MCF-7 cells that is only moderately efficient compared to the maximal 97% recovery observed at 2 mm/s [48]. Lengthening the 320 parallel channels to 20 or 30 mm (Figure 4.18b) adds resistance to the parallel channels and improves flow uniformity, but clearly the flow distributions in these devices still vary around the optimal 2 mm/s linear velocity [21].

To enable Z-configuration devices with very high throughputs that also retain flow uniformity, the inlet and outlet channels can be tapered (Figure 4.19) to systematically correct the parabolic profiles. If all parallel channel velocity terms in Eqs 4.12a–g are replaced with a constant linear velocity (uniform flow distribution), the equations can be reduced into a single equation that describes the geometry of Z-configuration networks with any number of parallel channels (N) that have uniform flow distribution [61]:

$$\frac{R_i}{A_i} = \frac{i}{(N+1-i)} \cdot \frac{R_{N+1-i}}{A_{N+1-i}} \tag{4.13}$$

Equation 4.13 provides a method of designing the inlet and outlet channel geometries. For the 250-channel Z-configuration shown in Figure 4.19, the geometry of the segment of the inlet channel joining the 1st and 2nd parallel channels ($i = 1$) is related to

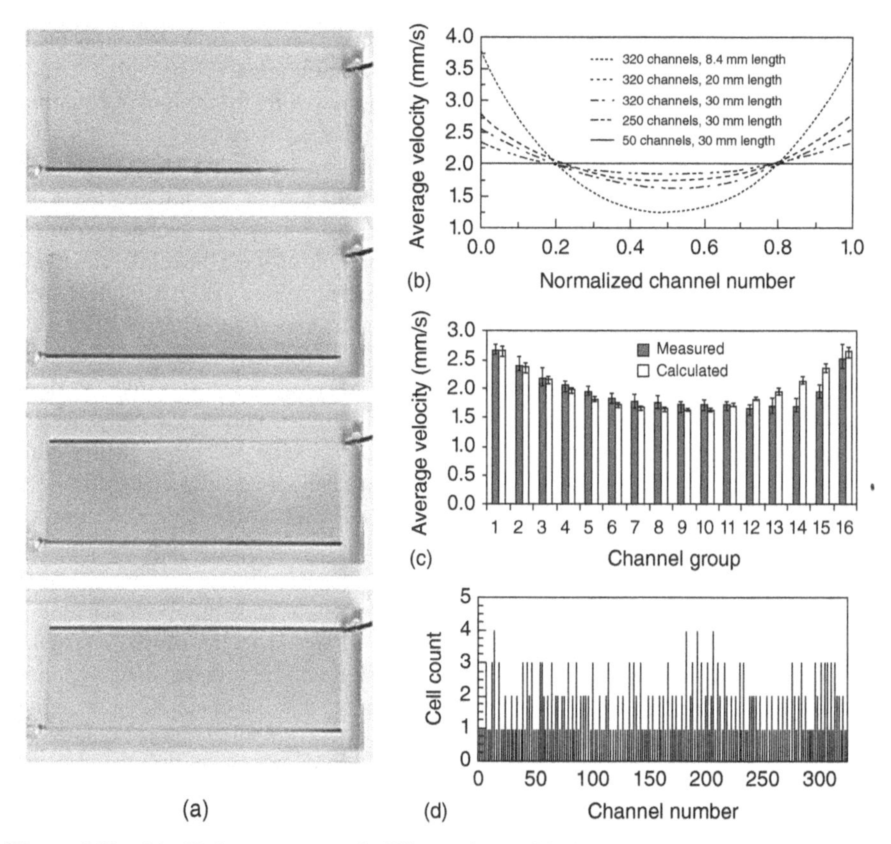

Figure 4.18 (a) Various stages of filling of a 320-channel *Z*-configuration device (20-mm-long parallel channels) with a dye solution. (b) Numerical simulation results showing the distribution of flow velocities for different configurations of the CTC selection beds arranged in a *Z*-configuration. (c) Average linear velocities of fluid in 16 groups of 20 adjacent sinusoidal, high-aspect-ratio microchannels based on the results shown in (a) (filled bars) and theoretical values obtained via the network analysis model (empty bars). (d) Distribution of cells selected in 20- mm-long microchannels. Kamande *et al.* [21]. Reproduced with permission of the American Chemical Society.

the geometry of the segment joining the 249th and 250th parallel channels ($i = 249$) by Eq. 4.13, which requires that the first segment is widened. Continuing this trend, tapered inlet and outlet channels are generated such as those shown in Figure 4.19 and provide uniform flow distributions, with very good agreement to CFD simulations as demonstrated in the field of fuel cells [61]. These optimized designs offer the ability to scale up *Z*-configuration systems well beyond 320 channels, enabling extremely high-throughput microfluidic devices while retaining optimal linear velocity throughout all parallel channels.

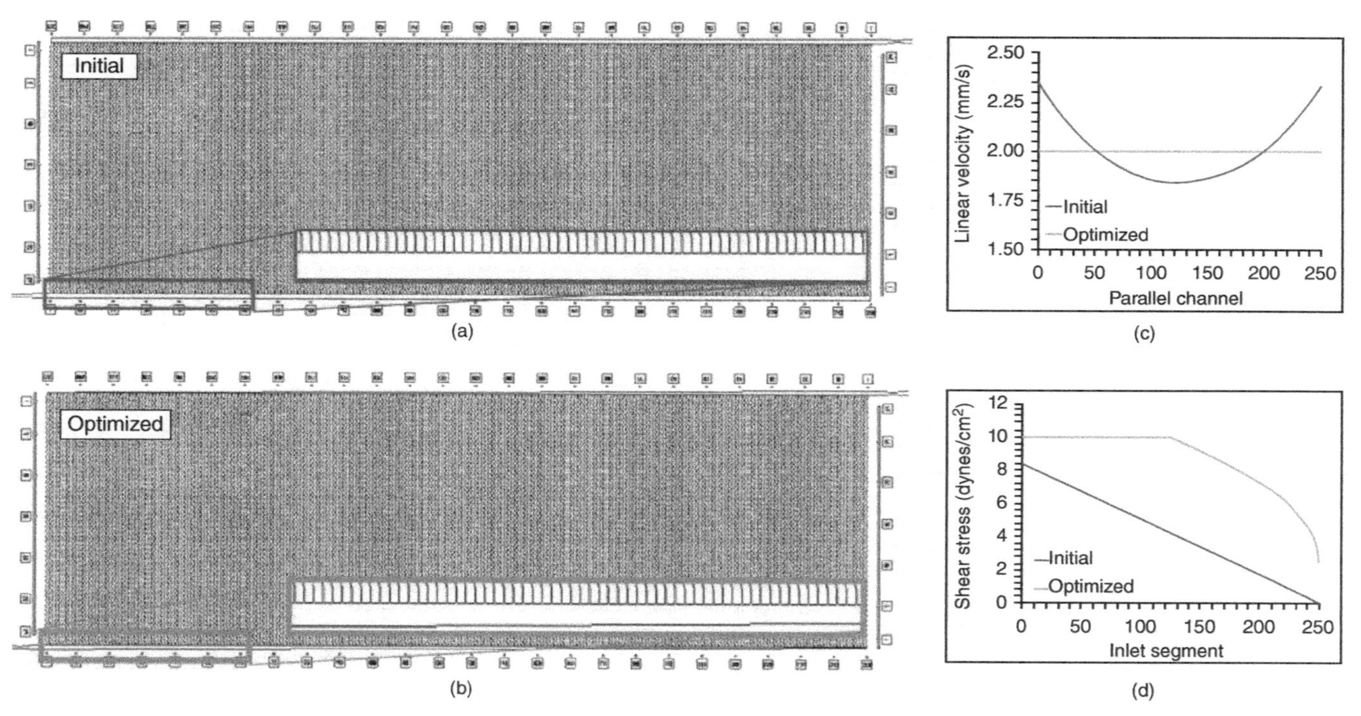

Figure 4.19 (a, b) AutoCAD renderings of Z-configuration cell isolation units with 250 parallel channels that utilize straight and tapered inlet and outlet channels. The Z-configuration with tapered inlets and outlets offers (c) constant linear velocity throughout the parallel array and (d) controlled shear stress throughout the tapered inlet and outlet channels to ensure mechanical stability of the CTCs.

4.3 CLINICAL APPLICATIONS OF SINUSOIDAL CTC MICROCHIP

A challenge for CTC assays is that some epithelial cancers do not show high clinical yields of CTCs even though the devices have been shown to demonstrate high recovery when using an immortalized cell line seeded into "normal" blood. For example, pancreatic ductal adenocarcinoma (PDAC), the fourth leading cause of cancer-related deaths worldwide [102–105], shows poor clinical yields when using EpCAM and positive selection formats. The CellSearch system has demonstrated clinical CTC yields for metastatic PDAC patients in the range of 0–1 per 7.5 ml of blood [106, 107]. However, a micropillar-based CTC fluidic device has generated good yields of CTCs for PDAC patients using anti-EpCAM antibodies as the positive selection marker [66].

The sinusoidal CTC chip was used to analyze EpCAM-positive CTCs from both locally resectable (nonmetastatic) and metastatic PDAC patient blood samples [21]. The data for this study are shown in Figure 4.20. Using the sinusoidal chip, a range of 34–95 (median = 51) CTCs/ml were selected for metastatic patients. The sinusoidal chip was also used to analyze blood from patients who had been determined to have locally resectable PDAC by CT scans; a range of 9–19 (median = 11) CTCs/ml for this group of patients was found. Pairwise Wilcoxon rank-sum tests suggested that there was a significant difference between metastatic patients and healthy donors ($p = 0.005$). Comparison between metastatic and locally resectable patients also indicated a statistically significant difference ($p = 0.023$) [21].

The purity for the CTC assay for the PDAC patients was calculated from the ratio of the number of CTCs selected to the total number of cells selected and was found to be 86%. The high purity levels resulted mainly from the high shear forces applied to the selected cells during sample processing and the postselection wash [45]. While the CTCs selected on the channel walls remained unaffected by the high shear forces due to the low K_D value between EpCAM and anti-EpCAM antibodies and the multipoint contact [48], the nonspecifically adsorbed leukocytes were more easily removed from the surface due to their lower adhesion force to the surface [45].

The sinusoidal CTC chip was also tested for selecting CTCs from other metastatic cancer patients [60]. Two prostate cancer CTCs are shown in Figure 4.21a that display the characteristics of CTCs including positive staining for DAPI (blue) and cytokeratins (red) with the lack of staining for leukocyte-specific antigen CD45 (green). Figure 4.21b shows a summary of the number of CTCs and contaminating WBCs isolated from blood samples from nine patients with metastatic prostate cancer. The number of CTCs detected ranged from 19 to 76 (median = 35) per milliliter of whole blood, whereas the number of contaminating WBCs was much lower, ranging from 2 to 8 (median = 7) per milliliter of blood. The purity was 89% [60].

Given the high purity of the CTC isolates using the sinusoidal CTC chip, molecular profiling of isolated CTCs was carried out on the metastatic prostate cancer patients [60]. For this example, mRNA was isolated from lysed CTCs followed by reverse transcription. The resulting cDNA was then amplified using polymerase chain reactions (PCRs) with six primer sets targeting GAPDH (glyceraldehyde-3-phosphate dehydrogenase; housekeeping gene control), PSMA

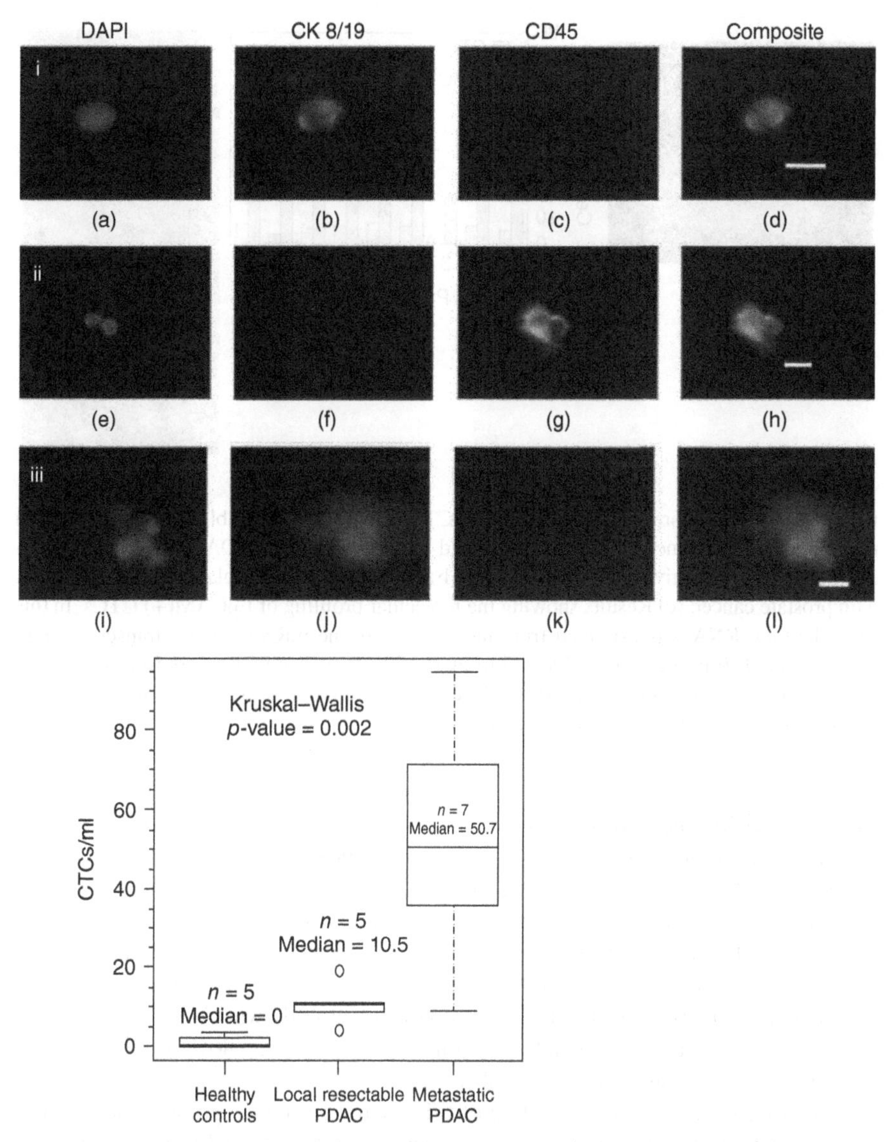

Figure 4.20 Selection and enumeration of CTCs via immunophenotyping. (a) Fluorescence images of various selected cells from a metastatic PDAC patient: (i) a CTC; (ii) two white blood cells; and (iii) a cluster of CTCs. (b, f, j) CTC marker for cytokeratin 8/19 (red) with b, j positive for this marker and f negative for this marker; and (c, g, k) leukocyte antigen marker CD45 (green) with c, k negative for this marker and g positive for this marker. Micrographs (i–l) are of an aggregate of six CTCs captured in the HT-CTC module. This aggregate showed positive for cytokeratins 8/19 (j) and negative for leukocyte marker CD45 (k). Bars are 10 μm. (Bottom) Also shown is a box plot from CTCs isolated from five healthy donors, five locally resectable PDAC patients, and seven metastatic PDAC patients. Kamande *et al.* [21]. Reproduced with permission of the American Chemical Society. (*See color plate section for the color representation of this figure.*)

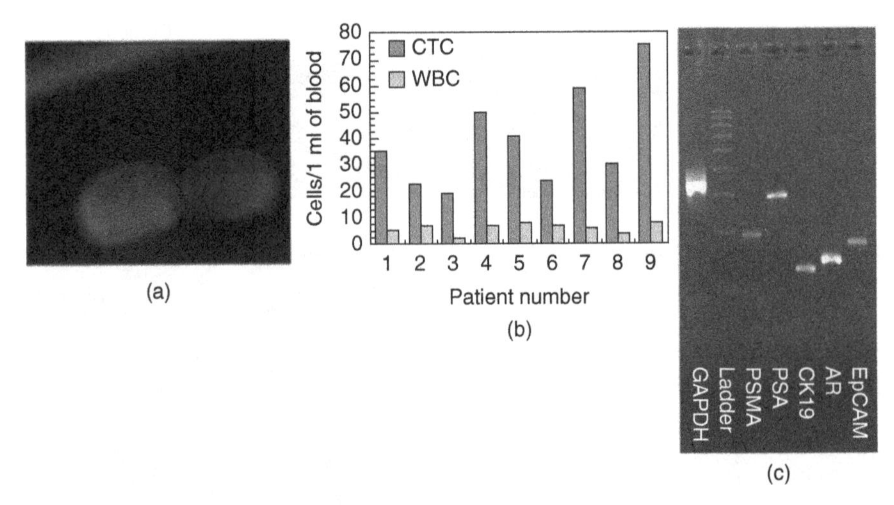

Figure 4.21 (a) Fluorescence image of two CTCs isolated from the blood of a patient with metastatic prostate cancer. CTCs demonstrated positive staining for DAPI (blue) and cytokeratins (red) and negative staining for CD45. (b) Summary of CTC isolation data for patients with prostate cancer. (c) Results showing the molecular profiling of EpCAM(+) CTCs. In this case, the total RNA was extracted from the CTCs with the mRNA reverse transcribed into cDNA using T20 primers and subjected to PCR with gene specific primers followed by gel electrophoresis. Hupert *et al.* [60]. Reproduced with permission of Springer. (*See color plate section for the color representation of this figure.*)

(prostate-specific membrane antigen), PSA (prostate-specific antigen), CK19, AR (androgen receptor), and EpCAM. As expected, PSA and EpCAM showed strong bands while there were only weak bands for the CK19 and PSMA genes.

In addition to locally resectable and metastatic PDAC cancer, the sinusoidal chip has been used to analyze blood samples from, colorectal cancer, ovarian cancer, and breast cancer. Based on the number of CTCs enumerated, the clinical sensitivity (number of patients with the disease properly identified) was found to be 100%, except for breast cancer in which the clinical sensitivity was 80%. In addition, the clinical specificity was found to be 100%.

In many cases, molecular profiling of CTCs can be used for guiding therapy. For example, *KRAS* mutations are frequently found in DNA extracted from cancer tissue and have been associated with a lack of response to epidermal growth factor receptor (EGFR)-targeted therapies [108]. Currently, it is recommended for colon and lung cancer patients that *KRAS* mutation testing is performed before initiating EGFR therapy to identify patients whom would or would not benefit from EGFR monoclonal antibody treatment [109]. It is important to note that *KRAS* oncogenes can be heterogeneous, and thus, different tumor margins must be examined to correctly identify mutated *KRAS* genes [110, 111]. A high concordance (92–96%) between primary tumors and metastatic sites were observed for *KRAS* mutations [112–114]. Liquid biopsies with CTCs were performed in metastatic colorectal cancer patients, and mutated *KRAS* oncogenes were found in ~40% of the CTCs, a similar frequency as that found in the primary tumor (~43%) [112–114].

A PCR/LDR (ligase detection reaction) method developed by Khanna *et al.* [115, 116] was used to detect point mutations in *KRAS* codons 12, 13, and 61 (see Figure 4.22a). LDR relies on the high specificity of the DNA ligase to discriminate base alterations in DNA [117]. In the first step of the assay, PCR amplifies the region of genomic DNA containing the mutations of interest. There are two primers that are designed to be complementary to the polymorphism (i.e., discriminating primer) and to a region directly adjacent (i.e., 5' phosphorylated common primer; see Figure 4.22a). Both primers hybridize to the target DNA, and only if there is a perfect match between them will the oligonucleotides be ligated by the ligase enzyme. A single-base mismatch at the junction inhibits ligation, and thus, single-base mutations can be distinguished [116, 117]. The presence of a mutation in an excess of

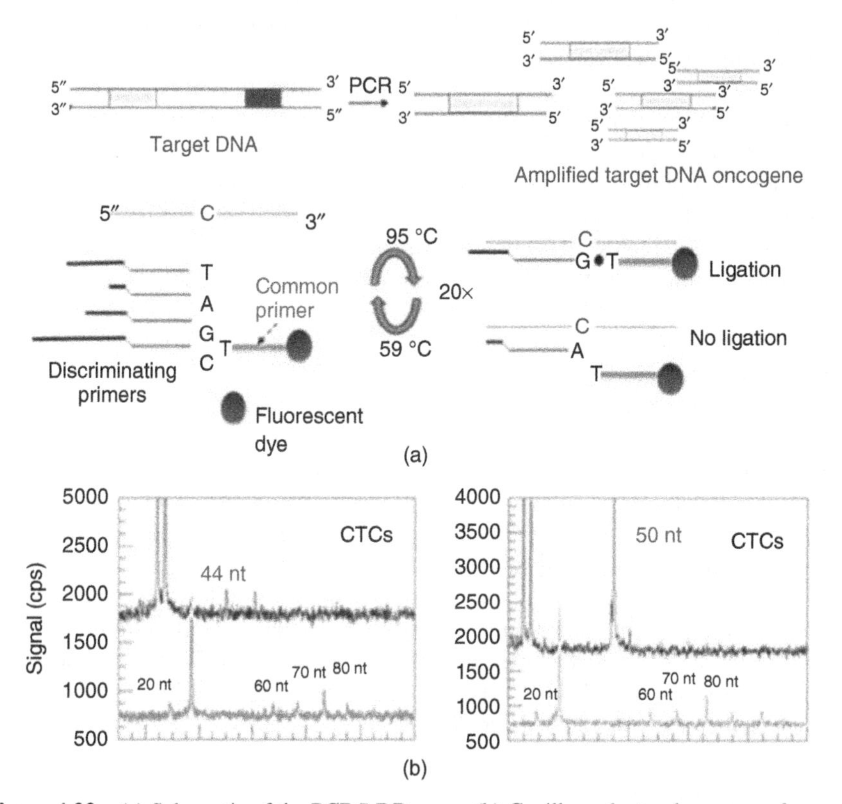

Figure 4.22 (a) Schematic of the PCR/LDR assay. (b) Capillary electropherograms for separated LDR products generated from CTCs isolated from a metastatic colorectal cancer blood sample. LDRs were carried out in 20 µl using a commercial Taq Ligase. LDR mixture contained: discriminating and common primers (4 nM each), DNA template 0.6–1 ng (3–5 fmol), 40 units of Taq DNA ligase. Thermocycling conditions were 94 °C for 1 min and 59 °C for 4 min cycled 20 times. Common primers were Cy5 labeled for detection.

TABLE 4.2 List of Sequences for Discriminating and Common Primers Used in the LDR Reaction

Mutation Site	Discriminating Primer 5′-3 (size)	Common Primer 5′-3′ (size)	Ligated Product Size
GGTGGC 35WT	TTTTTTTAAACTTGTGGTAGTTGGAGCTGG (30 nt)	Ph-TGGCTAGGCAAGAGTGCCT-Cy5 (20 nt)	50 (nt)
GATGGC G35A	TAAACTTGTGGTAGTTGGAGCTGA (24 nt)	Ph-TGGCGTAGGCAAGAGTGCCT-Cy5 (20 nt)	44 (nt)
GTTGGC G35T	TTTTTTTTTTTTAAACTTGTGGTAGTTGGAGCTGT (35 nt)	Ph-TGGCGTAGGCAAGAGTGCCT-Cy5 (20 nt)	55 (nt)
GGTGGC 34WT	TTTTTTTTTTTTTTTAAACTTGTGGTAGTTGGAGCTG (37 nt)	Ph-GTGGCGTAGGCAAGAGTGCCTTGACGATAC-Cy5 (30 nt)	67 (nt)
CGTGGC G34C	TTTTTTTTTAAACTTGTGGTAGTTGGAGCTC(31 nt)	Ph-GTGGCGTAGGCAAGAGTGCCTTGACGATAC-Cy5 (30 nt)	61 (nt)

wild-type alleles can be detected with improved specificity and sensitivity compared to Sanger sequencing [117]. In our assay, discriminating and common primers were designed to produce ligation products with different sizes (Table 4.2); they could be separated using capillary gel electrophoresis and the mutation identified based on the size of the product detected (see Figure 4.22b).

Electropherograms of LDR products generated from HT29 and LS180 gDNA showed peaks corresponding to 50 nt (34wt) and 67 nt (35wt) (electropherograms not shown). These two products indicated the presence of wild-type (wt) exon 1 codon 12 from the gDNA of these cell lines. These were the only products detected for HT29, while in the LS180 sample, a product of 44 nt was present as well. This was indicative of the G35A mutation being present, which is in agreement with the literature for this particular cell line [118]. LDRs with negative controls (no DNA in the reaction) showed no products within the electropherogram.

Reports indicate that mutated *KRAS* in colorectal cancer was found in codons 12 and 13 most often (95%) with ~80% occurring in codon 12 and 15% in codon 13. Other mutations in codon 61 occur less frequently. In codon 12, the mutations G35A and G35T are the most frequent [110]. From metastatic colorectal cancer patients, gDNA from CTCs was analyzed. One result is shown in Figure 4.22b where a patient showed a G35A mutation (Figure 4.22b, left electropherogram panel, product 44 nt, right panel for wt product). Overall, 75% metastatic colorectal cancer patients tested positive for *KRAS* mutations in codon 12.

The prevalence of *KRAS* mutations in PDAC is reported to be between 69 and 90% and represents the most frequent and earliest genetic alteration in this disease [51, 119]. In PDAC tissue, the most abundant mutations are G35A, G34C, and G35T [120]. For metastatic PDAC patients, 4 CTC samples out of 5 (80%) showed LDR products indicative of a mutated *KRAS*.

4.4 CONCLUSION

While there are a number of attractive and innovative technologies for the selection of CTCs using either biological or physical CTC properties (Table 4.1), implementation of any of these technologies into clinical research and eventually clinical practice requires a number of important operational characteristics such as high recovery, high throughput, favorable purity levels, and good clinical sensitivity and specificity. Many of the microfluidic technologies listed in Table 4.1 demonstrate most of these operational characteristics, but very few demonstrate favorable operational characteristics on all fronts, many showing less than favorable purity that can make molecular profiling problematic due to the high infiltration rate of contaminating leukocytes. The ability to generate CTC selection technologies that achieve high clinical yields of CTCs while also producing high purity of the selected CTC fraction will be necessary, especially as molecular profiling becomes more important in the clinical implementation of CTCs for guiding therapy through patient stratification based on genetic profiles.

The sinusoidal CTC chip can provide some important operational characteristics that will make it an attractive platform for clinical implementation, including high recovery (~98%), high throughput (15 ml/h), high clinical sensitivity and specificity (~100% and 99%, respectively, for metastatic disease), and, more importantly, high purity (>80%). The high purity will provide the ability to perform extensive molecular profiling on selected CTC fractions that can be instrumental in helping the clinician make therapeutic decisions on patients with metastatic disease. In addition, the high clinical yields as well as the high purity will improve the prospects of using clinical CTCs for next-generation DNA/RNA sequencing to search for recognized and unknown genetic alterations.

Finally, the CTC sinusoidal chip can be fabricated using a thermoplastic, making it appropriate for manufacturing via injection molding that can produce devices in a high production mode and at low cost, but with tight compliance providing high assay reproducibility.

ACKNOWLEDGMENTS

The authors thank Joyce Kamande, Mateusz Hupert, Kent Gordon, and Mira Pellerin for their help in assembling Table 4.1.

REFERENCES

1. Pachmann, U.A., Hekimian, K., Carl, S. *et al.* (2011) Comparing sequential steps for detection of circulating tumor cells: more specific or just less sensitive? *Wedmed Central Cancer*, **2**, WMC001490.
2. Clarke, M.F., Dick, J.E., Dirks, P.B. *et al.* (2006) Cancer stem cells – perspectives on current status and future directions: AACR workshop on cancer stem cells. *Cancer Research*, **66**, 9339–9344.
3. Gorges, T.M., Tinhofer, I., Drosch, M. *et al.* (2012) Circulating tumour cells escape from EpCAM-based detection due to epithelial-to-mesenchymal transition. *BMC Cancer*, **12**, 178–190.
4. Kasimir-Bauer, S., Hoffmann, O., Wallwiener, D. *et al.* (2012) Expression of stem cell and epithelial-mesenchymal transition markers in primary breast cancer patients with circulating tumor cells. *Breast Cancer Research*, **14**, R15–R23.

5. Raimondi, C., Gradilone, A., Naso, G. *et al.* (2011) Epithelial-mesenchymal transition and stemness features in circulating tumor cells from breast cancer patients. *Breast Cancer Research and Treatment*, **130**, 449–455.
6. Aktas, B., Tewes, M., Fehm, T. *et al.* (2009) Stem cell and epithelial-mesenchymal transition markers are frequently overexpressed in circulating tumor cells of metastatic breast cancer patients. *Breast Cancer Research*, **11**, R46–R54.
7. Mostert, B., Kraan, J., Vries, J.B. *et al.* (2011) Detection of circulating tumor cells in breast cancer may improve through enrichment with anti-cd146. *Breast Cancer Research and Treatment*, **127**, 33–41.
8. Smirnov, D.A., Zweitzig, D.R., Foulk, B.W. *et al.* (2005) Global gene expression profiling of circulating tumor cells. *Cancer Research*, **65**, 4993–4997.
9. Williams, A., Balic, M., Datar, R., and Cote, R. (2012) Size-based enrichment technologies for CTC detection and characterization, in *Minimal Residual Disease and Circulating Tumor cells in Breast Cancer* (eds M. Ignatiadis, C. Sotiriou, and K. Pantel), Springer, Berlin, Heidelberg, pp. 87–95.
10. Lim, L.S., Hu, M., Huang, M.C. *et al.* (2012) Microsieve lab-chip device for rapid enumeration and fluorescence in situ hybridization of circulating tumor cells. *Lab on a Chip*, **12**, 4388–4396.
11. Wang, S., Liu, K., Liu, J. *et al.* (2011) Highly efficient capture of circulating tumor cells by using nanostructured silicon substrates with integrated chaotic micromixers. *Angewandte Chemie International Edition*, **50**, 3084–3088.
12. Dharmasiri, U., Njoroge, S.K., Witek, M.A. *et al.* (2011) High-throughput selection, enumeration, electrokinetic manipulation, and molecular profiling of low-abundance circulating tumor cells using a microfluidic system. *Analytical Chemistry*, **83**, 2301–2309.
13. Stott, S.L., Hsu, C.-H., Tsukrov, D.I. *et al.* (2010) Isolation of circulating tumor cells using a microvortex-generating herringbone-chip. *Proceedings of National Academy of Sciences of United States of America*, **107**, 18392–18397.
14. Lin, H.K., Zheng, S., Williams, A.J. *et al.* (2010) Portable filter-based microdevice for detection and characterization of circulating tumor cells. *Clinical Cancer Research*, **16**, 5011–5018.
15. Hosokawa, M., Hayata, T., Fukuda, Y. *et al.* (2010) Size-selective microcavity array for rapid and efficient detection of circulating tumor cells. *Analytical Chemistry*, **82**, 6629–6635.
16. Bo, L., Tong, X., Zheng, S., Goldkorn, A. and Yu-Chong, T.. (2010) Micro Electro Mechanical Systems (MEMS), 2010 IEEE 23rd International Conference, pp. 935–938.
17. Xu, Y., Phillips, J.A., Yan, J.L. *et al.* (2009) Aptamer-based microfluidic device for enrichment, sorting, and detection of multiple cancer cells. *Analytical Chemistry*, **81**, 7436–7442.
18. Tan, S.J., Yobas, L., Lee, G.Y.H. *et al.* (2009) Microdevice for the isolation and enumeration of cancer cells from blood. *Biomedical Microdevices*, **11**, 883–892.
19. Thege, F.I., Lannin, T.B., Saha, T.N. *et al.* (2014) Microfluidic immunocapture of circulating pancreatic cells using parallel EpCAM and muc1 capture: characterization, optimization and downstream analysis. *Lab on a Chip*, **14**, 1775–1784.
20. Yu, X., He, R., Li, S. *et al.* (2013) Magneto-controllable capture and release of cancer cells by using a micropillar device decorated with graphite oxide-coated magnetic nanoparticles. *Small*, **9**, 3895–3901.
21. Kamande, J.W., Hupert, M.L., Witek, M.A. *et al.* (2013) Modular microsystem for the isolation, enumeration, and phenotyping of circulating tumor cells in patients with pancreatic cancer. *Analytical Chemistry*, **85**, 9092–9100.

22. Hou, S., Zhao, L., Shen, Q. *et al.* (2013) Polymer nanofiber-embedded microchips for detection, isolation, and molecular analysis of single circulating melanoma cells. *Angewandte Chemie International Edition*, **52**, 3379–3383.
23. Harouaka, R.A., Zhou, M.D., Yeh, Y.T. *et al.* (2014) Flexible micro spring array device for high-throughput enrichment of viable circulating tumor cells. *Clinical Chemistry*, **60**, 323–333.
24. Tan, W.H., Donovan, M.J., and Jiang, J.H. (2013) Aptamers from cell-based selection for bioanalytical applications. *Chemical Reviews*, **113**, 2842–2862.
25. Raimondi, C., Gradilone, A., and Gazzaniga, P. (2013) Controversies in circulating tumor cell count during therapy. *Expert Review of Molecular Diagnostics*, **13**, 499–507.
26. Lopez-Riquelme, N., Minguela, A., Villar-Permuy, F. *et al.* (2013) Imaging cytometry for counting circulating tumor cells: Comparative analysis of the CellSearch vs ImageStream systems. *APMIS*, **121**, 1139–1143.
27. Lin, H.C., Hsu, H.C., Hsieh, C.H. *et al.* (2013) A negative selection system PowerMag for effective leukocyte depletion and enhanced detection of EpCAM positive and negative circulating tumor cells. *Clinica Chimica Acta*, **419**, 77–84.
28. He, R.X., Zhao, L.B., Liu, Y.M. *et al.* (2013) Biocompatible TiO_2 nanoparticle-based cell immunoassay for circulating tumor cells capture and identification from cancer patients. *Biomedical Microdevices*, **15**, 617–626.
29. Chung, J., Issadore, D., Ullal, A. *et al.* (2013) Rare cell isolation and profiling on a hybrid magnetic/size-sorting chip. *Biomicrofluidics*, **7**, 054107.
30. Casavant, B.P., Mosher, R., Warrick, J.W. *et al.* (2013) A negative selection methodology using a microfluidic platform for the isolation and enumeration of circulating tumor cells. *Methods*, **64**, 137–143.
31. Sheng, W.A., Chen, T., Katnath, R. *et al.* (2012) Aptamer-enabled efficient isolation of cancer cells from whole blood using a microfluidic device. *Analytical Chemistry*, **84**, 4199–4206.
32. Santana, S.M., Liu, H., Bander, N.H. *et al.* (2012) Immunocapture of prostate cancer cells by use of anti-PSMA antibodies in microdevices. *Biomedical Microdevices*, **14**, 401–407.
33. Mostert, B., Kraan, J., Sieuwerts, A.M. *et al.* (2012) CD49f-based selection of circulating tumor cells (CTCS) improves detection across breast cancer subtypes. *Cancer Letters*, **319**, 49–55.
34. Kim, M.S., Sim, T.S., Kim, Y.J. *et al.* (2012) SSA-MOA: A novel CTC isolation platform using selective size amplification (SSA) and a multi-obstacle architecture (MOA) filter. *Lab on a Chip*, **12**, 2874–2880.
35. Jang, K., Tanaka, Y., Wakabayashi, J. *et al.* (2012) Selective cell capture and analysis using shallow antibody-coated microchannels. *Biomicrofluidics*, **6**, 044117.
36. Chen, K.C., Pan, Y.C., Chen, C.L. *et al.* (2012) Enumeration and viability of rare cells in a microfluidic disk via positive selection approach. *Analytical Biochemistry*, **429**, 116–123.
37. Takao, M. and Takeda, K. (2011) Enumeration, characterization, and collection of intact circulating tumor cells by cross contamination-free flow cytometry. *Cytometry Part A*, **79A**, 107–117.
38. Liu, Z.A., Fusi, A., Klopocki, E. *et al.* (2011) Negative enrichment by immunomagnetic nanobeads for unbiased characterization of circulating tumor cells from peripheral blood of cancer patients. *Journal of Translational Medicine*, **9**, 70.
39. Chen, C.L., Chen, K.C., Pan, Y.C. *et al.* (2011) Separation and detection of rare cells in a microfluidic disk via negative selection. *Lab on a Chip*, **11**, 474–483.
40. Zhao, M.X., Schiro, P.G., Kuo, J.S. *et al.* (2013) An automated high-throughput counting method for screening circulating tumor cells in peripheral blood. *Analytical Chemistry*, **85**, 2465–2471.

41. Schiro, P.G., Zhao, M.X., Kuo, J.S. *et al.* (2012) Sensitive and high-throughput isolation of rare cells from peripheral blood with ensemble-decision aliquot ranking. *Angewandte Chemie International Edition*, **51**, 4618–4622.
42. Svobodova, Z., Kucerova, J., Autebert, J. *et al.* (2014) Application of an improved magnetic immunosorbent in an ephesia chip designed for circulating tumor cell capture. *Electrophoresis*, **35**, 323–329.
43. Mottet, G., Perez-Toralla, K., Tulukcuoglu, E. *et al.* (2014) A three dimensional thermoplastic microfluidic chip for robust cell capture and high resolution imaging. *Biomicrofluidics*, **8**, 024109.
44. Autebert, J., Coudert, B., Bidard, F.C. *et al.* (2012) Microfluidic: an innovative tool for efficient cell sorting. *Methods*, **57**, 297–307.
45. Jackson, J.M., WItek, M.A., Hupert, M.L. *et al.* (2014) UV activation of polymeric high aspect ratio microstructures: ramifications in antibody surface loading for circulating tumor cell selection. *Lab on a Chip*, **14**, 106–117.
46. Dharmasiri, U., Balamurugan, S., Adams, A.A. *et al.* (2009) Highly efficient capture and enumeration of low abundance prostate cancer cells using prostate-specific membrane antigen aptamers immobilized to a polymeric microfluidic device. *Electrophoresis*, **30**, 3289–3300.
47. Chin, C.D., Linder, V., and Sia, S.K. (2012) Commercialization of microfluidic point-of-care diagnostic devices. *Lab on a Chip*, **12**, 2118–2134.
48. Adams, A.A., Okagbare, P.I., Feng, J. *et al.* (2008) Highly efficient circulating tumor cell isolation from whole blood and label-free enumeration using polymer-based microfluidics with an integrated conductivity sensor. *Journal of the American Chemical Society*, **130**, 8633–8641.
49. Madou, M.J. (2012) *Fundamentals of microfabrication and nanotechnology*, 3rd edn, CRC Press, Boca Raton, FL.
50. Rånby, B.G. and Rabek, J.F. (1975) *Photodegradation, Photo-oxidation, and Photostabilization of Polymers: Principles and Applications*, Wiley, London, New York.
51. Bryant, K.L., Mancias, J.D., Kimmelman, A.C., and Der, C.J. (2014) KRAS: feeding pancreatic cancer proliferation. *Trends in Biochemical Sciences*, **39**, 91–100.
52. Dharmasiri, U., Witek, M.A., Adams, A.A., and Soper, S.A. (2010) Microsystems for the capture of low-abundance cells, in *Annual Review of Analytical Chemistry* vol. 3 (ed E.S.Z.R.N. Yeung), pp. 409–431.
53. Lalkhen, A.G. and McCluskey, A. (2008) Clinical tests: sensitivity and specificity. *CEACCP*, **8**, 221–223.
54. Hristozava, T., Konschak, R., Budach, V., and Tinhofer, I. (2012) A simple multicolor flow cytometry protocol for detection and molecular characterization of circulating tumor cells in epithelial cancers. *Cytometry Part A*, **81A**, 489–495.
55. Kamande, J.; Jackson, J.M.; Witek, M.A.; Hupert, M.L.; Gordon, K.; Pellerin, M.; Soper, S.A., Materials and microfluidics: Enabling the clinical utility of circulating tumor cells, Chem. Soc. Rev., 2014, *British Journal of Cancer*, 2014, **111**, 817–822.
56. Went, P.T.H., Lugli, A., Meier, S. *et al.* (2004) Frequent EpCAM protein expression in human carcinomas. *Human Pathology*, **35**, 122–128.
57. Yang, L., Lang, J.C., Balasubramanian, P. *et al.* (2009) Optimization of an enrichment process for circulating tumor cells from the blood of head and neck cancer patients through depletion of normal cells. *Biotechnology and Bioengineering*, **102**, 521–534.

58. Hristozava, T., Konschak, R., Stromberger, C. *et al.* (2011) The presence of circulating tumor cells (CTCS) correlates with lymph node metastasis in nonresectable squamous cell carcinoma on the head and neck region (SCCHN). *Annals of Oncology*, **22**, 1878–1885.

59. Racila, E., Euhus, D., Weiss, A.J. *et al.* (1998) Detection and characterization of carcinoma cells in the blood. *Proceedings of National Academy of Sciences of United States of America*, **95**, 4589–4594.

60. Hupert, M.L., Jackson, J.M., Wang, H. *et al.* (2013) Arrays of high-aspect ratio microchannels for high-throughput isolation of circulating tumor cells (CTCS). *Microsystem Technologies*, **20**, 1815–1825.

61. Jackson, J.M., Hupert, M.L., and Soper, S.A. (2014) Discrete geometry optimization for reducing flow non-uniformity, asymmetry, and parasitic minor loss pressure drops in z-type configurations of fuel cells. *Journal of Power Sources*, **269**, 274–283.

62. Becker, H. and Gartner, C. (2000) Polymer microfabrication methods for microfluidic analytical applications. *Electrophoresis*, **21**, 12–26.

63. Becker, H. and Locascio, L.E. (2002) Polymer microfluidic devices. *Talanta*, **56**, 267–287.

64. Soper, S.A., Ford, S.M., Qi, S. *et al.* (2000) Polymeric microelectromechanical systems. *Analytical Chemistry*, **72**, 643A–651A.

65. Hupert, M.L., Guy, W.J., Llopis, S.D. *et al.* (2007) Evaluation of micromilled metal mold masters for the replication of microchip electrophoresis devices. *Microfluidics and Nanofluidics*, **3**, 1–11.

66. Nagrath, S., Sequist, L.V., Maheswaran, S. *et al.* (2007) Isolation of rare circulating tumour cells in cancer patients by microchip technology. *Nature*, **450**, 1235–1239.

67. Herne, T.M. and Tarlov, M.J. (1997) Characterization of DNA probes immobilized on gold surfaces. *Journal of the American Chemical Society*, **119**, 8916–8920.

68. Butler, J.E., Ni, L., Brown, W.R. *et al.* (1993) The immunochemistry of sandwich ELISAs – VI. Greater than 90% of monoclonal and 75% of polyclonal anti-fluorescyl capture antibodies (CAbs) are denatured by passive adsorption. *Molecular Immunology*, **30**, 1165–1175.

69. Soper, S.A., Kelly, K., McCarley, R. *et al.* (2000) High-aspect ratio micromachining in plastics: a logical choice for the fabrication of DNA analysis devices. *Abstracts Papers of American Chemical Society*, **219**, U111.

70. Vaidya, B., Soper, S.A., and McCarley, R.L. (2002) Surface modification and characterization of microfabricated poly(carbonate) devices: Manipulation of electroosmotic flow. *Analyst*, **127**, 1289–1292.

71. Xu, Y.C., Vaidya, B., Patel, A.B. *et al.* (2003) Solid-phase reversible immobilization in microfluidic chips for the purification of dye-labeled DNA sequencing fragments. *Analytical Chemistry*, **75**, 2975–2984.

72. McCarley, R.L., Vaidya, B., Wei, S.Y. *et al.* (2005) Resist-free patterning of surface architectures in polymer-based microanalytical devices. *Journal of the American Chemical Society*, **127**, 842–843.

73. Miller, A.A., Lawton, E.J., and Balwit, J.S. (1954) Effect of chemical structure of vinyl polymers on crosslinking and degradation by ionizing radiation. *Journal of Polymer Science*, **14**, 503–504.

74. Fox, R.B., Isaacs, L.G., and Stokes, S. (1963) Photolytic degradation of poly(methyl methacrylate). *Journal of Polymer Science: Part A*, **1**, 1079–1086.
75. Moore, J.A. and Choi Jin, O. (1991) Degradation of poly(methyl methacrylate), in *Radiation Effects on Polymers*, American Chemical Society, pp. 156–192.
76. Reiser, A. (1983) Developments in polymer photochemistry. *Journal of Polymer Science: Polymer Letters Edition*, **21**, 679–679.
77. Diepens, M. and Gijsman, P. (2007) Photodegradation of bisphenol a polycarbonate. *Polymer Degradation and Stability*, **92**, 397–406.
78. Humphrey, J.S. Jr. (1968) Photochemical reactions of aryl polycarbonate esters. *Polymer Preprints, American Chemical Society, Division of Polymer Chemistry*, **9**, 453–460.
79. Rivaton, A. (1995) Recent advances in bisphenol-a polycarbonate photodegradation. *Polymer Degradation and Stability*, **49**, 163–179.
80. Nagai, N., Okumura, H., Imai, T., and Nishiyama, I. (2003) Depth profile analysis of the photochemical degradation of polycarbonate by infrared spectroscopy. *Polymer Degradation and Stability*, **81**, 491–496.
81. Wei, S., Vaidya, B., Patel, A.B. *et al.* (2005) Photochemically patterned poly(methyl methacrylate) surfaces used in the fabrication of microanalytical devices. *Journal of Physical Chemistry B*, **109**, 16988–16996.
82. Roy, S., Yue, C.Y., Lam, Y.C. *et al.* (2010) Surface analysis, hydrophilic enhancement, ageing behavior and flow in plasma modified cyclic olefin copolymer (coc)-based microfluidic devices. *Sensors and Actuators B: Chemical*, **150**, 537–549.
83. Roy, S. and Yue, C.Y. (2011) Surface modification of coc microfluidic devices: a comparative study of nitrogen plasma treatment and its advantages over argon and oxygen plasma treatments. *Plasma Processes and Polymers*, **8**, 432–443.
84. Tsao, C.W., Hromada, L., Liu, J. *et al.* (2007) Low temperature bonding of PMMA and COC microfluidic substrates using uv/ozone surface treatment. *Lab on a Chip*, **7**, 499–505.
85. Dole, M. (1973) *Radiation Chemistry of Macromolecules*, Academic Press, New York.
86. Shadpour, H., Musyimi, H., Chen, J.F., and Soper, S.A. (2006) Physiochemical properties of various polymer substrates and their effects on microchip electrophoresis performance. *Journal of Chromatography*, **1111**, 238–251.
87. Gleghorn, J.P., Pratt, E.D., Denning, D. *et al.* (2010) Capture of circulating tumor cells from whole blood of prostate cancer patients using geometrically enhanced differential immunocapture (GEDI) and a prostate-specific antibody. *Lab on a Chip*, **10**, 27–29.
88. Ligrani, P.M. and Niver, R.D. (1988) Flow visualization of dean vortices in a curved channel with 40 to 1 aspect ratio. *Physics of Fluids*, **31**, 3605–3617.
89. Kuntaegowdanahalli, S.S., Bhagat, A.A.S., Kumar, G., and Papautsky, I. (2009) Inertial microfluidics for continuous particle separation in spiral microchannels. *Lab on a Chip*, **9**, 2973–2980.
90. Di Carlo, D., Irimia, D., Tompkins, R.G., and Toner, M. (2007) Continuous inertial focusing, ordering, and separation of particles in microchannels. *Proceedings of National Academy Sciences of United States of America*, **104**, 18892–18897.
91. Umpleby, H.C., Fermor, B., Symes, M.O., and Williamson, R.C.N. (1984) Viability of exfoliated colorectal-carcinoma cells. *British Journal of Surgery*, **71**, 659–663.
92. Baskurt, O.K. and Meiselman, H.J. (2003) Blood rheology and hemodynamics. *Seminars in Thrombosis and Hemostasis*, **29**, 435–450.
93. Asmolov, E.S. (1999) The inertial lift on a spherical particle in a plane Poiseuille flow at large channel Reynolds number. *Journal of Fluid Mechanics*, **381**, 63–87.

94. Segre, G. and Silberberg, A. (1961) Radial particle displacements in Poiseuille flow of suspensions. *Nature*, **189**, 209–210.
95. Chang, K.C. and Hammer, D.A. (1999) The forward rate of binding of surface-tethered reactants: effect of relative motion between two surfaces. *Biophysical Journal*, **76**, 1280–1292.
96. Battle, K.N., Jackson, J.M., Witek, M.A. *et al.* (2014) Solid-phase extraction and purification of membrane proteins using a UV-modified PMMA microfluidic bioaffinity MUSPE device. *Analyst*, **139**, 1355–1363.
97. Rao, C.G., Chianese, D., Doyle, G.V. *et al.* (2005) Expression of epithelial cell adhesion molecule in carcinoma cells present in blood and primary and metastatic tumors. *International Journal of Oncology*, **27**, 49–57.
98. Zhang, W., Hu, P., Lai, X., and Peng, L. (2009) Analysis and optimization of flow distribution in parallel-channel configurations for proton exchange membrane fuel cells. *Journal of Power Sources*, **194**, 931–940.
99. Kotz, K.T., Xiao, W., Miller-Graziano, C. *et al.* (2010) Clinical microfluidics for neutrophil genomics and proteomics. *Nature Medicine*, **16**, 1042–1047.
100. Kays, W.M. and Crawford, M.E. (1980) *Convective heat and mass transfer*, 2nd edn, McGraw-Hill, New York.
101. Pullagurla, S.R., Witek, M.A., Jackson, J.M. *et al.* (2014) Parallel affinity-based isolation of leukocyte subsets using microfluidics: application for stroke diagnosis. *Analytical Chemistry*, **86**, 4058–4065.
102. Riker, A., Libutti, S.K., and Bartlett, D.L. (1997) Advances in the early detection, diagnosis, and staging of pancreatic cancer. *Surgical Oncology*, **6**, 157–169.
103. Hariharan, D., Saied, A., and Kocher, H.M. (2008) Analysis of mortality rates for pancreatic cancer across the world. *HPB*, **10**, 58–62.
104. Hezel, A.F., Kimmelman, A.C., Stanger, B.Z. *et al.* (2006) Genetics and biology of pancreatic ductal adenocarcinoma. *Genes and Development*, **20**, 1218–1249.
105. Grasso, D., Garcia, M.N., and Iovanna, J.L. (2012) Autophagy in pancreatic cancer. *International Journal of Cell Biology*, **2012**, 760498–760504.
106. Cristofanilli, M., Budd, G.T., and Ellis, M.J. (2004) Circulating tumor cells, disease progression, and survival in metastatic breast cancer. *New England Journal of Medicine*, **351**, 781–791.
107. Hayes, D.F. and Smerage, J. (2008) Is there a role for circulating tumor cells in the management of breast cancer? *Clinical Cancer Research*, **14**, 3646–3650.
108. Amado, R.G., Wolf, M., Peeters, M. *et al.* (2008) Wild-type KRAS is required for panitumumab efficacy in patients with metastatic colorectal cancer. *Journal of Clinical Oncology*, **26**, 1626–1634.
109. National Comprehensive Cancer Network (2014) Clinical practice guidelines in oncology: colon cancer (version 3.2014). http://www.nccn.org/professionals/physician_gls/pdf/colon.pdf (accessed August 4, 2014).
110. Tan, C.D. (2012) X., Kras mutation testing in metastatic colorectal cancer. *World Journal of Gastroenterology*, **18**, 5171–5180.
111. Yen, L.C., Yeh, Y.S., Chen, C.W. *et al.* (2009) Detection of KRAS oncogene in peripheral blood as a predictor of the response to cetuximab plus chemotherapy in patients with metastatic colorectal cancer. *Clinical Cancer Research*, **15**, 4508–4513.

112. Santini, D., Loupakis, F., Vincenzi, B. *et al.* (2008) High concordance of KRAS status between primary colorectal tumors and related metastatic sites: Implications for clinical practice. *The Oncologist*, **13**, 1270–1275.
113. Knijn, N., Mekenkamp, L.J.M., Klomp, M. *et al.* (2011) KRAS mutation analysis: A comparison between primary tumours and matched liver metastases in 305 colorectal cancer patients. *British Journal of Cancer*, **104**, 1020–1026.
114. Watanabe, T., Kobunai, T., Yamamoto, Y. *et al.* (2011) Heterogeneity of KRAS status may explain the subset of discordant kras status between primary and metastatic colorectal cancer. *Diseases of the Colon and Rectum*, **54**, 1170–1178.
115. Favis, R. and Barany, F. (2000) Mutation detection in k-ras, BRCA1, BRCA2, and p53 using PCR/LDR and a universal DNA microarray. *Annals of the New York Academy of Sciences*, **906**, 39–43.
116. Khanna, M., Park, P., Zirvi, M. *et al.* (1999) Multiplex PCR/LDR for detection of k-ras mutations in primary colon tumors. *Oncogene*, **18**, 27–38.
117. Barany, F. (1991) Genetic disease detection and DNA amplification using cloned thermostable ligase. *Proceedings of National Academy of Sciences of United States of America*, **88**, 189–193.
118. Dunn, E.F., Iida, M., Myers, R.A. *et al.* (2011) Dasatinib sensitizes KRAS mutant colorectal tumors to cetuximab. *Oncogene*, **30**, 561–574.
119. Eser, S., Schnieke, A., Schneider, G., and Saur, D. (2014) Oncogenic KRAS signalling in pancreatic cancer. *British Journal of Cancer*, advance online publication. **111**, 817–822.
120. Zinsky, R., Servet, B., Holger, B. *et al.* (2010) Analysis of KRAS mutations of exon 2 codons 12 and 13 by snapshot analysis in comparison to common DNA sequencing. *Gastroenterology Research and Practice*, **2010**, 789363.
121. Yu, M., *et al.* (2012) RNA sequencing of pancreatic circulating tumour cells implicates WNT signaling in metastasis. *Nature*, **487**, 510–513.
122. Shim, S., *et al.* (2013) Dielectrophoresis has broad applicability to marker-free isolation of tumor cells from blood by microfluidic systems. *Biomicrofluidics*, **7**, 011808.
123. Shim, S., *et al.* (2013) Antibody-independent isolation of circulating tumor cells by continuous-flow dielectrophoresis. *Biomicrofluidics*, **7**, 011807.
124. Yang, F., *et al.* (2013) Dielectrophoretic separation of prostate cancer cells. *Technology in Cancer Research & Treatment*, **12**, 61–70.
125. Lu, J., *et al.* (2010) Isolation of circulating epithelial and tumor progenitor cells with an invasive phenotype from breast cancer patients. *International Journal of Cancer*, **126**, 669–683.
126. Allard, W. J., *et al.* (2004) Tumor Cells Circulate in the Peripheral Blood of All Major Carcinomas but not in Healthy Subjects or Patients With Nonmalignant Diseases. *Clinical Cancer Research*, **10**, 6897–6904.
127. Ozkumur, E., *et al.* (2013) Inertial focusing for tumor antigen-dependent and -independent sorting of rare circulating tumor cells. *Sci. Transl. Med.*, **5**, 179ra47.
128. Hughes, A.D., *et al.* (2012) Microtube Device for Selectin-Mediated Capture of Viable Circulating Tumor Cells from Blood. *Clin. Chem.*, **58**, 846–853.
129. Saliba, A.E., *et al.* (2010) Microfluidic sorting and multimodal typing of cancer cells in self-assembled magnetic arrays. *Proc. Natl. Acad. Sci. U. S. A.*, **107**, 14524–14529.
130. Issadore, D., *et al.* (2012) Ultrasensitive clinical enumeration of rare cells ex vivo using a micro-hall detector. *Sci. Transl. Med.*, **4**, 141ra192.

5

CELL SEPARATION USING INERTIAL MICROFLUIDICS

NIVEDITA NIVEDITA AND IAN PAPAUTSKY

Department of Electrical Engineering and Computing Systems, University of Cincinnati, Cincinnati, OH, USA

5.1 INTRODUCTION

Separation of cells is a critical sample preparation step in many diagnostic, therapeutic, and cell biology applications [1–6]. Conventional approaches such as fluorescently activated cell sorting (FACS) are the most widely used [7, 8] and offer high throughput and resolution. However, wide adoption of these methods in clinical settings is limited due to system complexity, need for user training, and high cost [4]. When selectivity is critical, as in detection of circulating tumor cells (CTCs), immunoselection is the preferred approach [9] although such labeling can adversely impact cell properties [10]. Nevertheless, high cost and low throughput limit the wide acceptance of these systems as well. Furthermore, recent evidence suggests that surface biomarkers recognized during the immunoselection may not be homogenously expressed [2, 11].

Rapid growth of microfluidics has brought considerable attention to new technologies for cell separations and sorting due to a number of promising advantages, including high throughput, high efficiency, low cost, and simplicity [12–20]. These microfluidic separation technologies can be categorized as either active or passive approaches. Active microfluidic systems use external actuators to exert magnetic [21], dielectrophoretic [22], or acoustic [23, 24] forces on cells to sort them based on a variety of markers such as size, charge, or polarizability. These active methods generally offer excellent separation efficiency but often have limited throughput

Circulating Tumor Cells: Isolation and Analysis, First Edition. Edited by Z. Hugh Fan.

and require complex sample preparation or sophisticated external control. Passive microfluidic systems based on hydrodynamic forces [25–27], size filtration [28, 29], and pinched flow fractionation [30, 31] are generally simpler and easier to operate. These label-free methods sort cells based on their physical properties and often significantly reduce operational complexity. Despite fewer viable sorting markers, the passive sorting approaches are generally preferable.

Inertial microfluidics is a rapidly emerging subfield of passive microfluidics that is an alternative to the high-cost and labor-intensive nature of current cell-sorting methods. Inertial microfluidics relies on manipulation of hydrodynamic forces for label-free and membrane-free sorting of cells. It is the interaction of these forces on cells in flowing fluid that leads to specific migration behavior depending on the cell size [25, 27], shape [4], and deformability [32, 33]. The approach has been demonstrated in microchannels of various geometries, such as straight [27, 34] and spiral channels [26]. These platforms generally offer excellent separation performance. For example, Zhou *et al.* [34] demonstrated high purity (>90%) and ultrahigh separation efficiency (>99%) at flow rates as high as ~100 µl/min. Our previous demonstration in a spiral channel has shown even better performance in terms of throughput (~1 ml/min) and efficiency (~90%), subject to an appropriate outlet system [26].

In this chapter, we show that inertial microfluidic devices can be used for separation of cells, including rare cells such as CTCs. Although most of the today's microfluidic separation techniques sacrifice throughput in order to improve separation efficiency, this is not the case in inertial microfluidics. The devices we discuss herein can provide both high separation efficiency and high throughput and, thus, show a tremendous promise for separation and isolation of rare circulating cells.

5.2 DEVICE FABRICATION AND SYSTEM SETUP

Microchannels were fabricated in polydimethysiloxane (PDMS, Sylgard 184, Dow Corning) with the standard soft lithography process. First, a master was formed in SU-8 photoresist (2075, Microchem Corp.). A mixture of PDMS base and curing agent (10:1 ratio) was then poured on the master. After degassing, PDMS was cured for 4 h on a 60 °C hotplate. The cured PDMS devices were peeled off, and inlet/outlet ports were punched with a 14-gauge syringe needle. PDMS was bonded to microscope glass slides using a handheld plasma surface treater (BD-20AC, Electro-Technic Products Inc.).

For microparticle or cell experiments, syringes loaded with particle or cell mixtures were connected to the devices using a 1/16″ peek tubing (Upchurch Scientific) with appropriate fittings (Upchurch Scientific). A syringe pump (NE-1000, New Era Pump Systems, Inc.) was used to control the flow. An inverted epifluorescence microscope (IX71, Olympus Inc.) equipped with a 12-bit high-speed charge-coupled device (CCD) camera (Retiga EXi, QImaging) was used for fluorescence and bright-field imaging. For streak velocimetry, at least 20 images were overlaid and pseudo-colored using Image J to demonstrate particle focusing streaks. To image trajectory of particles in bright field, the exposure time was set to 10 µs and 300 images were taken sequentially and stacked in Image J to establish a complete view of particle or cell motion.

A hemocytometer was used to count particles or cells in samples collected from each outlet to determine their concentration. Each sample was counted at least 3× to reduce potential errors. Before any of the methods is used, it is important to understand the underlying fundamentals of the inertial microfluidic devices and the physics of the inertial focusing, which we discuss in the following section.

5.3 INERTIAL FOCUSING IN MICROFLUIDICS

Inertial focusing has been used for high-throughput focusing and separation of cells in microscale channels [25]. The phenomenon describes migration of microparticles or cells across streamlines into equilibrium positions within the flow cross section as they travel downstream in a fluidic channel at a finite-channel Reynolds number (*Re*) [35, 36]. This migration is governed by two primary inertial lift forces [27, 35, 36]. The shear-gradient-induced lift force (F_s) and the wall-induced lift force (F_w) as illustrated in Figure 5.1a. As particles or cells travel in a microchannel, they experience a velocity difference on either side due to the parabolic velocity profile of the flow. This velocity difference generates a lift force, F_s, which acts to push particles or cells down the velocity gradient toward the channel wall. Near the channel wall, the compression of the flow builds the pressure and generates F_w, which pushes particles and cells away from the wall.

The balance of these two forces leads to equilibrium at specific cross-sectional positions. The net lift force (F_L) at these positions can be defined as

$$F_L = \rho C_L G^2 {a_p}^4 \tag{5.1}$$

where r is the fluid density, C_L is the lift coefficient which is a function of the particle position across the channel cross-section, G is the shear rate of the fluid, and a_p is the particle diameter. The average value of G for a Poiseuille flow is given by $G = 2U_f/D_h$, where U_f is the fluid velocity and D_h is the hydraulic diameter of the channel ($D_h = 4A/P$, where A is the area and P is the perimeter of the channel cross-section) [27].

Microchannel geometries dictate the equilibrium behavior in inertial focusing. Microchannels used in inertial microfluidics can be grouped in two major categories as straight and curved microchannels. In straight channels, particle focusing behavior further varies with the cross-sectional shape. In a straight channel with a circular cross section, Segre and Silberberg [37] first observed focusing of ~1-mm-diameter particles into an annulus at ~0.2*D* (diameter of the pipe) in a macroscale cylinder pipe (Figure 5.1b), which can be explained entirely by the balance of the two primary inertial lift forces [38]. While the balance of the two lift forces explains particle focusing in a channel with circular cross section, focusing in square and rectangular channels is more complex due to the radial asymmetry of the channel cross section.

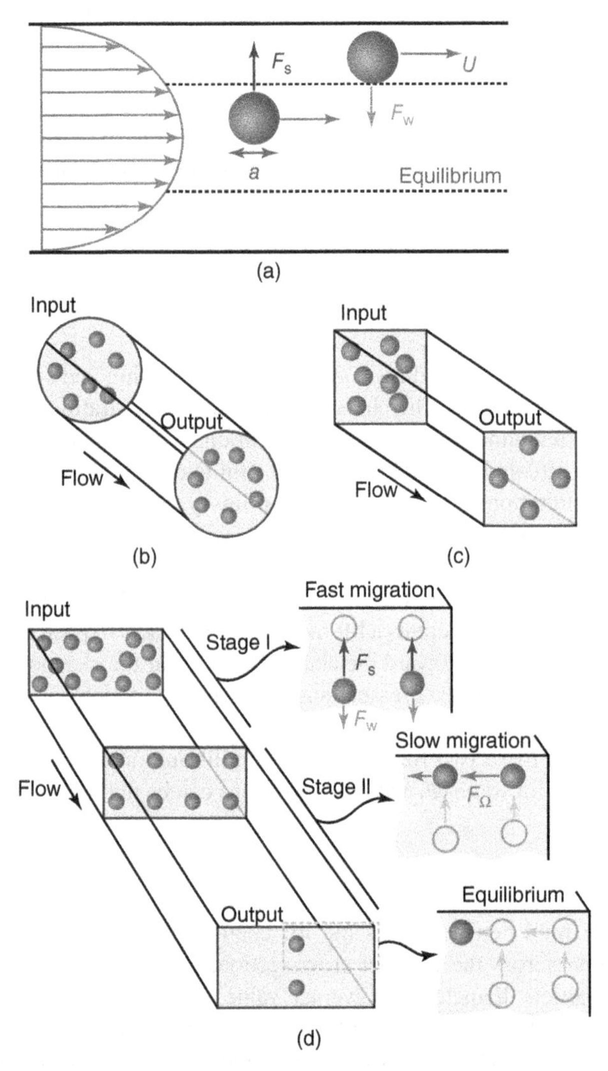

Figure 5.1 Schematic illustration of inertial focusing in different channel geometries. (a) Two inertial lift forces with opposite direction and orthogonal to the flow direction act to equilibrate microparticles near walls. (b) Inertial migration and final equilibrium positions in rectilinear microchannel with circular cross section. (c) Inertial migration and final equilibrium positions in rectilinear microchannel with square cross section. (d) Inertial migration and final equilibrium positions in rectilinear microchannel with rectangular cross section. (e) Inertial focusing in curved microchannel. Zhou and Papautsky [27]. Reproduced with permission of the Royal Society of Chemistry.

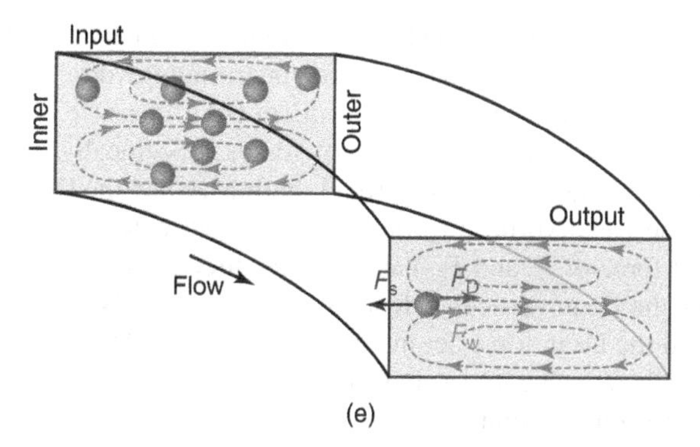

Figure 5.1 (*continued*)

In a straight channel with square cross section, previous research indicates four fully symmetric equilibrium positions at the center of each channel wall (Figure 5.1c) [39]. In a straight channel with rectangular cross section, the difference of velocity profile at the longer and shorter walls in rectangular channel reduces the focusing positions to two at the center of each longer wall [27]. This equilibrium behavior can be explained by the addition of a rotational lift force F_{Ω} in a two-stage migration model (Figure 5.1d). In the first stage, microparticles focus along the long wall where F_s balances F_w. Once this initial equilibrium is reached, particle motion is dominated by F_{Ω}, which plays an important role near channel wall (Figure 5.1d). As a result, particles migrate to the center of long sidewalls in this second stage. One should note that as the channel Reynolds number (*Re*) exceeds a certain threshold, particles can also equilibrate into four positions at the center of each wall in a rectangular channel [40].

Precise prediction of channel length for particle focusing is essential to the design of a straight inertial microfluidic device. Zhou and Papautsky [27] reported an in-depth investigation of inertial focusing in straight microchannels and showed that the complete focusing length (L) in a rectangular straight channel can be defined as

$$L = \frac{3\pi\mu D_h^2}{4\rho U_f a_p^3}\left(\frac{H}{C_L^-} + \frac{W}{C_L^+}\right), \quad W > H \tag{5.2}$$

where H is the height and W is the width of the channel cross section, C_L^- is the negative lift coefficient, and C_L^+ is the positive life coefficient. These two lift coefficients can be obtained from our recent work [27]. Equation 5.2 is directly applicable to low-aspect-ratio channels, which are the most common in microfluidic systems, where W is the longer channel dimension and H is the shorter one. For high-aspect-ratio channels, H and W must be swapped to represent the channel height. While Eq. 5.2 was derived for a rectangular microchannel, it can be easily applied to a square channel by setting $W = H$ or a round microchannel by setting $W = 0$ and H to diameter. Note that in a round channel, only the first stage of particle equilibration occurs, dominated by the shear-induced negative lift force; the result is the Segré annulus near the capillary walls. Thus, Eq. 5.2 is generally applicable for inertial focusing in straight microchannels.

In a curved or spiral microchannel, the curvilinear geometry introduces the Dean force, which leads to the formation of two counterrotating vortices (Dean vortices) in the cross section of the channel (Figure 5.1e) [41]. Presence of the Dean force disrupts equilibrium of the two inertial lift forces. Specifically, particles flowing near the top and the bottom walls are strongly influenced by the Dean force and thus migrate along the Dean vortices toward the inner and outer walls. Near the outer wall, the net lift force of F_s and F_w has the same direction as the Dean force. Thus, particles follow the Dean vortices independent of their size. Near the inner wall, the net lift force and the Dean force act in opposite directions, leading to possible force balance for particle focusing into a single position [42, 43]. In the following sections, we discuss the various straight and spiral microchannels, cell-sorting devices, and their various applications.

5.4 CANCER CELL SEPARATION IN STRAIGHT MICROCHANNELS

Inertial migration in straight microchannels shows great potential for isolation of CTCs from blood with high efficiency and purity. One example of a microfluidic device based on inertial migration is our MARCS (modulation of aspect ratio for complete separation) device [34]. Building on our two-stage model of inertial focusing in rectangular microchannels, the MARCS device consists of two channel segments with reversed aspect ratio (AR). This is schematically illustrated in Figure 5.2a. The upstream high-AR channel (segment I) is where randomly dispersed particles or cells at the input complete both stages of inertial migration and focus into two equilibrium positions at the centers of the channel sidewalls. The downstream channel (segment II) reverses the AR and is where equilibrium positions shift to the centers of the top and bottom walls. This shift in equilibrium positions causes cells to regain migration velocity toward the new equilibrium locations. Since the migration velocity U_m is strongly dependent on particle or cell diameter a_p, scaling as $U_m \propto a_p^3$ [35], the larger particles or cells migrate much faster than the smaller ones. Consequently, the downstream length necessary for the larger cells to reach equilibrium is much shorter than that of the smaller ones, and the slowly migrating smaller cells remain near channel sidewalls while the larger cells equilibrate near centers of the top and bottom walls. A three-outlet system at the microchannel end then permits a complete separation based on size alone, with the larger cells eluting in the center outlet and the smaller ones in the side outlets.

The complete separation concept was validated with a mixture of 20- and 9.94-μm-diameter particles. The results shown in Figure 5.2b demonstrate that separation occurs in three steps. First, as expected, the randomly distributed particles at the input gradually migrate to their equilibrium positions near sidewalls of the high-AR upstream channel, forming two streams for each particle. Streaks of the 20 μm diameter particles are clearly visible at ~20 mm downstream, although the smaller 9.94 μm diameter particles are indistinguishable due to proximity of the equilibrium positions. To illustrate the positions of the particle streams, we measured the lateral distance d_m of each stream from peak maxima in fluorescent intensity

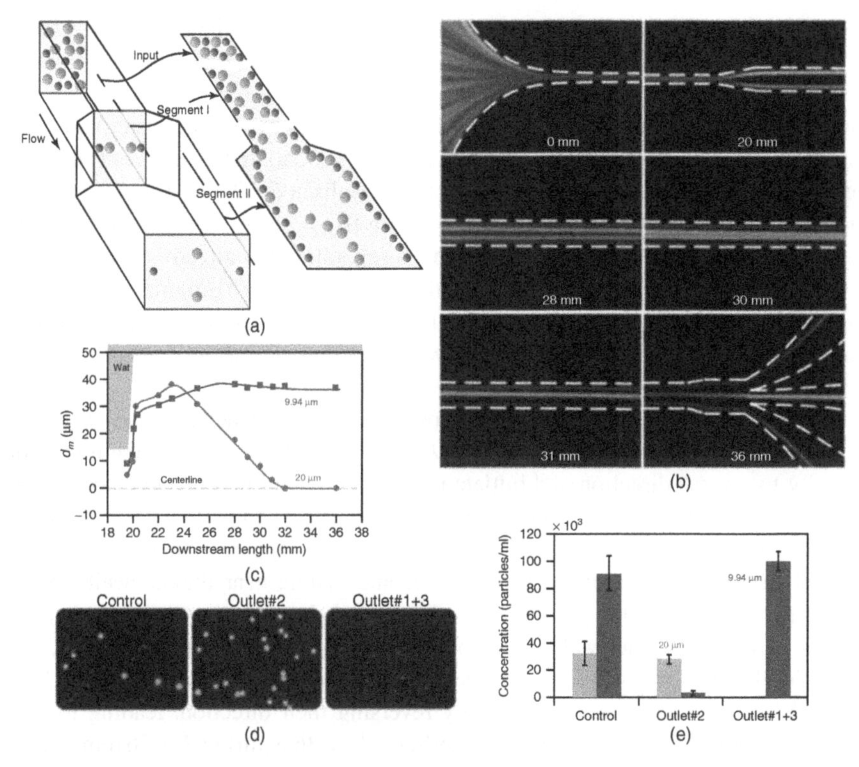

Figure 5.2 (a) Schematic of the device concept for size-based separation. Mixture of particles first flows through a high-AR channel (upstream segment) where they migrate to their equilibrium positions centered at the channel side walls. Once all particles fully focus into stable positions, the channel expands into a low-AR channel (downstream segment), which modifies the inertial lift and shifts the equilibrium positions to the center of the top and bottom walls. Due to strong dependence of migration velocity on particle size, large particles refocus much faster than small particles, leading to complete separation at the outlet. (b) Fluorescent images demonstrating progressive inertial migration of 20 μm (green) and 9.94 μm microparticles (red) leading to complete separation at the outlet. (c) Lateral distance d_m as a function of downstream length indicating the migration trajectory of each sized particles. (d) Fluorescent images of microparticles collected from inlet and outlets (e) Concentrations of microparticles from inlet sample and outlet samples indicate successful separation. Zhou *et al.* [34]. Reproduced with permission of the Royal Society of Chemistry. (*See color plate section for the color representation of this figure.*)

line scans (Figure 5.2c). The particle positions at the end of the upstream segment indicate that the smaller red particles are closer to the channel sidewalls, which is expected since they experience less wall-induced lift force ($F_w \propto a^3/\delta$, where δ is the minimum distance between particle and wall). The green particles were observed at $\delta = 1$ μm ($\delta = ½W\text{-}d_m\text{-}½ap$, $d_m = 5$ μm), while the smaller red particles were at $\delta = 2$ μm ($d_m = 9$ μm) at $L = 19.5$ mm downstream.

In the second step, the reversed microchannel AR leads to the increasing δ and a substantial reduction in F_w, which activates lateral migration under the influence of the shear-induced lift force F_s. The significant dependence of the migration velocity on particle diameter ($U_m \propto a_p^3$) leads the larger particles to migrate faster and cross trajectory of the smaller particles (red curve in Figure 5.2c). In this transient state, the larger 20-μm-diameter particles are closer to sidewalls at $\delta = 2$ μm ($d_m = 38$ μm), and the 9.94-μm-diameter particles are at $\delta = 12$ μm ($d_m = 33$ μm) at $L = 23$ mm downstream. These distinct positions suggest that a separation has already occurred; nevertheless the relatively small gap between the streams ($\Delta d_m = 5$ μm) makes it difficult to design an outlet system to fractionate the sample. This gap, however, is substantially amplified further downstream following the second trajectory crossing, as we discuss next.

As the large particles migrate to closer to channel sidewall, they begin to experience the rotation-induced lift force F_Ω, which ultimately causes the migration velocity to reverse direction and initiate movement toward microchannel centerline (Figure 5.2c). Thus, at $L > 23$ mm, the larger particles migrate in the direction opposite to that of the smaller particles. This is also apparent in Figure 5.2b as the red streams of the smaller particles become visible near the sidewalls. Since migration velocity of the smaller particles and their transient equilibrium position is closer to channel wall, they continue migration toward the sidewalls. For example, the smaller particles reach the maximum lateral position of $d_m = 38.1$ μm at $L = 28$ mm downstream before slowly reversing their direction, leading to very slow migration toward the channel centerline ($d_m = 36.8$ μm) at $L = 36$ mm, where the larger particles have already settled in their new equilibrium positions.

The second trajectory crossing occurs at $L \sim 24$ mm (Figure 5.2c) as the large particles migrate rapidly to the channel centerline, dominated by the rotation-induced lift force, creating large spacing between the two particle streams ($\Delta d_m = 37$ μm). This distance is more than 7× that of the spacing in the second step. Since the focusing length is strongly dependent on particle size, the smaller (9.94 μm) particles will not migrate to the width-centered equilibrium positions until ~60 mm downstream. The particle-free spacing (~20 μm) observed between the two particle streams without any further channel expansion permits an easy design of an outlet system to fractionate the follow and separate particles. Indeed, we demonstrated complete separation using a simple three-channel outlet, with both fluorescent and bright-field images confirming that small particles exit through the two side outlets while the large particles elute from the central outlet (Figure 5.2d).

Our results shown in Figure 5.2e demonstrate the excellent separation performance of the system. Samples from side outlets (#1 and #3) were combined due to symmetry, and the volumes of all samples were adjusted to that of the control for better comparison. Both of our fluorescent images and quantitative counts show that at input $Re = 40$ (flow rate of 100 μl/min), almost every large particle is collected from outlet #2, with efficiency $\eta = 99.3\%$ and purity = 90.8% for the 20-μm-diameter particles. Similarly, $\eta = 94.5\%$ and purity = 99.6% for the smaller 9.94-μm-diameter particles in side outlets.

One target application of the described microfluidic device is to process a blood sample to extract cells of interest such as CTCs. Since the whole blood is rather viscous (female hematocrit ~40%), a dilution is necessary to achieve Newtonian fluid and reduce the cell-to-cell interactions before processing through the device. Typically, a 0.5% hematocrit (Ht) is used to perform cell separations in an inertial microfluidic system [44, 45]. Yet, higher hematocrit (less dilute) blood is preferred to save the processing time. To determine the minimum dilution factor that offers acceptable performance, blood samples of diminishing hematocrit were introduced into the device. We took bright-field images near the outlet and measured the full width at half maximum (FWHM) of the intensity distribution across the channel width (Figure 5.3a). As the hematocrit decreases with increasing dilution factor, FWHM first drops rapidly and then converges to ~7 μm when hematocrit Ht < 1%. This suggests that the MARCS system would offer good focusing of cells at 50× dilution (Ht = 0.8%) (Figure 5.3b).

Application of our microfluidic system to separation of CTCs from the blood was demonstrated using human prostate epithelial (HPET) cells. HPET cells are derived from a high-grade (Gleason 9) prostate punch biopsy and recapitulate the histopathology of human prostate cancer *in vivo* [46] and can be present in the blood stream [47]. The target size range of HPET cells for this application was narrowed to 18–22 μm in diameter and cannot be separated with conventional flow cytometry due to their fragile nature (relatively high shear rates in the flow cytometer tend to lyse these

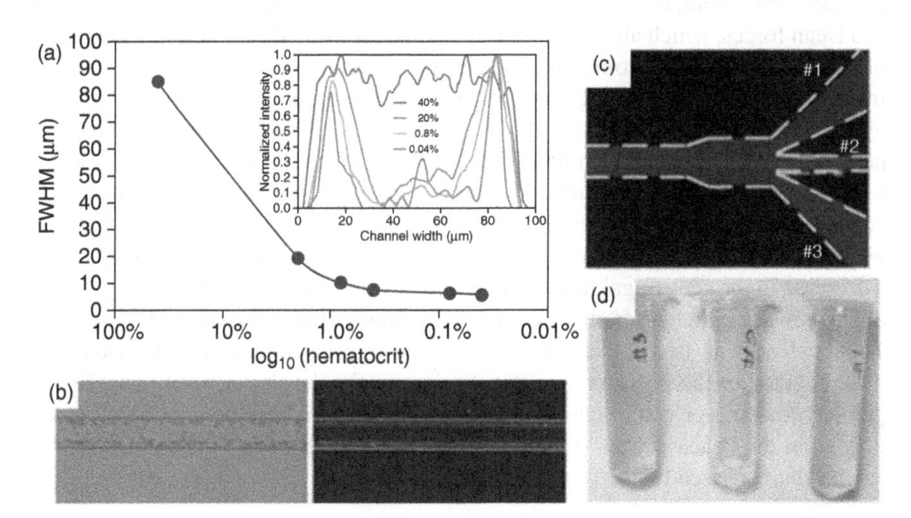

Figure 5.3 (a) FWHM as a function of blood hematocrit. The inset shows average intensity obtained from five measurements for each hematocrit value. (b) Bright-field image illustrating focusing of diluted blood (Ht = 0.8%) at ~35 mm downstream and color inversion of the 100-frame-stacked image. (c) Fluorescence image showing HPET cells flowing in the central outlet. (d) Image illustrating that most of RBCs were collected from side outlets (#1 and #3), which appear to be darker gray, while the central outlet (#2) is clear. Reproduced with permission of the Royal Society of Chemistry.

cells). The cells were stained with green live cell tracker dye (ER-Tracker™ Green, Invitrogen) to permit visualization. The fluorescent streak in the middle (Figure 5.3c) indicates HPET cells behave similar to 20-μm-diameter particles and exit from the central outlet. As expected, most of the RBCs flow out the channel from side exits, which is confirmed by sample collected from the three outlets. Samples in side outlets appear darker gray in Figure 5.3d due to presence of RBCs, while sample from the central outlet is clear (Figure 5.3d). The successful separation of HPET cells from human blood indicates that the device can be a promising solution for the isolation of CTCs considering its extremely high efficiency and purity.

In this section, we introduced the concept of using simple, straight microchannel design for potential separation of CTCs from human blood with high efficiency and purity. The demonstrated straight microchannel exhibits size-based separation of microparticles with efficiency $\eta > 99\%$ and purity >90%. Moreover, the device offers excellent performance over a wide range of flow conditions ($30 < Re < 80$). Most importantly, the complete separation of rare cells using HPET cell spiked blood has been achieved using the device. Such an outcome suggests that the approach is a promising alternative to isolation of CTCs. In the next section, we show how spiral microchannels can be used for separation at high throughput.

5.5 CANCER CELL SEPARATION IN SPIRAL MICROCHANNELS

As discussed earlier, cells in spiral channels experience a combination of inertial lift and Dean forces, which alter their focusing behavior [41]. This is due to the presence of Dean flow in these geometries, which manifests as the counterrotating vortices. In a curved channel of rectangular cross section, laminar Poiseuille flow is disrupted by the centrifugal force, shifting the maximum point of velocity distribution of the parabolic profile of the laminar flow from the center of the channel toward the concave wall (Figure 5.4) [43]. This shift causes a sharp velocity gradient to develop near the concave wall between the point of maximum velocity and the outer concave channel wall where the velocity is zero. The development of the sharp velocity gradient causes a decrease in the centrifugal force on the fluid near the concave wall, thereby increasing the pressure gradient from the concave wall toward the convex wall (Figure 5.4b). The local velocity near the channel walls is not significant enough to provide complete balance of the pressure gradient, leading to the development of instability of pressure known as Dean instability, which in turn leads to development of secondary flow from the concave wall to the convex wall of the channel. This secondary flow can be characterized by the Dean number, *De*.

$$De = Re\sqrt{\frac{D_{\mathrm{h}}}{2R}} \tag{5.3}$$

where D_{h} is the hydraulic diameter of the microchannel, *Re* is the Reynolds number, and *R* is the radius of the curvature of the convex surface of the curved channel. Hence, the strength of the secondary flow is strongly dependent on the dimensions of the channel and the radius of curvature. The secondary flow in spiral microchannels

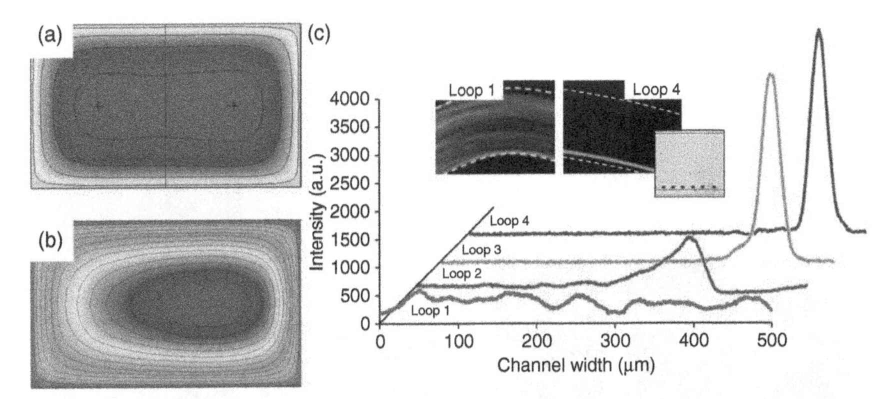

Figure 5.4 (a) CFD-ACE simulation of a curved rectangular channel, velocity and pressure distribution at the inlet where the flow is laminar and the curvature has not been introduced; (b) velocity and pressure distribution at the center of the curved channel, which indicated the shift of the center of maximum velocity toward the concave wall. (c) Intensity plot of 20-μm-diameter particles across the width of the channel at the end of each loop in the spiral (loop1 being the innermost loop and loop4 being the outermost). The two inset figures show the fluorescent images of the 20 μm polystyrene particles at the innermost loop (loop1) of the spiral device (500 μm × 110 μm) and at loop4 focused in a single stream near the inner channel wall. Nivedita and Papautsky [43]. Reproduced with permission of the American Institute of Physics. (*See color plate section for the color representation of this figure.*)

manifests itself in the form of two primary counterrotating vortices below a certain *De*. The flow of the vortices is directed from the center of the channel toward the concave wall and back to the center of the channel.

As discussed in the section of the physics of inertial focusing, spiral devices with rectangular channels use the inertial lift forces acting on cells within laminar flow, coupled with rotational Dean drag due to curvilinear microchannel geometry, to focus cells in single streams near the inner channel wall depending on the size of the cells/particles. The two primary counter rotating vortices displace the cells from the multiple focusing positions and focus them in a single stream near the inner channel wall/convex wall depending on their size, with the largest focusing closest to the inner wall. The net Dean force (F_D) experienced by the cells can be defined as

$$F_D = 5.4 \times 10^{-4} \pi \mu De^{1.63} a_p \tag{5.4}$$

where *De* is the dimensionless Dean number and μ is the viscosity of the fluid. Thus, the ratio of the net inertial lift force and the Dean force is strongly dependent on the cell size (a_p^3). Hence, the magnitude of the Dean force and the net inertial lift force governs the focusing position of the cells and the focusing length required.

The size dependence forms the basis of the design of spiral devices for size-based separation of particles and cells. As the cells focus in a single stream near the inner channel wall, with the largest being closest to it, an optimal bifurcation of the outlet

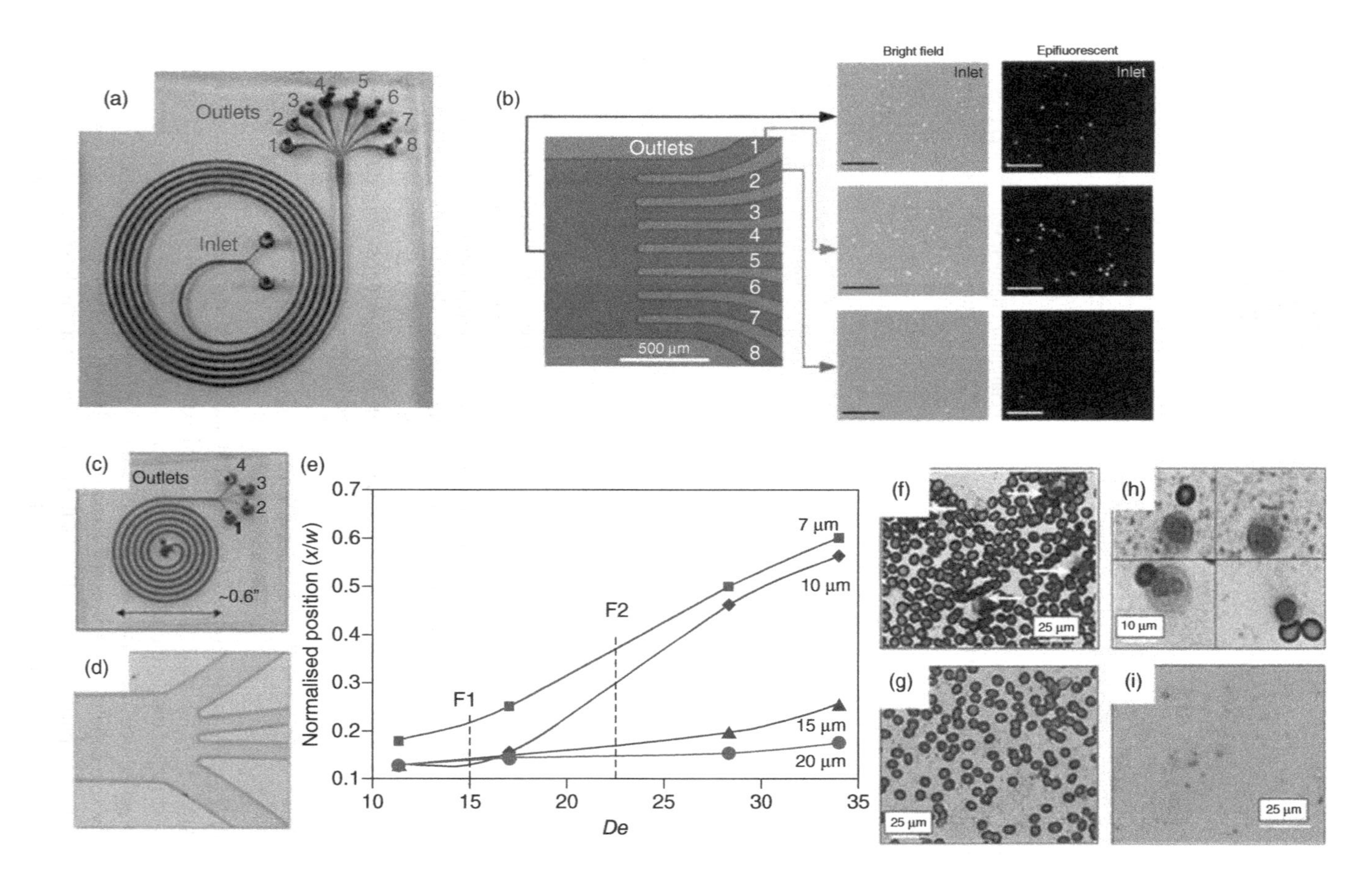
(a)
Outlets
1
2
3
4
5
6
7
8
Inlet
(b)
Outlets
1
2
3
4
5
6
7
8
500 μm
Bright field
Epifluorescent
Inlet
Inlet
(c)
Outlets
4
3
2
1
~0.6"
(d)
(e)
Normalised position (x/w)
0.7
0.6
0.5
0.4
0.3
0.2
0.1
10
15
20
25
30
35
De
F1
F2
7 μm
10 μm
15 μm
20 μm
(f)
25 μm
(g)
25 μm
(h)
10 μm
(i)
25 μm

Figure 5.5 (a) Photograph of the five-loop spiral microchannel with two inlets and eight outlets fabricated in PDMS. (b) Bright-field and epifluorescent images illustrating the distribution of the bigger~15-μm-diameter SH-SY5Y cells (pseudo-colored green) and the smaller ~8-μm-diameter C6 glioma cells at the inlet and the first two outlets of the spiral microchannel (scale bar 100 μm). (c) Image of the spiral sorting device. (d) Bright-field image of the outlet system.(e) Normalized focusing position of particles (x is the distance of the focused stream from the inner channel wall, and w is the width of the channel) as function of De. (f) Bright-field images of the stained blood samples after they were collected from each outlet of the design2 and centrifuged. Inlet has all the cells present. Arrows indicate the white blood cells. (g) outlets 2 and 3 have RBCs and platelets. (h) Outlet1 has majority of WBCs (neutrophils, eosinophils, and monocytes), some platelets, and very little RBCs; and (i) outlet 4 has only diluted plasma and platelets. Kuntaegowdanahalli *et al.* [26]. Reproduced with permission of the Royal Society of Chemistry. Nivedita and Papautsky [43]. Reproduced with permission of the American Institute of Physics.

system aids the collection of the separated cells. Figure 5.4c illustrates the focusing mechanism in a 500 μm × 110 μm Archimedean spiral device with the inner radius of 2 mm. The intensity line scan indicates subsequent focusing of a sample containing 20-μm-diameter polystyrene particles. At a flow rate of 900 μl/min, it has been observed that the particles focus in a tighter stream, almost a single-particle stream by the time they reach loop 4 (loop 1 being the innermost loop and loop 4 being the outermost loop of the spiral). The calculated FWHM at loop 1 was ~340.2 μm, which defines the intensity distribution of the particles almost across the whole width of the channel. As the sample moved downstream, the particles experienced Dean drag along with shear-induced lift force and wall-induced lift force causing them to slowly focus closer to the inner channel wall. At the end of loops 2, 3, and 4, the FWHM in each case was 68.9, 28.7, 24.1 μm, respectively, from the inner wall. The FWHM indicates the width of the focused stream of the particles and by the time the fourth loop ends, the width was small enough to conclude that the particles focused in a single stream, one after the other.

Spiral devices have been used for sorting a variety of cell types with high efficiency and throughput. They have recently been optimized to an area of <1 in.^{2}, which makes them even less expensive due to reduced material and manufacturing costs and small enough to be integrated on lab-on-chip analysis systems. The low aspect ratio of the channel cross section makes the device fabrication easy as compared to high-aspect-ratio channels. We have used spiral devices for sorting SH-SY5Y neuroblastoma and C6 glioma cells [26] (Figure 5.5a). Based on the size of these cells, the mixture was passed through a 500-μm-wide and 120-μm-high spiral microchannel at $De \sim 11.8$ to collect the large (~15 μm diameter) cells from outlet 1 and the small (~8 μm diameter) cells from outlet 2 in an eight-outlet system (Figure 5.5b). The SH-SY5Y neuroblastoma cells were labeled with Cell-Tracker™ Green (Invitrogen Corp.) for better visualization and to confirm separation. The PDMS-based devices were thoroughly flushed with PBS and antibiotics (1 × PSN) before running the cell mixture. This separation was achieved with >80% efficiency and 90% cell viability with a throughput at ~1 million cells/min. The devices were further optimized for size reduction (1/10) and higher efficiency of sorting (>95%), especially for applications such as sorting blood cells [43] and cancer stem cell (CSC) sorting [48].

Since CTCs are present in the blood, separation of cellular blood components (namely RBCs and WBCs) has been an important aspect to investigate. To sort blood cells, polystyrene beads were first used to confirm focusing positions and flow parameters [43]. At the flow rate of 1.8 ml/min, 10-, 15-, and 20-μm-diameter particles (representing WBCs) eluted in the first outlet and the 7.32-μm-diameter particles (representing RBCs) eluted in the second and third outlets (Figure 5.5e). Hence, when 500× diluted blood sample (Ht ~ 0.1%) was run through spiral devices with 500 μm × 110 μm channel cross section at 1.8 ml/min (Figure 5.5f), WBCs focused closer to the inner channel wall owing to their larger size and RBCs focused closer to the center of the channel. The cells collected from the first outlet had 95% of WBCs and 6% of RBCs, with a few platelets (Figure 5.5h). The sample collected from the second and third outlets had 94% RBCs (Figure 5.5g). The fourth outlet contained platelets only (Figure 5.5i). These devices showed separation efficiency of ~95%, with high throughput of $\sim 10^7$ RBCs/ml and $\sim 10^5$ cells/WBCs (which corresponds to ~3.6 μl/min of whole blood). Thus, the device achieves a complete separation of leukocytes and erythrocytes. In the next section, we discuss the use of these devices for cancer cell sorting.

Sorting of HPET cells from other cancer cells has been reported in spiral devices (Figure 5.6a) [49]. HPET cells were successfully sorted and separated at a flow rate of ~2 ml/min (Figure 5.6b). The smaller cells (~15 μm) were eluted in the second outlet, and the smallest (~10 μm) cells were eluted in the third outlet from the inner wall (Figure 5.6c). The larger HPET cells (>20 μm diameter) were eluted in the first outlet closest to the channel inner wall (Figure 5.6d). Because these cells are highly fragile, the flow rate was optimized to prevent them from lysing in the device, leading to ~100% recovery rate at the output. The devices were also used to separate other cell lines, including DU-145 (derived from brain metastasis) and LNCaP (derived from left supraclavicular lymph node), from HPET cells. Recently, Han and coworkers [48] demonstrated the use of spiral microchannels with trapezoidal cross section for label-free isolation of CTCs. They used a spiral channel with one inlet and two outlets in the center of the spiral device with the height of the channel cross section varying from 80 μm at the inner channel wall and 130 μm at the outer channel wall (Figure 5.6e). This trapezoidal geometry enhances the separation resolution allowing higher purity of separation. The sample used to demonstrate CTC isolation was constructed by spiking ~500 tumor cells (MCF-7, T24 and MDA-MB-231) into 7.5 ml of blood. This sample was then brought down to optimal concentration (~×0.5) by RBC lysing and resuspension. The sample was then introduced into the spiral sorting system, following which the samples collected at the two outlets were identified using immunofluorescence using surface markers such as CK+/CD45–. The CTCs focused closer to the inner channel wall due to the balance of the Dean drag and the hydrodynamic force and the WBCs and platelets got trapped in the core of the primary Dean vortices near the outer channel wall. On an average, >80% sorting efficiency

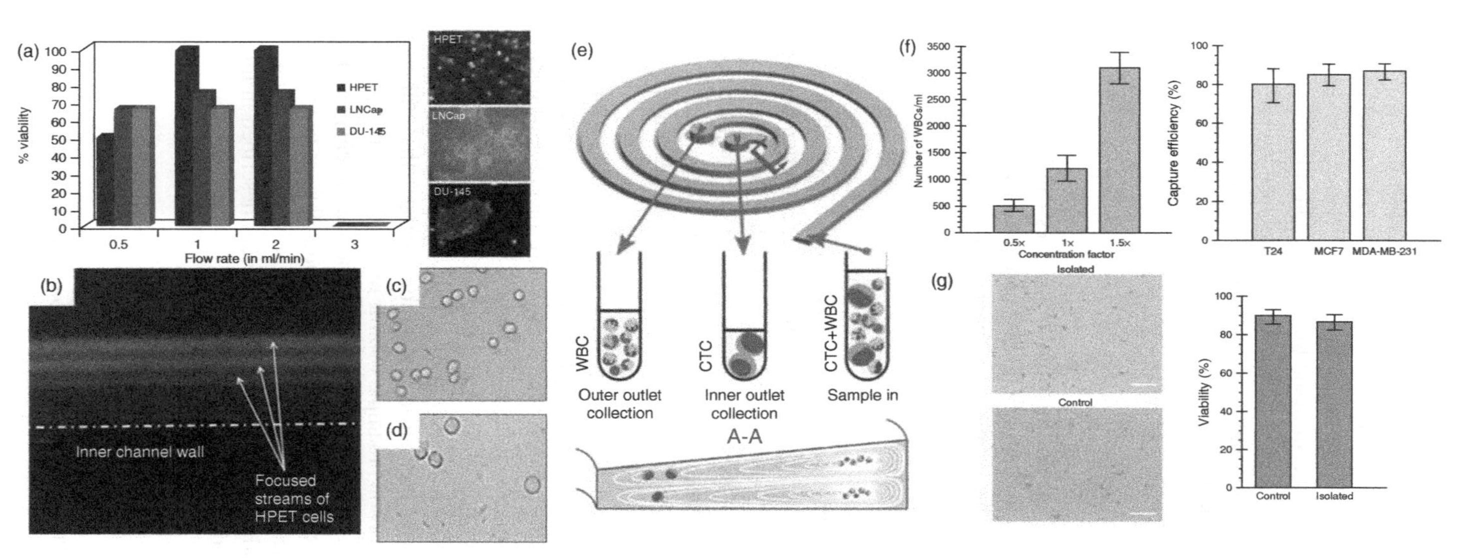

Figure 5.6 (a) Viability plot for each of the HPET, LNCap, and DU-145 cell lines in spiral sorting device (b) Fluorescent image of the focused streams of HPET cells. Phase contrast images of HPET cells at outlets 2 and 3 (c) and outlet1 (d). (e) Schematic of the principle of CTC isolation and enrichment by spiral microchannel with trapezoidal cross section. (f) The effect of the WBC concentration on the performance of spiral microchannel and the final purity and the histogram plot indicating separation efficiency for different cell lines (~>80%). (g) Phase contrast micrographs of control (unsorted) and sorted MDA-MB-231 cells stained using the trypan blue dye indicating high cell viability. Warkiani *et al.* [48]. CC BY-NC 3.0. Published by The Royal Society of Chemistry.

was observed for each of the cell lines (Figure 5.6f and g). The frequency of CTC isolation from each of the samples ranged from 3 to 125 cells/ml.

In this section, we describe the basic operating principle of inertial spiral sorting devices. Spiral devices discussed here have been effectively used for sorting a variety of cell types including CSCs from tissue biopsies, blood cells for point-of-care devices for blood analysis, and neuroblastoma cells. Their planar and low-aspect-ratio design reduces the fabrication time and does not create clogging issues, which are a main concern in high-aspect-ratio channels.

5.6 CONCLUSIONS

In this chapter, we show that inertial microfluidic devices can be used for separation of cells, including rare cells such as CTCs. These devices provide both higher sorting efficiency and higher throughput as compared to other passive microfluidic sorting techniques. For example, dielectrophoresis (DEP) [21, 22] offers high separation efficiency (~92%) but low throughput (on the order of ~1 μL/min). Alternatively, hydrodynamic filtration [28] offers as much as 30× enrichment of cells, at a slightly better throughput of ~20 μL/min (for a 10× diluted whole blood). Another popular technique, pinched flow fractionation [31] also offers high sorting efficiency (>90%) but at even lower throughput of <50 μl/h. The same can be said for another technique, magnetophoresis [50, 51], which has efficiency of 97% but throughput only up to 0.33 μl/min. Ultimately, most of the today's microfluidic separation techniques sacrifice throughput in order to improve separation efficiency. The inertial microfluidic devices address this trade-off and, hence, show promise for separation and isolation of rare circulating cells.

The inertial devices discussed in this chapter can be fabricated not only in PDMS but also in thermoplastics such as poly(methyl methacrylate) and cyclic olefin copolymer. This offers the flexibility of using the wide range of polymer microfabrication methods, such as injection molding, hot embossing, laser ablation, and even roll-to-roll processing. Cascading of these systems for separation of more than two to four distinct cell types should be achievable with further optimization. Sample dilution is another challenge of the inertial separation approach that still needs to be further explored, but the reduction in sample volume required for cell sorting compensates for the dilution requirement. The devices described in the chapter are broadly applicable and could be extended to other cell types and cell sizes as long as the cell ranges are maintained. Due to the passive nature of the separation mechanism, its simple planar structure, and low costs, the inertial devices described in this chapter are easy to use and are disposable.

REFERENCES

1. Toner, M. and Irimia, D. (2005) Blood-on-a-chip. *Annual Review of Biomedical Engineering*, **7**, 77–103.

2. Paterlini-Brechot, P. and Benali, N.L. (2007) Circulating tumor cells (CTC) detection: clinical impact and future directions. *Cancer Letters*, **253**, 180–204.
3. Osborne, G.W. (2011) Recent advances in flow cytometric cell sorting. *Methods in Cell Biology*, **102**, 533–556.
4. Gossett, D.R., Weaver, W.M., Mach, A.J. *et al.* (2010) Label-free cell separation and sorting in microfluidic systems. *Analytical and Bioanalytical Chemistry*, **397**, 3249–3267.
5. Steeg, P.S. (2006) Tumor metastasis: mechanistic insights and clinical challenges. *Nature Medicine*, **12**, 895–904.
6. Den Toonder, J. (2011) Circulating tumor cells: the Grand Challenge. *Lab on a Chip*, **11**, 375–377.
7. Givan, A.L. (2011) Flow cytometry: an introduction. *Methods in Molecular Biology*, **699**, 1–29.
8. Greve, B., Kelsch, R., Spaniol, K. *et al.* (2012) Flow cytometry in cancer stem cell analysis and separation. *Cytometry. Part A*, **81A**, 284–293.
9. Criscitiello, C., Sotiriou, C., and Ignatiadis, M. (2010) Circulating tumor cells and emerging blood biomarkers in breast cancer. *Current Opinions in Oncology*, **22**, 552–558.
10. Adams, J.D. and Soh, H.T. (2009) Perspectives on utilizing unique features of microfluidics technology for particle and cell sorting. *JALA*, **14**, 331–340.
11. Rupp, A., Rupp, C., Keller, S. *et al.* (2011) Loss of EpCAM expression in breast cancer derived serum exosomes: role of proteolytic cleavage. *Gynecologic Oncology*, **122**, 437–446.
12. Lee, W.C., Bhagat, A.A.S., Huang, S. *et al.* (2011) High-throughput cell cycle synchronization using inertial forces in spiral microchannels. *Lab on a Chip*, **11**, 1359–1367.
13. Hur, S.C., Tse, H.T.K., and Di Carlo, D. (2010) Sheathless inertial cell ordering for extreme throughput flow cytometry. *Lab on a Chip*, **10**, 274–280.
14. Choi, S., Song, S., Choi, C., and Park, J. (2009) Microfluidic self-sorting of mammalian cells to achieve cell cycle synchrony by hydrophoresis. *Analytical Chemistry*, **81**, 1964–1968.
15. Nagrath, S., Sequist, L.V., Maheswaran, S. *et al.* (2007) Isolation of rare circulating tumour cells in cancer patients by microchip technology. *Nature*, **450**, 1235–1239.
16. Nilsson, J., Evander, M., Hammarström, B., and Laurell, T. (2009) Review of cell and particle trapping in microfluidic systems. *Analytica Chimica Acta*, **649**, 141–157.
17. Cheng, X., Irimia, D., Dixon, M. *et al.* (2007) A microfluidic device for practical label-free CD4+ T cell counting of HIV-infected subjects. *Lab on a Chip*, **7**, 170–178.
18. Wu, Z., Willing, B., Bjerketorp, J. *et al.* (2009) Soft inertial microfluidics for high throughput separation of bacteria from human blood cells. *Lab on a Chip*, **9**, 1193–1199.
19. Pratt, E.D., Huang, C., Hawkins, B.G. *et al.* (2011) Rare cell capture in microfluidic devices. *Chemical Engineering Science*, **66**, 1508–1522.
20. Stott, S.L., Hsu, C., Tsukrov, D.I. *et al.* (2010) Isolation of circulating tumor cells using a microvortex-generating herringbone-chip. *Proceedings of National Academy of Sciences of United States of America*, **107**, 18392–18397.
21. Gijs, M. A. M. (2004) Magnetic bead handling on-chip: new opportunities for analytical applications. *Microfluid. Nanofluid.*, **1**, 22–40.

22. Borgatti, M., Bianchi, N., Mancini, I. *et al.* (2008) New trends in non-invasive prenatal diagnosis: applications of dielectrophoresis-based Lab-on-a-chip platforms to the identification and manipulation of rare cells. *International Journal of Molecular Medicine*, **21**, 3–12.
23. Agarwal, G. and Livermore, C. (2011) Chip-based size-selective sorting of biological cells using high frequency acoustic excitation. *Lab on a Chip*, **11**, 2204–2211.
24. Jeong, J.S., Lee, J.W., Lee, C.Y. *et al.* (2011) Particle manipulation in a microfluidic channel using acoustic trap. *Biomedical Microdevices*, **13**, 779–788.
25. Di Carlo, D. (2009) Inertial microfluidics. *Lab on a Chip*, **9**, 3038–3046.
26. Kuntaegowdanahalli, S.S., Bhagat, A.A.S., Kumar, G., and Papautsky, I. (2009) Inertial microfluidics for continuous particle separation in spiral microchannels. *Lab on a Chip*, **9**, 2973–2980.
27. Zhou, J. and Papautsky, I. (2013) Fundamentals of inertial focusing in microchannels. *Lab on a Chip*, **13**, 1121–1132.
28. Yamada, M. and Seki, M. (2005) Hydrodynamic filtration for on-chip particle concentration and classification utilizing microfluidics. *Lab on a Chip*, **5**, 1233–1239.
29. Nam, K. and Eddington, D.T. (2010) Size-based separation of microparticles in a multilayered microfluidic device. *Journal of Microelectromechanical Systems*, **19**, 375–383.
30. Takagi, J., Yamada, M., Yasuda, M., and Seki, M. (2005) Continuous particle separation in a microchannel having asymmetrically arranged multiple branches. *Lab on a Chip*, **5**, 778–784.
31. Yamada, M., Nakashima, M., and Seki, M. (2004) Pinched flow fractionation: continuous size separation of particles utilizing a laminar flow profile in a pinched microchannel. *Analytical Chemistry*, **76**, 5465–5471.
32. Hou, H.W., Bhagat, A.A.S., Lin Chong, A.G. *et al.* (2010) Deformability based cell margination – a simple microfluidic design for malaria-infected erythrocyte separation. *Lab on a Chip*, **10**, 2605–2613.
33. Hur, S.C., Mach, A.J., and Di Carlo, D. (2011) High-throughput size-based rare cell enrichment using microscale vortices. *Biomicrofluidics*, **5**, 022206.
34. Zhou, J., Giridhar, P.V., Kasper, S., and Papautsky, I. (2013) Modulation of aspect ratio for complete separation in an inertial microfluidic channel. *Lab on a Chip*, **13**, 1919–1929.
35. Bhagat, A.A.S., Kuntaegowdanahalli, S.S., and Papautsky, I. (2008) Enhanced particle filtration in straight microchannels using shear-modulated inertial migration. *Physics of Fluids*, **20**, 101702.
36. Bhagat, A.A.S., Kuntaegowdanahalli, S.S., and Papautsky, I. (2009) Inertial microfluidics for continuous particle filtration and extraction. *Microfluidics and Nanofluidics*, **7**, 217–226.
37. Segré, G. and Silberberg, A. (1961) Radial particle displacements in poiseuille flow of suspensions. *Nature*, **189**, 209–210.
38. Matas, J., Morris, J.F., and Guazzelli, É. (2004) Inertial migration of rigid spherical particles in Poiseuille flow. *Journal of Fluid Mechanics*, **515**, 171–195.
39. Di Carlo, D., Irimia, D., Tompkins, R.G., and Toner, M. (2007) Continuous inertial focusing, ordering, and separation of particles in microchannels. *Proceedings of the National Academy of Sciences of the United States of America*, **104**, 18892–18897.
40. Amini, H., Lee, W., and Di Carlo, D. (2014) Inertial microfluidic physics. *Lab on a Chip*, **14**, 2739–2761.

41. Bhagat, A.A.S., Kuntaegowdanahalli, S.S., and Papautsky, I. (2008) Continuous particle separation in spiral microchannels using dean flows and differential migration. *Lab on a Chip*, **8**, 1906–1914.
42. Bhagat, A.A.S., Kuntaegowdanahalli, S.S., Kaval, N. *et al.* (2010) Inertial microfluidics for sheath-less high-throughput flow cytometry. *Biomedical Microdevices*, **12**, 187–195.
43. Nivedita, N. and Papautsky, I. (2013) Continuous separation of blood cells in spiral microfluidic devices. *Biomicrofluidics*, **7**, 54101.
44. Mach, A.J. and Di Carlo, D. (2010) Continuous scalable blood filtration device using inertial microfluidics. *Biotechnology and Bioengineering*, **107**, 302–311.
45. Bhagat, A.A.S., Hou, H.W., Li, L.D. *et al.* (2011) Pinched flow coupled shear-modulated inertial microfluidics for high-throughput rare blood cell separation. *Lab on a Chip*, **11**, 1870–1878.
46. Kasper, S. (2008) Exploring the origins of the normal prostate and prostate cancer stem cell. *Stem Cell Reviews*, **4**, 193–201.
47. Kruck, S., Gakis, G., and Stenzl, A. (2012) Circulating and disseminated tumor cells in the management of advanced prostate cancer. *Advances in Urology*, **135281**, 5 pages.
48. Warkiani, M.E., Guan, G., Luan, K.B. *et al.* (2014) Slanted spiral microfluidics for the ultra-fast, label-free isolation of circulating tumor cells. *Lab on a Chip*, **14**, 128–137.
49. Nivedita, N., Giridhar, P.V., Kasper, S., and Papautsky, I. (2011) *Sorting Human Prostate Epithelial (HPET) Cells in an Inertial Microfluidic Device*. 15th International Conference on Miniaturized Systems for Chemistry and Life Sciences 2011, MicroTAS 2011 **2**, 1230–1232.
50. Pamme, N. and Manz, A. (2004) On-chip free-flow magnetophoresis: continuous flow separation of magnetic particles and agglomerates. *Analytical Chemistry*, **76**, 7250–7256.
51. Pamme, N. and Wilhelm, C. (2006) Continuous sorting of magnetic cells via on-chip free-flow magnetophoresis. *Lab on a Chip*, **6**, 974–980.

6

MORPHOLOGICAL CHARACTERISTICS OF CTCs AND THE POTENTIAL FOR DEFORMABILITY-BASED SEPARATION

Simon P. Duffy

Department of Mechanical Engineering, University of British Columbia, Vancouver, British Columbia, Canada

Hongshen Ma

Department of Mechanical Engineering, University of British Columbia, Vancouver, British Columbia, Canada; Vancouver Prostate Centre, Vancouver General Hospital, Vancouver, British Columbia, Canada; Department of Urologic Science, University of British Columbia, Vancouver, British Columbia, Canada

6.1 INTRODUCTION

Circulating tumor cell (CTC) separation based on affinity capture of epithelial cell surface markers has demonstrated the potential to use CTC enumeration to predict disease outcomes in multiple cancers. However, these enrichment strategies suffer from the lack of a universally expressed cell surface marker that can distinguish CTC from leukocytes, the primary contaminant cell type. As a result, significant population of CTCs may be missed due to expression of cell surface biomarkers below the detection threshold. Specifically, events associated with CTC dissemination may result in a downregulation of epithelial biomarkers, causing these strategies to overlook key

Circulating Tumor Cells: Isolation and Analysis, First Edition. Edited by Z. Hugh Fan.

subpopulations of CTCs. Alternatively, CTCs can potentially be distinguished from leukocytes based on their biophysical properties, such as size and deformability. Separation based solely on size has been studied extensively and has shown promising results. However, this approach is hindered by the significant overlap in the size range of CTCs and leukocytes, which could be addressed using deformability-based separation where deformability is a proxy for morphological differences between CTCs and leukocytes. This chapter begins by describing the limitations of affinity-capture-based CTC separation methods and the known morphological differences between CTCs and leukocytes. We then briefly review the mechanism for CTC separation based on biophysical parameters with a focus on filtration methods that enable CTC separation based on cellular deformability, as well as the problems imposed by clogging. Finally, this chapter concludes by introducing two emerging mechanisms for deformability-based separation that could potentially eliminate clogging.

6.2 LIMITATIONS OF ANTIBODY-BASED CTC SEPARATION METHODS

The isolation of CTCs represents a potentially major advance in cancer research and treatment because it enables a less invasive means to sample highly relevant tumor cells shed from the primary tumor without the need for more invasive procedures, such as biopsies and surgeries. Conventional enrichment strategies, such as those used in the CellSearch® system (see Chapter 19 for details), involve capturing CTCs using magnetically labeled antibodies against the epithelial cell surface protein epithelial cell adhesion molecule (EpCAM). The labeled cells can then be enriched by magnetic attraction, while the contaminant hematological cells are removed by washing. The count of CTCs established in this manner using the CellSearch system has been shown to correlate with the survival rate for patients with metastatic breast, prostate, and colorectal cancers [1–4]. However, despite the established relationship between CTC number and clinical outcome, the CTC detection rate for advanced patients in these three cancers are only 25–50% [1, 3–5]. This low detection rate may be largely attributed to the lack of universally expressed cell surface markers on CTCs.

CTC separation schemes based on affinity capture of the protein EpCAM were born from observations that EpCAM is found on epithelial cells that make up structural components of tissues and organs, but not in normal circulating cells in the blood stream [6]. However, the expression of this protein varies significantly between tumor types and between patients. In fact, the development of more sensitive affinity-capture strategies revealed significant CTC populations exist with levels of EpCAM expression below the detection limit of CellSearch (hereto referred to as EpCAM-low CTCs) [7]. Additionally, within an individual tumor, the level of EpCAM expression can be highly variable, permitting EpCAM-low tumor cells to give rise to EpCAM-low CTCs [8, 9]. Therefore, the presence of EpCAM-low CTCs may limit the success of affinity-capture-based enrichment methods.

The expression for EpCAM may also be reduced as a result of events associated with metastasis. Specifically, while carcinoma tumors exhibit many of the features of the progenitor epithelial cell, including apical–basal cell polarity and intercellular adhesion, these tumor cells may undergo an epithelial-to-mesenchymal transition (EMT). EMT is a phenomenon, observed during metastasis, whereby epithelial cells undergo changes associated with loss of cell polarity and increased migratory behavior, and it is thought that EMT contributes to the dissemination of CTCs. One of the results of this process is a phenotypic switch within the tumor cell where the expression for epithelial cell markers, such as EpCAM, is reduced [10–14]. EpCAM is also known to be particularly susceptible to downregulation, as diminished EpCAM expression is associated with an EMT-like state [15]. Not surprisingly, reduced EpCAM expression in patient tumor cells has been found to correlate with poor survival [16, 17]. A recent study further confirms that EpCAM expression is dynamic in disseminated tumor cells and its expression correlates to distinct proliferative stages [8]. Therefore, enrichment strategies that rely on epithelial cell surface markers, such as EpCAM, may fail to capture the subpopulations of CTC that is most likely to disseminate into the bloodstream.

As a result of the considerable challenges associated with using affinity capture of EpCAM to select CTCs, there has been significant interest in developing other methods to discriminate CTCs from hematological cells. Specifically, we consider the potential to distinguish CTCs from leukocytes based on their morphological characteristics, which enables using biomechanical methods to separate CTCs from leukocytes.

6.3 MORPHOLOGICAL AND BIOPHYSICAL DIFFERENCES BETWEEN CTCs AND HEMATOLOGICAL CELLS

Normal blood cells of healthy individuals consist of erythrocytes, leukocytes, and thrombocytes, or equivalently, red blood cells, white blood cells, and platelets. Red blood cells are discoid shaped, 8 μm in diameter, and 2–3 μm thick. They are devoid of a nucleus and are therefore highly deformable to permit them to transit freely through the microvasculature of the body. Platelets also do not have a nucleus and are even smaller blood cells (1–4 μm). White blood cells have a nucleus and range from 7 to 20 μm in diameter. These cells are physically most similar to epithelial cells, and therefore, they are the primary type of contaminant cells in CTC separation and identification processes.

White blood cells are generally spherical in shape when circulating within the bloodstream, without any obvious cell polarity but may adopt a more polar morphology when activated. In contrast, epithelial cells conform primarily to squamous (square), cuboidal (cube), or columnar (elongated) shapes. These cells typically line the surface of the body and its cavities with the apical surface of the cell contacting the outside or cavity lumen while the basal surface contacts the underlying connective tissue.

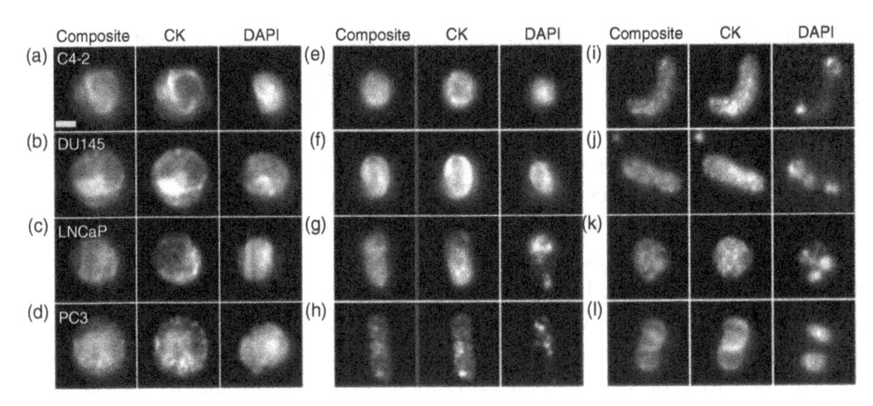

Figure 6.1 Difference in size, shape, and nuclear volume of cultured cancer cells and CTCs. Cultured cancer cells (a–d) and CTCs from patients with prostate cancer (e–l) were stained for cytokeratin (green) and with DAPI nuclear stain (pink). The cells are imaged at the same scale, indicated by the yellow line that represents a length of 5 μm. (*See color plate section for the color representation of this figure.*)

Tumor cells derived from the epithelium share many of the characteristics of their progenitor epithelial cells but increasingly deviate from this morphological plan. In fact, CTCs display a range of cell sizes and shapes that distinguish them from the parental epithelial cells (Figure 6.1). The process of dissemination to the circulatory system imposes significant selective pressures that likely cause CTCs to be morphologically distinct from their parental cells as not all disseminated cells are likely to have the fitness to survive the challenging environment of the circulatory system. CTCs are often assumed to be larger than white blood cells, of which neutrophils and lymphocytes are the most common. These cells have a mean diameter of ~8.5 and 6–7 μm, respectively [18, 19]. The assumption that CTCs are larger than white blood cells largely derives from the observed diameter of cultured tumor cells (e.g., HeLa, LNCaP, MCF-7), which are 13.4 μm in diameter on average and can be as large as 15–20 μm [19–22] (Figure 6.2). Cultured cells may be appropriate to model CTCs from non-small-cell lung cancer or breast cancer that have a larger mean diameter of ~16 or 13 μm, respectively [20, 23]. However, CTCs of prostate and colon cancer are significantly smaller than cultured tumor cells, with estimates of mean cell size ranging 8–11 μm [20, 24]. Variation in cell size of CTCs combined with variability in cell size among white blood cells [18, 24] contributes to poor resolution of these cells on the basis of size alone. For this reason, biomechanical separation strategies that enrich CTCs based on size alone, such as hydrodynamic chromatography [25], inertial flow [23, 26, 27], dielectrophoresis (DEP) [28], and filtration of fixed cell samples [29], are likely to underperform when enriching for CTCs from patient blood, compared with model samples created by doping cultured cancer cells into the blood from healthy donors. This issue could be partly mitigated by using cultured cancer cells such as L1210 mouse lymphoma cells that have a smaller mean diameter (10 μm) and could be a closer representation of the size of patient-derived CTCs [30–32].

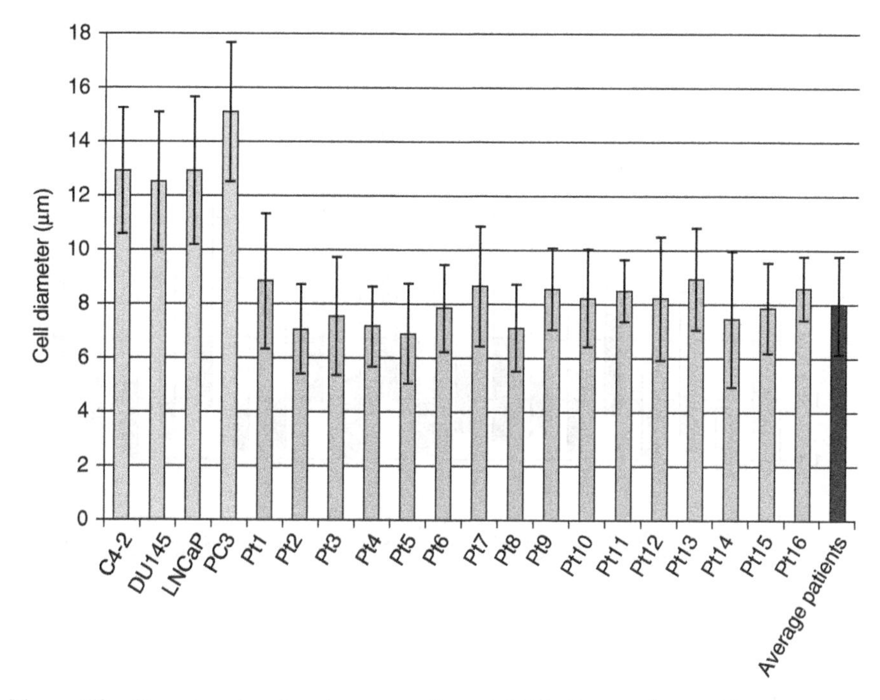

Figure 6.2 Diameter of cultured cancer cells and CTCs from patients with prostate cancer. CTCs are significantly smaller than cultured cells ($p < 0.001$), with an average diameter of 7.97 μm. This is in contrast to the cultured tumor cells that have an average diameter of 10.38 μm.

While adherent tumor cells conform to a polarized columnar morphology, CTCs become more spherical or round following detachment from the primary tumor [33]. In this way, they may be morphologically similar to white blood cells. However, it is important to note that the majority of white blood cells have a strictly spherical morphology, and it may be possible to distinguish CTCs based on their characteristic elongated shape (Figure 6.3a). This eccentricity of cell shape may be a consequence of apoptotic events associated with CTC dissemination [34, 35]. In fact, by quantification of cell elongation, cultured cancer cells were found to have a generally spherical shape while CTCs were significantly more elongated [24]. Interestingly, cytomorphological abnormality of CTCs has been correlated to poor clinical outcome in metastatic breast, colorectal, and prostate cancers [36].

Another distinguishing cytomorphological characteristic is the nuclear-to-cytoplasm (N:C) ratio. Cancer cells often develop an enlarged nucleus as a consequence of chromosomal abnormality. The N:C ratio of CTCs may vary significantly between individual CTCs. This ratio is generally even higher in CTCs than in cultured cancer cells (Figure 6.3b), which would be anticipated to have a greater N:C ratio than leukocytes [33, 37]. A potential biomechanical implication of greater N:C ratio is the increased cell rigidity and an associated decreased deformability.

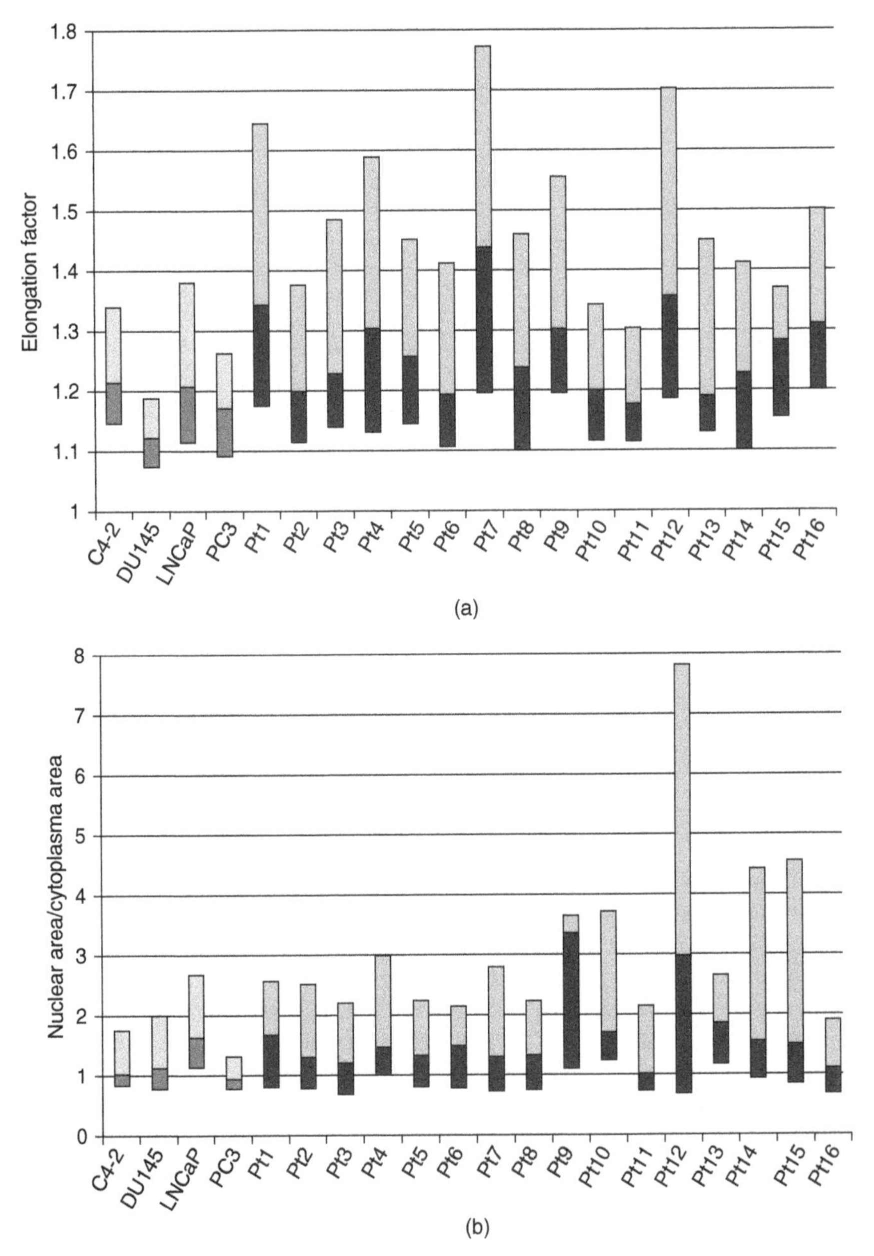

Figure 6.3 Cytological characteristics of CTCs. (a) An elongation factor was determined by bisecting the cell with perpendicular major and minor axes. The elongation factor represents the ratio of major and minor axes. CTCs exhibit an attenuated elongation, even compared to cultured tumor cells. (b) The nuclear-to-cytoplasmic (N:C) exhibited a high degree of intra- and interpatient variation but the median N:C ratio of CTCs was greater than that of cultured cancer cells.

A previous micropipette aspiration study performed by Guliak and colleagues determined that the nuclei of articular chondrocytes were twice as viscous as the cytoplasm and 3–4 times stiffer [38]. A more quantitative study of neutrophils indicated that the nucleoplasm is 10 times more viscous than the cytoplasm [39]. Similarly, Maniotis *et al.* [40] observed that the nucleus of endothelial cells behaved as if it were nine times stiffer than the cytoplasm in response to micropipette aspiration. The combination of greater N:C ratio and greater rigidity of the nucleus relative to cytoplasm suggests that CTCs are significantly more rigid than leukocytes, and the cell deformability represents an important biomechanical characteristic by which to distinguish these cell phenotypes.

6.4 HISTORICAL AND RECENT METHODS IN CTC SEPARATION BASED ON BIOPHYSICAL PROPERTIES

Seal [41] was the first person to leverage the biophysical differences between CTCs and hematological cells to enrich CTCs using differential centrifugation in the late 1950s. While similar approaches have been used in recent studies [42], the general consensus of the field is that differential centrifugation is insufficiently selective for enriching CTCs because of the exceeding rarity of these cells. When CTCs are present in the blood, the projected ratio of CTCs to leukocytes is typically $1{:}10^5$ to $1{:}10^6$. Therefore, an enrichment factor of at least 10^3 to 10^4 is required in order to reasonably deplete the leukocyte count sufficiently to individually identify and analyze CTCs from the remaining leukocytes.

In 1964, Seal proposed leveraging the differences in cell size to enrich CTCs from hematological cells using a simple filter. An initial demonstration of this principle was performed using perforated plastic tape as a micropore filter. The results showed the capture of apparently malignant cells in the blood of metastatic cancer patients [43]. The fabrication of micropore filters was later improved by using energetic neutrons to generate precise 8 μm diameter pores. Using these microstructures, Vona *et al.* [44] demonstrated that the microfiltration strategy can detect a single tumor cell spiked into 1 ml of blood. This approach was later developed into multiple commercial products, including ScreenCell®, RareCell®, ClearCell®, all of which demonstrated the ability to isolate tumor cells doped into the blood, as well as CTCs from the patient's blood [45–47].

Recent advances in microfluidics improved the capabilities to fabricate structures at the scale of single cells and precisely control fluid flow. Leveraging these capabilities, a number of microfluidic strategies have been developed to separate CTCs from blood cells. These methods are broadly classified into three groups: dielectrophoresis, hydrodynamic chromatography, and mechanical filtration. DEP (see Chapter 18 for details) discriminates CTCs from hematological cells based on cell size, nuclear morphology, and membrane morphology, which collectively contribute to the dielectric properties of the cell [19, 48–51]. DEP achieves a high capture efficiency (>90%) for cultured tumor cells but is currently limited in both enrichment (<100 over white blood cells) and throughput (<1 ml/h) [50, 52, 53].

Hydrodynamic chromatography involves flowing cells past obstacles that cause cells of different sizes to diffract along different flow paths. This strategy allows for very sensitive separation based on size and can be operated at very high flow rates; however, enrichment is often poor (<100 over leukocytes in many cases) [27, 54–58]. Traditional microfluidic strategies often operate in the low Reynolds number ($Re \ll 1$) regime where inertial effects can be ignored. Recent studies have shown that when the Reynolds number is on the order of 1, where both inertia and viscosity are finite, forces arising from inertial effects, including drag forces from Dean flows, shear gradient lift forces, and wall-effect lift forces, can be balanced against each other to enable size-based separation [23, 26, 58–60]. Hydrodynamic CTC enrichment techniques based on these inertial effects have been recently reported with an impressive capacity to capture cultured tumor cells. These techniques are discussed in greater detail elsewhere (e.g., Chapter 5).

Mechanical filtration is the process of deforming a cell sample through microstructures to capture cells based on a combination of size and deformability. CTC enrichment techniques developed using microfluidic devices include weir [31, 61], pillar [30, 62], and pore [63, 64] types of microstructures. Similar to conventional micropore filtration, microfluidic-filtration-based strategies have been demonstrated with the ability to effectively capture patient-derived CTCs [29, 47, 65–68] (see Chapter 7 for details).

A prerequisite for many filtration methods is the pretreatment of patients' blood samples using a cell fixative. Fixing the sample is necessary to allow the cells to stay intact throughout the filtration process, which may damage or lyse the cells and clog the filter with cellular debris. Fixing the cells presents several limitations: First, fixing the cells rigidifies the cell membrane and destroys any differences in the internal composition of the cell. Therefore, subsequent separation processes can only distinguish CTCs based on size. As discussed earlier, there may be significant overlap between the size of CTCs and leukocytes [20, 36] and that deformability-based separation may be a more selective process. Second, fixing the cells increases the adsorption of cells in the filter matrix, which greatly increases the potential for background cells to adsorb and contaminate the filter microstructure. In fact, this property will cause red blood cells to clog the filter, even though they are significantly smaller than CTCs. As a result, red blood cells are typically lysed prior to processing [44, 45]. Red-blood-cell lysis protocols operate on the premise that red blood cells are more fragile than nucleated cells. However, CTCs are also likely to be fragile cells, as they undergo apoptosis upon dissemination from the primary tumor [34, 35]. This suggests that pretreatment to lyse red blood cells may also lyse or damage a significant fraction of the CTCs. Finally, fixing the cells renders them nonviable, severely limiting the potential downstream characterization that can be applied to these cells, such as propagation by cell tissue culture and in xenograft models, which can provide a tremendous amount of information on the metastatic potential of these cells. Specifically, following engraftment of tumors onto mice, it has been demonstrated that EpCAM immunoaffinity capture techniques can enrich these cells and the captured cells can be expanded in culture for 12–15 days [69]. This is a valuable model, but the CTCs with the greatest metastatic potential may not express EpCAM as a consequence of EMT [15]. Conversely, depletion of hematological cells from patient-derived blood by RosetteSep® CTC enrichment has been shown to isolate CTCs that can generate tumors when transplanted into mice [70].

Filtration strategies have also been employed to enrich unfixed CTCs, albeit at a lower capture efficiency, and human CTCs have also been cultured for 15 days, within the filter matrix [71]. The capture of viable CTCs presents the opportunity to perpetuate the tumors by grafting them into mice. However, a limitation of this culture system is that individual relevant CTCs, such as those displaying metastatic characteristics, are not easily extracted from the filter matrix. Filtration-based CTC enrichment strategies have not yet isolated enough cells to transplant from tumors to mice. Therefore, there is a need for a filter-based strategy to enrich a sufficient number of viable CTCs that cell proliferation can be achieved in culture and by transplanting into mice. This filtration system would ideally efficiently capture CTCs from whole blood, without the need for red blood cell lysis and cell fixation, and the enriched CTCs would be easily extracted from the enrichment device.

6.5 MICROFLUIDIC RATCHET FOR DEFORMABILITY-BASED SEPARATION OF CTCs

6.5.1 Microfluidic Ratchet Mechanism

The primary limitation to conventional CTC filtration strategies is the need to fix the cell sample before processing, which renders the cells incapable of deforming through the filter matrix and results in separation of the cells on the basis of size alone. Cell fixation also contributes to clogging of the sorting matrix as well as poor selectivity for CTCs because of the significant overlap in cell size between CTCs and white blood cells [20, 36]. To address these issues, our research group has developed a technique for deformability-based separation of CTCs, known as the microfluidic ratchet mechanism. The key advantage of this mechanism is its ability to avoid clogging to enable highly selective deformability-based CTC separation directly from whole blood.

The microfluidic ratchet mechanism is based on the deformation of single cells through tapered constrictions. When the opening of the constriction is smaller than the diameter of the cell, the deformation force exhibits a directional asymmetry where pushing the cell through the constriction along the direction of the taper requires less force than against the direction of the taper (Figure 6.4a) [30]. Therefore, an oscillatory flow of an appropriate magnitude would selectively transport cells unidirectionally (Figure 6.4b) [29, 30, 72]. Cell separation based on deformability is achieved when one cell type is transported unidirectionally, while another cell type is completely blocked by the constrictions. For the latter cells, the oscillatory flow provides a constant source of mechanical agitation that prevents cells from adhering to the filter microstructures to prevent clogging.

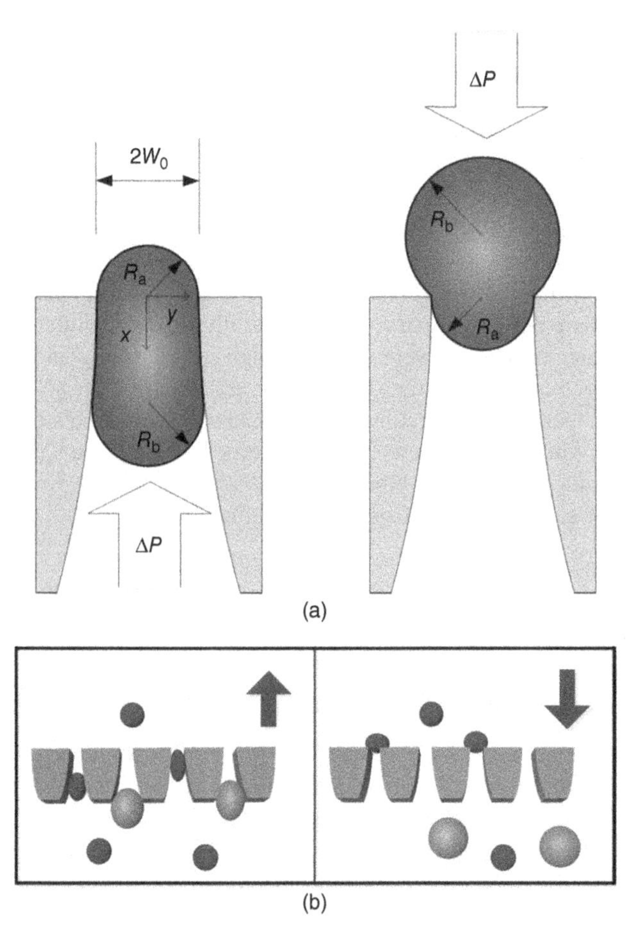

Figure 6.4 Microfluidic ratchet mechanism. (a) When cells are deformed through a funnel-shaped constriction, the applied pressure required for their transit is dependent on the difference between the radius of the leading edge (R_a) and the trailing edge (R_b) of the cell. Therefore, since R_b is constrained by the funnel constriction when transiting in the forward direction, less pressure is required for deforming the cell in the forward direction than in the reverse. (b) By oscillating the applied pressure back and forth, large cells fail to transit the channel but to not clog the filter pore. Conversely, despite the oscillation of pressure, cells that deform through the constriction do not transit backward through the funnel pore.

6.5.2 Design of the Microfluidic Ratchet

The microfluidic device for CTC enrichment based on the microfluidic ratchet mechanism consists of an inlet, where cells are introduced; a central separation area, where cells are sorted based on size and deformability; and two outlets – a "leukocyte" outlet that collects discarded white blood cells and a collection outlet that collects the enriched cells (Figure 6.5a). Oscillation microchannels also lead into the separation area and generate the oscillatory flow.

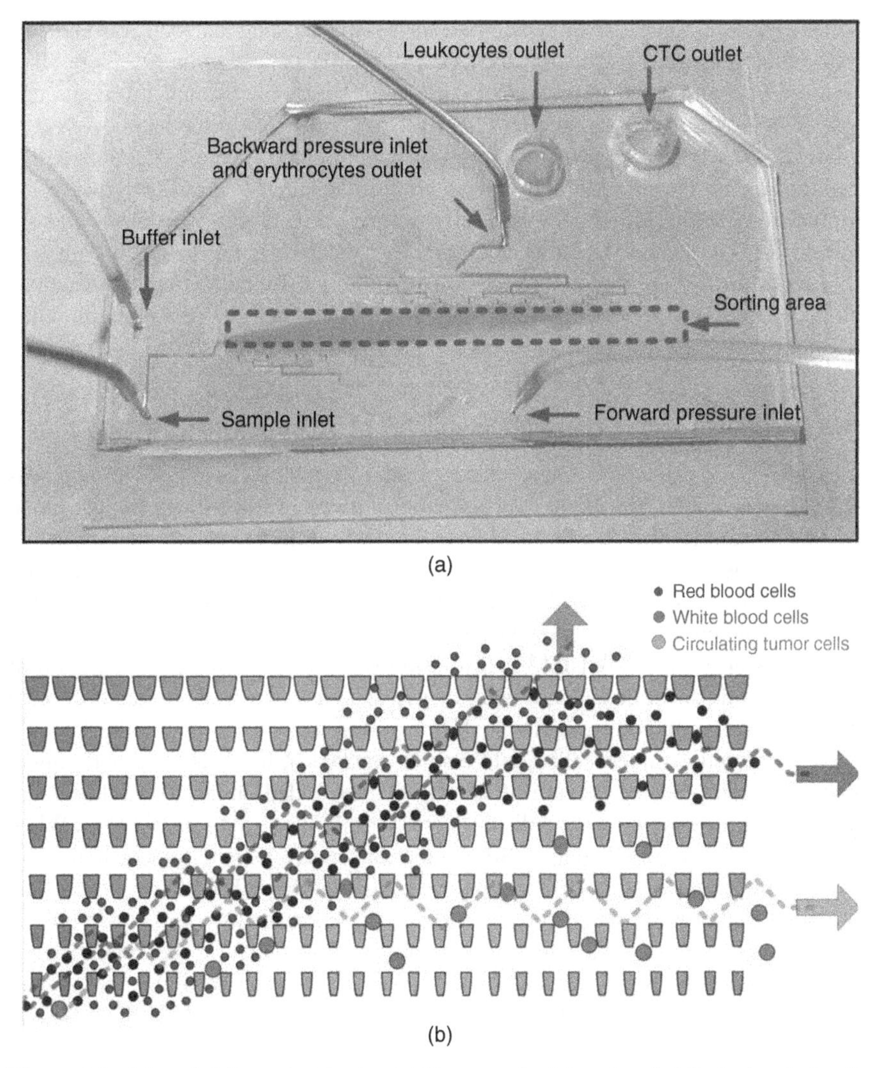

Figure 6.5 Microfluidic ratchet device. (a) Photograph of the microfluidic device during the processing of whole blood. Blood is directed to the sample inlet and migrates to the sorting area, where CTCs are separated from white blood cells and each cell time is directed to a different outlet. (b) The sorting is achieved by aligning funnel-shaped micropores into rows, where each row has an incrementally smaller pore size. Thus, the different cell types are vertically displaced and CTCs can be captured from the rows with pore sizes less than 6 μm.

The separation area consists of a two-dimensional array of microscale funnel constrictions that are arranged in 32 rows. Each row consists of 2048 individual constrictions, and the constriction pore size for each row is incrementally smaller from the bottom to the top of the sorting area. Cells are introduced into the bottom of the sorting area, where they experience oscillatory flow as well as a leftward applied pressure, such that they migrate in a stepwise diagonal path through the sorting matrix and are vertically displaced, based on their size and deformability (Figure 6.5b). The cells transit funnel pores, ranging in size from 18 to 2 μm, and those cells that transit pores smaller than 6 μm are redirected to the leukocyte outlet while cells that fail to transit these pores are redirected to the collection outlet. Erythrocytes are the smallest and the most deformable cells, so they migrate to the terminal row. White blood cells generally deform through the 6 μm pores and are directed into the leukocyte outlet. Conversely, cancer cells fail to reach or transit the 6 μm pores and are directed to the collection outlet. By collecting cell fractions below the threshold row, a cell suspension can be obtained that is highly enriched for cancer cells.

6.5.3 Validation of the Microfluidic Ratchet Mechanism

UM-UC13 (UC13) bladder cancer cells represent a good model to test these devices for CTC enrichment. Similar to CTCs, their size range overlaps that of white blood cells, while their rigidity is 5–10-fold greater [73]. In order to visually identify white blood cells, whole blood from healthy donors was stained fluorescent blue with Hoechst 33342 dye. Conversely, cultured UC13 cells were independently stained fluorescent green with Calcein-AM Green dye. By mixing these two stained cell populations at a ratio of 1 UC13 cell to 1000 white blood cells, the separation of leukocytes and cancer cells could be clearly observed as the cells migrated through the sorting matrix and collected in the device outlets.

The differently stained cells could also be tracked in the leukocyte and collection outlets. The specific capture of cancer cells depended on the forward pressure employed when oscillating the cells back and forth in the sorting matrix [66]. When a higher pressure was applied, only 80% of the cancer cells were redirected to the collection outlet (Figure 6.6a). However, by reducing the forward pressure, greater than 90% of the cells could be captured. In contrast, at a lower forward pressure, a greater number of white blood cells contaminated the collection outlet (Figure 6.6b). Therefore, the system can be operated at a low forward pressure to maximize the number of CTCs captured or at a high forward pressure to enhance the purity of the CTC isolated, at the expense of the capture efficiency. Notably, however, even at the lowest forward pressure, this system achieved nearly 10^4-fold enrichment of cancer cells. This is a major improvement over other microfluidic strategies and is comparable, in selectivity, with immunomagnetic enrichment strategies, such as CellSearch.

The rate at which blood flows through the microfluidic ratchet is an important consideration. A higher flow rate may allow more rapid sample processing but may also result in a fouling of the sorting mechanism as well as physical damage to the cells within the sorting matrix. With this in mind, a flow rate of ~1 ml of whole blood per hour has been deemed optimal for the enrichment of living cancer cells. However, two important advantages of this device are its small size (each device is approximately the size of a credit card) and that the device can be operated perpetually, without user intervention. Therefore, the devices can be operated in parallel to dramatically increase the rate of sample processing. To date, up to four devices have

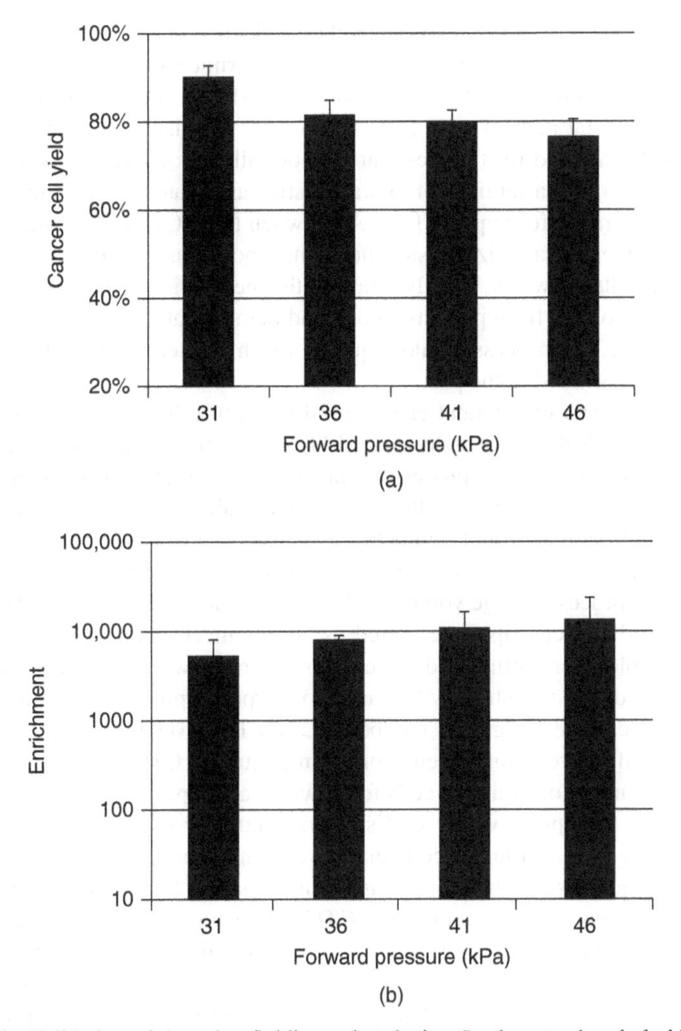

Figure 6.6 Validation of the microfluidic ratchet device. Leukocytes in whole blood were labeled by Hoechst dye and UM-UC13 cancer cells by Calcein-AM Green. When the two cell populations were mixed and sorted by microfluidic ratchet, (a) tumor cell capture efficiency and (b) tumor sample enrichment were assessed over a range of applied forward pressure.

been operated simultaneously in parallel without any reduction in the effectiveness of cancer cell enrichment.

6.5.4 Viability of Cells Enriched by the Microfluidic Ratchet Mechanism

While the CellSearch system has been credited with the ability to guide therapeutic decisions based on CTC enumeration, it has still not been concretely established that CTCs have the capacity to initiate metastasis, the capacity to act as a seed to spread a tumor from one part of the body to another. Evidence that CTCs may act as seed for cancer metastasis has come from xenograft experiments in mice. In these experiments, cultured human breast cancer cells (MDA-MB-231) were implanted into the breast mammary fat pads of mice [74]. CTCs were then isolated from the blood of the mice and examined for features that are normally associated with tumor metastasis. The CTCs had a number of characteristics associated with metastasis, such as a greater tolerance for hypoxia [74]. Also, when these CTCs were implanted into healthy mice, they metastasized faster and formed more massive tumors than the original cultured cells. However, to truly establish the metastatic potential of CTCs, these cells must be isolated from patients' blood and demonstrate the ability to form new tumors in mice. This process would require a much greater number of living CTCs than have been isolated in the past.

Immunomagnetic enrichment can be used to capture living CTCs, but this process may miss cells that have undergone EMT, an important step in tumor metastasis. Microfiltration strategies can process a relatively large volume of blood and capture CTCs, but the fixation of these cells reduces their viability. Also, the CTCs become trapped in the filter matrix and cannot be easily removed from the device. In contrast to these conventional strategies, the microfluidic ratchet mechanism can be operated perpetually to process a large volume of blood and selectively capture CTCs. When UC13 cultured cancer cells were sorted using this mechanism, >98% of the cells remained viable after sorting and the cells proliferated with no observable change in growth rate or cell morphology (Figure 6.7). Perpetual processing of patient blood may allow for capture of a greater number of CTCs. However, because the cells maintain their proliferative ability, even short-term culture of CTCs could exponentially increase the number of available cells for downstream experimentation.

The ability to capture viable CTCs offers a number of additional advantages. Because CTCs represent tumor cells that have been released from a primary tumor, they can be used to assess the characteristics of the patient tumor without the need for a tissue biopsy. Invasion assay of living CTCs can examine how effectively the cell can migrate from the blood stream and penetrate into the tissues, thereby providing an indication of the aggressiveness of the tumor [74]. Some tumors express biomarkers, such as the secretion of prostate serum antigen (PSA) by prostate tumor cells. By suspension of CTCs in a culture medium and monitoring the secretion of PSA, this tumor marker can be sensitively assessed [75]. Together, the analysis of living cells can reveal a great deal about the primary tumor and can be used as a surrogate tumor marker to rapidly evaluate the efficacy of therapeutic interventions [1, 76] or to evaluate the efficacy of new drugs in clinical trials [77].

6.6 RESETTABLE CELL TRAP FOR DEFORMABILITY-BASED SEPARATION OF CTCs

6.6.1 Resettable Cell Trap Mechanism

The microfluidic ratchet mechanism performs deformability-based CTC separation by selective depletion of leukocytes. A converse approach to achieve the same goal

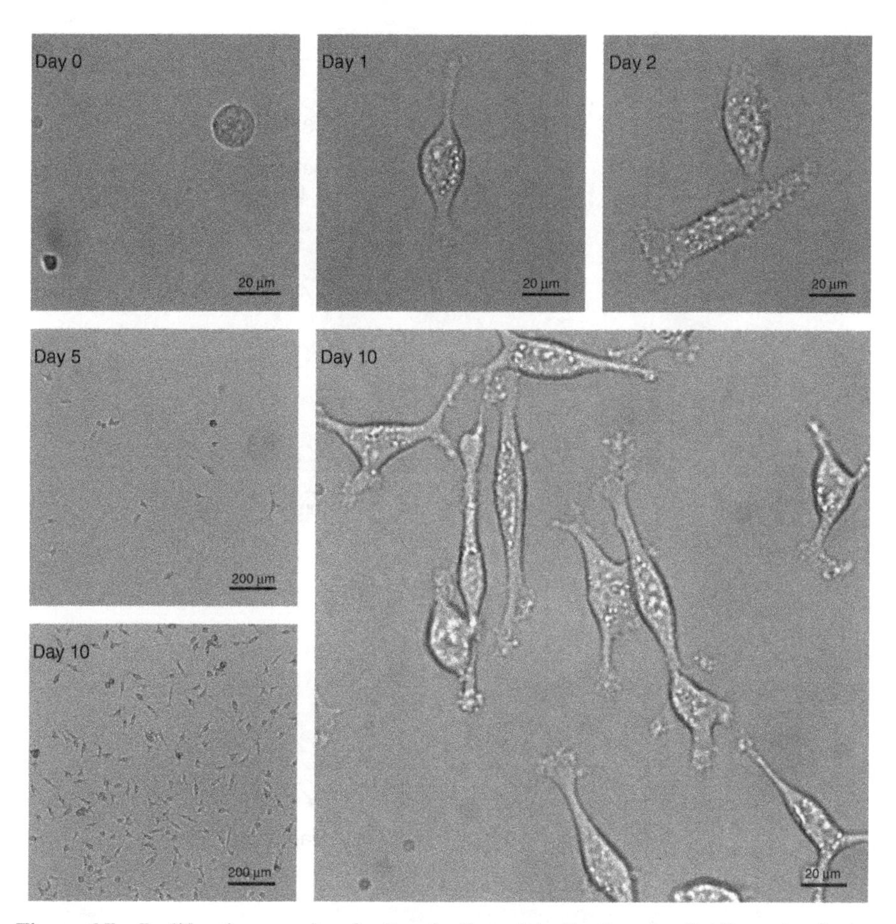

Figure 6.7 Proliferative capacity of cultured cells enriched by the microfluidic ratchet device. Cancer cells that are enriched by the microfluidic ratchet device appear to demonstrate the same ability to proliferate, over 10 days, as normal cultured cells. While they are initially circularized due to treatment with trypsin, prior to sorting, the cancer cells re-establish their normal elongated shape as they grow in culture.

is to selectively capture CTCs. The resettable cell trap mechanism is a weir filter microstructure that performs this function and uses a flexible microstructure to periodically remove trapped cells to prevent clogging [31]. Specifically, blood is infused into individual cell traps, arranged in series, which specifically adsorb cancer cells while directing hematological cells to the waste reservoir (Figure 6.8a). The trap can then be opened to release the enriched cancer cells into the collection reservoir.

The fabrication of such devices is made possible by advances in multilayer soft lithography that allow the mechanism to be generated in two layers. The top layer has a textured surface, which defines the cell traps. The second layer is a control layer that forms a flexible diaphragm with the first layer that is capable of opening and

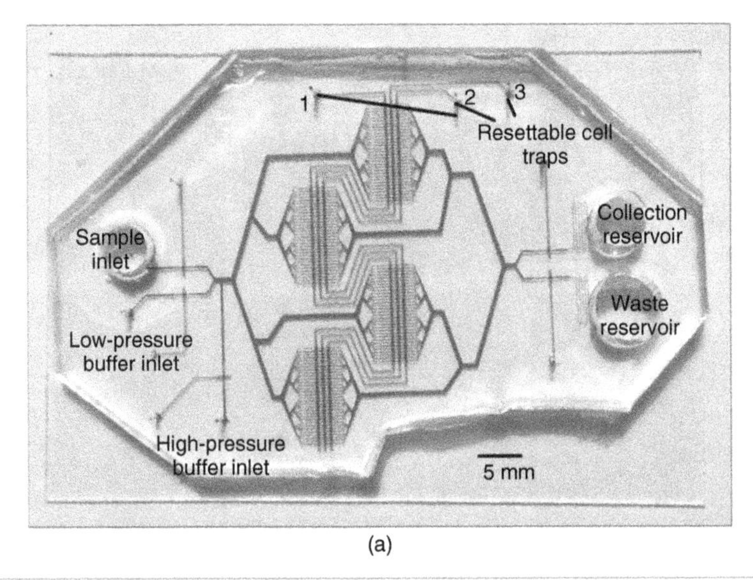

(a)

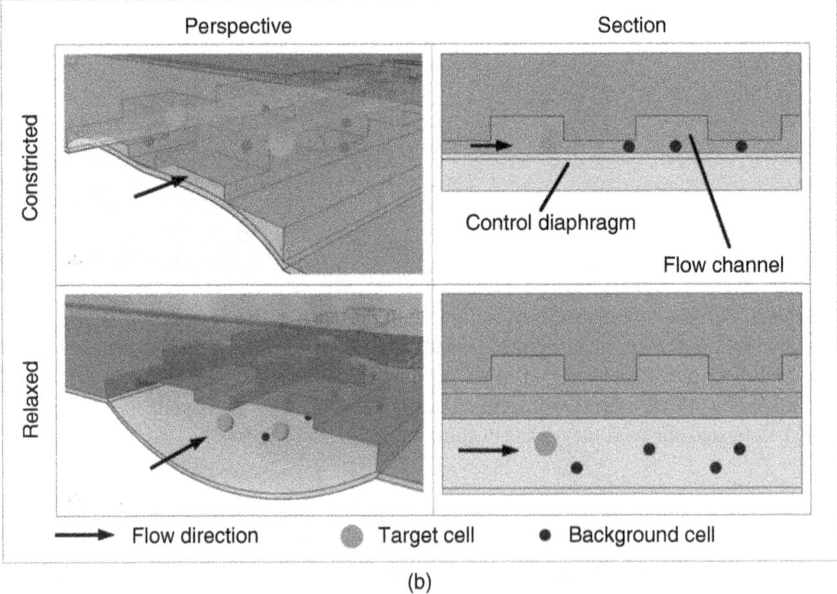

(b)

Figure 6.8 Resettable cell trap microfluidic device. (a) Whole blood is applied to a sample inlet and is directed into an array of parallel cell traps for sorting. White blood cells are flushed from the traps into the waste reservoir while CTCs are released from the trap into the collection reservoir. (b) The cell trap operates by the inflation of a flexible diaphragm that can constrict to specifically capture CTCs while permitting transit of white blood cells (top panel). The trap can then be relaxed to release all of the remaining adsorbed cells (bottom panel). (*See color plate section for the color representation of this figure.*)

closing the cell traps. This device has two states of operation: an open state where all cells can transit the microchannel unobstructed and a semiclosed state where smaller and more deformable hematological cells are permitted to transit, while the more rigid cancer cells are retained (Figure 6.8b). The flexible membrane of the control layer can alter the height of the channel relative to the textured surface, and a central fin functions as a mechanical stop to ensure that the channel does not close entirely. By alternating between open and semiclosed states, as well as applying relatively high fluid pressure during the semiclosed state, the device can process a sample to selectively retain larger and more rigid cancer cells.

Cell separation is performed in this device in three phases: In the *filtration phase*, the cell traps are closed sufficiently to retain cancer cells while permitting transit of leukocytes. During this filtration phase, the collection outlet is closed while the waste outlet is open. Thus, the leukocytes are collected in the waste chamber. In the *purging phase*, the sample inlet is closed and the device is flushed with buffer under low pressure to release any entrapped leukocytes in the cell trap. Finally, during the *collection phase*, the device is flushed with buffer under high pressure with the cell trap is opened, and only the outlet to the collection chamber is open. During this phase, the entrapped cancer cells are flushed into the outlet chamber.

As other filtration mechanisms, the resettable cell trap is a binary filter that captures cells based on a combination of size and deformability. However, the resettable cell trap has two major advantages over conventional filtration strategies. First, the filtration pore size can be dynamically tuned for optimal capture of cancer cells, based on the specific biomechanical properties of cell population. This capability is achieved by varying the degree of deflection for the control layer membrane to achieve a semiclosed state with a specific height, relative to the textured surface. Second, in contrast to conventional filtration systems, the resettable cell trap is highly resistant to clogging. During the purge phase, low pressure is applied to the trap while it is in a closed state, which removes the leukocytes that are likely to clog the device. Furthermore, by alternating between the filtration and the purge phases, contaminating leukocytes can be more stringently removed from the microchannel. Also, by opening the trap during the collection phase, any remaining contaminating cells are purged from the microchannel prior to the next enrichment cycle.

6.6.2 Validation of the Resettable Cell Trap Device

UM-UC13 (UC13) bladder cancer cells are used as a model for CTCs, as described in Section 6.5.3, because they overlap in size with white blood cells and the primary distinguishing characteristic is their greater cell rigidity. White blood cells are stained with fluorescent Hoechst 33342 dye (blue) in whole blood, and the UC13 cells are stained with Calcein-AM Green fluorescent, so that the two cell types can be visually discriminated. When the UC13 is mixed with whole blood at a ratio of 1 UC13 to 1000 leukocytes, the efficiency of U13 capture as well as the degree of enrichment for these cells can be determined by microscopic analysis of the waste and collection outlets.

Because the cell trap mechanism is dynamically tunable, the conditions can be optimized to vary the selectivity for target cell capture. Using UC13 cells, suspended in phosphate-buffered saline (PBS), the conditions were determined for consistent capture of 95% of the UC13 cells. Similarly, when UC13 cells were doped into whole blood, diluted 1:1 with PBS, the system consistently captured ~94% of the target cells with a modest enrichment of 183-fold [73]. However, in addition to the fact that this system can be optimized for maximum target cell capture, it was observed that pipetting the sample isolated from the collection outlet back into the device inlet, such that the sample was enriched three times, resulted in an enrichment factor of >1800-fold and >90% capture efficiency for UC13 cells (Figure 6.9). This observation resulted in a re-engineering of the system such that three cell traps were oriented in series within a single microchannel. This variation of the resettable cell trap device is capable of consistently achieving ~94% capture of UC13 and ~10^3-fold enrichment of UC13 [26]. Given that this device does not appear to sacrifice capture efficiency for sample selectivity, there is a potential for adding even more cell traps in series to obtain a higher degree of cancer cell enrichment.

6.6.3 Application of the Resettable Cell Trap for CTC Enrichment

The resettable cell trap device permits highly sensitive and selective capture of CTCs from patient-derived blood samples. While conventional microfiltration strategies

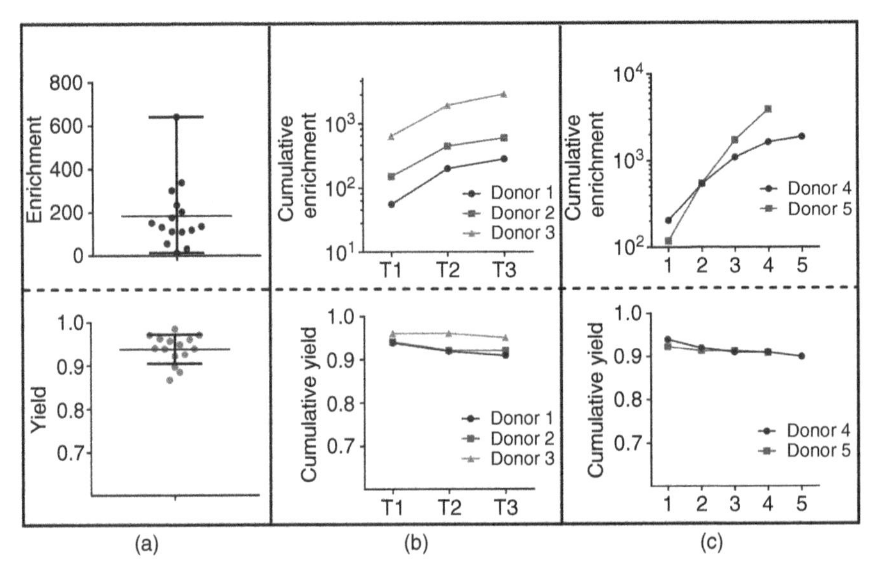

Figure 6.9 Enrichment and yield of UM-UC13 cancer cells doped into whole blood and processed using the resettable cell trap device. The graphs indicate the degree of cancer cell enrichment and yield (a) following filtration of cells through a single cell trap, (b) following filtration through three sequential cell traps and (c) filtration through four or five sequential cell traps.

have demonstrated impressive capabilities to enrich cultured cancer cells [44, 45, 63], a static pore geometry may not be appropriate for the enrichment of CTCs due to the differences in CTC cell size. The resettable cell trap can enrich CTCs based on both size and deformability and has the added advantage of a capability to dynamically vary the pore size for optimal CTC capture. The resettable cell trap is also more sensitive than immunoaffinity systems, such as CellSearch. This is possibly because of the capture efficiency of the device, but perhaps also because the resettable cell trap may capture CTCs that do not express the EpCAM antigen at a sufficiently high level to permit efficient immunoaffinity capture [7]. As a preliminary investigation of the efficacy of the resettable cell trap for CTC enrichment and identification, blood was obtained from six patients with metastatic castrate-resistant prostate cancer, for whom a count of only 0–4 CTCs/7.5 ml blood was identified by CellSearch. Blood from these patients was processed by resettable cell trap and stained for an intact nucleus (DAPI), cytokeratin (CK), and leukocyte CD45 biomarkers. Using the accepted definition of CTCs ($DAPI^+CK^+CD45^-$), five of the patients were observed to have >150 CTCs/7.5 ml blood [65]. The improved sensitivity of this system represents a significant advance because it may permit earlier detection of CTCs in patients with metastatic cancer.

6.7 SUMMARY

The separation of CTCs using affinity capture of the cell surface marker EpCAM suffers from antigenic bias where they may fail to capture CTCs with reduced expression for EpCAM through EMT [78, 79]. Advances in microfabrication have enabled the development of label-free technologies that may enrich CTCs on the basis of their distinct biomechanical characteristics. Many of these strategies, such as hydrodynamic chromatography [25], inertial flow [23, 26, 27], dielectrophoresis [28], and filtration of fixed cell samples [29], enrich CTCs on the basis of size alone. These systems are effective in capturing cultured cancer cells that tend to be significantly larger than leukocytes, but patient-derived CTCs may be considerably smaller than cultured cancer cells and may overlap with leukocytes in their size distribution [24].

An important characteristic that distinguishes CTCs from leukocytes is their reduced deformability arising from greater nucleus-to-cytoplasm ratio. Microfluidic filtration strategies aiming to enrich CTCs based on a combination of size and deformability are limited by clogging, which significantly reduces their selectivity and ability to extract the separated cells. The microfluidic ratchet and resettable cell trap mechanisms are examples of technologies designed to overcome this challenge. The microfluidic ratchet employs funnel-shaped microconstrictions that capture CTCs but enable leukocytes migrate unidirectionally under oscillatory flow. The oscillatory flow constantly agitates the sample to prevent clogging and adsorption that may reduce filter selectivity and prevent extraction of target cells. The resettable cell trap mechanism entraps the target cell within microstructured pockets generated in the textured surface of the microfluidic channel. After capturing the target cells, this device can be then purged and the captured cells can be released in such a way that cells are not retained within the microchannel to contribute to clogging.

In conclusion, the separation of CTCs based on their biophysical properties, specifically cellular deformability, has the potential to overcome the antigenic bias of CTC separation based on cell surface markers. Therefore, in contrast to affinity capture methods, biophysical separation could potentially provide access to a greater population of CTCs. This enhanced sensitivity may permit identification of the more relevant CTCs in metastasis and may thereby aid in the evaluation of disease status. Furthermore, the ability to capture viable CTCs in a format that permits their downstream characterization promises to provide valuable insight into the relationship between these cells and tumor metastasis. Together with the established clinical value of CTC enumeration, the characterization of CTCs may provide a more accurate assessment of disease status in cancer patients and may give rise to new strategies to personalize treatment and evaluate drug efficacy.

REFERENCES

1. Cristofanilli, M., Budd, G.T., Ellis, M.J. *et al.* (2004) Circulating tumor cells, disease progression, and survival in metastatic breast cancer. *New England Journal of Medicine*, **351**, 781–791.
2. Krivacic, R.T., Ladanyi, A., Curry, D.N. *et al.* (2004) A rare-cell detector for cancer. *Proceedings of National Academy of Sciences of United States of America*, **101**, 10501–10504.
3. De Bono, J.S., Scher, H.I., Montgomery, R.B. *et al.* (2008) Circulating tumor cells predict survival benefit from treatment in metastatic castration-resistant prostate cancer. *Clinical Cancer Research Official Journal of American Association of Cancer Research*, **14**, 6302–6309.
4. Cohen, S.J., Punt, C.J.A., Iannotti, N. *et al.* (2008) Relationship of circulating tumor cells to tumor response, progression-free survival, and overall survival in patients with metastatic colorectal cancer. *Journal of Clinical Oncology: Official Journal of the American Society of Clinical Oncology*, **26**, 3213–3221.
5. Riethdorf, S., Fritsche, H., Müller, V. *et al.* (2007) Detection of circulating tumor cells in peripheral blood of patients with metastatic breast cancer: a validation study of the CellSearch system. *Clinical Cancer Research: Official Journal of American Association of Cancer Research*, **13**, 920–928.
6. Went, P.T., Lugli, A., Meier, S. *et al.* (2004) Frequent EpCam protein expression in human carcinomas. *Human Pathology*, **35**, 122–128.
7. Ozkumur, E., Shah, A.M., Ciciliano, J.C. *et al.* (2013) Inertial focusing for tumor antigen-dependent and -independent sorting of rare circulating tumor cells. *Science Translational Medicine*, **5**, 179ra47.
8. Driemel, C., Kremling, H., Schumacher, S. *et al.* (2013) Context-dependent adaption of EpCAM expression in early systemic esophageal cancer. *Oncogene*, **33**, 4904–4915.

9. Yoshida, G.J. and Saya, H. (2014) EpCAM expression in the prostate cancer makes the difference in the response to growth factors. *Biochemical and Biophysical Research Communications*, **443**, 239–245.
10. Aktas, B., Tewes, M., Fehm, T. *et al.* (2009) Stem cell and epithelial-mesenchymal transition markers are frequently overexpressed in circulating tumor cells of metastatic breast cancer patients. *Breast Cancer Research*, **11**, R46.
11. Kasimir-Bauer, S., Hoffmann, O., Wallwiener, D. *et al.* (2012) Expression of stem cell and epithelial-mesenchymal transition markers in primary breast cancer patients with circulating tumor cells. *Breast Cancer Research*, **14**, R15.
12. Armstrong, A.J., Marengo, M.S., Oltean, S. *et al.* (2011) Circulating tumor cells from patients with advanced prostate and breast cancer display both epithelial and mesenchymal markers. *Molecular Cancer Research*, **9**, 997–1007.
13. Joosse, S.A., Hannemann, J., Spötter, J. *et al.* (2012) Changes in keratin expression during metastatic progression of breast cancer: impact on the detection of circulating tumor cells. *Clinical Cancer Research: Official Journal of American Association of Cancer Research*, **18**, 993–1003.
14. Raimondi, C., Gradilone, A., Naso, G. *et al.* (2011) Epithelial-mesenchymal transition and stemness features in circulating tumor cells from breast cancer patients. *Breast Cancer Research and Treatment*, **130**, 449–455.
15. Vannier, C., Mock, K., Brabletz, T., and Driever, W. (2013) Zeb1 regulates E-cadherin and Epcam (epithelial cell adhesion molecule) expression to control cell behavior in early zebrafish development. *Journal of Biological Chemistry*, **288**, 18643–18659.
16. Kimura, H., Kato, H., Faried, A. *et al.* (2007) Prognostic significance of EpCAM expression in human esophageal cancer. *International Journal of Oncology*, **30**, 171–179.
17. Songun, I., Litvinov, S.V., van de Velde, C.J.H. *et al.* (2005) Loss of Ep-CAM (CO17-1A) expression predicts survival in patients with gastric cancer. *British Journal of Cancer*, **92**, 1767–1772.
18. Downey, G.P., Doherty, D.E., Schwab, B. *et al.* (1990) Retention of leukocytes in capillaries: role of cell size and deformability. *Journal of Applied Physiology Bethesda Md 1985*, **69**, 1767–1778.
19. Gascoyne, P.R.C., Shim, S., Noshari, J. *et al.* (2013) Correlations between the dielectric properties and exterior morphology of cells revealed by dielectrophoretic field-flow fractionation. *Electrophoresis*, **34**, 1042–1050.
20. Coumans, F.A.W., van Dalum, G., Beck, M., and Terstappen, L.W.M.M. (2013) Filter characteristics influencing circulating tumor cell enrichment from whole blood. *PLoS ONE*, **8**, e61770.
21. Song, H., O'Connor, K.C., Lacks, D.J. *et al.* (2003) Monte Carlo simulation of LNCaP human prostate cancer cell aggregation in liquid-overlay culture. *Biotechnology Progress*, **19**, 1742–1749.
22. Shi, Y., Ryu, D.D., and Ballica, R. (1993) Rheological properties of mammalian cell culture suspensions: Hybridoma and HeLa cell lines. *Biotechnology and Bioengineering*, **41**, 745–754.
23. Sollier, E., Go, D.E., Che, J. *et al.* (2014) Size-selective collection of circulating tumor cells using vortex technology. *Lab on a Chip*, **14**, 63–77.

24. Park, S., Ang, R.R., Duffy, S.P. *et al.* (2014) Morphological differences between circulating tumor cells from prostate cancer patients and cultured prostate cancer cells. *PLoS ONE*, **9**, e85264.
25. Karabacak, N.M., Spuhler, P.S., Fachin, F. *et al.* (2014) Microfluidic, marker-free isolation of circulating tumor cells from blood samples. *Nature Protocols*, **9**, 694–710.
26. Hou, H.W., Warkiani, M.E., Khoo, B.L. *et al.* (2013) Isolation and retrieval of circulating tumor cells using centrifugal forces. *Scientific Reports*, **3**, 1259.
27. Hyun, K.-A., Kwon, K., Han, H. *et al.* (2013) Microfluidic flow fractionation device for label-free isolation of circulating tumor cells (CTCs) from breast cancer patients. *Biosensors and Bioelectronics*, **40**, 206–212.
28. Shim, S., Stemke-Hale, K., Tsimberidou, A.M. *et al.* (2013) Antibody-independent isolation of circulating tumor cells by continuous-flow dielectrophoresis. *Biomicrofluidics*, **7**, 011807–011812.
29. Lin, H.K., Zheng, S., Williams, A.J. *et al.* (2010) Portable filter-based microdevice for detection and characterization of circulating tumor cells. *Clinical Cancer Research*, **16**, 5011–5018.
30. McFaul, S.M., Lin, B.K., and Ma, H. (2012) Cell separation based on size and deformability using microfluidic funnel ratchets. *Lab on a Chip*, **12**, 2369–2376.
31. Gerhardt, T., Woo, S., and Ma, H. (2011) Chromatographic behaviour of single cells in a microchannel with dynamic geometry. *Lab on a Chip*, **11**, 2731–2737.
32. Breier, A., Stefanková, Z., Barancík, M., and Tribulová, N. (1994) Time dependence of [3H]-vincristine accumulation by L1210 mouse leukemic cells. Effect of P-glycoprotein overexpression. *General Physiology and Biophysics*, **13**, 287–298.
33. Marrinucci, D., Bethel, K., Bruce, R.H. *et al.* (2007) Case study of the morphologic variation of circulating tumor cells. *Human Pathology*, **38**, 514–519.
34. Méhes, G., Witt, A., Kubista, E., and Ambros, P.F. (2001) Circulating breast cancer cells are frequently apoptotic. *American Journal of Pathology*, **159**, 17–20.
35. Larson, C.J., Moreno, J.G., Pienta, K.J. *et al.* (2004) Apoptosis of circulating tumor cells in prostate cancer patients. *Cytometry Part J*, **62**, 46–53.
36. Ligthart, S.T., Coumans, F.A.W., Bidard, F.-C. *et al.* (2013) Circulating tumor cells count and morphological features in breast, colorectal and prostate cancer. *PLoS ONE*, **8**, e67148.
37. Lazar, D.C., Cho, E.H., Luttgen, M.S. *et al.* (2012) Cytometric comparisons between circulating tumor cells from prostate cancer patients and the prostate-tumor-derived LNCaP cell line. *Physical Biology*, **9**, 016002.
38. Guilak, F., Tedrow, J.R., and Burgkart, R. (2000) Viscoelastic properties of the cell nucleus. *Biochemical and Biophysical Research Communications*, **269**, 781–786.
39. Dong, C., Skalak, R., and Sung, K.L. (1991) Cytoplasmic rheology of passive neutrophils. *Biorheology*, **28**, 557–567.
40. Maniotis, A.J., Chen, C.S., and Ingber, D.E. (1997) Demonstration of mechanical connections between integrins, cytoskeletal filaments, and nucleoplasm that stabilize nuclear structure. *Proceedings of National Academy of Sciences of United States of America*, **94**, 849–854.
41. Seal, S.H. (1959) Silicone flotation: a simple quantitative method for the isolation of free-floating cancer cells from the blood. *Cancer*, **12**, 590–595.

42. Rosenberg, R., Gertler, R., Friederichs, J. *et al.* (2002) Comparison of two density gradient centrifugation systems for the enrichment of disseminated tumor cells in blood. *Cytometry*, **49**, 150–158.
43. Seal, S.H. (1964) A sieve for the isolation of cancer cells and other large cells from the blood. *Cancer*, **17**, 637–642.
44. Vona, G., Sabile, A., Louha, M. *et al.* (2000) Isolation by size of epithelial tumor cells: a new method for the immunomorphological and molecular characterization of circulating tumor cells. *American Journal of Pathology*, **156**, 57–63.
45. Desitter, I., Guerrouahen, B.S., Benali-Furet, N. *et al.* (2011) A new device for rapid isolation by size and characterization of rare circulating tumor cells. *Anticancer Research*, **31**, 427–441.
46. Ma, Y.-C., Wang, L., and Yu, F.-L. (2013) Recent advances and prospects in the isolation by size of epithelial tumor cells (ISET) methodology. *Technology in Cancer Research & Treatment*, **12**, 295–309.
47. Tan, S.J., Lakshmi, R.L., Chen, P. *et al.* (2010) Versatile label free biochip for the detection of circulating tumor cells from peripheral blood in cancer patients. *Biosensors and Bioelectronics*, **26**, 1701–1705.
48. Wang, X.-B., Huang, Y., Gascoyne, P.R.C. *et al.* (1994) Changes in Friend murine erythroleukaemia cell membranes during induced differentiation determined by electrorotation. *Biochimica et Biophysica Acta BBA – Biomembranes*, **1193**, 330–344.
49. Yang, J., Huang, Y., Wang, X.B. *et al.* (1999) Cell separation on microfabricated electrodes using dielectrophoretic/gravitational field-flow fractionation. *Analytical Chemistry*, **71**, 911–918.
50. Alazzam, A., Stiharu, I., Bhat, R., and Meguerditchian, A.-N. (2011) Interdigitated comb-like electrodes for continuous separation of malignant cells from blood using dielectrophoresis. *Electrophoresis*, **32**, 1327–1336.
51. Gascoyne, P.R.C., Noshari, J., Anderson, T.J., and Becker, F.F. (2009) Isolation of rare cells from cell mixtures by dielectrophoresis. *Electrophoresis*, **30**, 1388–1398.
52. Moon, H.-S., Kwon, K., Kim, S.-I. *et al.* (2011) Continuous separation of breast cancer cells from blood samples using multi-orifice flow fractionation (MOFF) and dielectrophoresis (DEP). *Lab on a Chip*, **11**, 1118–1125.
53. Huang, S.-B., Wu, M.-H., Lin, Y.-H. *et al.* (2013) High-purity and label-free isolation of circulating tumor cells (CTCs) in a microfluidic platform by using optically-induced-dielectrophoretic (ODEP) force. *Lab on a Chip*, **13**, 1371–1383.
54. Carlo, D.D. (2009) Inertial microfluidics. *Lab on a Chip*, **9**, 3038–3046.
55. Huang, L.R., Cox, E.C., Austin, R.H., and Sturm, J.C. (2004) Continuous particle separation through deterministic lateral displacement. *Science*, **304**, 987–990.
56. Liu, Z., Huang, F., Du, J. *et al.* (2013) Rapid isolation of cancer cells using microfluidic deterministic lateral displacement structure. *Biomicrofluidics*, **7**, 011801.
57. Sun, J., Li, M., Liu, C. *et al.* (2012) Double spiral microchannel for label-free tumor cell separation and enrichment. *Lab on a Chip*, **12**, 3952–3960.
58. Moon, H.-S., Kwon, K., Hyun, K.-A. *et al.* (2013) Continual collection and re-separation of circulating tumor cells from blood using multi-stage multi-orifice flow fractionation. *Biomicrofluidics*, **7**, 014105.
59. Hur, S.C., Mach, A.J., and Di Carlo, D. (2011) High-throughput size-based rare cell enrichment using microscale vortices. *Biomicrofluidics*, **5**, 022206.

60. Bhagat, A.A.S., Kuntaegowdanahalli, S.S., and Papautsky, I. (2008) Continuous particle separation in spiral microchannels using Dean flows and differential migration. *Lab on a Chip*, **8**, 1906–1914.
61. Ji, H.M., Samper, V., Chen, Y. *et al.* (2008) Silicon-based microfilters for whole blood cell separation. *Biomedical Microdevices*, **10**, 251–257.
62. Lu, B., Xu, T., Zheng, S., *et al.* (2010). *Parylene Membrane Slot Filter for the Capture, Analysis and Culture of Viable Circulating Tumor Cells*. 2010 IEEE 23rd International Conference on Micro Electro Mechanical Systems (MEMS), pp. 935–938.
63. Zheng, S., Lin, H., Liu, J.-Q. *et al.* (2007) Membrane microfilter device for selective capture, electrolysis and genomic analysis of human circulating tumor cells. *Journal of Chromatography. A*, **1162**, 154–161.
64. Zheng, S., Lin, H.K., Lu, B. *et al.* (2011) 3D microfilter device for viable circulating tumor cell (CTC) enrichment from blood. *Biomedical Microdevices*, **13**, 203–213.
65. Qin, X., Park, S., Duffy, S.P. *et al.* (2015) Size and deformability based separation of circulating tumor cells from castrate resistant prostate cancer patients using resettable cell traps. *Lab on a Chip*, **15** (10), 2278–2286.
66. Park, S., Jin, C., and Ang, R. *et al.* (2014) *Deformability Based Separation of Circulating Tumor Cells from Patients with Castrate Resistant Prostate Cancer*. Proceedings of MicroTAS Conference, San Antonio, Texas, USA.
67. Huang, T., Jia, C.-P. *et al.* (2014) Highly sensitive enumeration of circulating tumor cells in lung cancer patients using a size-based filtration microfluidic chip. *Biosensors and Bioelectronics*, **51**, 213–218.
68. Lim, L.S., Hu, M., Huang, M.C. *et al.* (2012) Microsieve lab-chip device for rapid enumeration and fluorescence in situ hybridization of circulating tumor cells. *Lab on a Chip*, **12**, 4388–4396.
69. Helzer, K.T., Barnes, H.E., Day, L. *et al.* (2009) Circulating tumor cells are transcriptionally similar to the primary tumor in a murine prostate model. *Cancer Research*, **69**, 7860–7866.
70. Hodgkinson, C.L., Morrow, C.J., Li, Y. *et al.* (2014) Tumorigenicity and genetic profiling of circulating tumor cells in small-cell lung cancer. *Nature Medicine*, **20**, 897–903.
71. Benali-Furet, N., Ye, F., Ezaoui, S. *et al.* (2012) Ex vivo expansion of CTCs in culture isolated by size from patients with melanoma. *Journal of Clinical Oncology*, **30**, abstract e21046.
72. Maheswaran, S. and Haber, D.A. (2015) Ex vivo culture of ctcs: an emerging resource to guide cancer therapy. *Cancer research*, **75** (12), 2411–2415.
73. Beattie, W., Qin, X., Wang, L., and Ma, H. (2014) Clog-free cell filtration using resettable cell traps. *Lab on a Chip*, **14**, 2657–2665.
74. Ameri, K., Luong, R., Zhang, H. *et al.* (2010) Circulating tumour cells demonstrate an altered response to hypoxia and an aggressive phenotype. *British Journal of Cancer*, **102**, 561–569.
75. Alix-Panabières, C., Schwarzenbach, H., and Pantel, K. (2012) Circulating tumor cells and circulating tumor DNA. *Annual Review of Medicine*, **63**, 199–215.
76. Massard, C., Chauchereau, A., and Fizazi, K. (2009) The quest for the "bony Grail" of detecting circulating tumour cells in patients with prostate cancer. *Annals of Oncology*, **20**, 197–199.

77. Reid, A.H.M., Attard, G., Danila, D.C. *et al.* (2010) Significant and sustained antitumor activity in post-docetaxel, castration-resistant prostate cancer with the CYP17 inhibitor abiraterone acetate. *Journal of Clinical Oncology: Official Journal of the American Society of Clinical Oncology*, **28**, 1489–1495.
78. Paterlini-Brechot, P. and Benali, N.L. (2007) Circulating tumor cells (CTC) detection: clinical impact and future directions. *Cancer Letters*, **253**, 180–204.
79. Willipinski-Stapelfeldt, B., Riethdorf, S., Assmann, V. *et al.* (2005) Changes in cytoskeletal protein composition indicative of an epithelial-mesenchymal transition in human micrometastatic and primary breast carcinoma cells. *Clinical Cancer Research: Official Journal of American Association of Cancer Research*, **11**, 8006–8014.

7

MICROFABRICATED FILTER MEMBRANES FOR CAPTURE AND CHARACTERIZATION OF CIRCULATING TUMOR CELLS (CTCs)

ZHENG AO, PH.D.

Sheila and David Fuente Graduate Program in Cancer Biology, Department of Pathology and Laboratory Medicine, Dr. John T. Macdonald Foundation Biomedical Nanotechnology Institute, University of Miami Miller School of Medicine, Miami, FL, USA

RICHARD J. COTE, M.D., FRCPATH, FCAP

Department of Pathology and Laboratory Medicine, Department of Biochemistry and Molecular Biology, Dr. John T. MacDonald Foundation Biomedical Nanotechnology Institute, University of Miami Miller School of Medicine, Miami, FL, USA

RAM H. DATAR, M.PHIL., PH.D.

Department of Pathology and Laboratory Medicine, Department of Biochemistry and Molecular Biology, Dr. John T. MacDonald Foundation Biomedical Nanotechnology Institute, University of Miami Miller School of Medicine, Miami, FL, USA

ANTHONY WILLIAMS, PH.D.

Department of Surgery, Section of Urology, University of Chicago – Pritzker School of Medicine, Chicago, IL, USA and Department of Pathology and Laboratory Medicine, University of Miami Miller School of Medicine, Miami, FL, USA

7.1 INTRODUCTION

Metastasis disease accounts for 90% of cancer-related mortality. Thus, interrogating metastatic process not only provides valuable clinical prognosis information but

Circulating Tumor Cells: Isolation and Analysis, First Edition. Edited by Z. Hugh Fan.

also sheds light on discovering cancer-targeted therapy. Metastatic process is usually perceived as a multistep process including tumor cell invasion into neighboring tissue, detachment from tumor lesions, migration via blood or lymphatics, arrest, and eventually intravasation at a distant location to form secondary lesions. The migration of tumor cells via blood vessels is a critical link in this metastasis cascade that was not being studied extensively until recently. These tumor cells found in circulation are named circulating tumor cells (CTCs). The challenge to study CTCs is that CTCs are rare in population in circulation, often found to be as few as one to a few hundred per milliliter of blood. On the other hand, millions of leukocytes and billions of erythrocytes are present in the same volume of blood. Thus, technologies to isolate these cells with high efficiency are required in order to capture and further characterize these cells. One property that has recently been widely explored to distinguish CTCs from the background cells is that CTCs, as a subpopulation or tumor cells, are inherently larger in size compared to surrounding normal blood components. Thus, microfiltration technologies have been developed based on this size difference to isolate CTCs from cancer patients' peripheral blood.

7.2 SIZE-BASED ENRICHMENT OF CIRCULATING TUMOR CELLS

The size-based isolation of CTCs was initially explored by using track-etched polycarbonate filters. These polycarbonate filters are fabricated using ^{235}U fission fragments to bombard polycarbonate membranes, which are then subjected to warm sodium hydroxide etching to generate pores with a uniform size [1]. Using this type of technology, the technique known as "isolation by size of epithelial tumor cells," or ISET, was developed to isolate CTCs from cancer patients' peripheral blood. [2] After collection of peripheral blood from a patient, the sample was drawn through an ISET filter by negative pressure. Using a cut-off of 8 μm, larger CTCs were retained on the filter, while the majority of the leukocytes and erythrocytes passed through it. Postfiltration, the filter was subjected to immunohistochemistry (IHC) so that CTCs could be distinguished from the rest of the normal blood components retained on the filter. Although effective and tested in several disease settings (including breast cancer, lung cancer, pancreatic cancer cutaneous melanoma, and uveal melanoma [3–7]), track-etched filters can be limited due to the randomized deposition of pores, which can lead to less effective filtration area and fused pores (Figure 7.1a). This can potentially yield lower capture efficiency and higher clogging events.

To alleviate this concern, our group has developed a parylene-based filter with uniformly deposited pores generated with oxygen plasma etching in reactive ion etching (RIE) technique [8]. Briefly, parylene C was deposited to 10 μm thickness, which was then masked with photoresist material (Cr/Au or AZ 9260). The parylene C membrane was then etched with RIE, and pores were generated before the final step to strip off the photoresist with acetone. By employing this novel technique, our group has manufactured filters with 40,000 pores with 8 μm diameter, evenly distributed in a 6 mm × 6 mm filtration area [8] (Figure 7.1b). This pore deposition

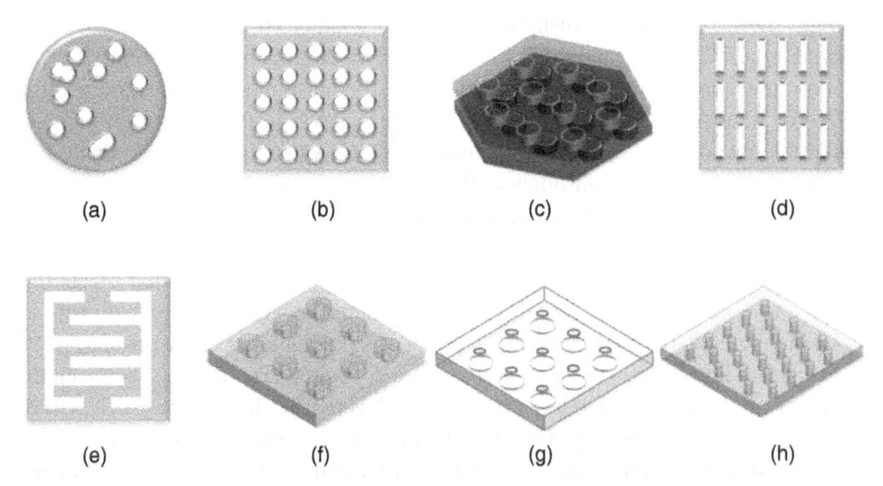

Figure 7.1 Brief illustration of existing microfabricated filter membrane structures for capture and characterization of CTCs. (a) Track-etched polycarbonate filter, (b) parylene C filter with evenly distributed pore arrays, (c) 3D bilayer membrane filter, (d) slot-shaped pore filter membrane, (e) flexible micro spring array (FMSA), (f) three-dimensional palladium filter, (g) microfilter with conical-shaped holes, and (h) VyCap microsieves.

in a controllable fashion will both increase the effective filtration area and ensure uniform filtration pore sizes on the whole filter.

To use this parylene-based filter for CTC enumeration, a blood sample needs to be first diluted 1:1 with phosphate-buffered saline (PBS) and fixed with an end concentration of 1% formalin for 10 min. Postfixation, the blood sample is positively pressured through the filter using a bench-top syringe pump at a rate of 200 ml/h. Following sample filtration, an equal volume of PBS is filtered to further clean the filter and reduce unwanted leukocytes from the filter. The filter can be then subjected to immunofluorescence (IF) or IHC and other downstream analyses. We have tested this filter in several disease settings including breast, prostate, colorectal, and bladder cancers. In a pilot cohort of 57 samples collected from these four cancer types, our filter demonstrated better performance in sensitivity compared to the CellSearch™ platform (see Chapter 19 for details) for the blood draws collected from the same time points in all four disease settings [9]. Currently, our efforts are focused on using this microfiltration platform to characterize CTCs from a Phase 2 clinical trial in patients with relapsed or refractory metastatic bladder cancer clinical trial to monitor therapeutic responses [10]. Additionally, employing this microfabricated filter, urine samples can also be examined to detect urothelial carcinoma of the bladder as an alternative to the standard urine cytology with lower cost and potentially better performance [11].

Although this novel design of microfilter with well-controlled pore deposition is effective in enumeration of CTCs, the prefixation step of sample with 1% formalin, although mild, prohibits downstream functional characterization of CTCs such as establishing CTC culture and drug sensitivity. Also, the fixation can potentially lead to suboptimal conditions for RNA profiling of CTCs due to the chemical modification of RNA by fixatives and to the loss or degradation of RNA during the fixation process. This problem can be more evident in single CTC RNA profiling because single-cell analysis usually requires high-quality RNA. Thus, retrieval of live CTCs from blood can be vital to enable these types of analysis. In order to accommodate this need of live CTC capture, novel designs of microfilters focus on decreasing shear pressure during the filtration process to maintain CTC viability. One design that meets this need is a double-layered 3D membrane structure that our group has developed. As shown in (Figure 7.1c), on the top layer, pores of 9 μm diameter are deposited, while on the bottom layer, pores of 8 μm diameter are deposited. The distance between the double layers is engineered to be 6.5 μm. Thus, during filtration, larger CTCs will be trapped at the edge of the gap. This design allows us to capture viable CTCs. Using a model system for demonstration, cultured tumor cells spiked into healthy donor's blood could be retrieved and maintained in culture for at least 2 weeks [12].

Another design of the microfilter that can be employed for viable CTC capture is achieved by modifying the pore geometry. Using a design of "slot" pores instead of round pores, CTCs can be captured without prefixation due to the larger fill factor of the new design that reduces the shear pressure during filtration. By employing this capture platform, cultured tumor cells in 1 ml of healthy donor's blood can be recovered with 90% capture efficiency and 90% cell viability as validated by measuring telomerase activity by quantitative real-time polymerase chain reaction (qRT-PCR) [13] (Figure 7.1d). In another design of filter, Harouaka *et al.* [14] reported a flexible micro spring array (FMSA) device (Figure 7.1e). This design of filter increases the porosity of the filter to 30–50% and thus decreases the risk of clogging further. (The parylene microfilter technology is protected by the following US patents: 8,815,092; 8,551,425; 8,288,170; 8,114,289; 7,846,743; 7,846,393)

Other filter designs have also been developed with the aim of decreasing shear pressure on CTCs during filtration so as to allow viable CTC capture. In one design, reported by Yusa *et al.*, a three-dimensional palladium filter was developed. This filter consisted of two layers, where on the lower layer an 8 μm sized pore was deposited and on the top layer a 30 μm sized pore was deposited surrounding the lower layer pore forming a pocket shape (Figure. 7.1f). By using this design, a blood sample can be processed without fixation and the blood moves through the filter by gravity with no pressure implemented in the process [15].

Tang *et al.* have developed another design of microfilter for CTC capture. In this design, traditional cylinder-shaped holes are replaced with conical-shaped holes (Figure 7.1g). With this conical design, the pressure is slightly increased between the smaller pore facing the top of the filter and larger pore facing the bottom of the filter. This additional pressure, although small, can facilitate leukocytes to escape the pores and decrease non-CTC cell retention on the filter, thus increasing capture purity [16].

Another microfilter being explored for CTC isolation is the VyCap microsieve. Fabricated using silicon nitride, these microsieves contain evenly distributed pores. By spiking cultured tumor cells into healthy donor's blood, 5 μm pore size is determined to generate the highest recovery rate using the microsieve, as well as manifesting less leukocytes retention as compared to track-etched filters [17].

7.3 COMPARISON BETWEEN SIZE-BASED CTC ISOLATION AND AFFINITY-BASED ISOLATION

The two prevalent types of CTC isolation technologies are affinity-based CTC capture and size-based CTC capture. While affinity-based capture of CTCs relies on CTC expression of cancer-specific cell membrane antigen, size-based CTC capture is antigen agnostic. The most common antigen used in affinity-based CTC capture in epithelial cancer is EpCAM (epithelial cell adhesion molecule). Although EpCAM is uniquely expressed on tumor cells and not on normal blood components, and thus can be used to capture CTCs from peripheral blood with high specificity, concerns have been raised that EpCAM-based capture can lead to loss of certain subtypes of CTCs with low or no EpCAM expression. This can be due to a process known as epithelial–mesenchymal transition (EMT) during metastasis. During EMT, cells from epithelial origin can shed off epithelial phenotype and gain a mesenchymal phenotype, which increases cell invasion capability and facilitates tumor cells to extravasate into the circulation system. Thus, tumor cells, which have undergone EMT, may lose certain epithelial phenotypes, including EpCAM expression. Recently, Zhang *et al.* [18] have reported that a certain subtype of CTCs isolated from breast cancer that has metastatic competency to the brain is EpCAM negative. Also, by using the microsieve described in Section 7.2, EpCAM-negative tumor cells can be detected from CellSearch waste [19]. Additionally, a direct comparison study of CellSearch and ISET for CTC detection performed by Farace *et al.* [20] indicated that ISET technology outperformed CellSearch in metastatic prostate and lung cancer samples, whereas the CellSearch platform captured more CTCs in metastatic breast cancer samples as compared with ISET technology. Also, as discussed in Section 7.2, data from a pilot cohort study indicated that our microfilter technology with evenly distributed pores has a better performance compared to CellSearch in metastatic breast, prostate, colorectal, and bladder cancers [9]. Additionally, with the necessity of allowing antigen–antibody binding events to happen, affinity-based technology is usually more time-consuming (e.g., 3–4 h process time for CellSearch per sample) as compared to microfiltration technology (e.g., 3–5 min process time for our microfilter or ISET technology per sample). On the other hand, several groups have recently reported “small” CTCs in affinity-based capture. Coumans *et al.* [21] reported that the number of CTCs isolated by CellSearch, whether their cytokeratin area is larger or smaller than $4\,\mu m \times 4\,\mu m$, is predictive of survival of castration-resistant prostate cancer patients.

To conclude, we have summarized the molecular and functional characterization assays of CTC enabled by microfabricated filter membrane platforms in Table 7.1. To address this issue, emerging technologies for enrichment of CTCs employ strategies to use either several antibodies recognizing a combination of antigen expression on CTCs [18] or, as the strategy employed by CTC-iChip, use a combination

TABLE 7.1 Molecular and Functional Characterization of CTCs Downstream of Capture by Microfabricated filter Membranes

	Genomic Analysis	Gene Expression Analysis	Functional Characterization
ISET	PCR [2], FISH [23]	qRT-PCR [3, 6], single-cell qRT-PCR [24], Multiplex IF [23]	—
Parylene C-Microfilter	PCR [8], FISH [11]	—	—
Slot pore microfilter	—	Single-cell qRT-PCR [13]	Viable CTC capture, CTC culture in model system [13]
3D parylene C microfilter	—	—	Viable CTC capture [12]
FMSA	—	Multiplex IF [14]	Viable CTC capture, CTC culture, drug sensitivity test in model system [26]
3D palladium filter	FISH, DNA sequencing [15]	qRT-PCR [15]	Viable CTC capture [15]
Conical-shaped microfilter	—	—	Viable CTC capture [16]

of hydrodynamic size-based sorting to debulk the whole blood and then use an affinity-based strategy to positively capture CTCs or negatively deplete leukocytes [22]. Further studies to molecularly and functionally characterize CTCs need to be performed to determine whether CTCs neglected by either type of technology are of clinical and/or biological importance.

7.4 CHARACTERIZATION OF CTCs CAPTURED BY MICROFILTERS

Although CTC number can be of prognosis value in cancer patient management, further characterization can provide additional information for both clinical management of cancer and biological studies to better understand metastatic processes. Here, we briefly discuss the potential downstream characterization of CTCs after filtration-based enrichment technologies described in Section 7.2.

7.4.1 Genomic Analysis of CTCs Enriched by Microfabricated Filter Membrane

Analyzing CTC genomics can provide prognostic information for cancer patient management, reveal drug targets on CTCs, and reveal CTC lineage tracing from primary tumor. Genomic analysis can be performed on CTCs from different perspectives. One technology that is being used to study gene amplification and chromosomal rearrangements in CTCs is DNA fluorescence *in situ* hybridization (FISH). A membrane filter can serve as a convenient platform for both on-chip capture and on-chip FISH characterization because most filters are made of chemically inertial materials, which are often compatible with the FISH protocol. Our group (unpublished data), as well as Nakanishi's group [15], has performed HER2 FISH assay on tumor cells spiked into healthy donor's blood and retrieved by microfilter. Also, using an ISET filter, CTCs were detected from nonsmall-cell lung cancer (NSCLC) patients, and ALK (anaplastic lymphoma kinase) rearrangement was examined in CTCs by FISH. CTCs from ALK-positive NSCLC patients were also positive for ALK rearrangement with a mesenchymal phenotype indicating potential clonal selection from primary tumor [23].

In addition to DNA FISH, PCR can also serve as an effective tool to study DNA mutations in CTCs. Using laser capture microdissection (LCM), CTCs can be isolated from a filter membrane for PCR analysis. Vona *et al.* [2] performed LCM to dissect Hep3B cells captured on ISET filters and mapped p53 deletion by PCR. Also, our group has demonstrated that, through filter electrolysis, DNA can be extracted from tumor cells captured on a filter for PCR analysis [8]. Additionally, Yusa et al. demonstrated that, by PCR-based whole genome amplification, DNA isolated from tumor cells captured by a 3D palladium filter can be used for sequencing analysis of KRAS mutation [15].

7.4.2 Gene Expression Analysis of CTC Enriched by Microfabricated Filter Membrane

In addition to genomic analysis of CTCs, gene expression of CTCs can also provide valuable information on clinical aspects such as patient prognosis and treatment response monitoring, as well as assist biological understating of metastatic process such as tumor tropism and tumor dormancy. qRT-PCR can be employed to interrogate gene expression profile of CTCs. As demonstrated by Pinzani and colleagues [3], CTCs isolated from an ISET filter using LCM can be analyzed by qRT-PCR for HER2 gene expression, which showed that results from 7 of the 44 patients analyzed were strongly correlated to HER2 gene amplification in corresponding primary tumors. Additionally, in the cutaneous melanoma disease setting, De Giorgi *et al.* [6] used qRT-PCR to analyze tyrosinase expression in CTCs from direct filter lysate.

A common drawback of microfiltration technology, as well as other existing technologies, is that they often capture nontumor cells. Thus, to study nontumor-specific gene expression profiles, further isolation of CTCs is required. LCM is effective in pooled cell analysis. However, if single-cell gene expression profiling is to be achieved, a less harsh isolation technology is desirable, such as micropipetting. Chen *et al.* [24] reported that, by using micropipetting to isolate CTCs captured on membrane filter, 38 CTCs from eight prostate cancer patients were analyzed for 84 EMT-related and reference genes.

Additionally, CTC gene expression can also be examined at the protein level. As already integrated in most of the CTC enrichment platforms, IF and IHC are usually employed to distinguish CTCs from nontumor cells captured on membrane filter. Using multispectrum imaging, additional protein markers can be interrogated. As mentioned earlier, using an ISET filter, CTCs detected from NSCLC patients present a mesenchymal phenotype, which is validated by addition of vimentin staining, cytokeratin, and CD45 staining for CTC identification [23].

7.4.3 Functional Characterization of CTCs Enriched by Microfabricated Filter Membrane

As an emerging field in CTC research, functional characterization of CTCs can potentially provide information on drug sensitivity of the patient's primary tumor and metastasis as demonstrated by Yu *et al.* [25], who used CTC cultures established from blood samples of metastatic breast cancer patients. Thus, exploring the idea of *in vivo* drug sensitivity test using a minimal invasive blood draw can potentially benefit the patient treatment through the concept of "personalized therapy."

The prerequisite to perform functional characterization of CTCs is to achieve live CTC capture. As described in Section 7.2, first generation of track-etched filter or round pore filter is not suitable for this kind of task since prefixation is required for cells to survive the increasing shear pressure during filtration process. Thus, several modifications on filter structure or pore geometry have been attempted and proved to be successful in capturing CTCs viably and maintaining their ability to proliferate. For example, as reported by Gallant et al., FMSA device can be used to retrieve viable HCT-116 cancer cells spiked in blood, and drug sensitivity tests can be performed on these cells. After injecting these cells into SCiD mice, the FMSA-derived cells can form tumors, and these mouse models can serve as a drug sensitivity test platform [26].

7.5 CONCLUSION

As a fast expanding area of research, platforms for the detection, enumeration, and analysis of CTCs are being developed as tools to assist in comprehensive patient disease management. As microfabricated filter membranes are fast evolving, with high porosity and novel structures, to enable viable CTC capture, better purity of capture, and reduced risks of clogging, other technologies are also being developed to combine the concept of size-based isolation with affinity-based isolation. With the advance of capture technology development and emerging single-cell analysis tool, the future directions of CTC tests will not only focus on rendering reliable and automated CTC number output as a reference for prognosis but also provide us with a comprehensive snapshot of cancer genomics, transcriptomics, proteomics, and a portfolio of drug sensitivity, all mapped from a minimally invasive blood draw.

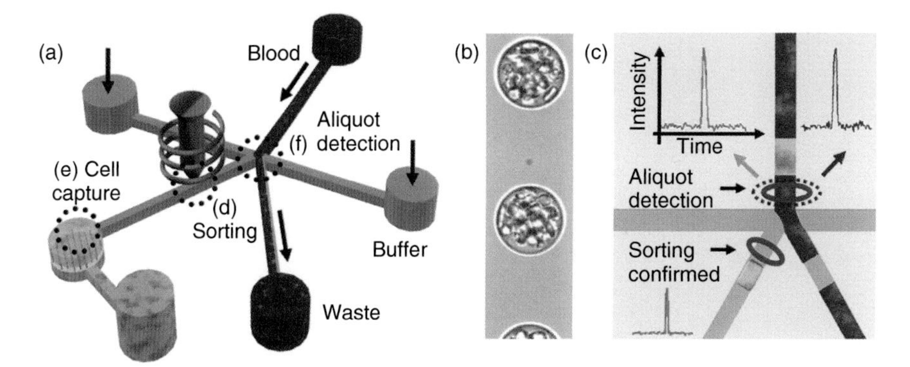

Figure 3.3 (a) Overview of the microfluidic chip. (b) A high-speed camera image of a continuous stream of aliquots. (c) Sorting by laser-induced fluorescence. (See page 56 for full caption.)

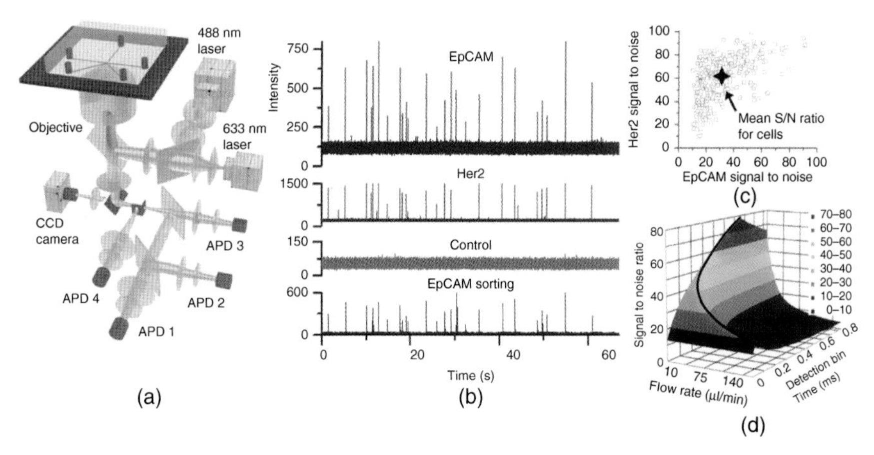

Figure 3.4 (a) The setup for aliquot ranking. (b) A segment of the APD traces. (c) The S/N ratio for each single CTC. (d) The S/N ratio for EpCAM-labeled single CTCs. (See page 58 for full caption.)

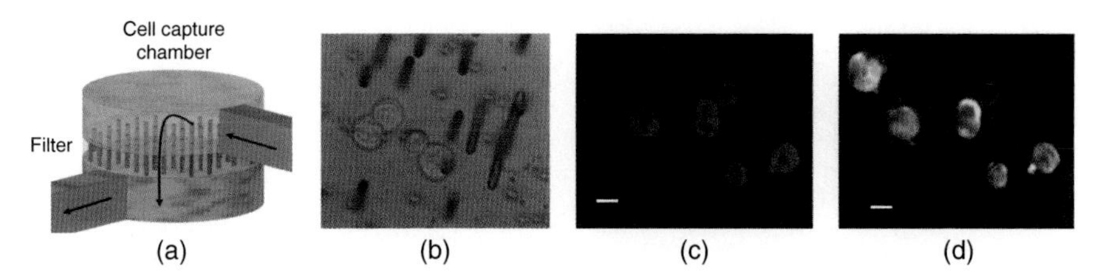

Figure 3.7 (a) The structure of the cell capture chamber. (b) The bright field images of four MCF-7 cells isolated by eDAR. (c) & (d) Fluorescence images of the captured CTCs. (See page 64 for full caption.)

Circulating Tumor Cells: Isolation and Analysis, First Edition. Edited by Z. Hugh Fan.

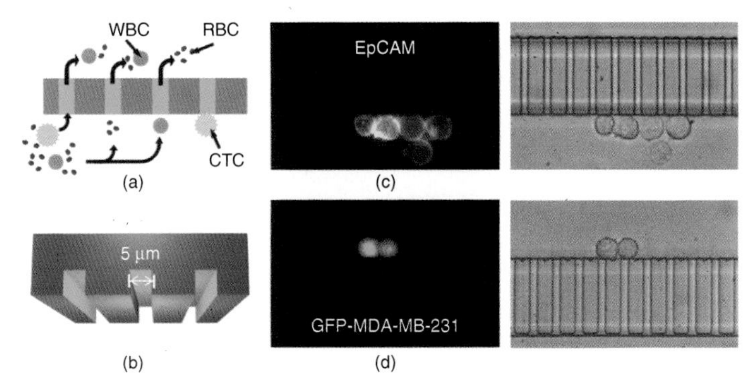

Figure 3.8 (a) Structure of microslits for the on-chip purification. (b) The 3D model of the microslits. (c) & (d) Fluorescence (left) and bright field (right) images of five MCF-7 cells and two MDA-MB-231 cells. (See page 65 for full caption.)

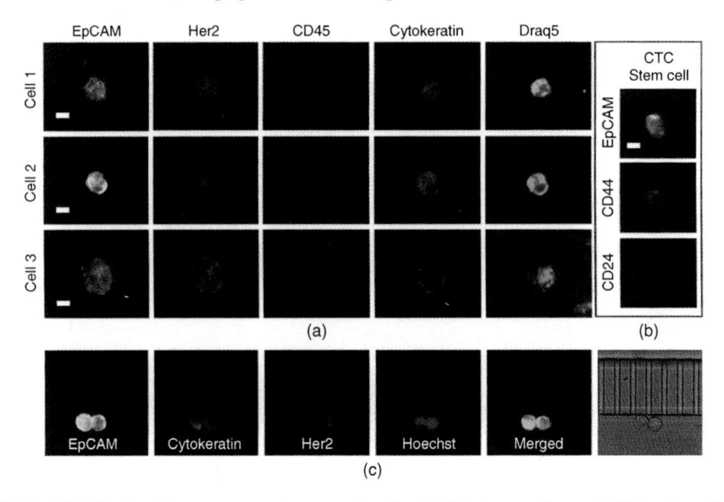

Figure 3.9 Multicolor fluorescence images for the CTCs trapped on eDAR microchip. (See page 66 for full caption.)

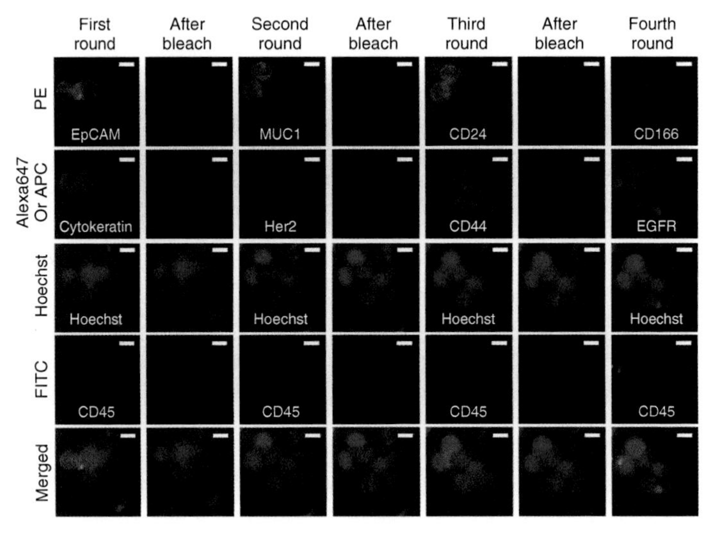

Figure 3.14 Sequential immunostaining and photobleaching results. (See page 74 for full caption.)

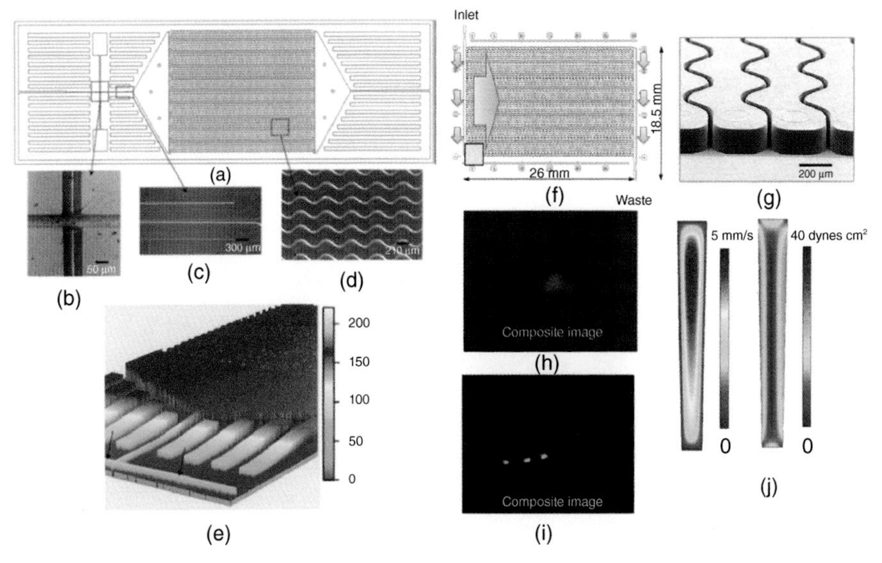

Figure 4.1 (Left, a–e) Schematics of the first-generation, high-throughput CTC selection device. (Right, f–j) Second-generation sinusoidal device. (See page 91 for full caption.)

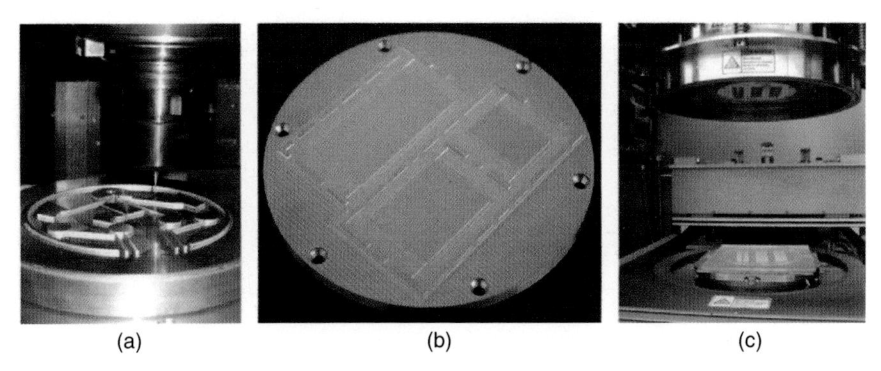

Figure 4.2 (a) High-precision micromilling. (b) A completed brass master mold. (c) Repeated microreplication. (See page 92 for full caption.)

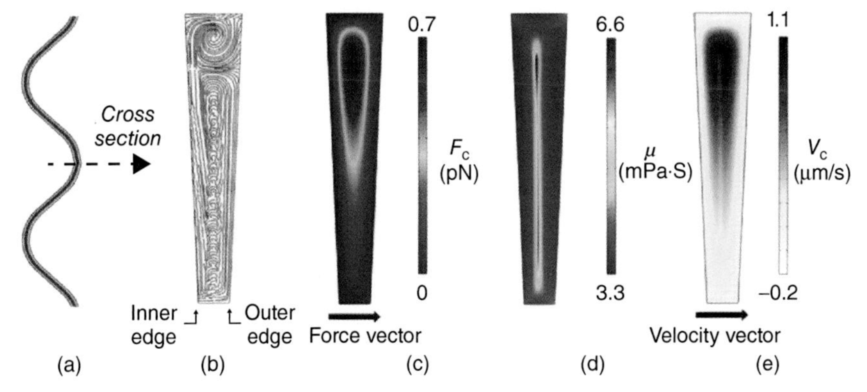

Figure 4.10 (a) Longitudinal velocity profile of blood flow. (b) Cross-sectional velocity streamlines of weak Dean flow. (c) Cross-sectional centrifugal forces. (d) Cross section of non-Newtonian viscosity profile. (e) cross-sectional centrifugal velocities. (See page 101 for full caption.)

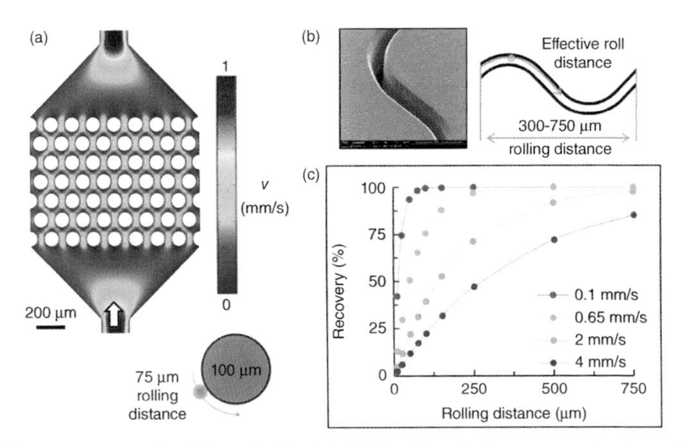

Figure 4.13 (a) Schematic of flow fields. (b) SEM of the high-aspect-ratio sinusoid channel. (c) Recovery of CTCs. (See page 106 for full caption.)

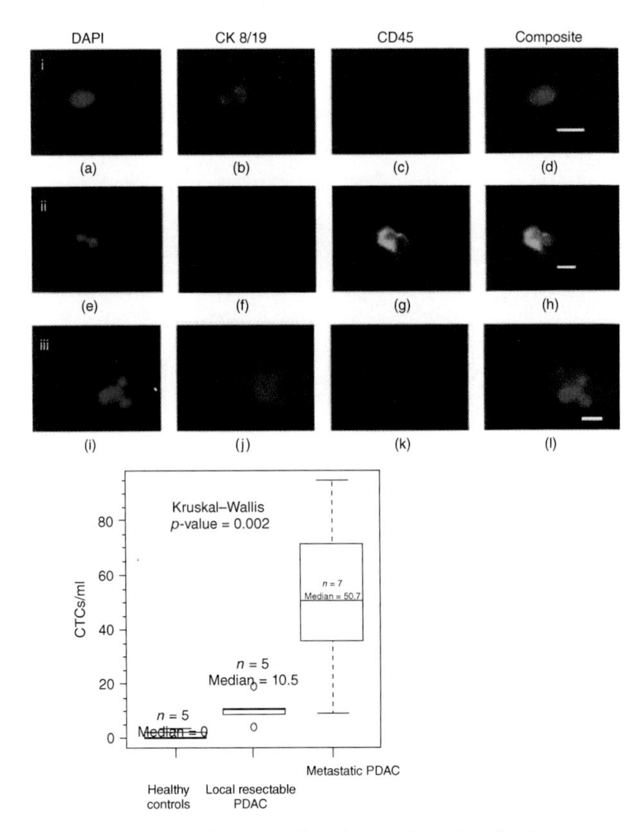

Figure 4.20 Top: Fluorescence images of various selected cells from a metastatic PDAC patient. Bottom: CTCs isolated from CTCs from both locally resectable (nonmetastatic) and metastatic PDAC patient blood samples. (See page 115 for full caption.)

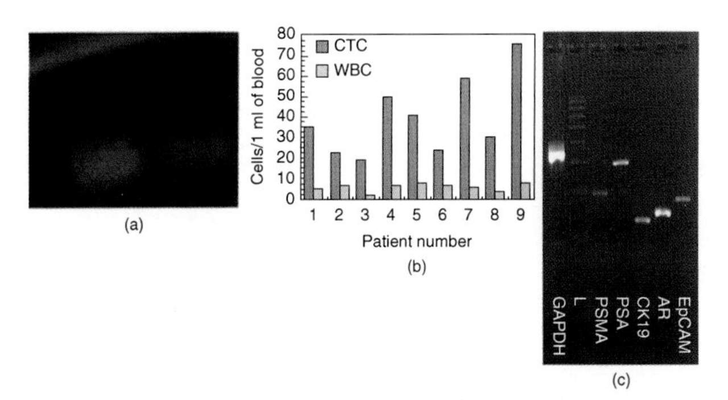

Figure 4.21 (a) Fluorescence image. (b) Summary of CTC isolation data. (c) Molecular profiling of EpCAM(+) CTCs. (See page 116 for full caption.)

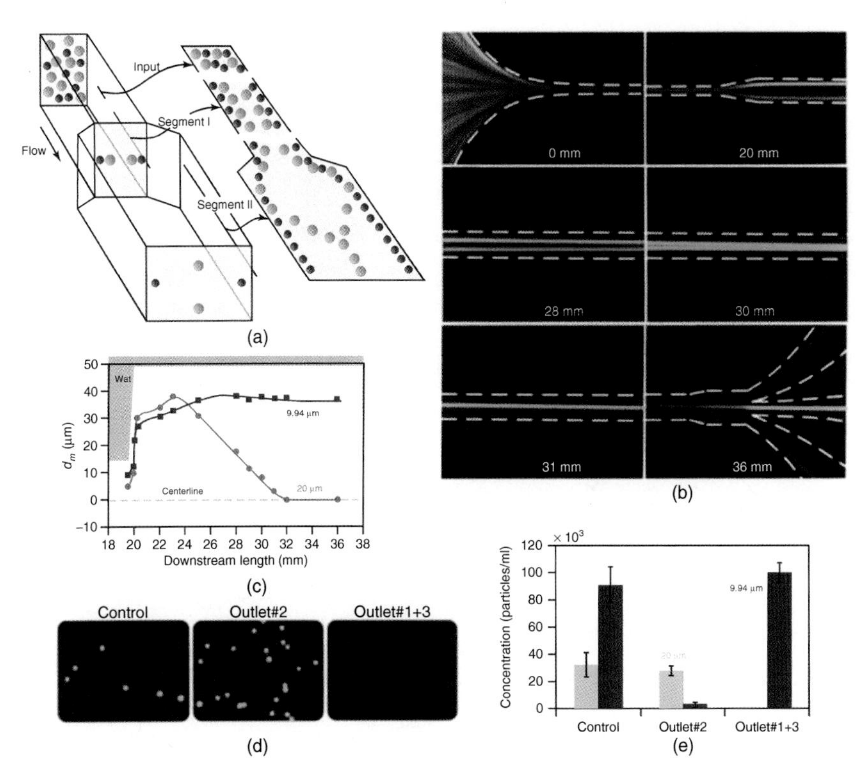

Figure 5.2 (a) Schematic of the device concept for size-based separation. (b) Fluorescent images demonstrating progressive inertial migration. (c) d_m as a function of downstream length. (d) Fluorescent images of microparticles collected from inlet and outlets. (e) Concentrations of microparticles from inlet sample and outlet samples. (See page 133 for full caption.)

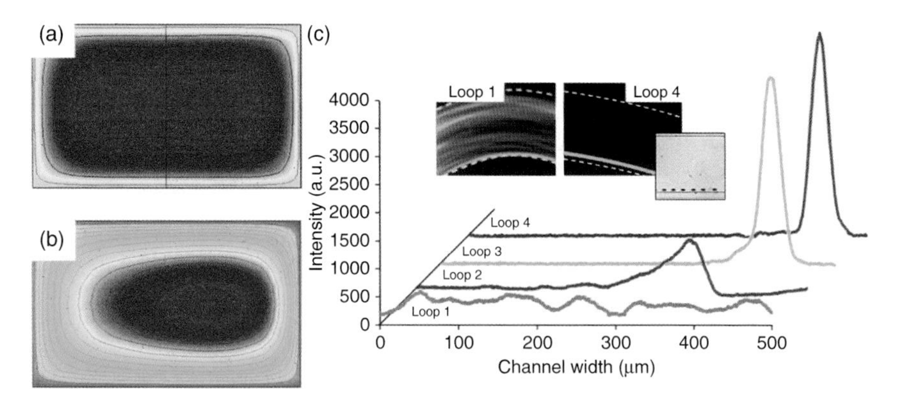

Figure 5.4 (a) CFD-ACE simulation of a curved rectangular channel. (b) Velocity and pressure distribution. (c) Intensity plot of 20-μm-diameter particles across the width of the channel. (See page 137 for full caption.)

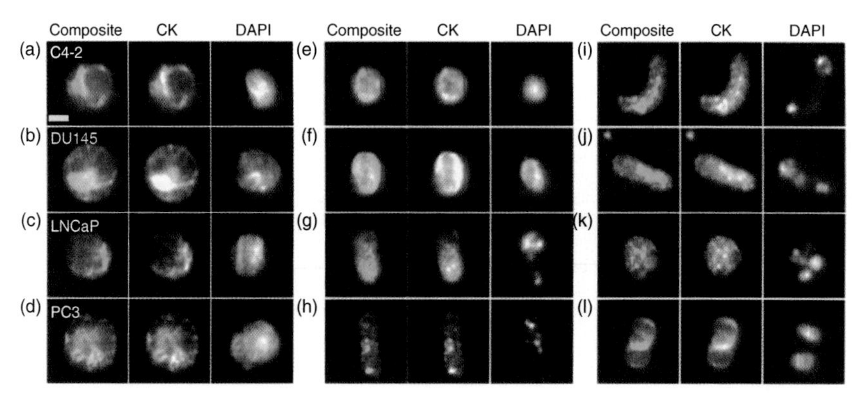

Figure 6.1 Difference in size, shape, and nuclear volume of cultured cancer cells and CTCs. (See page 150 for full caption.)

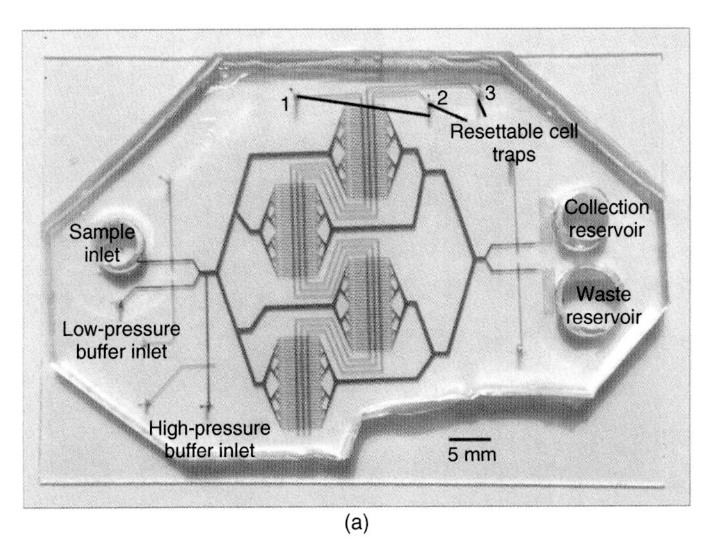

(a)

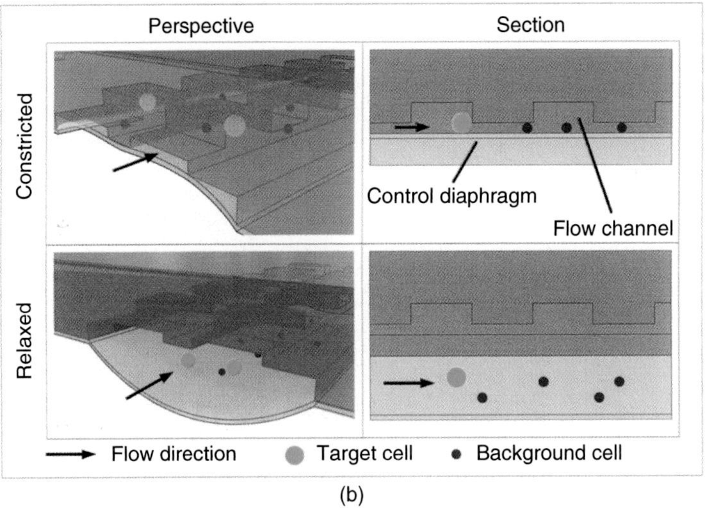

(b)

Figure 6.8 Resettable cell trap microfluidic device. (See page 162 for full caption.)

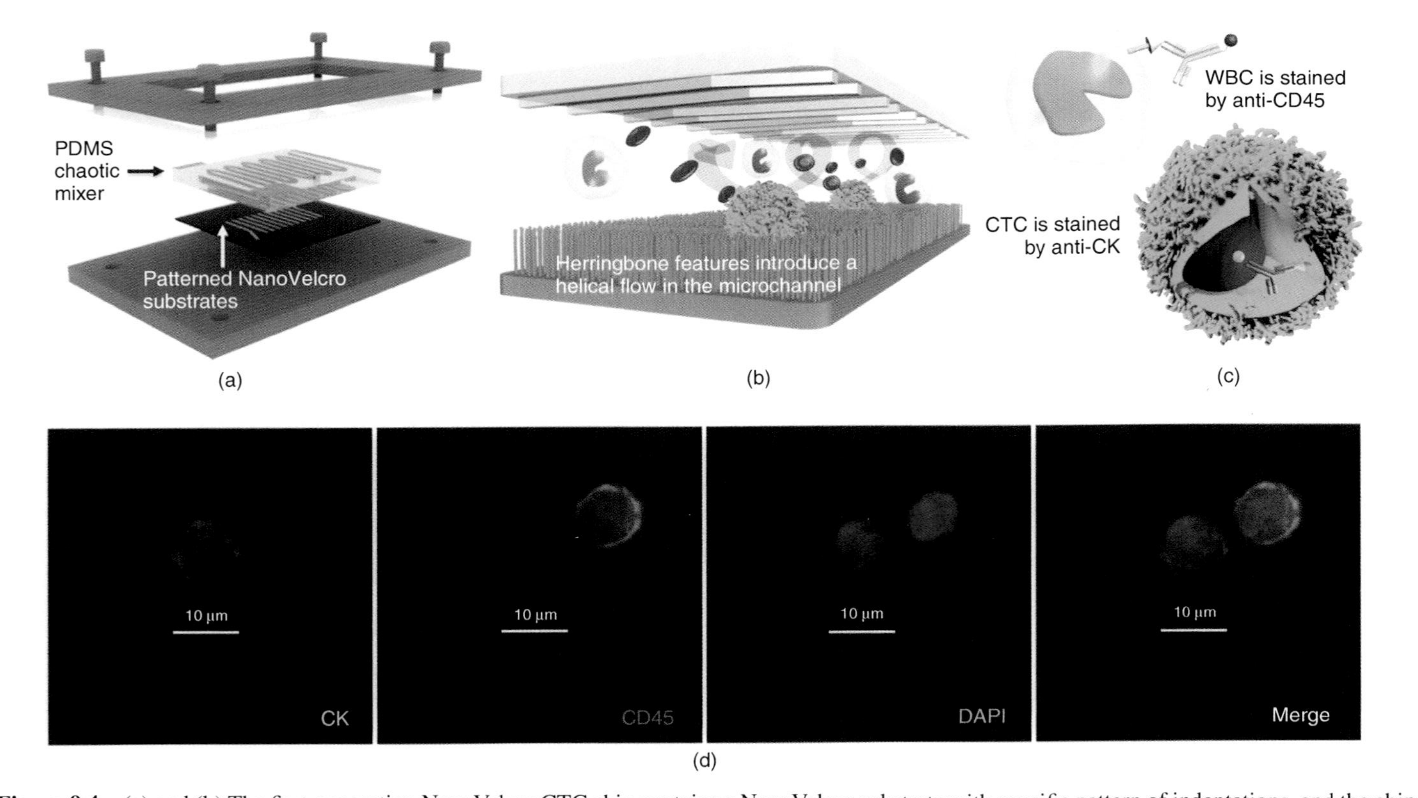

Figure 9.4 (a) and (b) The first-generation NanoVelcro CTC chip contains a NanoVelcro substrate with specific pattern of indentations, and the chip is covered with a PDMS chaotic mixer. (c) The three-color ICC protocol. (d) Fluorescence micrographs of one captured prostate cancer CTC. (See page 211 for full caption.)

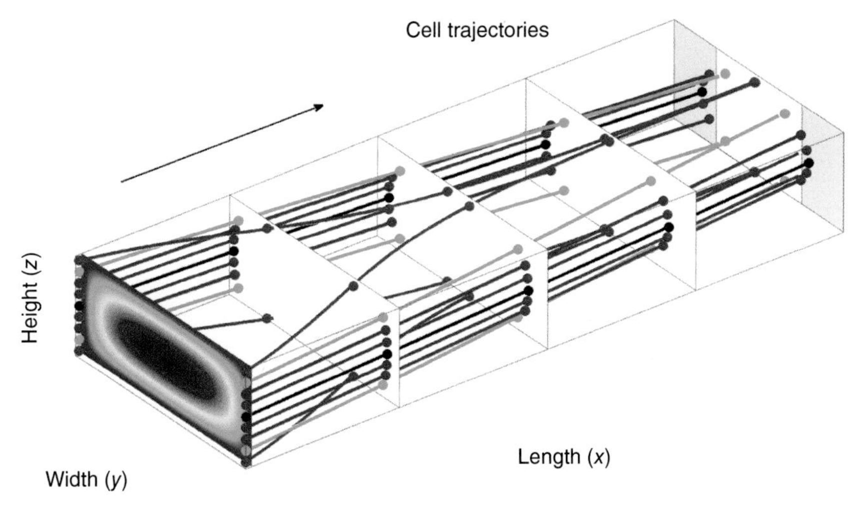

Figure 10.3 Simulated trajectories of identical particles undergoing acoustophoresis. (See page 232 for full caption.)

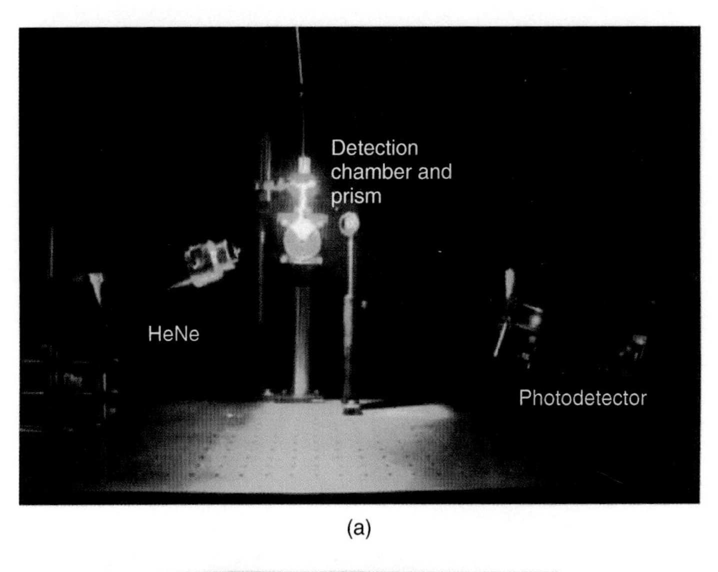

(a)

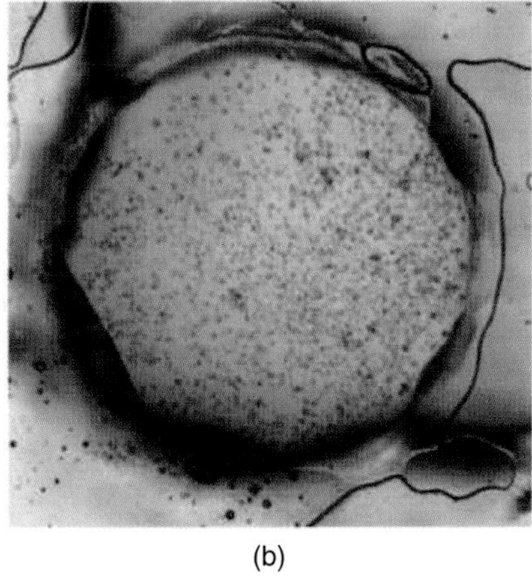

(b)

Figure 11.4 (a) The optical detection setup. (b) Pigmented melanoma cells in the microcuvette.

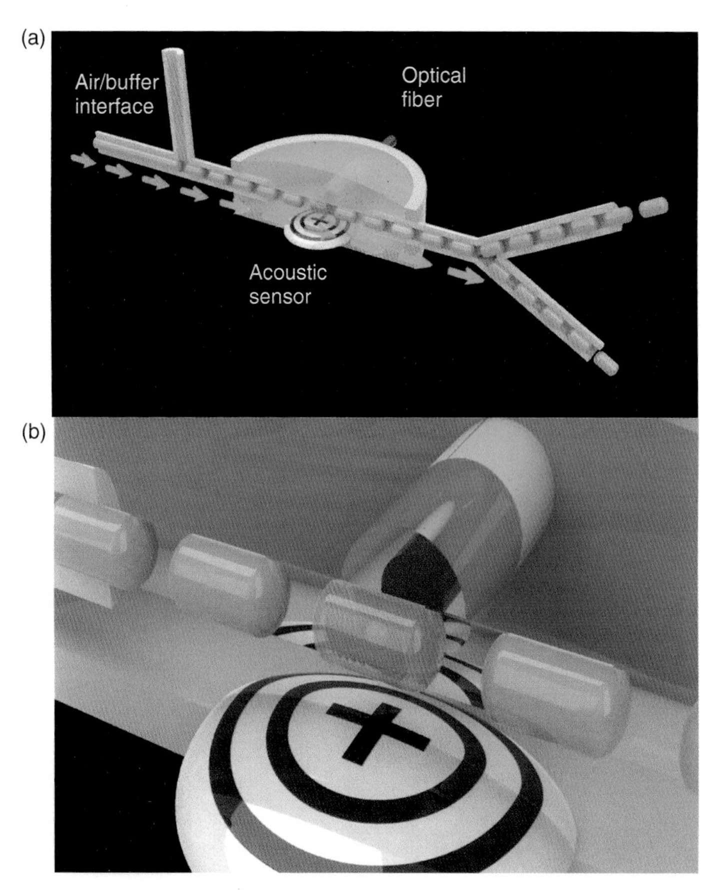

Figure 11.5 (a) The photoacoustic flowmeter. (b) The irradiated droplet containing a CMC. (See page 257 for full caption.)

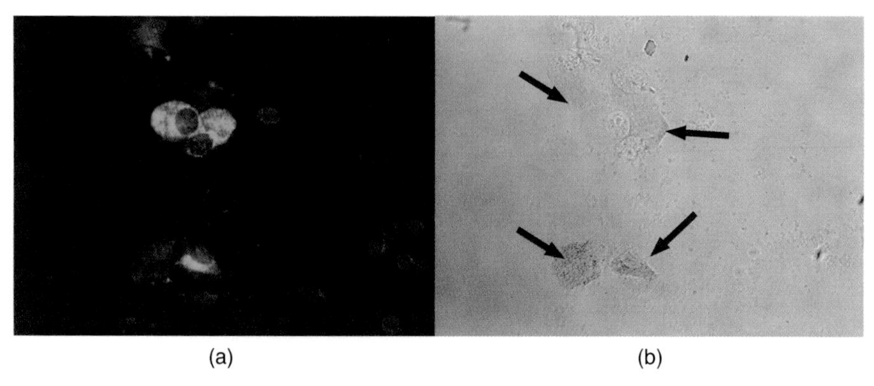

Figure 11.10 (a) Cultured melanoma cells after photoacoustic capture and immunohistochemical staining. (b) Bright-field microscope images of pigmented structures. (See page 261 for full caption.)

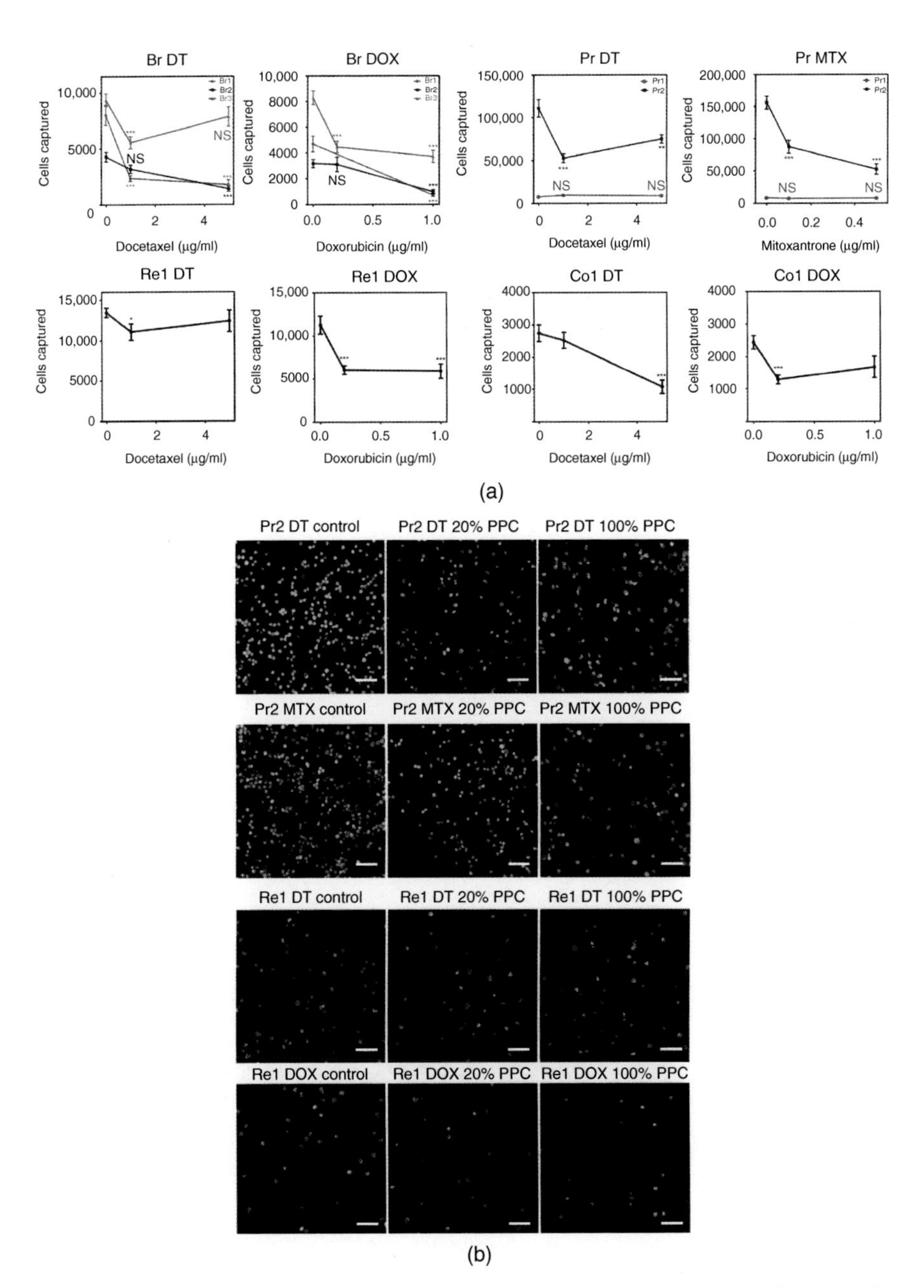

Figure 12.3 (a) CTC counts of patient samples treated with chemotherapeutic drugs. (b) Example micrographs of two patient samples. (See page 274 for full caption.)

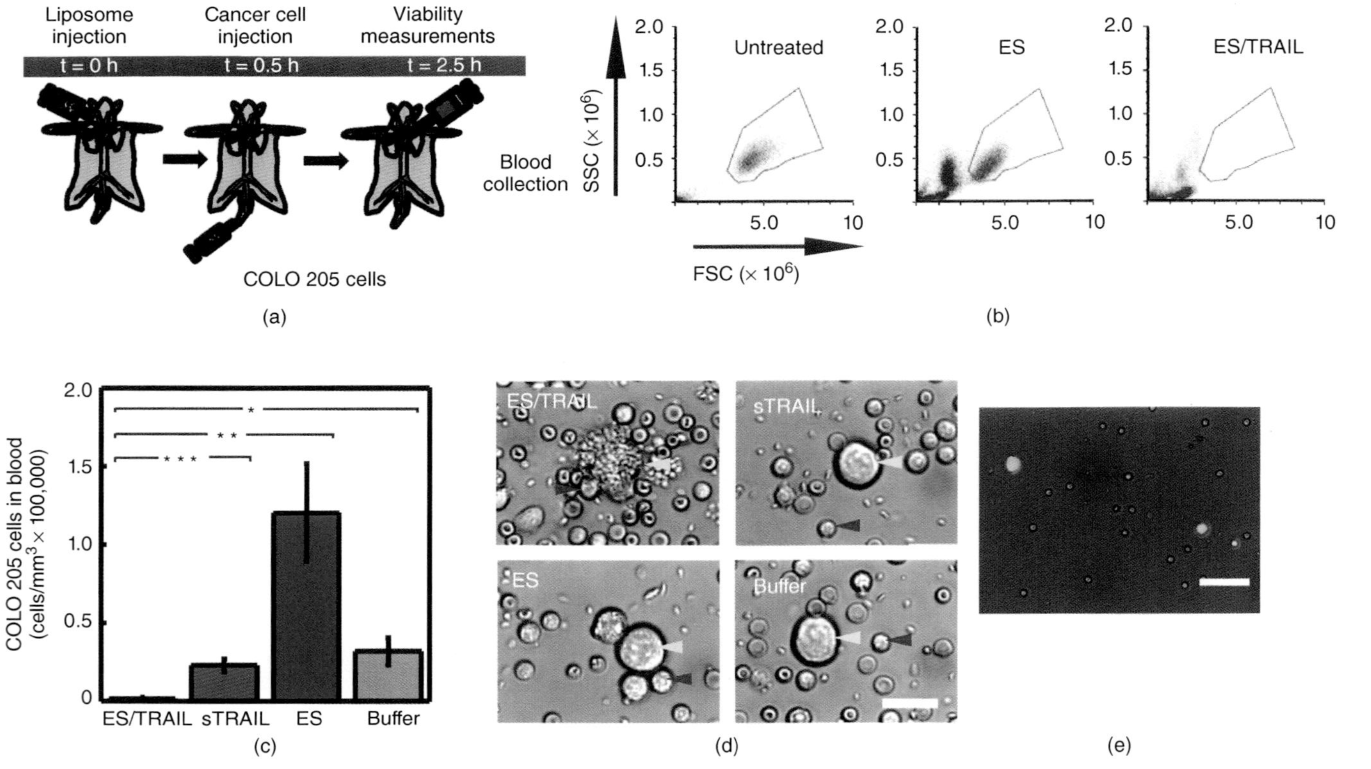

Figure 12.4 (a) Schematic of procedure for *in vivo* liposome experiments. (b) Flow cytometry of COLO205 cells *in vitro*. (c) Number of viable cells recovered from blood of mice. (d) Representative micrographs of cells recovered from mouse blood. (e) Leukocytes functionalized with fluorescent ES/TRAIL liposomes recovered during cardiac puncture. (See page 276 for full caption.)

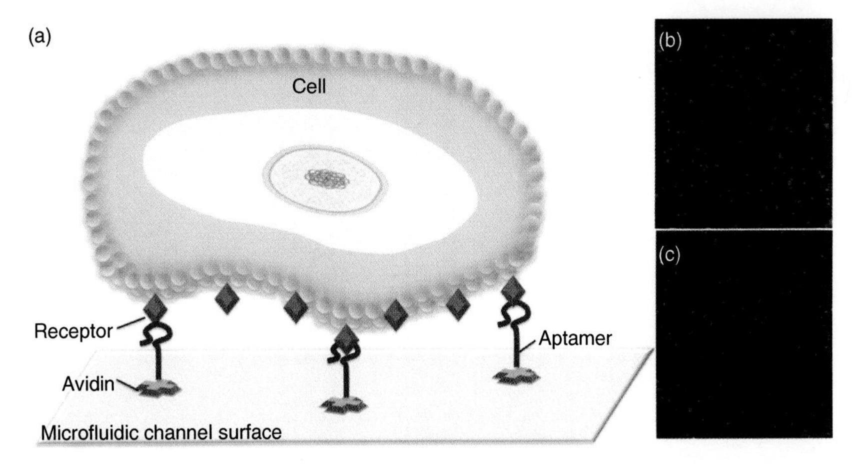

Figure 13.3 (a) Schematic of surface modification and cell capture. (b) Representative image of target and control cells. (c) Image after the capture experiment. (See page 294 full caption.)

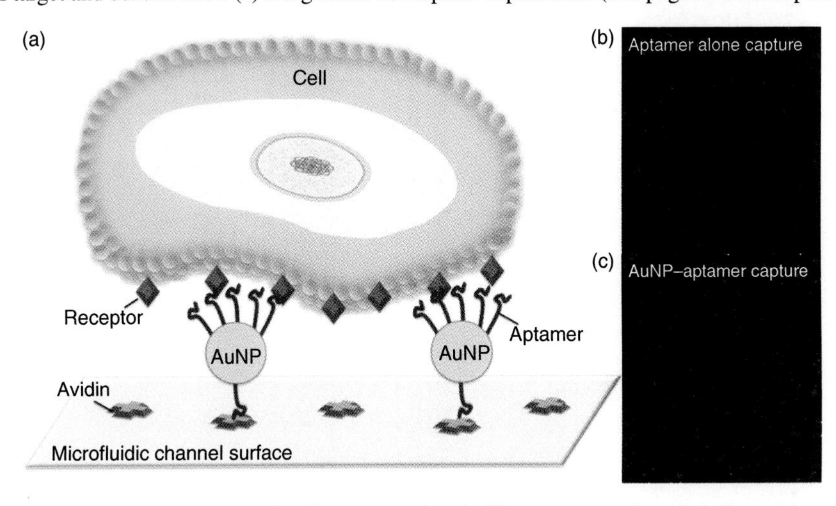

Figure 13.4 (a) Schematic of cell capture using AuNP–aptamers. (b and c) Comparison in the capture efficiency. (See page 295 for full caption.)

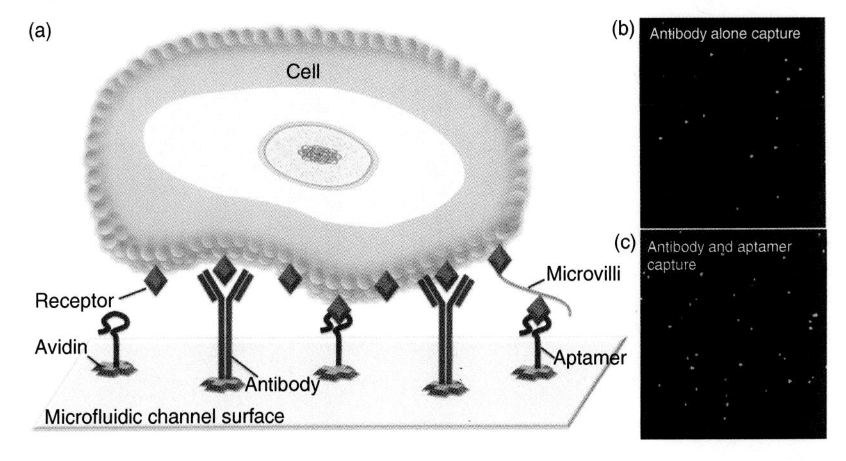

Figure 13.5 (a) Schematic of an ensemble of antibodies and aptamers and the multivalent interactions of one cell with the ensemble. (b and c) Comparison in the capture efficiency of target cells. (See page 296 for full caption.)

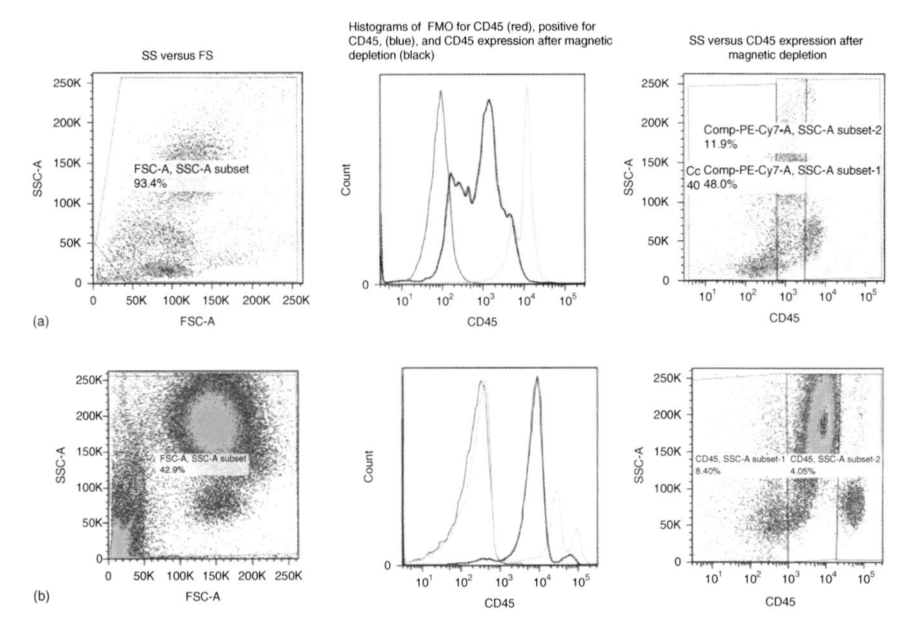

Figure 14.4 Flow cytometry plots of normal blood (a) and metastatic breast cancer patient blood (b). (See page 308 for full caption.)

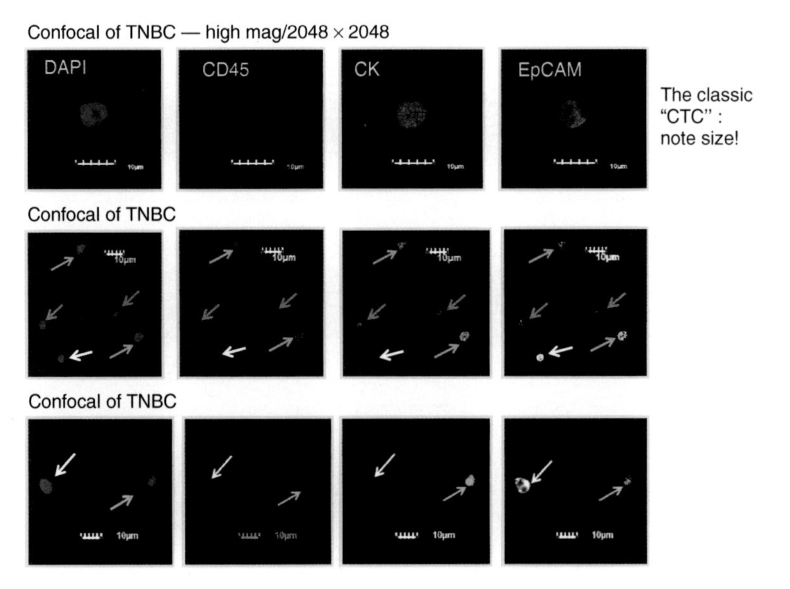

Figure 14.6 Confocal images of enriched samples from a metastatic breast cancer patient. (See page 310 for full caption.)

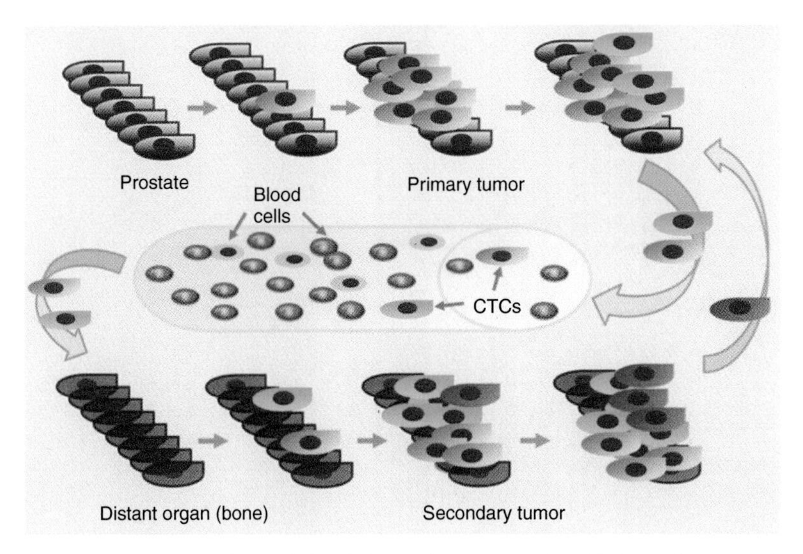

Figure 16.1 Illustration of the origin of circulating tumor cells (CTCs) and metastasis. (See page 330 for full caption.)

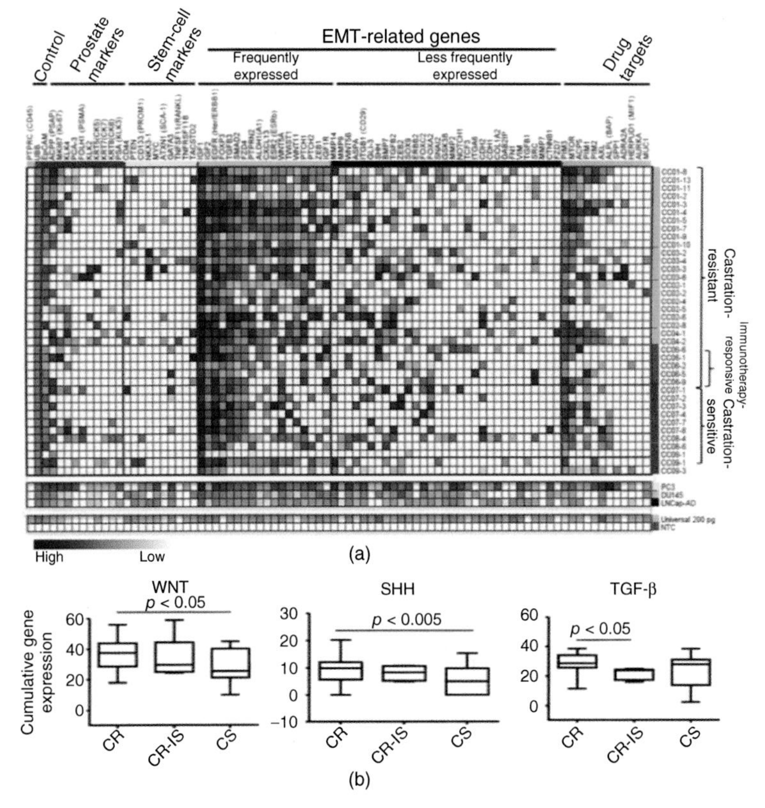

Figure 16.6 Elevated expression of EMT-related genes in CTC castration-resistant prostate cancer patients. (See page 340 for full caption.)

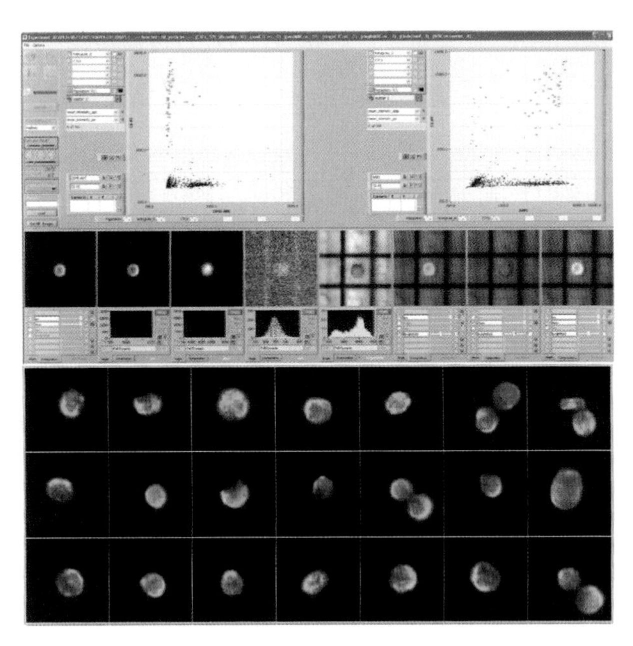

Figure 18.1 *DEPArray™ system CellBrowser™*. (See page 370 for full caption.)

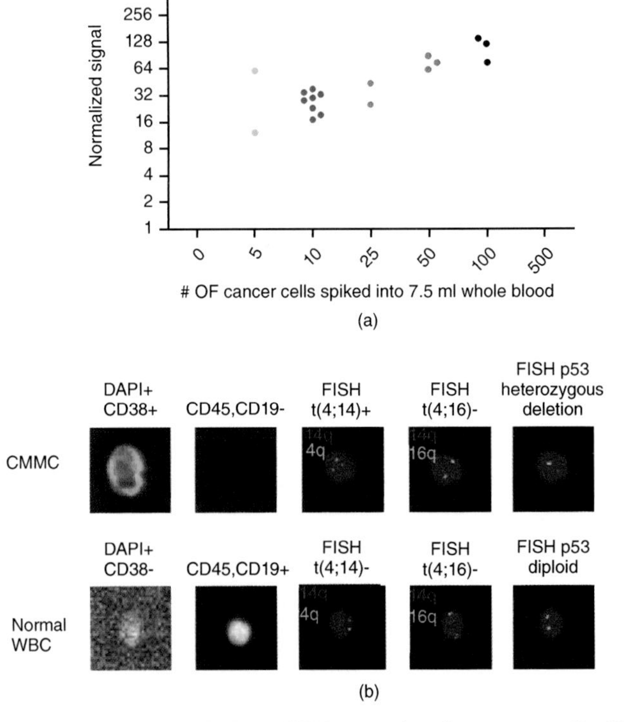

Figure 19.2 (a) Detection of EGFR T790M mutation in cancer cells H1975 using CELLSEARCH® system. (b) Cytogenetic analysis of circulating multiple myeloma cells. (See page 389 for full caption.)

REFERENCES

1. Fleischer, R.L., Alter, H.W., Furman, S.C. *et al.* (1972) Particle track etching. *Science*, **178**, 255–263.
2. Vona, G., Sabile, A., Louha, M. *et al.* (2000) Isolation by size of epithelial tumor cells: a new method for the immunomorphological and molecular characterization of circulating tumor cells. *The American Journal of Pathology*, **156** (1), 57–63.
3. Pinzani, P., Salvadori, B., Simi, L. *et al.* (2006) Isolation by size of epithelial tumor cells in peripheral blood of patients with breast cancer: correlation with real-time reverse transcriptase–polymerase chain reaction results and feasibility of molecular analysis by laser microdissection. *Human Pathology*, **37** (6), 711–718.
4. Hofman, V., Bonnetaud, C., Ilie, M.I. *et al.* (2011) Preoperative circulating tumor cell detection using the isolation by size of epithelial tumor cell method for patients with lung cancer is a new prognostic biomarker. *Clinical Cancer Research*, **17** (4), 827–835.
5. Khoja, L., Backen, A., Sloane, R. *et al.* (2011) A pilot study to explore circulating tumour cells in pancreatic cancer as a novel biomarker. *British Journal of Cancer*, **106** (3), 508–516.
6. De Giorgi, V., Pinzani, P., Salvianti, F. *et al.* (2010) Application of a filtration-and isolation-by-size technique for the detection of circulating tumor cells in cutaneous melanoma. *Journal of Investigative Dermatology*, **130** (10), 2440–2447.
7. Mazzini, C., Pinzani, P., Salvianti, F. *et al.* (2014) Circulating tumor cells detection and counting in uveal melanomas by a filtration-based method. *Cancers*, **6** (1), 323–332.
8. Zheng, S., Lin, H., Liu, J.Q. *et al.* (2007) Membrane microfilter device for selective capture, electrolysis and genomic analysis of human circulating tumor cells. *Journal of Chromatography. A*, **1162** (2), 154–161.
9. Lin, H.K., Zheng, S., Williams, A.J. *et al.* (2010) Portable filter-based microdevice for detection and characterization of circulating tumor cells. *Clinical Cancer Research*, **16** (20), 5011–5018.
10. Hahn N. (2015) Phase 2 study of docetaxel +/− OGX-427 in patients with relapsed or refractory metastatic bladder cancer. ClinicalTrials.gov [Internet]. Bethesda (MD): National Library of Medicine (US). 2011 [2014.3.3]. http://clinicaltrials.gov/ct2/show/NCT01780545. NLM Identifier: NCT 01780545 (accessed 24 November 2015).
11. Birkhahn, M., Mitra, A.P., Williams, A.J. *et al.* (2013) A novel precision-engineered microfiltration device for capture and characterisation of bladder cancer cells in urine. *European Journal of Cancer*, **49** (15), 3159–3168.
12. Zheng, S., Lin, H.K., Lu, B. *et al.* (2011) 3D microfilter device for viable circulating tumor cell (CTC) enrichment from blood. *Biomedical Microdevices*, **13** (1), 203–213.
13. Lu, B., Xu, T., Zheng, S., *et al.* (2010) *Parylene Membrane Slot Filter for the Capture, Analysis and Culture of Viable Circulating Tumor Cells*. Micro Electro Mechanical Systems (MEMS), 2010 IEEE 23rd International Conference, pp. 935–938, IEEE.
14. Harouaka, R.A., Zhou, M.D., Yeh, Y.T. *et al.* (2014) Flexible micro spring array device for high-throughput enrichment of viable circulating tumor cells. *Clinical Chemistry*, **60** (2), 323–333.

15. Yusa, A., Toneri, M., Masuda, T. *et al.* (2014) Development of a new rapid isolation device for circulating tumor cells (CTCs) using 3D palladium filter and its application for genetic analysis. *PLoS ONE*, **9** (2), e88821.
16. Tang, Y., Shi, J., Li, S. *et al.* (2014) Microfluidic device with integrated microfilter of conical-shaped holes for high efficiency and high purity capture of circulating tumor cells. *Scientific Reports*, **4**, 6052.
17. Coumans, F.A., van Dalum, G., Beck, M., and Terstappen, L.W. (2013) Filter characteristics influencing circulating tumor cell enrichment from whole blood. *PLoS ONE*, **8** (4), e61770.
18. Zhang, L., Ridgway, L.D., Wetzel, M.D. *et al.* (2013) The identification and characterization of breast cancer CTCs competent for brain metastasis. *Science Translational Medicine*, **5** (180), 180ra48.
19. Barradas, A. and Terstappen, L.W. (2013) Towards the biological understanding of CTC: capture technologies, definitions and potential to create metastasis. *Cancers*, **5** (4), 1619–1642.
20. Farace, F., Massard, C., Vimond, N. *et al.* (2011) A direct comparison of CellSearch and ISET for circulating tumour-cell detection in patients with metastatic carcinomas. *British Journal of Cancer*, **105** (6), 847–853.
21. Coumans, F.A.W., Doggen, C.J.M., Attard, G. *et al.* (2010) All circulating EpCAM+ CK+ CD45− objects predict overall survival in castration-resistant prostate cancer. *Annals of Oncology*, **21** (9), 1851–1857.
22. Ozkumur, E., Shah, A.M., Ciciliano, J.C. *et al.* (2013) Inertial focusing for tumor antigen–dependent and–independent sorting of rare circulating tumor cells. *Science Translational Medicine*, **5** (179), 179ra47.
23. Pailler, E., Adam, J., Barthélémy, A. *et al.* (2013) Detection of circulating tumor cells harboring a unique ALK rearrangement in ALK-positive non–small-cell lung cancer. *Journal of Clinical Oncology*, **31** (18), 2273–2281.
24. Chen, C.L., Mahalingam, D., Osmulski, P. *et al.* (2013) Single-cell analysis of circulating tumor cells identifies cumulative expression patterns of EMT-related genes in metastatic prostate cancer. *The Prostate*, **73** (8), 813–826.
25. Yu, M., Bardia, A., Aceto, N. *et al.* (2014) Ex vivo culture of circulating breast tumor cells for individualized testing of drug susceptibility. *Science*, **345** (6193), 216–220.
26. Gallant, J.N., Matthew, E.M., Cheng, H. *et al.* (2013) Predicting therapy response in live tumor cells isolated with the flexible micro spring array device. *Cell Cycle*, **12** (13), 2132.

8

MINIATURIZED NUCLEAR MAGNETIC RESONANCE PLATFORM FOR RARE CELL DETECTION AND PROFILING

SANGMOO JEONG, CHANGWOOK MIN, HUILIN SHAO,
CESAR M. CASTRO, RALPH WEISSLEDER, AND HAKHO LEE

Center for Systems Biology, Massachusetts General Hospital, Harvard Medical School, Boston, MA, USA

8.1 INTRODUCTION

Circulating tumor cells (CTCs), which are shed from the primary tumor and then circulate through the bloodstream, have emerged as a promising surrogate for tissue-based markers, enabling noninvasive, real-time disease detection [1, 2]. Various investigations confirmed the potential of CTCs as an early indicator of tumor expansion and metastasis [3]; CTCs have also been used to predict disease progression, response to treatment, relapse, and overall survival [4–6].

Despite great potential of CTCs, their clinical use for detection and characterization of cancer still remains a significant technical challenge [7]. The task typically requires exquisite sensitivity to detect small number (10–100) of cells amid a vast background of host cells (e.g., $>10^6$ leukocytes in 1 ml blood). The most commonly used CTC assay, CellSearch (see Chapter 19 for the details), isolates CTCs from peripheral blood based on positive selection of epithelial surface marker (EpCAM, epithelial cell adhesion molecule) expression and differentiation from leukocytes via the absence of CD45 [4]. Although EpCAM is considered a key molecular marker for CTCs, it has been shown that EpCAM is not highly expressed in ~40% of cancers and completely absent in ~20% [8]. EpCAM can be further downregulated in

Circulating Tumor Cells: Isolation and Analysis, First Edition. Edited by Z. Hugh Fan.

cells undergoing epithelial–mesenchymal transition (EMT) during increased cell proliferation [5, 9]. Sensitive CTC assays beyond the EpCAM-based methods are thus needed to promote rare cancer cell studies.

We have established a novel detection technology, termed micronuclear magnetic resonance (μNMR), for rapid and highly sensitive biomarker detection [10–13]. The μNMR exploits nuclear magnetic resonance (NMR) to detect cells labeled with magnetic nanoparticles (MNPs): samples containing MNP-labeled cells exhibit faster relaxation of NMR signals due to local magnetic fields created by MNPs [14]. The detection principle based on the magnetic interaction allows μNMR to be performed with minimal sample purification steps, which minimizes cell loss and simplifies assay procedures. We have systematically improved the μNMR as a versatile platform for sensitive detection on a wide range of targets, including nucleic acids [15, 16], proteins [10], microvesicles [17], bacteria [18–20], and mammalian cells [8]. Most recently, the platform has been optimized for rapid detection and profiling of CTCs [21, 22]. In this chapter, we review the μNMR technology with emphasis on its technical developments and translational work on CTC detection.

8.2 μNMR TECHNOLOGY

MNPs generate local magnetic fields that accelerate the relaxation of ^{1}H nuclear spins in NMR condition. Between principal spin relaxation mechanisms, namely, the longitudinal (R_1) and the transverse (R_2) ones, the effect of MNPs is more pronounced on the R_2 process. Samples containing MNP-labeled cells display larger R_2 relaxation rate than nonlabeled ones (Figure 8.1), which allows for noninvasive cellular detection. The "R_2-enhancing" capacity of MNPs is defined as transverse relaxivity, r_2 [23], and R_2 can be expressed as

$$R_2 = R_w + r_2 \cdot \left(\frac{N_p}{V}\right) \tag{8.1}$$

where R_w is the relaxation rate of the background (water), V is the NMR detection volume, and N_p is the total number of MNPs in V. When cancer cells are labeled with MNPs, the associated R_2 change ($\Delta R_2 = R_2 – R_w$) is given as

$$\Delta R_2 = r_2 \cdot \left(\frac{N_p}{V}\right) = r_2 \cdot \left(n \cdot \frac{N_c}{V}\right) \tag{8.2}$$

where n is the number of MNPs per cell and N_c is the total number of cells in V. Note that a higher ΔR_2 for a given number of target cells indicates more sensitive detection and profiling capability of μNMR. Based on Eq. 8.2, three approaches have been pursued to improve the μNMR sensitivity: (i) synthesizing MNPs with high r_2; (ii) efficient labeling MNPs on cells, which increases the number of MNPs per cell (n); (iii) miniaturizing a NMR probe, which decreases the detection volume (V). The following sections describe how each approach has been designed and implemented.

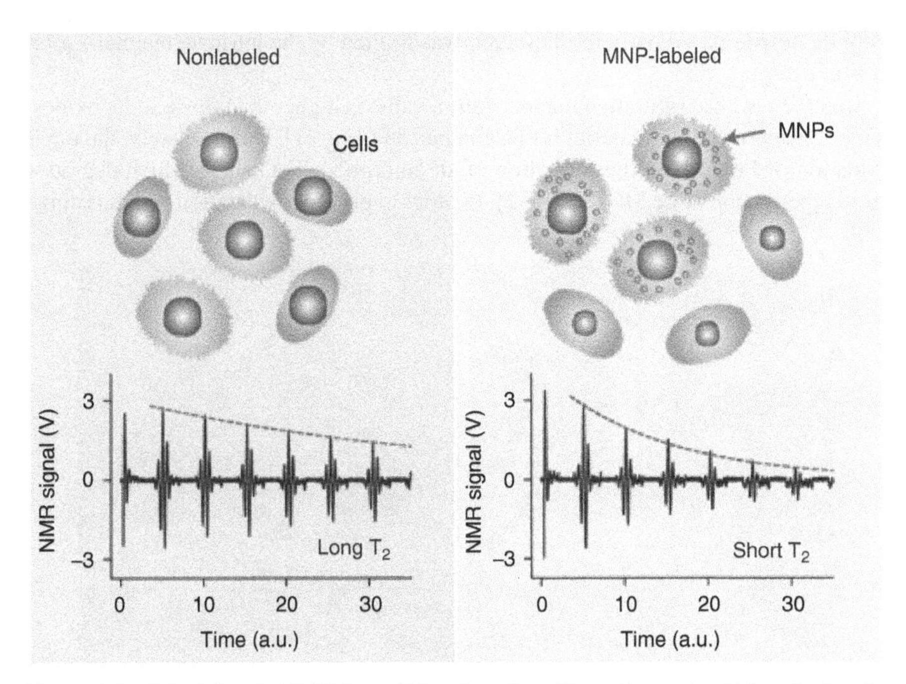

Figure 8.1 Principle of μNMR for cellular detection. Magnetic nanoparticles shorten the transverse relaxation time of water protons. The NMR signal from the magnetically labeled samples (b) decays faster than that from the nonlabeled samples (a). The shortening of relaxation time provides the sensing mechanism of μNMR. Lee *et al.* [10]. Reproduced with permission of Nature Publishing Group.

8.2.1 Magnetic Nanoparticles with High Transverse Relaxivity

The transverse relaxivity (r_2) of an MNP increases with the size (d) and magnetization (M) of the particle, as shown in Eq. 8.3 [24, 25].

$$r_2 \sim d^2 \cdot M^2 \tag{8.3}$$

In an effort to make MNPs with larger size and higher magnetization, ferrite-based (Fe_xO_y) MNPs have been explored first. In addition to their excellent stability and ease of synthesis, the ferrite-based MNPs can be doped with ferromagnetic elements including manganese (Mn), cobalt (Co), or nickel (Ni) to improve M [26–28]. We have synthesized manganese (Mn)-doped magnetite ($MnFe_2O_4$), as the composition is known to have the highest M due to the high spin quantum number (5/2) of Mn^{2+}. Moreover, a seed-mediated growth method was employed to make larger $MnFe_2O_4$ particles (Figure 8.2a) and thereby further increase r_2 [11]. The particles with a diameter ≤16 nm were highly monodisperse and superparamagnetic at room temperature. Their r_2 values reached up to 420/s·mM [Fe + Mn], which was >eight times higher than that of conventional ferrite MNPs (Fe_3O_4 or γFe_2O_3). The r_2 value from the ferrite-based particles, however, was limited by the intrinsic magnetization of $MnFe_2O_4$.

Iron (Fe), whose saturation magnetization value is higher than those of its oxides, can be a superior MNP material for an enhanced r_2 [30, 31]. Unfortunately, the rapid oxidation of Fe leads to the formation of an amorphous FeO shell, which degrades the magnetization of Fe MNPs [31, 32]. In order to prevent oxidization and maximize

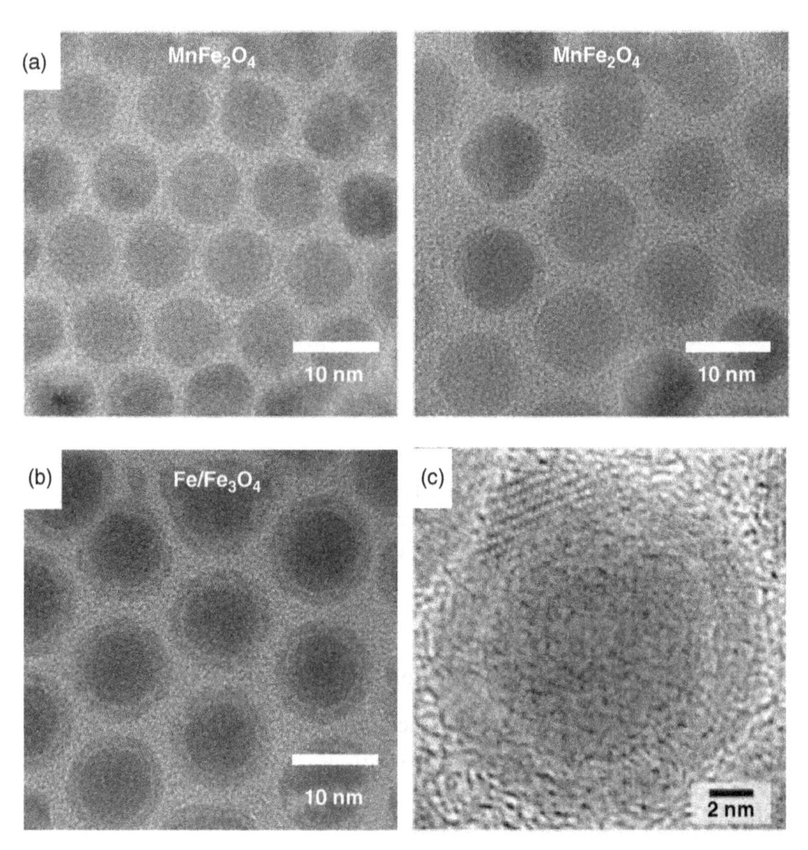

Figure 8.2 Magnetic nanoparticles synthesized for high transverse relaxivity (r_2). (a) Transmission electron microscope (TEM) images of manganese-doped magnetite ($MnFe_2O_4$) particles. Smaller particles (left) were used as a seed to grow bigger particles (right). Adapted from Lee and Weissleder [11]. (b) TEM image of Fe-core and Fe_3O_4-shell (Fe/Fe_3O_4) particles. Yoon *et al.* [29]. Reproduced with permission of John Wiley and Sons, Inc. (c) High-resolution TEM image of an $Fe/MnFe_2O_4$ shell particle. Note that the shell is polycrystalline with a grain size of ~3 nm. (d) Plot of r_2 as a function of magnetization and diameter of MNPs. The dotted line indicates the bulk magnetization value of $MnFe_2O_4$. Because of their stronger magnetization, Fe-core MNPs achieved higher r_2 than ferrite-based MNPs. $Fe/MnFe_2O_4$ MNPs assumed the highest r_2. CLIO, cross-linked iron oxide nanoparticle; MION, monocrystalline iron oxide nanoparticle.

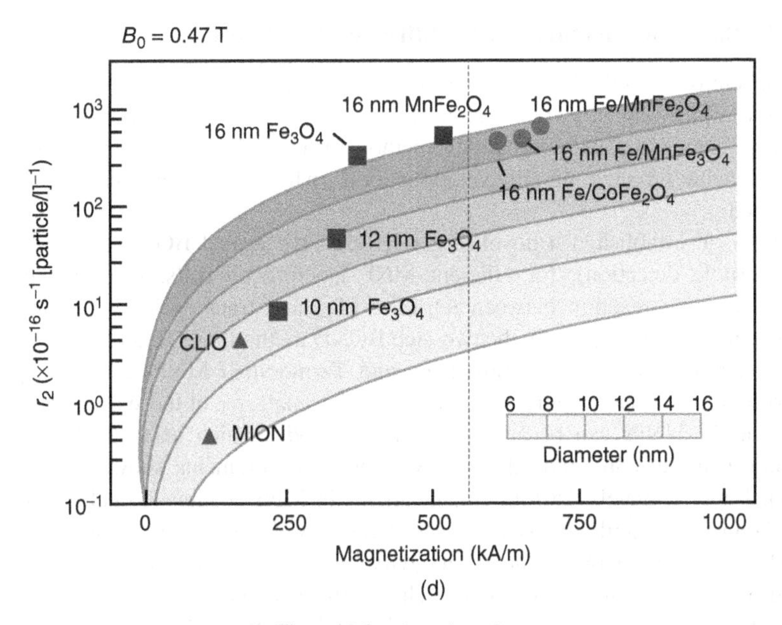

Figure 8.2 *(continued)*

the magnetization of Fe MNPs, we have developed a novel core–shell MNP [29]. This hybrid particle consists of an elemental Fe core and a crystalline ferrite (Fe_3O_4) shell. The structure has the following advantages. First, the crystalline shell protects the core from oxidation. Thus, the overall magnetic moment is stable over time. Second, the shell is overcoated rather than oxidizing the core, leaving the Fe-core intact. Third, the shell could be metal-doped, which permits further tailoring of the particles' magnetic properties. Figure 8.2b shows the transmission electron micrograph (TEM) image of Fe-core/Fe_3O_4-shell (Fe/Fe_3O_4) MNPs. A high-resolution TEM image of Fe/$MnFe_2O_4$ (Figure 8.2c) confirmed the polycrystalline nature of the oxide shell with a grain size of ~3 nm.

Metal-doped magnetite particles and core–shell hybrid structures assumed significantly enhanced r_2 (Figure 8.2d). Note that the magnetization of a particle increases with the particle size. This could be attributed to the reduced effect of a magnetically disordered surface layer in larger particles. In the nanoscale regime, the magnetization (M) of an MNP can be described as

$$M = \frac{M_0(r-d)^3}{r^3} \tag{8.4}$$

where M_0 is the bulk magnetization, r is the particle radius, and d is the thickness of disordered surface layer [33]. Enlarging particle size, therefore, increased r_2 faster than expected.

8.2.2 Bioorthogonal Strategy for Efficient MNP Labeling

A conventional method for MNP labeling uses MNPs preconjugated with target-specific affinity ligands [10], which often requires extensive optimization of both affinity ligands and the conjugation method for each new target. The approach is not amenable for scale-up clinical applications, wherein multiple markers are often analyzed.

We have established a novel targeting strategy, termed BOND (bioorthogonal nanoparticle detection), for efficient MNP labeling on cells. BOND exploits a chemoselective reaction between tetrazine (Tz) and trans-cyclooctene (TCO), as shown in Figure 8.3a [35]. In the two-step BOND method, TCO-modified antibodies are first bound to cellular biomarkers, and Tz-modified MNPs are subsequently coupled to the TCO counterparts. Because of the small size of these coupling agents, multiple Tz-MNPs can be attached to an antibody, which leads to denser MNP loading (Figure 8.3b). Indeed, flow cytometry measurements showed that BOND yielded approximately 15-fold enhancements in MNP loading on cells, compared to labeling with antibody–MNP direct conjugates (Figure 8.3c). A higher number of MNPs per cell increases the r_2 relaxivity (Eq. 8.2), which improves the detection sensitivity of μNMR (Figure 8.3d). The platform combining μNMR and BOND has been shown to be broadly applicable for both extracellular and intracellular biomarker detection and was used for clinical applications [8, 34, 36].

8.2.3 Miniaturized NMR Probe

The signal-to-noise ratio (SNR) of NMR is proportional to the excitation magnetic field over the sample volume produced by unit current [37]. A larger excitation magnetic field per unit current is typically generated by a smaller diameter coil, which has motivated the development of various "microcoils" as an NMR probe. A conventional design wherein a coil is wound around a capillary tube, however, results in a relatively low filling ratio due to the thick capillary walls [38, 39].

The μNMR system achieved a high filling ratio by embedding a microcoil into a microfluidic system [11]. In this approach, the microcoil was wound around a polyethylene tube (1.5 mm diameter) and molded in polydimethylsiloxane (PDMS). After PDMS curing, the microcoil was tightly bound to the PDMS mold, but the tube was easily extracted allowing a microfluidic channel to be fabricated. The structure allowed samples to fill the entire coil bore for NMR detection, which resulted in more than 350% enhancement in NMR signal level (Figure 8.4a). In addition, the PDMS–coil block could be easily integrated with other microfluidic components (e.g., pumps, valves, or filters), enabling NMR sample preparation and detection on a single chip (Figure 8.4b) [17].

The miniaturized NMR probe (diameter: 1 mm, length: 2 mm) allowed a portable permanent magnet (NdFeB) to be used to generate NMR magnetic fields. However, there was one challenge to use a permanent magnet for routine clinical applications: temperature-dependent fluctuation of the magnetic field. For example, 1 °C increase in temperature reduces the magnetic field approximately 0.1% from its initial value,

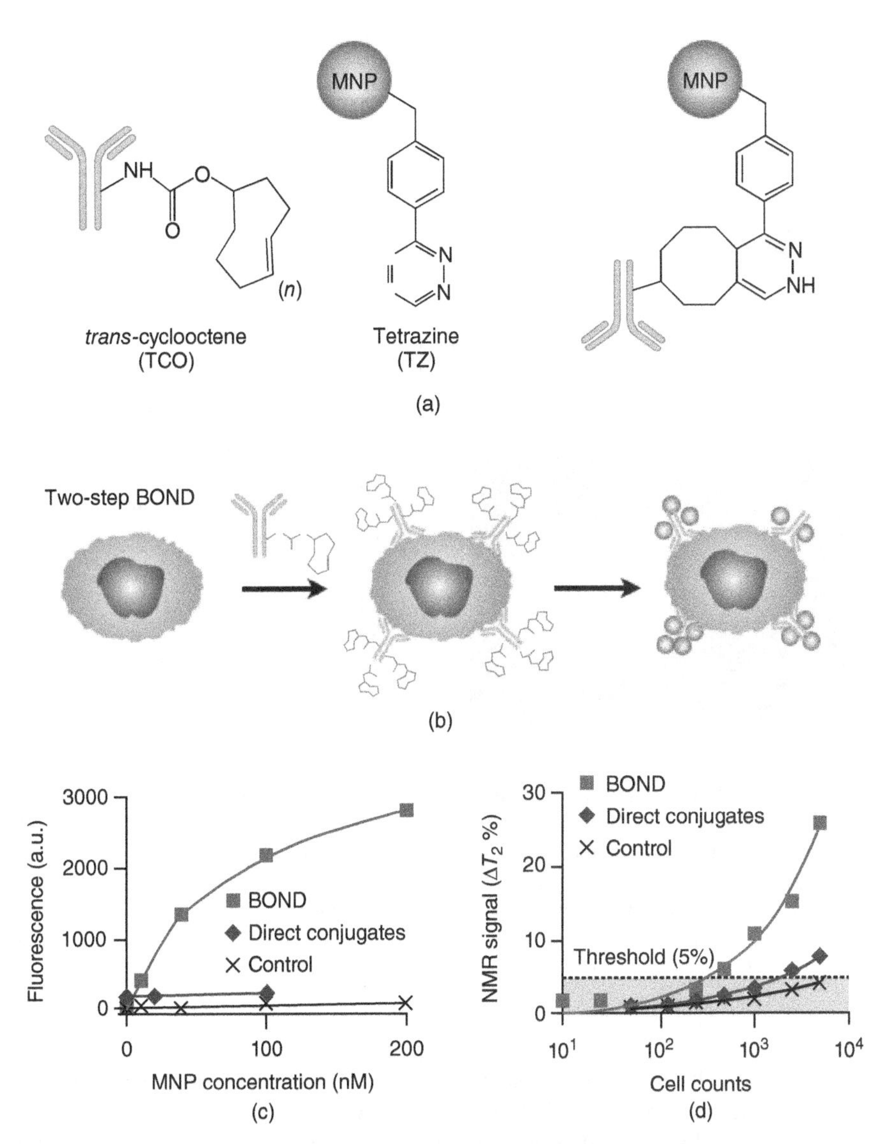

Figure 8.3 Bioorthogonal nanoparticle detection (BOND). (a) Reaction schematic. Trans-cyclooctene (TCO) is conjugated with antibody, and tetrazine (Tz) with MNP. TCO-Tz coupling is based on the Diels–Alder cycloaddition. (b) Cells are prelabeled with TCO–antibodies and targeted with Tz-MNPs. The antibody provides sites for multiple MNP binding. (c, d) Compared to the direct targeting with MNP–antibody conjugates, BOND method achieves higher MNP loading on target cells as confirmed by fluorescent (c) and μNMR (d) measurements. Adapted from Haun *et al.* [34].

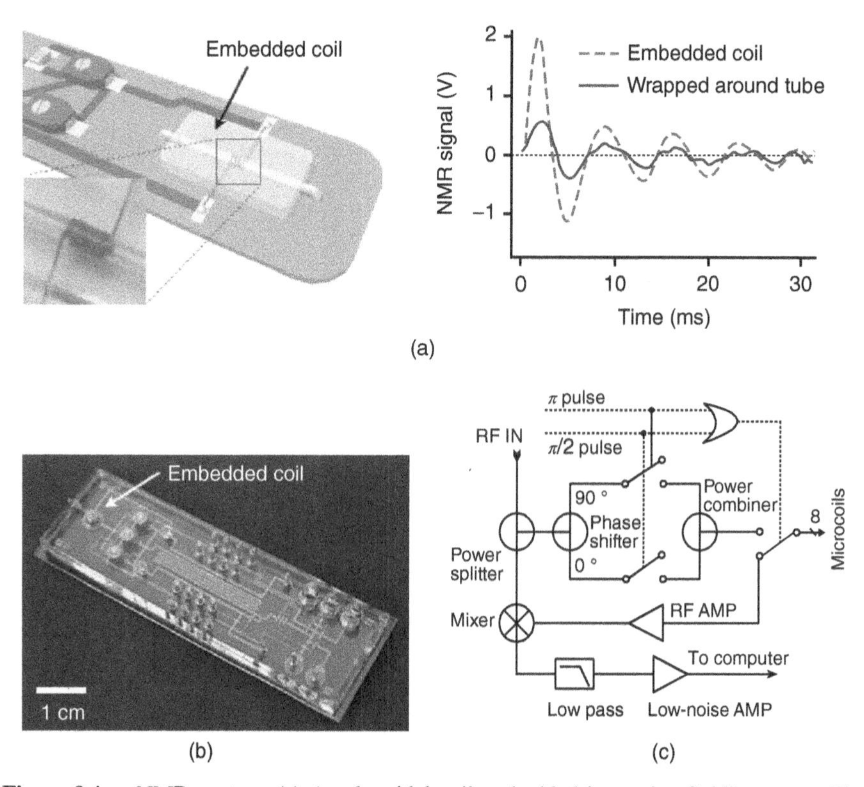

Figure 8.4 μNMR system. (a) A solenoidal coil embedded in a microfluidic system. The entire bore of the coil is available to samples, which maximizes the filling factor (≈1). (Right) The embedded coil design enhanced the NMR signal by more than 350%, compared to a similar coil wrapped around a tube. (b) A solenoidal coil was integrated with a sophisticated microfluidic system. MNP labeling, sample washing and NMR detection can be performed on-chip. (c) Schematic of the μNMR electronic components. Adapted from Lee and Weissleder [11]; Lee *et al.* [10]; Shao *et al.* [17].

which changes the magnetic resonance frequency (f_0) accordingly. Such shift of f_0 could place the NMR signals beyond the processing range (higher than cutoff frequency of the low-pass filter in the circuit diagram in Figure 8.4c), distorting the measured signal. The μNMR solved this problem with a new design of software: a self-adjusting algorithm [13]. As summarized in Figure 8.5a, it scans a wide range of frequencies to find f_0 before each NMR measurement. For the efficient scan of frequency, this algorithm is equipped with two methods: a coarse and a fine tracker. For the first measurement, the μNMR system conducts both tracking methods; for the subsequent measurements, it only uses the fine tracker based on the f_0 used in the immediate previous measurement. This self-adjusting algorithm significantly improved the reliability of the μNMR measurements. For example, without using the self-adjusting, the measured R_2 changed up to 200% from its initial value.

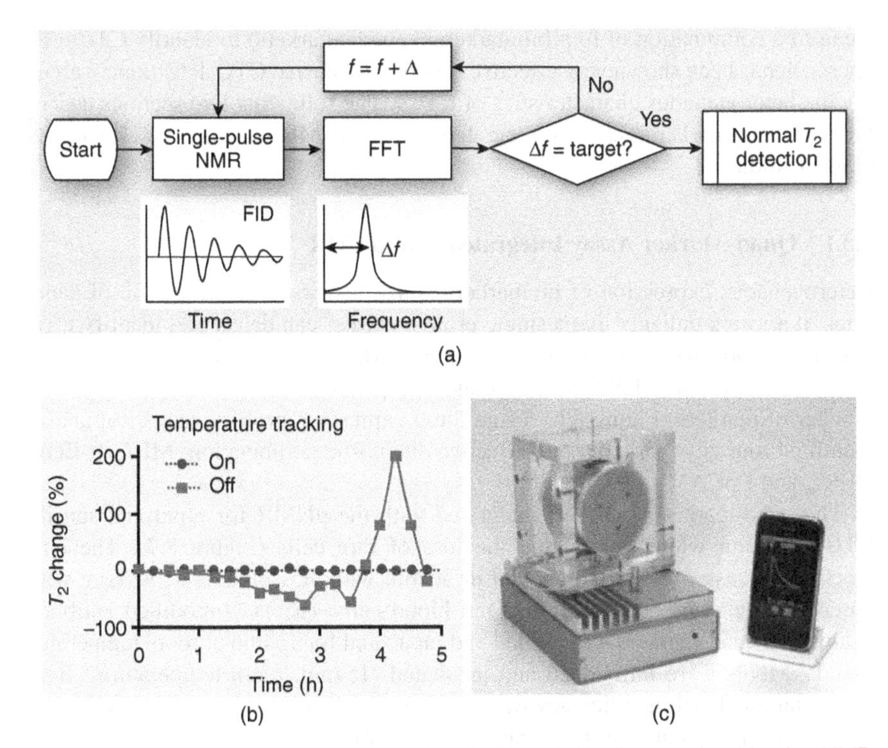

Figure 8.5 μNMR designed for clinical applications. (a) Self-adjusting algorithm in μNMR to compensate a temperature-dependent fluctuation of magnetic field from the permanent magnet. FID, free induction decay; FFT, fast Fourier transformation. (b) Plot of T_2 value changes from the same sample measurement with and without the self-adjusting algorithm. (c) μNMR system for point-of-care operation, featuring automatic system configuration and a smartphone interface. Issadore *et al.* [13]. Reproduced with permission of the Royal Society of Chemistry.

With the self-adjusting algorithm on, the relative R_2 fluctuation was less than 1% (Figure 8.5b).

The μNMR system was packaged for clinical use by integrating portable electronics with a graphical user interface. The electronics were assembled using off-the-shelf parts, which reduced the system cost to less than $200. The programmable interface allowed remote system control and data management with a mobile terminal, such as smartphones or tablet computers (Figure 8.5c). With such user-friendly characteristics, the μNMR system has been used in various translational research.

8.3 CLINICAL APPLICATION OF μNMR FOR CTC DETECTION AND PROFILING

Counting and molecularly profiling CTCs are increasingly recognized to facilitate early cancer detection as well as treatment monitoring [40]. In the μNMR system, we used a combination of four biomarkers ("quad" markers) to identify CTCs. This approach has been shown very effective for comprehensive CTC detection, overcoming the heterogeneous characteristics of CTCs. The following two sections describe how the quad-marker assay was integrated into the μNMR setup and how it catalyzed clinical studies.

8.3.1 Quad-Marker Assay Integrated with μNMR

Heterogeneous expression of biomarkers is a well-known characteristic of cancer. Thus, it appears unlikely that a single protein marker can detect and identify CTCs. In order to find a panel of appropriate protein markers for an effective CTC detection, fine needle aspirates (FNA) from a cohort of patients ($n = 50$) were profiled for a number of markers (Figure 8.6). From these expression profiles, statistical analysis identified four key markers as an effective diagnostic combination: MUC-1, EGFR, HER2, and EpCAM [8].

The quad-marker assay was integrated with the μNMR for rapid and sensitive CTC detection while minimizing the loss of rare cells (Figure 8.7). The entire labeling process could be completed in 30 min without isolation steps: (i) a whole blood sample was lysed to remove red blood cells; (ii) TCO-modified antibodies against the quad-markers were added and incubated for 15 min at room temperature; (iii) Tz-MNPs were introduced and incubated (15 min, room temperature). It was found that the labeling efficiency was maximized when the cells were fixed between red blood cell lysis and antibody labeling steps [21].

The clinical study based on 15 patients and 10 normal controls showed the effectiveness of the quad-marker μNMR for CTC detection [21]. As shown in Figure 8.8a, the quad-marker μNMR demonstrated the presence of CTCs in 13 patients, while CellSearch, a gold standard for CTC detection, identified CTCs in only one patient. Another study also compared the detection rates of the quad-marker μNMR and CellSearch: the former detected about 38% of tumor cells spiked in 7 ml blood samples, whereas the latter detected about 9.1% (Figure 8.8b).

This superior sensing capability of the quad-marker μNMR is mainly attributed to the multimarker detection scheme [21]. With the same μNMR setup, a quad-marker assay demonstrated sensitivity more than twice as high as a single-marker (EpCAM-based) assay (Figure 8.8c). The multimarker detection scheme is particularly important for the detection of CTCs in cancers undergoing EMT or in EpCAM-negative cancers. Figure 8.9 compares the μNMR signal with the single-marker (EpCAM-only) and the quad-marker assay in 12 different types of epithelial cancer cells. Regardless of EpCAM expression levels, the quad-marker method showed the highest μNMR signal for all cancer cells.

8.3.2 Comparison of Biomarkers in CTC and Bulk Tumor Cell

To establish CTCs as a noninvasive surrogate for tissue biopsies, we investigated the correlation between CTCs and their matched distal metastases. The molecular profiles of CTCs in peripheral blood and bulk tumors in fine-needle aspirates were

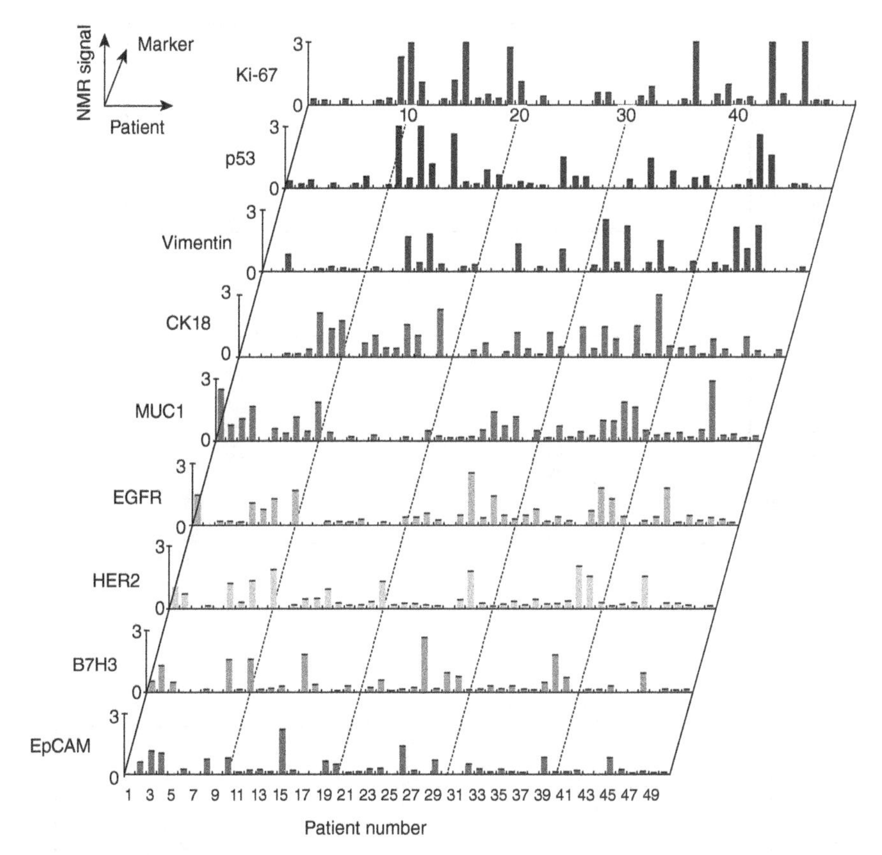

Figure 8.6 Profiling of tumor cells using the μNMR platform. Single fine-needle aspirates were obtained from patients and were tagged via the BOND strategy for μNMR detection. The profiling results indicate a high degree of heterogeneity in protein expression both across the different patient samples and even with the same tumor. Haun *et al.* [8]. Reproduced with permission of The American Association for the Advancement of Science.

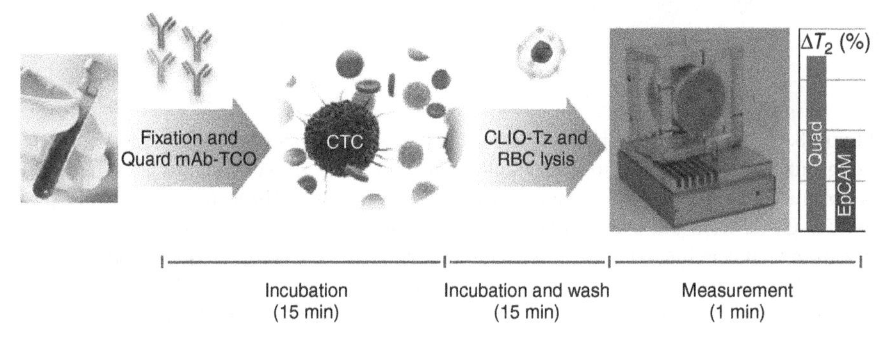

Figure 8.7 Quad-marker μNMR. TCO-modified antibodies are added to the whole blood to target CTCs. Red blood cells are then lysed, and Tz-modified MNPs were added for MNP labeling. The preparation process requires only 30 min. Ghazani *et al.* [21], pp. 388–395. Reproduced with permission of Elsevier.

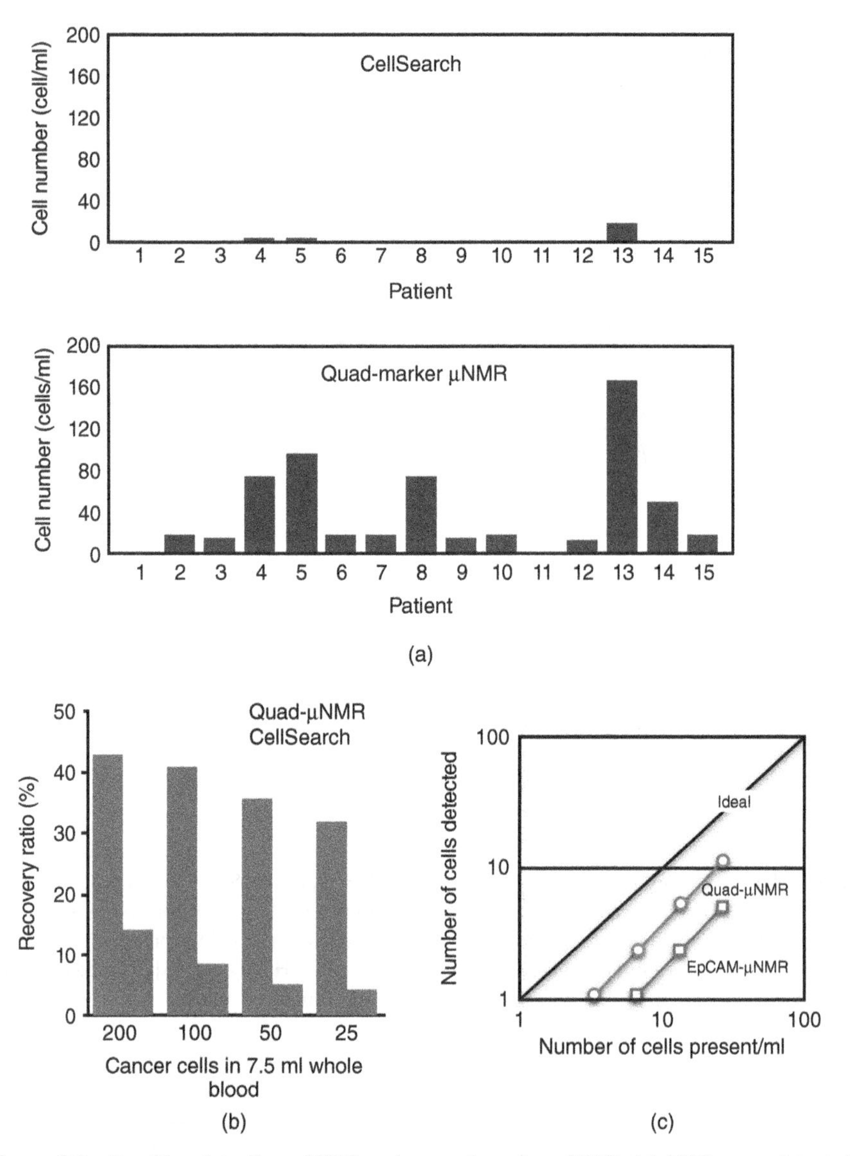

Figure 8.8 Sensitive detection of CTCs using quad-marker μNMR. (a) CTCs were detected using quad-marker μNMR (top) and CellSearch (bottom), a gold standard assay, from the peripheral blood of advanced-stage ovarian cancer patients. Thirteen patients were positive for CTCs from the quad-marker μNMR measurements, but only one patent was positive from CellSearch assay. (b) Comparison of detection sensitivities of quad-marker μNMR and CellSearch using spiked cells in whole blood. With quad-μNMR, the average recovery rate was 38% across the various cell concentrations assessed. Similar experiments with CellSearch showed an average recovery rate of 9.1%. (c) Comparison of detection sensitivities of quad-marker and EpCAM μNMR. The quad-marker assay outperformed the single-marker one in the same μNMR setup. Adapted from Ghazani *et al.* [21], pp. 388–395. Reproduced with permission of Elsevier.

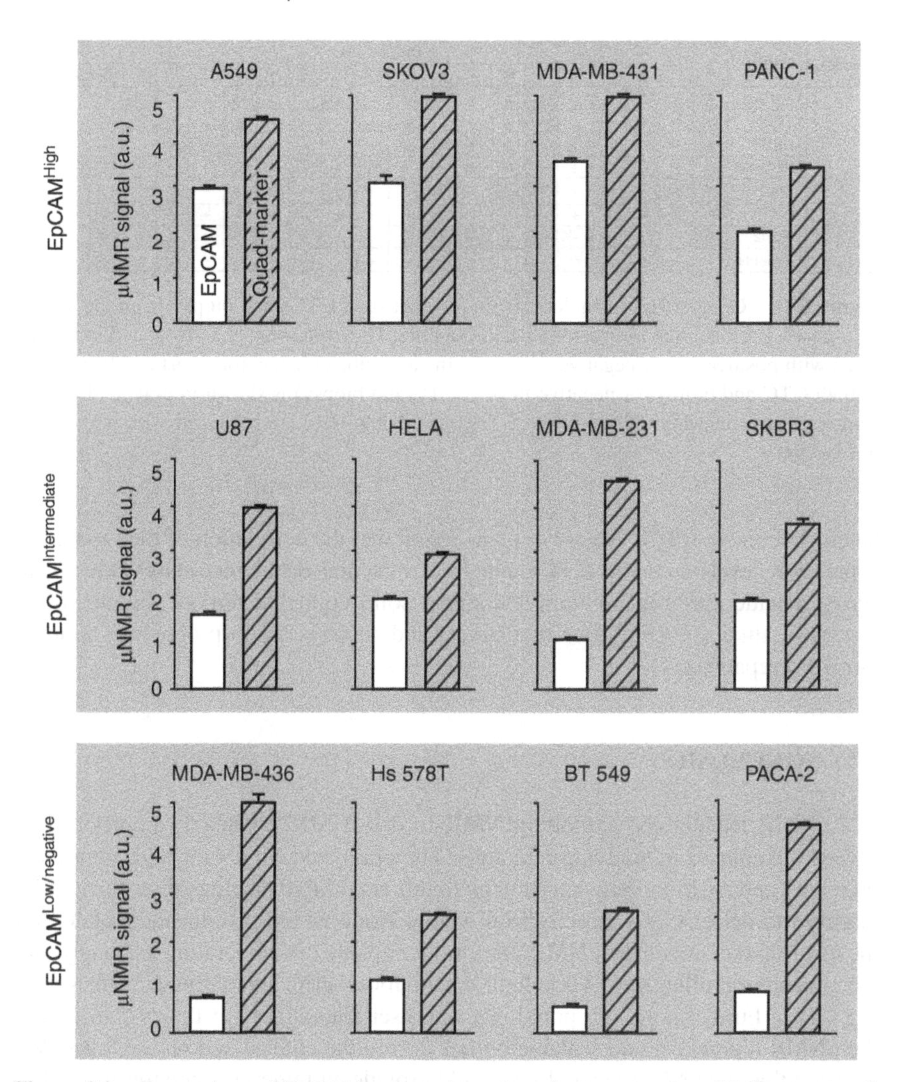

Figure 8.9 Comparison of EpCAM-only and quad-marker assay using μNMR. Across all cancer cell lines with different EpCAM expression levels, the quad-marker assay (striped bars) showed higher NMR signal than the EpCAM-only assay (solid bars). All measurements were performed in triplicate. The error bar indicates the standard deviation. Adapted from Ghazani *et al.* [21], pp. 388–395. Reproduced with permission of Elsevier.

analyzed with the quad-marker μNMR assay [21]. In the quad-marker set, we replaced MUC-1 with vimentin, a key EMT marker expressed in CTCs [41–43]. The expression level of vimentin was generally 50% lower in patients with worsening clinical trajectories than in patients with stable or improving trajectories, even though the expression levels of EGFR, EpCAM, and HER2 were considerably

Sample	Marker	Patient no.																				
		1	2	3	5	6	7	8	9	10	11	13	14	15	16	17	18	23	29	32	33	34
CTC	EpCAM	+	–	+	+	–	+	+	+	+	+	–	+	–	–	+	+	+	–	+	+	–
	EGFR	+	–	+	+	–	+	+	+	+	+	–	–	–	–	+	+	+	–	+	+	–
	HER2	+	+	+	–	+	+	+	+	+	+	–	–	–	+	+	+	+	–	+	+	+
	Vimentin	–	+	+	–	+	+	+	+	+	+	–	+	–	+	+	+	+	–	+	+	+
Biopsy	EpCAM	+	–	+	+	+	+	+	–	+	+	+	+	+	+	+	–	+	–	–	–	–
	EGFR	+	–	+	+	+	+	–	–	+	+	+	+	–	+	+	+	+	–	+	–	–
	HER2	+	–	+	+	+	+	–	+	+	+	+	+	–	–	+	+	+	+	+	–	–
	Vimentin	+	–	+	–	+	–	–	–	–	–	–	–	–	–	+	–	+	+	+	+	–

Figure 8.10 Comparison of molecular profiles between CTCs and biopsy from the site of metastases. Each column represents a patient sample. The expression of a given marker is indicated with positive (+) or negative (–) signs. In each subject, concordant test result (positive in both CTC and biopsy, or negative in both CTC and biopsy) is shown in gray, and discordant test result in white. Ghazani *et al.* [22], pp. 1009–1017. Reproduced with permission of Elsevier.

heterogeneous in CTCs. Notably, it was found that the correlation of the biomarker expression levels between CTCs and their matched distal metastases was weak across a wide spectrum of intra-abdominal solid epithelial tumors (Figure 8.10). Extensive studies over various biomarkers and subjects are in progress to validate those discrepancies.

8.4 CONCLUSION

The μNMR technology, exploiting NMR to detect MNPs-labeled samples, offers several advantages in studying rare cells. The optimized MNPs for high transverse relaxivity and the bioorthogonal strategy for efficient MNP labeling allow the μNMR platform to detect CTCs directly from whole blood within 30 min. Integrated with the quad-marker assay, the μNMR platform is capable of enumerating and analyzing the molecular profiles of CTCs at high sensitivities – critical requirements for studying CTCs. Furthermore, the portability and user-friendliness of the system render the μNMR a practical and scalable option for routine clinical and research use. We envision that the μNMR technology would provide an important step toward understanding cancer biology and personalized cancer therapies.

REFERENCES

1. Nagrath, S., Sequist, L.V., Maheswaran, S. *et al.* (2007) Isolation of rare circulating tumour cells in cancer patients by microchip technology. *Nature*, **450**, 1235–1239.
2. Lang, J.M., Casavant, B.P., and Beebe, D.J. (2012) Circulating tumor cells: getting more from less. *Science Translational Medicine*, **4**, 141ps13.
3. Chaffer, C.L. and Weinberg, R.A. (2011) A perspective on cancer cell metastasis. *Science*, **331**, 1559–1564.

4. Cristofanilli, M., Budd, G.T., Ellis, M.J. *et al.* (2004) Circulating tumor cells, disease progression, and survival in metastatic breast cancer. *New England Journal of Medicine*, **351**, 781–791.
5. Maheswaran, S., Sequist, L.V., Nagrath, S. *et al.* (2008) Detection of mutations in EGFR in circulating lung-cancer cells. *New England Journal of Medicine*, **359**, 366–377.
6. Pierga, J.Y., Bidard, F.C., Mathiot, C. *et al.* (2008) Circulating tumor cell detection predicts early metastatic relapse after neoadjuvant chemotherapy in large operable and locally advanced breast cancer in a phase II randomized trial. *Clinical Cancer Research*, **14**, 7004–7010.
7. Yu, M., Stott, S., Toner, M. *et al.* (2011) Circulating tumor cells: approaches to isolation and characterization. *Journal of Cell Biology*, **192**, 373–382.
8. Haun, J.B., Castro, C.M., Wang, R. *et al.* (2011) Micro-NMR for rapid molecular analysis of human tumor samples. *Science Translational Medicine*, **3**, 71ra16.
9. Armstrong, A.J., Marengo, M.S., Oltean, S. *et al.* (2011) Circulating tumor cells from patients with advanced prostate and breast cancer display both epithelial and mesenchymal markers. *Molecular Cancer Research*, **9**, 997–1007.
10. Lee, H., Sun, E., Ham, D., and Weissleder, R. (2008) Chip-NMR biosensor for detection and molecular analysis of cells. *Nature Medicine*, **14**, 869–874.
11. Lee, H. and Weissleder, R. (2009) Rapid detection and profiling of cancer cells in fine-needle aspirates. *Proceedings of the National academy of Sciences of the United States of America*, **106**, 12459–12464.
12. Lee, H., Yoon, T.J., and Weissleder, R. (2009) Ultrasensitive detection of bacteria using core-shell nanoparticles and an NMR-filter system. *Angewandte Chemie International Edition in English*, **48**, 5657–5660.
13. Issadore, D., Min, C., Liong, M. *et al.* (2011) Miniature magnetic resonance system for point-of-care diagnostics. *Lab on a Chip*, **11**, 2282–2287.
14. Gillis, P. and Koenig, S.H. (1987) Transverse relaxation of solvent protons induced by magnetized spheres: application to ferritin, erythrocytes, and magnetite. *Magnetic Resonance in Medicine*, **5**, 323–345.
15. Liong, M., Hoang, A.N., Chung, J. *et al.* (2013) Magnetic barcode assay for genetic detection of pathogens. *Nature Communications*, **4**, 1752.
16. Chung, H.J., Castro, C.M., Im, H. *et al.* (2013) A magneto-DNA nanoparticle system for rapid detection and phenotyping of bacteria. *Nature Nanotechnology*, **8**, 369–375.
17. Shao, H., Chung, J., Balaj, L. *et al.* (2012) Protein typing of circulating microvesicles allows real-time monitoring of glioblastoma therapy. *Nature Medicine*, **18**, 1835–1840.
18. Budin, G., Chung, H.J., Lee, H., and Weissleder, R. (2012) A magnetic Gram stain for bacterial detection. *Angewandte Chemie International Edition in English*, **51**, 7752–7755.
19. Chung, H.J., Reiner, T., Budin, G. *et al.* (2011) Ubiquitous detection of Gram-positive bacteria with bioorthogonal magnetofluorescent nanoparticles. *ACS Nano*, **5**, 8834–8841.
20. Liong, M., Fernandez-Suarez, M., Issadore, D. *et al.* (2011) Specific pathogen detection using bioorthogonal chemistry and diagnostic magnetic resonance. *Bioconjugate Chemistry*, **22**, 2390–2394.
21. Ghazani, A.A., Castro, C.M., Gorbatov, R. *et al.* (2012) Sensitive and direct detection of circulating tumor cells by multimarker micro-nuclear magnetic resonance. *Neoplasia*, **14**, 388–395.

22. Ghazani, A.A., McDermott, S., Pectasides, M. *et al.* (2013) Comparison of select cancer biomarkers in human circulating and bulk tumor cells using magnetic nanoparticles and a miniaturized micro-NMR system. *Nanomedicine*, **9**, 1009–1017.
23. Gossuin, Y., Gillis, P., Hocq, A. *et al.* (2009) Magnetic resonance relaxation properties of superparamagnetic particles. *WIREs Nanomedicine and Nanobiotechnology*, **1**, 299–310.
24. Brooks, R.A., Moiny, F., and Gillis, P. (2001) On T2-shortening by weakly magnetized particles: the chemical exchange model. *Magnetic Resonance in Medicine*, **45**, 1014–1020.
25. Gillis, P., Moiny, F., and Brooks, R.A. (2002) On T2-shortening by strongly magnetized spheres: a partial refocusing model. *Magnetic Resonance in Medicine*, **47**, 257–263.
26. Lee, J.H., Huh, Y.M., Jun, Y.W. *et al.* (2007) Artificially engineered magnetic nanoparticles for ultra-sensitive molecular imaging. *Nature Medicine*, **13**, 95–99.
27. Sun, S., Zeng, H., Robinson, D.B. *et al.* (2004) Monodisperse MFe2O4 (M= Fe, Co, Mn) nanoparticles. *Journal of the American Chemical Society*, **126**, 273–279.
28. Jun, Y., Lee, J., and Cheon, J. (2008) Chemical design of nanoparticle probes for high-performance magnetic resonance imaging. *Angewandte Chemie International Edition in English*, **47**, 5122–5135.
29. Yoon, T.J., Lee, H., Shao, H., and Weissleder, R. (2011) Highly magnetic core-shell nanoparticles with a unique magnetization mechanism. *Angewandte Chemie International Edition in English*, **50**, 4663–4666.
30. Huber, D.L. (2005) Synthesis, properties, and applications of iron nanoparticles. *Small*, **1**, 482–501.
31. Qiang, Y., Antony, J., Sharma, A. *et al.* (2006) Iron/iron oxide core-shell nanoclusters for biomedical applications. *Journal of Nanoparticle Research*, **8**, 489–496.
32. Seo, W.S., Lee, J.H., Sun, X. *et al.* (2006) FeCo/graphitic-shell nanocrystals as advanced magnetic-resonance-imaging and near-infrared agents. *Nature Mater.*, **5**, 971–976.
33. Jun, Y., Seo, J., and Cheon, J. (2008) Nanoscaling laws of magnetic nanoparticles and their applicabilities in biomedical sciences. *Accounts of Chemical Research*, **41**, 179–189.
34. Haun, J.B., Devaraj, N.K., Hilderbrand, S.A. *et al.* (2010) Bioorthogonal chemistry amplifies nanoparticle binding and enhances the sensitivity of cell detection. *Nature Nanotechnology*, **5**, 660–665.
35. Devaraj, N.K., Upadhyay, R., Haun, J.B. *et al.* (2009) Fast and sensitive pretargeted labeling of cancer cells through a tetrazine/trans-cyclooctene cycloaddition. *Angewandte Chemie International Edition in English*, **48**, 7013–7016.
36. Haun, J.B., Devaraj, N.K., Marinelli, B.S. *et al.* (2011) Probing intracellular biomarkers and mediators of cell activation using nanosensors and bioorthogonal chemistry. *ACS Nano*, **5**, 3204–3213.
37. Hoult, D.I. and Richards, R.E. (1976) The signal-to-noise ratio of the nuclear magnetic resonance experiment. *Journal of Magnetic Resonance*, **24**, 71–85.
38. Massin, C., Vincent, F., Homsy, A. *et al.* (2003) Planar microcoil-based microfluidic NMR probes. *Journal of Magnetic Resonance*, **164**, 242–255.
39. Eroglu, S., Gimi, B., Roman, B. *et al.* (2003) NMR spiral surface microcoils: design, fabrication, and imaging. *Concepts in Magenetic Resonance*, **17**, 1–10.
40. Yu, M., Bardia, A., Wittner, B.S. *et al.* (2013) Circulating breast tumor cells exhibit dynamic changes in epithelial and mesenchymal composition. *Science*, **339**, 580–584.

41. McInroy, L. and Määttä, A. (2007) Down-regulation of vimentin expression inhibits carcinoma cell migration and adhesion. *Biochemical and Biophysical Research Communications*, **360**, 109–114.
42. Barriere, G., Riouallon, A., Renaudie, J. *et al.* (2012) Mesenchymal characterization: alternative to simple CTC detection in two clinical trials. *Anticancer Research*, **32**, 3363–3369.
43. Connelly, L., Barham, W., Onishko, H.M. *et al.* (2011) NF-kappaB activation within macrophages leads to an anti-tumor phenotype in a mammary tumor lung metastasis model. *Breast Cancer Research*, **13**, R83.

9

NANOVELCRO CELL-AFFINITY ASSAY FOR DETECTING AND CHARACTERIZING CIRCULATING TUMOR CELLS

Millicent Lin, Anna Fong, and Sharon Chen

Department of Molecular and Medical Pharmacology, Crump Institute for Molecular Imaging (CIMI), California NanoSystems Institute (CNSI), University of California, Los Angeles, CA, USA

Yang Zhang

Biomedical Engineering Program, The University of Texas at El Paso, El Paso, TX, USA

Jie-fu Chen

Posadas Lab, Urologic Oncology Research Program, Samuel Oschin Comprehensive Cancer Institute, Cedars-Sinai Medical Center, Los Angeles, CA, USA

Paulina Do and Morgan Fong

Department of Molecular and Medical Pharmacology, Crump Institute for Molecular Imaging (CIMI), California NanoSystems Institute (CNSI), University of California, Los Angeles, CA, USA

Shang-Fu Chen

Department of Life Science, National Taiwan University, Taipei City, Taiwan (ROC)

Pauline Yang

Department of Molecular and Medical Pharmacology, Crump Institute for Molecular Imaging (CIMI), California NanoSystems Institute (CNSI), University of California, Los Angeles, CA, USA

An-Jou Liang

Department of Life Science, National Taiwan University, Taipei City, Taiwan (ROC)

Circulating Tumor Cells: Isolation and Analysis, First Edition. Edited by Z. Hugh Fan.

QINGYU LI, MIN SONG, SHUANG HOU, AND HSIAN-RONG TSENG

Department of Molecular and Medical Pharmacology, Crump Institute for Molecular Imaging (CIMI), California NanoSystems Institute (CNSI), University of California, Los Angeles, CA, USA

9.1 INTRODUCTION

9.1.1 Circulating Tumor Cells

Circulating tumor cells (CTCs) are rare cells shed from the primary tumor that travel through the circulatory system [1], then spread to different tissues of the body, where they reside and ultimately begin to proliferate as metastasis [2–4]. Multiple types of tumors can result from the invasion of epithelial tumor cells to additional organs distant from the primary ones. Patients with metastatic diseases may undergo biopsy, the gold standard for measuring the presence and extent of tumor. However, the procedure is invasive, and the spread of malignant cells can follow a tumor biopsy due to the lack of cohesion of malignant cells and the ease of detachment of cancer cells from a tumor. CTCs in peripheral blood can be regarded as a “liquid biopsy” [5] for effectively monitoring the progression of the disease and in determining the use of different treatments. The CTC characterization can act as a marker to track the real-time progress of the tumors, gauging the effectiveness of the patient’s treatment. This can provide an indication of patients’ tumor evolution and change patients’ predicted outcomes [5, 6]. The number of CTCs present in patients can be a promising aspiration in terms of the probability of their survival rates and the outcomes of their treatments.

9.1.2 Current CTC Capture Methods

Significant efforts have been made to develop a diversity of technologies for CTC capture. These methods use various principles to capture CTCs, with the goal of doing so with greater efficiency. The most widely used CTC detection assays are summarized as follows:

- **Immunomagnetic separation:** This approach uses capture-agent-labeled magnetic beads to target cell surface markers for the positive selection of CTCs [7–11] or CD45 for the negative depletion [12–14] of white blood cells (WBCs). Currently, CellSearch™ (see Chapter 19 for the details), an assay based on the positive selection mechanism, is the only Food and Drug Administration (FDA)-approved technology for CTC enumeration to predict patient outcome for metastatic breast [10], prostate [11], and colorectal [15] cancers. The method enriches cells using magnetic beads coated with antibodies against epithelial cell adhesion molecule (EpCAM). Over the past 10 years of clinical practice, oncologists have come to the conclusion that the widespread use of CellSearch assay is constrained by its limited sensitivity [16]. To further improve detection speed and efficiency, several sophisticated technologies have recently been developed, including the VerIFAST [7], magnetic sifter [8], MagSweeper [17], and IsoFlux [18] systems. In parallel, Massachusetts General Hospital team's iChip [14] assay has gained recognition from the field for its application of the negative depletion mechanism. iChip performs this task by using antibodies against CD45 and CD15 to immunomagnetically label leukocytes present in whole blood, allowing those cells to be separated from the sample. However, the purity of CTCs acquired by this technology remains insufficient for further biomolecular analysis.

- **Gradient centrifugation:** CTCs can be separated from other hematopoietic components based on density differences. The Ficoll [19] procedure uses density gradient liquids to separate whole blood into heavier particles (erythrocytes and neutrophils) and lighter particles (mononuclear cells, tumor cells, and plasma). If the whole blood is not centrifuged immediately, it tends to mix with the gradient, resulting in high contamination with WBCs, potential loss of CTCs, and general failure of separation. Oncoquick [20, 21], which generates a layered separation with the help of a porous barrier to prevent contamination with the gradient below, was developed to address this issue. Oncoquick depletes mononuclear cells with greater efficiency than Ficoll, but it suffers from frequent loss of CTCs due to migration of cells into the plasma layer.
- **Flow cytometry:** Flow cytometry [22–24] involves a process in which cells are suspended in a fluid that passes through an electronic detection device. It is capable of analyzing thousands of cells within a small sample volume in a small time frame and can perform several measurements on single cells in a heterogeneous mixture. Although flow cytometry is one of the most mature technologies for detecting and isolating subpopulations of cells, it does not meet the gold standard set by pathologists due to an inability to provide sufficient morphological information of CTCs. An assay [25, 26] known as ensemble-decision aliquot ranking (eDAR) is being developed in hopes of improving upon the current technology.
- **Microfluidic chips:** Several microfluidic-based technologies [27–34] have been developed to move forward in advancing CTC detection and isolation. The advantage of this technology are its capacity for automatic programming, flexibility in performing a large number of samples, and capacity for further molecular analysis. The disadvantage to this technology is the difficulty in assessing microscopy imaging. Due to the vertical depths of 3D devices (e.g., micropillars [27, 28] or herringbones [29, 30]), depth of field issues on the captured CTCs can arise. As a result, multiple cross-sectional scans are required to avoid out-of-focus or superimposed micrographs, a process that is time-consuming and often generates massive image files.
- **CTC filters:** Because the surrounding blood cells are generally smaller than CTCs, using filters [35–39] to separate whole blood can eliminate the majority

of peripheral cells. The parylene-based membrane microfilter device [36] uses two parylene membrane layers and a photolithography-defined gap to minimize stress, yielding viable cells for further molecular analysis with high efficiency. The advantage of this technology lies in its unbiased detection of CTCs from all types of cancer, including both epithelial and mesenchymal cell types. However, the greatest limitation of this technology is its sensitivity to size: questions have been raised regarding the loss of CTCs smaller than the filter's pores, a valid concern given that the size of CTCs is extremely variable.

- **High-throughput microscopy:** High-throughput microscopy is an imaging technique that uses a high-resolution microscope with the ability to capture images at a high speed to detect CTCs. The blood sample can be stained with fluorescent markers to identify CTCs, which leads to more accurate detection. This method is relatively difficult because it does not have an enrichment step, leaving CTCs in a large medium of other blood cells. Recently, a team of researchers from the Scripps Research Institute has enumerated [40, 41] label-free cells of interest by a cell density- and size-based amplified microscopy-based assay known as fiber array scanning technology (FAST), which generates "high-definition" images for detection, enumeration, and molecular analysis of CTCs. However, these findings were determined with a limited sample size; thus, the prognostic and clinical significance of the technology cannot be confirmed until further studies have been performed.
- **Other methods:** Combined assays for CTC detection have been widely used, incorporating multiple methods to achieve optimal cell recovery, such as processing [42, 43] whole blood by density gradient centrifugation and immunomagnetic isolation in parallel with the use of a microfluidic device to sort out cells of interest. TelomeScan [44, 45] is able to detect tumor cells by infecting cells with telomerase-specific, replication-selective adenovirus, which incorporates green fluorescence protein into cells through the process of replication. Nucleic-acid-based methods [46–48] have been widely used to exploit the characterization of CTCs, utilizing primers designed to target the specific gene of interest. However, the limitations derived from such technologies include lack of stability and reproducibility due to contamination, leading to false analysis.

Although existing technologies have been proven to detect CTCs with some efficiency, persisting challenges include (i) establishing a translational pipeline for researchers and clinicians to work jointly toward clinical validation and FDA approval, and, more importantly, (ii) recovering CTCs with the purity, viability, and molecular integrity required for subsequent molecular and functional analyses, which can provide insight into tumor biology during the critical window when treatment intervention can significantly improve a patient's prognosis. However, technologies invented to further exploit the biological characterization of CTCs have been challenged by the low abundance of CTCs (a few to hundreds per milliliter of whole blood) among a large number of erythrocytes ($\sim 10^9$ per milliliter of whole blood) [40, 49, 50], and the difficulty of separating CTCs among the background population of other hematopoietic components in the bloodstream. The fact that CTCs occur in low abundance has impeded the understanding of cancer evolution.

9.1.3 The Evolution of NanoVelcro Cell-Affinity Assays

In contrast to existing CTC detection methods, the UCLA team has developed the "NanoVelcro" cell-affinity assay, a novel technology in which capture-agent-coated nanostructured substrates [51–54] are used to efficiently immobilize viable CTCs. The mechanism of this assay works similarly to that of Velcro, in which increased surface area between two substrates enhances binding. Through continuous evolution over the course of several years, three generations of NanoVelcro chips, each holding a specific clinical utility, have been established. A graphical representation of the evolution of the NanoVelcro technology appears in Figure 9.1.

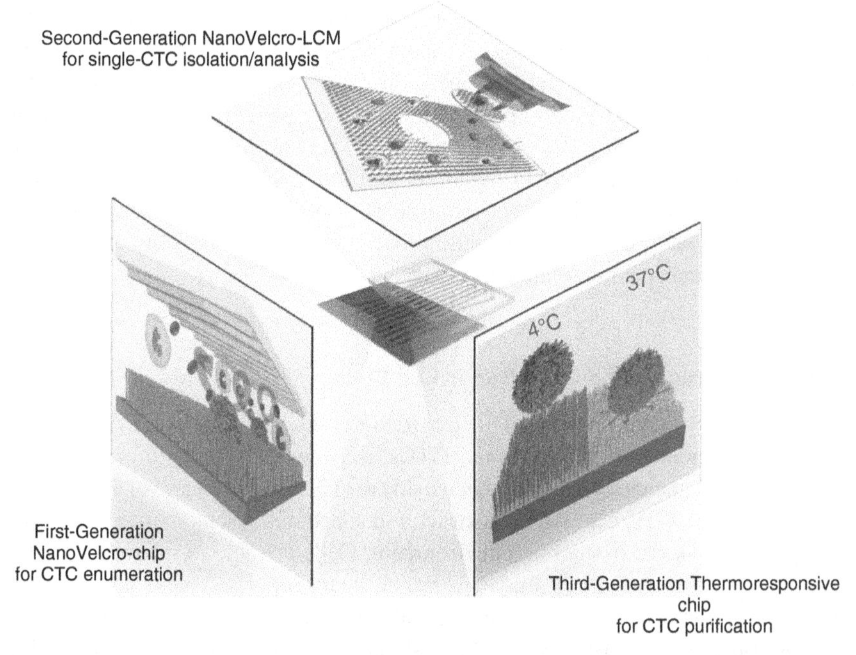

Figure 9.1 The evolution of the three generations of NanoVelcro CTC chips. The first-generation chip, composed of a silicon nanowire substrate (SiNS) chip with an overlaid microfluidic chaotic mixer, provides a high-sensitivity method that outperforms the FDA-approved CellSearch™ in CTC enumeration from clinical blood samples. The second-generation polymer nanosubstrate chip specializes in single CTC isolation, which can then be subjected to single-cell genotyping. The third-generation thermoresponsive NanoVelcro chip utilizes a capture-and-release mechanism at 37 °C and 4 °C, respectively; the temperature-induced conformational change alters the accessibility of the captured cells and allows for rapid CTC purification with desired viability and cellular integrity.

The first-generation NanoVelcro chip [55–57], consisting of a silicon nanowire substrate (SiNS) and an overlaid microfluidic chaotic mixer, was designed for CTC enumeration. The SiNS enhances local topographic interactions with nanoscale cell-surface components (e.g., microvilli) to increase CTC affinity, just as the textured surfaces of two Velcro strips stick together. The herringbone pattern of the serpentine chaotic mixing channel induces vertical flow of blood, which significantly increases contact frequency between the substrate and the cells and ensures that the blood that passes through is mixed well. Side-by-side analytical validation studies using clinical blood samples suggested that the sensitivity of the first-generation NanoVelcro chip outperforms that of CellSearch assay.

The second-generation NanoVelcro chip [58, 59], which combines a transparent polymer nanosubstrate in conjunction with laser capture microdissection (LCM) techniques, was developed for single-CTC isolation. The individually isolated CTCs can subsequently be subjected to single-CTC genotyping (e.g., Sanger sequencing and next-generation sequencing, NGS) to confirm their role as tumor liquid biopsy.

The third-generation NanoVelcro chip [60] uses temperature-dependent conformational changes of thermoresponsive polymer brushes grafted onto SiNS to effectively capture and release CTCs at 37 °C and 4 °C, respectively. The polymer brushes alter the accessibility of the capture agent on SiNS depending on the temperature of the system, allowing for rapid purification of CTCs while preserving their viability and molecular integrity. This causes minimum contamination of the surrounding WBCs and negligible disruption to viability and function of CTCs. This generation of the NanoVelcro chip holds great importance in allowing for further biomolecular analysis of viable CTCs, which could provide significant insight to give patients individualized and specific treatments.

9.1.4 Nanostructured Substrates for Cell Biology

The two main obstacles underlying the technologies described in the previous section include insufficient purity of detected CTCs and low CTC capture efficiency. To address these obstacles, efforts have been made to explore the concept of using nanostructures to aid in CTC capture. Nanostructured substrates assemble many characteristics presented in the tissue microenvironment. Cells are able to interact with their surroundings through contact guidance, which is an important factor that accounts for cell migration. Cell fates can be manipulated by incorporating chemical, mechanical, and topographical aspects into the designs of cell biointerfaces. Nanograftings embedded on the substrate are believed to alter cell fate by enhancing the adhesion of cells, as stated that cells grow in the direction of the grafting axis. In the pursuit of understanding how nanosubstrates affect cells, many researchers have developed substrates that mimic the features characterized by the extracellular matrix (ECM) [61–63]. Studies have shown that the embedded nanostructures promote communication between the cell and the microenvironment, inducing stimuli to accurately control cell functions. Serial designs of nanostructured substrates inspired by the idea of cell–material interaction have been put to use in capturing and isolating CTCs.

9.2 PROOF-OF-CONCEPT DEMONSTRATION OF NANOVELCRO CELL-AFFINITY SUBSTRATES

9.2.1 Stationary NanoVelcro CTC Assay

The NanoVelcro cell-affinity substrate [51–54] operates (Figure 9.2a) in a similar manner to Velcro, in which tangling between the two textured surfaces of a Velcro fastener leads to enhanced binding. SiNS, which allow for Velcro-like interactions [64] with nanoscale cell-surface components, were used in the proof-of-concept demonstration of the NanoVelcro cell-affinity assay [51]. Anti-EpCAM, the capture agent of interest, was grafted onto SiNS to confer specificity to a CTC-specific cell-affinity assay. Cell-capture affinity of the anti-EpCAM grafted SiNS was elevated when compared to that of an unstructured (i.e., flat silicon) substrate (Figure 9.2b). The anti-EpCAM grafted SiNS was fabricated through three continuous steps (Figure 9.2c): (i) introduction of densely packed silicon nanowires (100–200 nm in diameter) onto a silicon wafer, (ii) silane treatment and covalent conjugation of streptavidin onto SiNS, and (iii) grafting of biotinylated anti-EpCAM onto the streptavidin-coated SiNS. Comprehensive optimization for performing cell capture on whole blood samples in a stationary device setting was completed. Scanning electron microscopy (SEM) characterization of CTCs on SiNS (Figure 9.2d) revealed many interdigitated cell-surface components, supporting the working mechanism of the NanoVelcro cell-affinity substrate. Capture efficiency of NanoVelcro cell-affinity substrates using artificial blood samples was determined to be approximately 40–70%.

9.2.2 General Applicability of NanoVelcro CTC Substrates

In addition to SiNS, various other nanomaterials have been incorporated into NanoVelcro substrates using different fabrication approaches. Horizontally oriented TiO_2 nanofibers can be deposited onto glass slides using a combination of electrospinning and calcination processes. Enhanced CTC capture efficiency was observed for the resulting TiO_2 nanofiber-embedded NanoVelcro substrates [53] after anti-EpCAM conjugation, with TiO_2 nanofiber density affecting CTC-capture performance. Alternatively, organic conducting polymer (i.e., poly(3,4-ethylenedioxythiophene), PEDOT) nanodots [52] can be deposited onto ITO-coated glass substrates using an electrochemical method. Carboxylic groups on PEDOT backbones aid conjugation with anti-EpCAM, and enhanced capture performance, determined by nanodot size and density, was observed for the nanodots-embedded NanoVelcro substrates. Recently, a new approach [54], which combines chemical oxidative polymerization and a modified polydimethylsiloxane (PDMS) transfer printing technique, was established to introduce highly regular PEDOT nanorods onto glass substrates for CTC capture. The outstanding electrical transport properties, inherent biocompatibility, and manufacturing flexibility of PEDOT suggest that these PEDOT-based NanoVelcro substrates can be integrated with downstream electrical sensing and phenotyping.

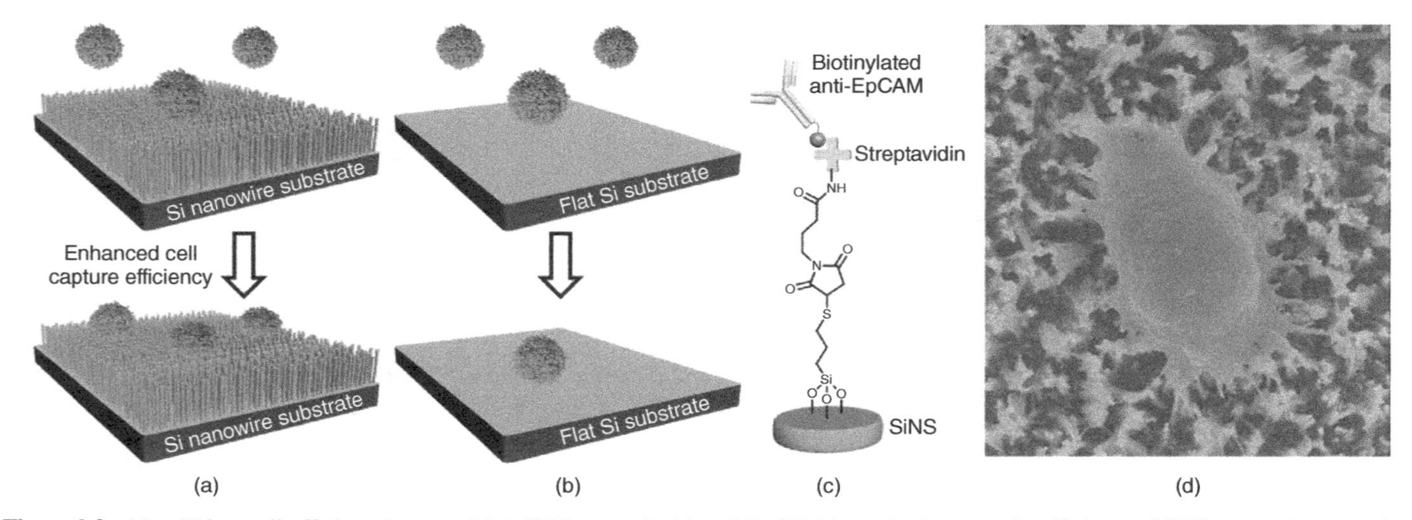

Figure 9.2 NanoVelcro cell-affinity substrates. (a) A SiNS covered with anti-EpCAM is used to improve the efficiency of CTC capture in comparison to (b) a flat silicon substrate that is covered with anti-EpCAM. SiNS increases efficiency compared to an unstructured silicon substrate due to Velcro-like interactions between the anti-EpCAM-coated SiNS and the surfaces of the target cells. (c) To increase CTC detection, biotinylated anti-EpCAM is added to the SiNS. (d) Images of the MCF7 cells captured on the SiNS using SEM.

Aside from these attempts, extensive research has been devoted to testing different nanostructured materials [65] for capturing CTCs and other types of rare cells. Since the proof-of-concept publication [51] in 2009, nanostructured materials, for example, layer-by-layer (LbL)-assembled nanostructures [66], gold cluster on silicon nanowires [67], Fe_3O_4 nanoparticles [68], polymer nanotubes [69], TiO_2 nanoparticles [70], dendrimers [71], graphene oxide nanosheets [72], transparent MnO_2 nanospheres [73], transferrin-conjugated nanostructured substrates [74], leukocyte-inspired particles [75], and other nanomaterials [76, 77] have demonstrated enhanced cell-capture performance when deposited on cell-capture substrates. Several of these works are highlighted in Figure 9.3. The works summarized support the general applicability of NanoVelcro cell-affinity assays and their potential for cell-sorting applications.

9.3 FIRST-GENERATION NANOVELCRO CHIPS FOR CTC ENUMERATION

9.3.1 Device Configuration of First-Generation NanoVelcro Chips

The UCLA team pioneered the concept of using NanoVelcro cell-affinity substrates [51–54] to enhance cell-capture performance by increasing the local topographic interactions between the cell-surface components and the nanosubstrates. The first-generation NanoVelcro chip [55] (Figure 9.4a) was designed on the basis of cell–substrate contact frequency by incorporating an overlaid PDMS with a chaotic mixing channel [78] onto the grafted substrate. A chip holder made of polyacrylate was used to immobilize the two components, with the integration of anti-EpCAM introduced onto the device prior to the cell-capture experiment. The shear force of the herringbone microstructures fixed in the roof of the chaotic mixer inhibits the separation of cells from the surface components (Figure 9.4b) as blood containing CTCs is passed through the chip, considerably enhancing contact frequency between the CTCs and the capture substrate while preserving the cell population. The optimal cell-capture conditions were determined using artificial cancer-cell-spiked blood under different parameters. Side-by-side comparison studies demonstrated that the NanoVelcro chip outperformed CellSearch assay, exhibiting a recovery rate of greater than 85%. In parallel, the use of an antibody cocktail in combination with cytomorphological characteristics enables specific detection of target cells as well as exclusion of false positives. The enrichment of three-color immunocytochemistry (ICC) protocol [79] (Figure 9.4c), along with footprint size of the cells, allows staining of DAPI, anti-CD45, and anti-CK in parallel to discriminate labeled target CTC cells (DAPI+/CK+/CD45-, size $>6\,\mu m$) from nonspecifically trapped WBCs (DAPI+/CK-/CD45+, size $<12\,\mu m$). Micropillar-based chips obtain the vertical depth that other microfluidic platforms lack in order to prevent out-of-focus or superimposed images of immobilized CTCs.

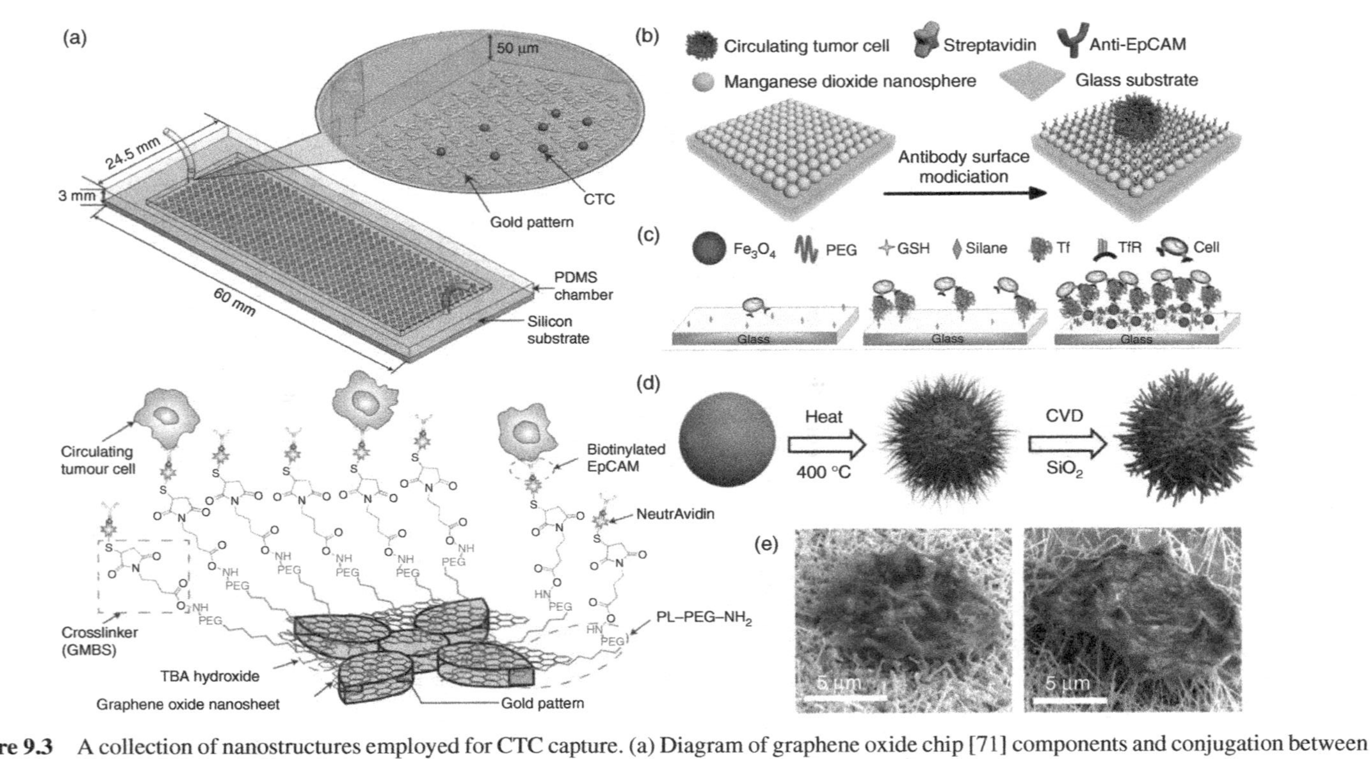

Figure 9.3 A collection of nanostructures employed for CTC capture. (a) Diagram of graphene oxide chip [71] components and conjugation between nanosheets and EpCAM antibodies. GMBS cross-linker binds to PL-PEG-NH2 on nanosheets, which adhere to the gold pattern. NeutrAvidin connects to the GMBS and biotinylated EpCAM. Zhang *et al*. [71]. Reproduced with permission of Nature Publishing Group. (b) Diagram of cell capture [72] using a cell-capture agent on the manganese dioxide surface. Yoon *et al*. [72]. Reproduced with permission of John Wiley and Sons, Inc. (c) Diagram of HCT116 cell capture [67] on SFG, Tf-SFG, and Tf-NMSFG following a 5 min incubation period. Park *et al*. [67]. Reproduced with permission of John Wiley and Sons, Inc. (d) Schematic diagram of the synthesis [73] of leukocyte-inspired particles. Huang *et al*. [73]. Reproduced with permission of John Wiley and Sons, Inc. To replicate the surface of the microvilli of leukocytes, nanofibers were coated with anti-EpCAM. (e) Magnified SEM

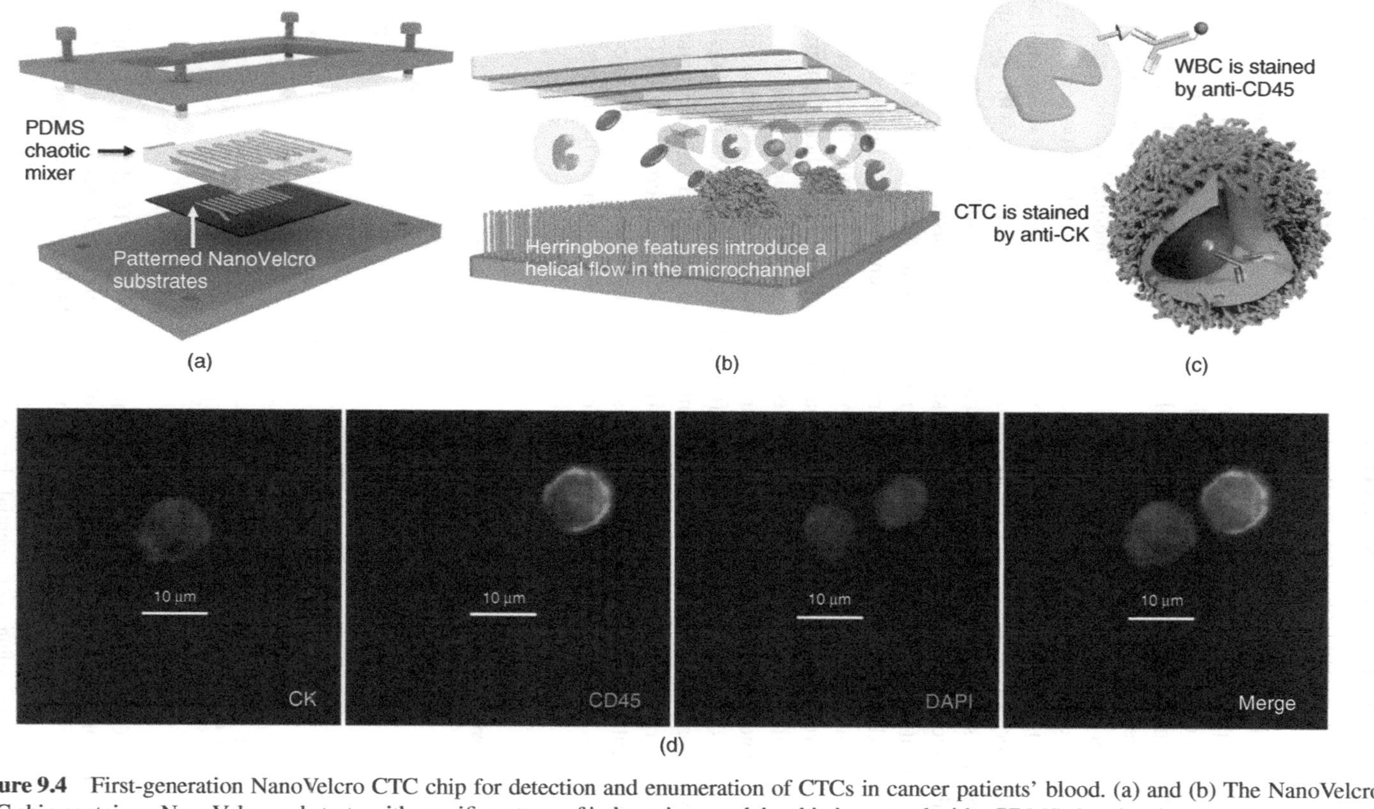

Figure 9.4 First-generation NanoVelcro CTC chip for detection and enumeration of CTCs in cancer patients' blood. (a) and (b) The NanoVelcro CTC chip contains a NanoVelcro substrate with specific pattern of indentations, and the chip is covered with a PDMS chaotic mixer. (c) The three-color ICC protocol allows captured CTCs ($DAPI^+/CK^+/CD45^-$, sizes >6 µm) to be differentiated from WBCs ($DAPI^+/CK^-/CD45^+$, sizes <12 µm). (d) A fluorescence micrograph of one captured prostate cancer CTC along with a nonspecifically captured WBC. (*See color plate section for the color representation of this figure.*)

9.3.2 Clinical Utility of First-Generation NanoVelcro Chips

A side-by-side analytical validation study [55] between the first-generation NanoVelcro chips and the current standard CellSearch assay demonstrated superior CTC capture results by the first-generation NanoVelcro chips. Utilizing blood samples taken from patients in varying stages of prostate cancer (PC), the first-generation NanoVelcro chips demonstrated finer sensitivity and a much broader dynamic range in CTC enumeration in 17 of the 26 clinical blood samples obtained (Figure 9.5a). Fluorescence micrographs of a single CTC isolated from 1.0 ml of a prostate cancer patient's blood sample are illustrated in Figure 9.4d. Oncologists have realized that the clinical utility of CellSearch assay is constrained by its high blood consumption (7.5 ml), poor sensitivity, and reduced dynamic range. Thus, we envision that the great performance observed for first-generation NanoVelcro chips will open up opportunities for (i) better monitoring of disease progression and treatment response and (ii) detecting CTCs at a relatively early stage of the disease.

Modifications [56] to improve the first-generation NanoVelcro CTC chip have yielded a highly sensitive yet inexpensive CTC enumeration assay that can provide invaluable information for the monitoring and treatment of cancer progression. Uro-Oncology teams at both Ronald Reagan UCLA Medical Center and Cedars-Sinai Medical Center have performed joint validation studies [56] with 40 prostate cancer patients; of the patients recruited, 23 had metastatic disease and 8 had localized disease. CTCs were isolated from all 40 patients. Over the course of different treatments, follow-up measurements were documented in these patients. As a direct response to the increased sensitivity of the NanoVelcro chips, CTCs of patients treated with corresponding therapies saw a considerable reduction in count. Over the course of up to 460 days, the index patient was given multiple therapies

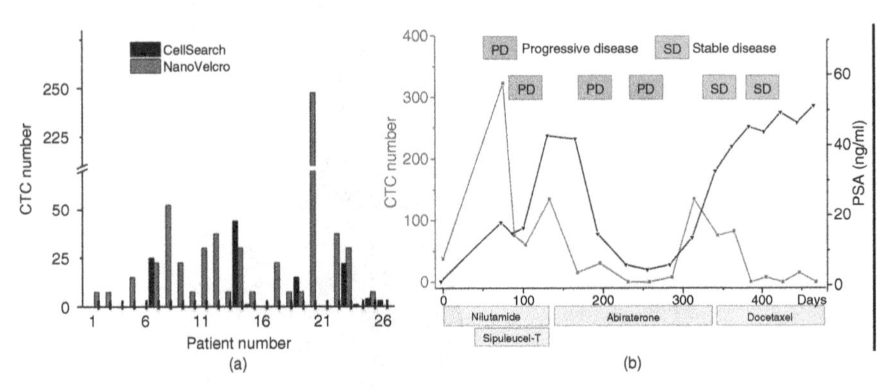

Figure 9.5 Clinical significance of first-generation NanoVelcro CTC chip. (a) A graph of CTC count comparison between the original NanoVelcro chips and CellSearch™ assay, based on samples from 26 prostate cancer patients. Wang *et al.* [55]. Reproduced with permission of John Wiley and Sons, Inc. (b) A graph of a prostate cancer patient's responses to treatment and progression by recording the sequential change in CTC and PSA. Lu *et al.* [56]. Reproduced with permission of Elsevier.

and had variable responses (Figure 9.5b) to which subsequent sequential analyses were performed. Specifically during nilutamide and sipuleucel-T treatment, CTC counts were consistent with both initial responses and subsequent failures. However, disease stabilization with docetaxel therapy introduced a CTC-PSA (prostate-specific antigen) discrepancy. Although imaging of the patient's bone revealed stabilization of the disease and a low CTC count, PSA progression was still noted. Under such circumstances with prostate cancer, the NanoVelcro chips have showcased the increased reliability of CTCs in serving as a biomarker as opposed to PSA.

Because the CellSearch assay has been used for multiple types of solid tumors, NanoVelcro chips – which are also based on the same anti-EpCAM capture agent – should be able to provide similar utility in oncology clinics. Studies toward applying NanoVelcro chips for different types of cancer (e.g., breast, lung, and pancreatic cancers, as well as melanoma) are making progress.

9.3.3 An Alternative Capture Agent, Aptamer

Anti-EpCAM antibodies have commonly been used in the capture of CTCs, but the use of such antibodies poses several issues: it requires a great sum of financial support, and the stability of the antibodies is unpredictable. These issues hinder research in that resources are not viable enough or obtainable at a reasonable price. However, the recent discovery of aptamers will allow a more efficient and accurate way to tag CTCs (Figure 9.6). Aptamers are single-stranded oligonucleotides with function similar to that of antibodies because they are able to differentiate between both other molecules and cells. Aptamers are produced by the process called *in vitro* cell-SELEX (systematic evolution of ligands by exponential enrichment). In this process, and through polymerase chain reaction (PCR), the ssDNA or ssRNA molecules that have a high binding affinity to the target cell or protein are selected and amplified. These molecules together then make the aptamer probe. The advantage of this technique is that not all the characteristics of the target cells have to be known in order for the aptamer to efficiently recognize the target cells.

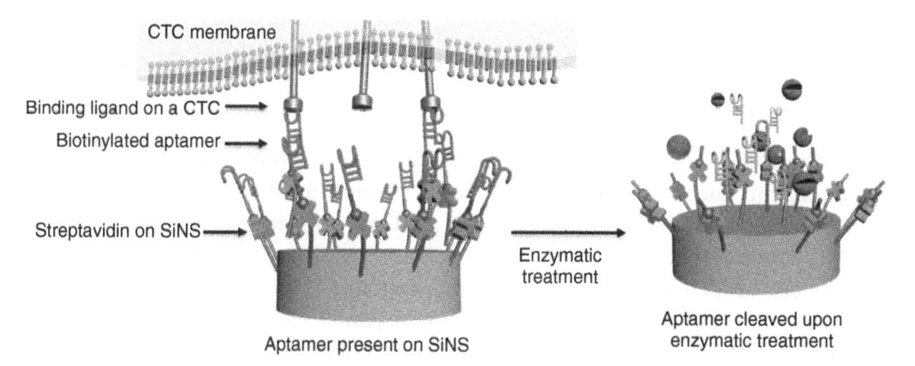

Figure 9.6 Capture and release of CTCs in the presence of aptamer-based capture agents. A visual representation of the interactions involved in the capture of the CTC, and the use of enzymes to release NSCLC CTCs from the SiNS grafted with aptamers.

Through the combined use of the "NanoVelcro" cell-affinity substrates with an aptamer-based capture agent, such as cell-SELEX [80], two single-stranded DNA aptamers were generated using A549 non-small-cell lung cancer (NSCLC) cells. In the same manner as Velcro, the interaction of the aptamer with the CTC inhibits further migration of the CTC. Created for CTC enumeration, the first model of this NanoVelcro chip has been improved and further developed for single-CTC isolation. The new-generation NanoVelcro chip has proven [57] to be highly efficient not only in the capture of NSCLC CTCs but also in the recovery of nanosubstrate-immobilized NSCLC CTCs upon treatment of a nuclease solution.

Originally, the NanoVelcro Chip was grafted with anti-EpCAM, but because of the issues previously stated, anti-EpCAM was replaced with DNA aptamers Ap-1 and Ap-2. After using *in vitro* cell-SELEX to positively select for A549 NSCLC cells, biotin groups were covalently bonded to the 5′ terminus of the aptamers to allow streptavidin-mediated conjugation onto SiNS. Then, a genetically engineered endonuclease (i.e., benzonase nuclease) was added to digest the DNA aptamer and left only the NSCLC cells. Finally, a sample of NSCLC CTCs was analyzed with the NanoVelcro Chip, and the results from the two cycles of the sample running through the chip showed that the purity of the CTCs increased.

A new technique [54] for CTC detection was recently developed to capture the CTCs without using any agents. Instead, this technique uses a glass slide, and depending on the roughness and etching on the slide, the affinities to CTCs will be different. By knowing the optimal roughness, the researchers will be able to detect the presence of CTCs. Although this technique has potential, more research needs to be done on the affinities between the glass slide and the different types of cancer cells to avoid having blood cells and cancer cells with similar affinity. Overlaps occur as a result of diversity in cell types, so techniques have been developed to address this issue and further increase the accuracy of detection. There are many ways to detect CTCs, and the use of aptamers with NanoVelcro Chips show great promise because of its efficiency and accuracy.

9.4 SECOND-GENERATION NANOVELCRO-LMD TECHNOLOGY FOR SINGLE CTC ISOLATION

The introduction of the first-generation NanoVelcro chip, which imitates the conditions of microenvironment, is able to resolve the risk of biopsy resulted from monitoring the disease progression of CTCs, creating a risk-free clinical use that is accessible to a wide range of cancer patients. The second-generation NanoVelcro-LMD technology [58, 59] was introduced to explore the molecular changes in human cancer. It is based on a transparent nanosubstrate covered with poly(lactic-*co*-glycolic) acid (PLGA) nanofibers coupled with laser microdissection (LMD) techniques. Mutational analyses (e.g., reverse transcription polymerase chain reaction (RT-PCR), Sanger sequencing, whole exome sequencing) can be performed by isolating single tumor cells captured on the PLGA nanofibers using LMD.

9.4.1 Preparation of PLGA NanoVelcro Chips

Electrospun PLGA nanofibers were embedded onto a commercial LMD slide (slide with predeposited 1.2 μm thick poly(phenylene) sulfide, PPS membrane, Figure 9.7a) to generate the PLGA NanoVelcro substrates [58, 59]. To enable identification of CTCs from whole blood samples, covalent conjugation of specific biotinylated antibodies (anti-CD146) onto the streptavidin-coating PLGA nanofibers (Figure 9.7b and c) was then employed. Biodegradable polymers such as poly(lactic acid) are known for their ease of reproducibility and manipulation in achieving optimal cell response through topographic designs of the surface. The capture efficiency of the nanostructured PLGA chip has significantly improved when compared to conventional PLGA chips.

9.4.2 NanoVelcro-LMD Technology and Mutational Analysis

Clinical validation of the utility of NanoVelcro-LMD technology was performed by isolating [58] single melanoma CTCs (Figure 9.7d and e) and detecting $BRAF^{V600E}$ from peripheral blood samples obtained from several stage-IV melanoma patients previously characterized to harbor $BRAF^{V600E}$. Present in 60% of melanomas, $BRAF^{V600E}$ is a signature oncogenic mutation that has been targeted by BRAF inhibitors (e.g., vemurafenib). Using a melanoma-specific capture agent (anti-CD146), individually captured circulating melanoma cells (CMCs, a subcategory of solid-tumor CTCs) were harvested for subsequent Sanger sequence analysis to detect the $BRAF^{V600E}$ mutation. The resulting data (Figure 9.7f) exhibited a robust signal-to-noise ratio while fluctuations in signal-to-noise ratios were often observed when sequencing biopsied melanoma tissue.

Distinctive oncogenic mutations from different solid tumors can also be genotyped using the same single-CTC method. For instance, in the presence of anti-EpCAM, this approach successfully isolated individual CTCs from both artificial and clinical blood samples. Specifically, a $KRAS^{G12V}$ mutation (Figure 9.7g), consistent with those found in patients' tumors, was revealed upon genotyping of single CTCs isolated from blood samples from patients with pancreatic cancer.

The isolated, captured CTCs with mutations were confirmed [81] by our pancreatic CTC identification immunostaining protocol. Immunostaining combined with LMD allows for the molecular characterization of CTCs, giving us an understanding into the biology of metastasis and potential molecular targets for therapeutics. The microfluidic NanoVelcro technology demonstrated capture of pancreatic CTCs and may prove useful in the development of CTCs as a biomarker in pancreatic cancer.

9.4.3 NanoVelcro-LCM Technology and Whole Exome Sequencing

In order to address the frequent loss of cells [58] incurred as a result of static charge during the collection processes, LCM has replaced LMD technology [59]. This modified version of LMD technology utilizes an infrared laser, liquefying the LCM cap into a "sticky finger" that grips the laser-dissected NanoVelcro substrate

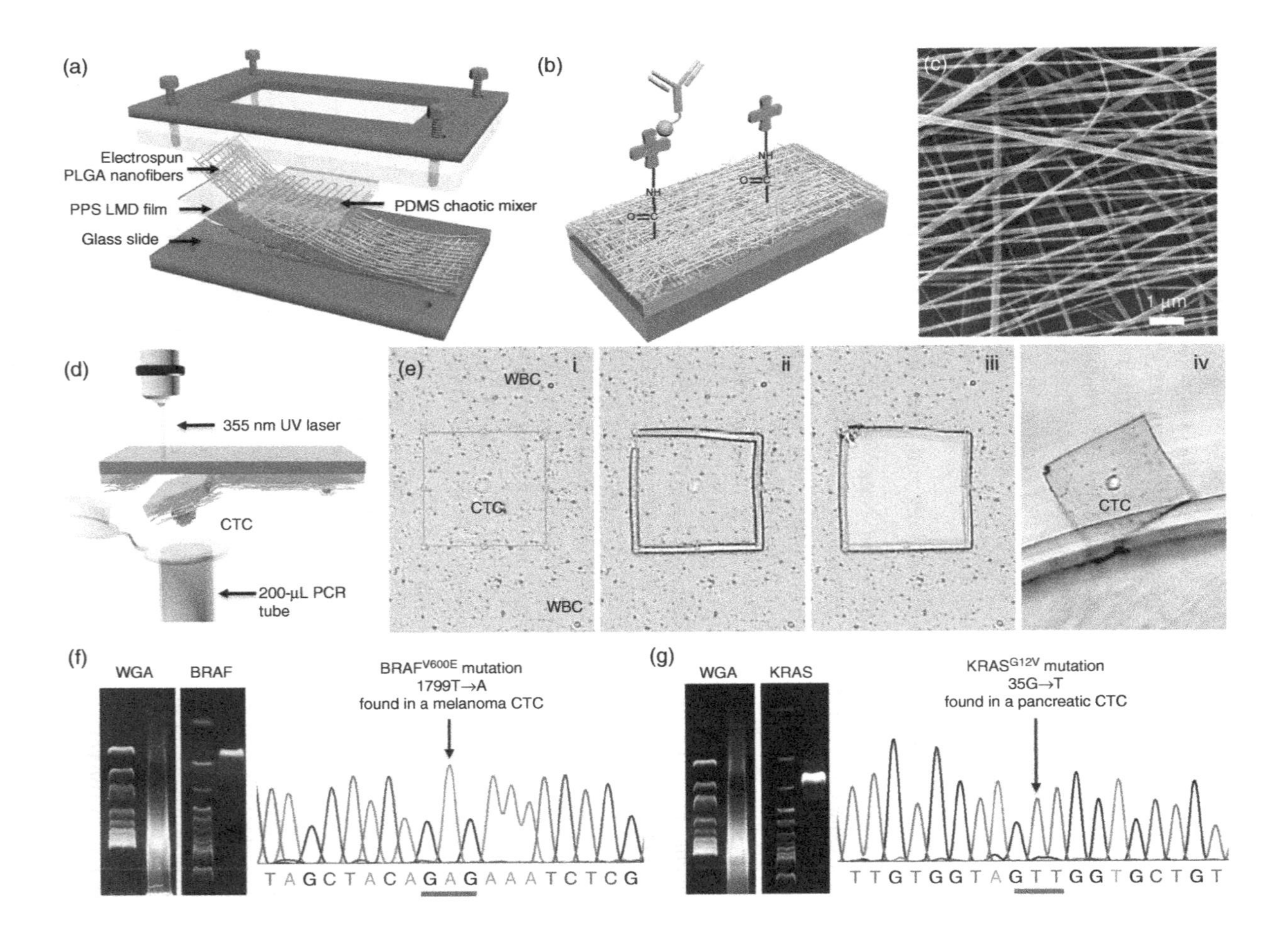

(a)
Electrospun PLGA nanofibers
PPS LMD film
Glass slide
PDMS chaotic mixer
(b)
NH
O=C
(c)
1 μm
(d)
355 nm UV laser
CTC
200-μL PCR tube
(e)
i
ii
iii
iv
WBC
CTC
WBC
CTC
(f)
WGA
BRAF
BRAF^{V600E} mutation
1799T→A
found in a melanoma CTC
T A G C T A C A G A G A A A T C T C G
(g)
WGA
KRAS
KRAS^{G12V} mutation
35G→T
found in a pancreatic CTC
T T G T G G T A G T T G G T G C T G T

Figure 9.7 Second-generation NanoVelcro CTC chip for single-CTC isolation, followed by mutational analyses. (a) The PDMS chaotic mixer is layered on top of a NanoVelcro chip that contains PLGA nanofibers. (b) For the binding of biotinylated capture agents (i.e., anti-EpCAM for prostate and pancreatic cancer), streptavidin is conjugated to PLGA nanofibers. (c) An image of the electrospun PLGA nanofibers by using SEM. (d) The graphic illustration of LMD-based single-CTC isolation. (e) The process to isolate single CTCs consists of (i) identification of CTC, (ii) isolation of the selected CTC using laser dissection, followed by (iii), and (iv) discharge of CTC from the silicon substrate into a 200 μl PCR tube. (f) Results of single-CTC WGA and gel electrophoresis after amplification in PCR with BRAF-specific primer. Through Sanger sequencing, further affirmation is gained because of the display of melanoma CTCs exhibiting the unique $BRAF^{V600E}$ mutation. (g) Pancreatic CTCs and the $KRAS^{G12V}$ mutation present.

(Figure 9.8a). Accordingly, the isolation of single CTCs obtained from advanced prostate cancer patients can be used for NGS [82] to identify mutations on a more expansive mutational level (Figure 9.8b and c). Typically, prostate cancer is known to carry a 10–15-year survival rate [83]. Given the fact that tumor heterogeneity evolves over time [84], there are concerns that historic tissue obtained upon diagnosis may not reflect the biology of a newly developed metastatic lesion. Serial samplings of metastatic lesions, however, are nearly impossible, because the majority [85] of prostate cancer patients have only bone metastasis, and the biopsy of bone is both invasive and technically difficult.

To determine the feasibility of monitoring tumor biology by CTCs, a streamlined process [59] was performed starting from CTC enrichment and isolation to individual CTC genomic analysis by whole exome sequencing (Exome-Seq). Single CTCs and a pool of WBCs (as a negative control) were isolated from a blood sample from a Cedars-Sinai Medical Center (CSMC) patient using the second-generation NanoVelcro chip in conjunction with LCM technology. After whole genome amplification (WGA) followed by quality check, the amplified genomic DNA was sequenced by standard exon-capture targeted sequencing. Results indicated that 25–80% of the targeted exome regions were sequenced with a mean coverage of 29–48× and that no chromosomal loss occurred during the isolation and sequencing processes. More shared mutations existed among CTCs than between CTCs and WBCs, as shown in Figure 9.8c. The similarity of CTCs and differences between CTCs and the WBC control verified the feasibility of using the NanoVelcro-LCM platform to capture pure CTCs for Exome-Seq.

The capability of NanoVelcro Microchip expands beyond CTC detection. It can also be used to isolate Circulating Fetal Nuclear Cells (CFNCs) [86] from maternal blood with similar methods. Individually isolated CFNCs were subjected to genetic analyses with FISH, microarray, and STR-based genomic fingerprinting assays, resulting in sufficient capture of CFNCs for each expectant mother. The results indicate that the NanoVelcro technology expands CTCs and has the potential for further applications.

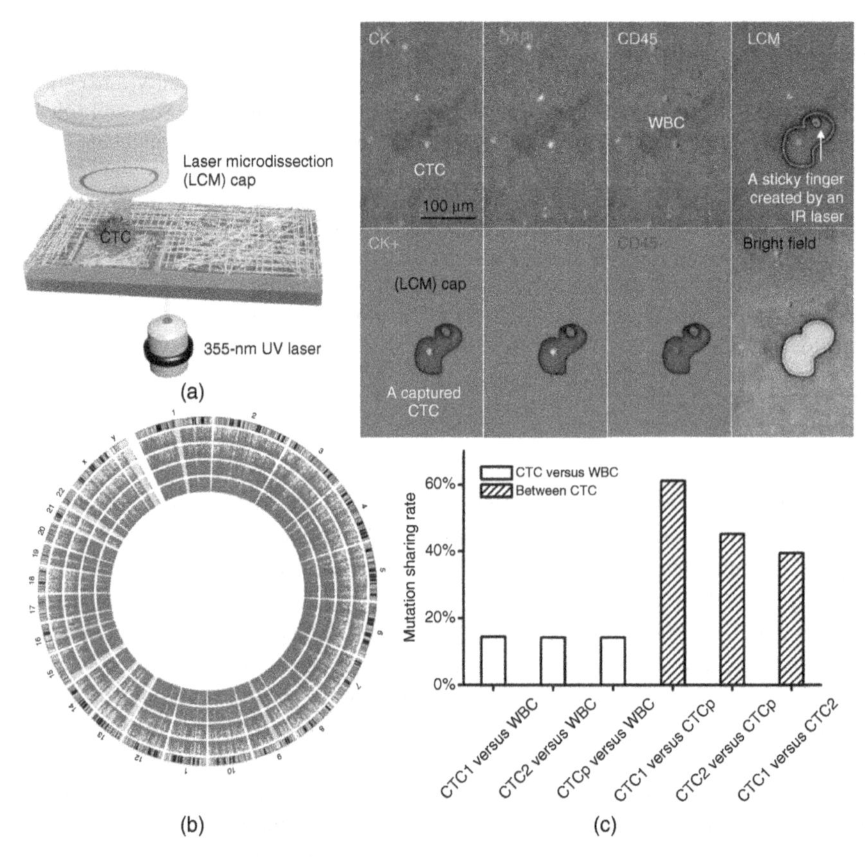

Figure 9.8 Second-generation NanoVelcro CTC chip for single CTC isolation, followed by whole exome sequencing. (a) Using the LCM system for single-CTC isolation requires recognition of CTC, identification of UV dissection route and IR sticky finger positions, UV laser dissection, and collection of the identified CTC on the LCM cap. (b) The Circos plots symbolize Exome-Seq's coverage areas. From the outside to inside, the rings represent CTC-2, CTC-1, pooled CTCs (CTCp), and WBC. The ring on the outside is the human reference genome's karyotype. (c) The mutual mutations between CTCs and WBCs are compared with mutations that CTCs have in common. Zhao *et al.* [59]. Adapted with permission of John Wiley and Sons, Inc.

Beyond single-CTC exome sequencing, continuous efforts will be devoted to exploring second-generation NanoVelcro-LCM/LMD in conjunction with advanced NGS, including whole genome sequencing, transcriptomic analysis by RNA sequencing, and epigenetic studies. In fact, the technology makes single-celled CTC isolation in PC patients possible [87], allowing for further whole genome sequencing. Determinants are used to compare tumor cells and CTCs against WBCs and normal adjacent tissue: analyses such as copy number variations, single nucleotide variations (SNVs), and structural variations (SVs) support genomic surrogacy of CTCs in PC patients. The information will help to better understand tumor heterogeneity and clonal evolution, as well as will provide real-time monitoring of patients' disease progression and response to specific treatments.

9.5 THIRD-GENERATION THERMORESPONSIVE NANOVELCRO CHIPS

The increasing use of molecular characterization and functional analysis of CTCs creates an urgent need to isolate CTCs with higher efficiency, better cell quality, and lower technical demand. Although the second-generation NanoVelcro-LCM technology [58, 59], single-CTC isolation exhibited exceptional precision, the labor intensity of the technology degraded CTC viability. Thus, the third-generation thermoresponsive NanoVelcro chip [60] rectifies these issues and can successfully capture and release CTCs at 37 and 4 °C, respectively, via grafting thermoresponsive polymer brushes [88] (i.e., poly(*N*-isopropylacrylamide, PIPAAm) onto the SiNS (Figure 9.9). This unique idea takes advantage of the fact that the polymer brushes undergo temperature-dependent conformational changes, thus effectively altering the accessibility of the capture agent on the NanoVelcro chip, allowing for rapid CTC purification with the desired viability and molecular integrity. In particular, the introduction of the biotin groups to the polymer brush amino groups conjugates the CTC-capture agent (e.g., anti-EpCAM) to form biotin-P-SiNWS. A temperature of 37 °C then exposes the polymer brushes' anti-EpCAM and hydrophobic domains to allow CTC capture. Subsequent CTC release occurs at 4 °C where the hydrophobic domains of the polymer brushes become hydrophilic as a result of a conformational change of the polymer brushes that shrouds the anti-EpCAM. As a result, improved CTC-capture efficiency is demonstrated by the thermoresponsive NanoVelcro chip without disturbing the surrounding WBCs; the successful release of a bulk of the captured cells at 4 °C achieved an approximate 90% viability.

In a recently published paper, researchers have been able to demonstrate [89] analysis in EGFR mutations through the operational system. To set the optimal parameters (three rounds of heating/cooling cycles at the speed of 0.5 ml/h), the team used artificial H1975 anti-EpCAM-positive NSCLC cells to determine the most favorable conditions for CTC capture and release. The data showed higher capture and release efficiency than non-EpCAM-expressing cells (i.e., HeLa and WBCs). Mutational analysis [58] of EGFR point mutations can be further enhanced with the aid of double CTC purification. The data resulted in 85% cell viability for double purification, while still maintaining CTC molecular integrity. Moreover, the researchers presented sequencing data [89] of purified CTCs from the seven NSCLC patients' blood samples in conjunction with their tumor tissues to prove the feasibility of the thermoresponsive purification system.

Headway is currently being made to permit rapid CTC purification from whole blood samples and facilitate downstream CTC characterization through the development of a user interface for the third-generation thermoresponsive NanoVelcro chips. Moreover, a diverse array of applications, such as individual-specific *in vitro*

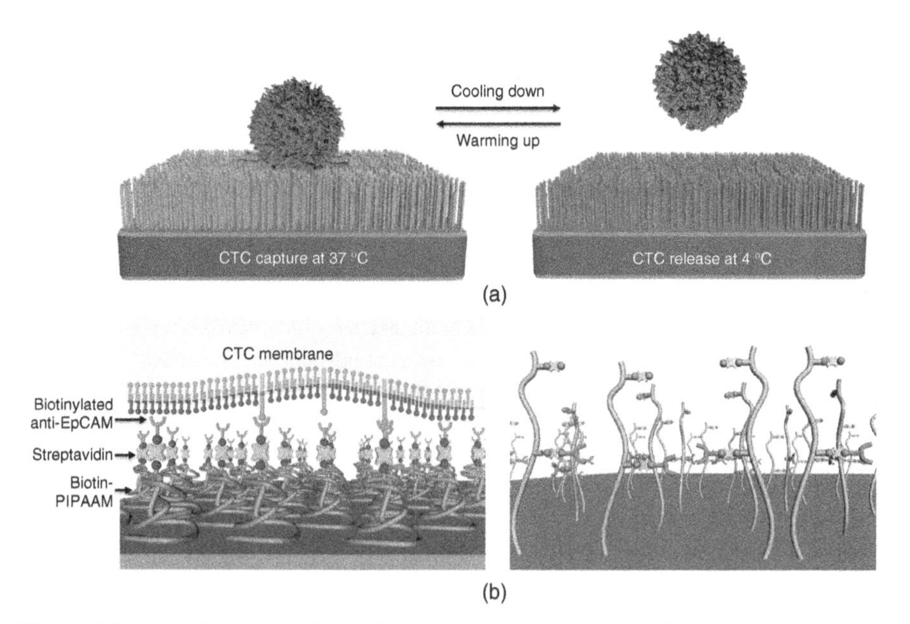

Figure 9.9 Operation mechanism of third-generation thermoresponsive NanoVelcro chip. (a) The thermoresponsive NanoVelcro system that is used to capture and discharge CTCs at 37 °C (left) and 4 °C (right). (b) The polymer brush configuration changes based on temperature, and the shifts will change anti-EpCAM's approach toward the capture and discharge of CTCs on the NanoVelcro.

screening for cancer and designing CTC-derived cancer lines, will be made possible via *ex vivo* expansion (e.g., culture) of these viable CTCs, bringing the possibility of personalized medicine within reach.

9.6 CONCLUSIONS AND FUTURE PERSPECTIVES

In the last 10 years, many different fields, such as bioengineering, materials science, chemistry, cancer biology, and oncology, have participated in the exploration in CTC research, which includes CTC detection, isolation, and characterization. This interest from various fields has led to substantial technological advancement in CTC detection. However, there is still much progress to be made. Persistent efforts are needed to obtain translational data and to develop more accurate and efficient treatment options to help cancer patients. Besides CTC detection, there is also an urgent need to develop CTC isolation and purification platforms with high efficiency so that further progress can also be made in successive molecular and functional analyses. These CTC-derived molecular signatures and functional readouts will ultimately be able to shed light on tumor biology that can provide physicians with invaluable information that can substantially impact patient morbidity.

The three generations of NanoVelcro CTC chips are based on the unique NanoVelcro functioning system, and researchers at UCLA have successfully proven that these chips are adept in detecting, isolating, and purifying CTCs from blood samples. A variety of capture agents were used to capture the CTCs from various different tumors, including melanoma as well as prostate, pancreatic, lung, and breast cancers. To further confirm the CTC's position as tumor liquid biopsy, the isolated CTCs were then put through successive molecular analyses (e.g., Sanger sequencing and NGS).

Some of the contents presented in this chapter have been published elsewhere in Accounts of Chemical Research [90]. Moreover, this chapter contains a wide variety of current research using this technology and compiles further applications for this technology. Currently, the viability of culturing purified CTCs is being investigated; if such methods are successful, more progress can be made toward personalizing treatments for cancer patients. Continuous efforts in research will be dedicated to applying these findings from research laboratories to oncology clinics.

ACKNOWLEDGMENT

This work was supported by National Institute of Health (R21-CA151159, R33-CA157396, P50-CA92131, R33-CA174562, and PO1-CA168585), Prostate Cancer Foundation (a Creativity Award), and UCLA Jonsson Comprehensive Cancer Center (an Impact Award), and National Natural Science Foundation of China (30900650/H1615, 81372501/H1615).

REFERENCES

1. Kling, J. (2012) Beyond counting tumor cells. *Nature Biotechnology*, **30**, 578–580.
2. Kaiser, J. (2010) Medicine. Cancer's circulation problem. *Science*, **327**, 1072–1074.
3. Bernards, R. and Weinberg, R.A. (2002) A progression puzzle. *Nature*, **418**, 823.
4. Criscitiello, C., Sotiriou, C., and Ignatiadis, M. (2010) Circulating tumor cells and emerging blood biomarkers in breast cancer. *Current Opinion in Oncology*, **22**, 552–558.
5. van de Stolpe, A., Pantel, K., Sleijfer, S. *et al.* (2011) Circulating tumor cell isolation and diagnostics: toward routine clinical use. *Cancer Research*, **71**, 5955–5960.
6. Krebs, M.G., Hou, J.M., Ward, T.H. *et al.* (2010) Circulating tumour cells: their utility in cancer management and predicting outcomes. *Therapeutic Advances in Medical Oncology*, **2**, 351–365.
7. Casavant, B.P., Guckenberger, D.J., Berry, S.M. *et al.* (2013) The VerIFAST: an integrated method for cell isolation and extracellular/intracellular staining. *Lab on a Chip*, **13**, 391–396.
8. Earhart, C.M., Hughes, C.E., Gaster, R.S. *et al.* (2014) Isolation and mutational analysis of circulating tumor cells from lung cancer patients with magnetic sifters and biochips. *Lab on a Chip*, **14**, 78–88.
9. Ozkumur, E., Shah, A.M., Ciciliano, J.C. *et al.* (2013) Inertial focusing for tumor antigen-dependent and -independent sorting of rare circulating tumor cells. *Science Translational Medicine*, **5**, 179ra147.

10. Cristofanilli, M., Budd, G.T., Ellis, M.J. *et al.* (2004) Circulating tumor cells, disease progression, and survival in metastatic breast cancer. *New England Journal of Medicine*, **351**, 781–791.
11. Shaffer, D.R., Leversha, M.A., Danila, D.C. *et al.* (2007) Circulating tumor cell analysis in patients with progressive castration-resistant prostate cancer. *Clinical Cancer Research*, **13**, 2023–2029.
12. Lara, O., Tong, X., Zborowski, M., and Chalmers, J.J. (2004) Enrichment of rare cancer cells through depletion of normal cells using density and flow-through, immunomagnetic cell separation. *Experimental Hematology*, **32**, 891–904.
13. Tong, X., Yang, L., Lang, J.C. *et al.* (2007) Application of immunomagnetic cell enrichment in combination with RT-PCR for the detection of rare circulating head and neck tumor cells in human peripheral blood. *Cytometry. Part B, Clinical Cytometry*, **72**, 310–323.
14. Karabacak, N.M., Spuhler, P.S., Fachin, F. *et al.* (2014) Microfluidic, marker-free isolation of circulating tumor cells from blood samples. *Nature Protocols*, **9**, 694–710.
15. Sastre, J., Maestro, M.L., Puente, J. *et al.* (2008) Circulating tumor cells in colorectal cancer: correlation with clinical and pathological variables. *Annals of Oncology*, **19**, 935–938.
16. Attard, G., Swennenhuis, J.F., Olmos, D. *et al.* (2009) Characterization of ERG, AR and PTEN gene status in circulating tumor cells from patients with castration-resistant prostate cancer. *Cancer Research*, **69**, 2912–2918.
17. Talasaz, A.H., Powell, A.A., Huber, D.E. *et al.* (2009) Isolating highly enriched populations of circulating epithelial cells and other rare cells from blood using a magnetic sweeper device. *Proceedings of the National Academy of Sciences of the United States of America*, **106**, 3970–3975.
18. Harb, W., Fan, A., Tran, T. *et al.* (2013) Mutational analysis of circulating tumor cells using a novel microfluidic collection device and qPCR assay. *Translational Oncology*, **6**, 528–538.
19. Noble, P.B. and Cutts, J.H. (1967) Separation of blood leukocytes by Ficoll gradient. *Canadian Veterinary Journal*, **8**, 110–111.
20. Rosenberg, R., Gertler, R., Friederichs, J. *et al.* (2002) Comparison of two density gradient centrifugation systems for the enrichment of disseminated tumor cells in blood. *Cytometry*, **49**, 150–158.
21. Gertler, R., Rosenberg, R., Fuehrer, K. *et al.* (2003) Detection of circulating tumor cells in blood using an optimized density gradient centrifugation. *Recent Results in Cancer Research*, **162**, 149–155.
22. Allan, A.L., Vantyghem, S.A., Tuck, A.B. *et al.* (2005) Detection and quantification of circulating tumor cells in mouse models of human breast cancer using immunomagnetic enrichment and multiparameter flow cytometry. *Cytometry. Part A*, **65**, 4–14.
23. Hu, Y., Fan, L., Zheng, J. *et al.* (2010) Detection of circulating tumor cells in breast cancer patients utilizing multiparameter flow cytometry and assessment of the prognosis of patients in different CTCs levels. *Cytometry. Part A*, **77**, 213–219.
24. He, W., Wang, H., Hartmann, L.C. *et al.* (2007) In vivo quantitation of rare circulating tumor cells by multiphoton intravital flow cytometry. *Proceedings of the National Academy of Sciences of the United States of America*, **104**, 11760–11765.
25. Zhao, M., Schiro, P.G., Kuo, J.S. *et al.* (2013) An automated high-throughput counting method for screening circulating tumor cells in peripheral blood. *Analytical Chemistry*, **85**, 2465–2471.

26. Schiro, P.G., Zhao, M., Kuo, J.S. *et al.* (2012) Sensitive and high-throughput isolation of rare cells from peripheral blood with ensemble-decision aliquot ranking. *Angewandte Chemie International Edition in English*, **51**, 4618–4622.
27. Gleghorn, J.P., Pratt, E.D., Denning, D. *et al.* (2010) Capture of circulating tumor cells from whole blood of prostate cancer patients using geometrically enhanced differential immunocapture (GEDI) and a prostate-specific antibody. *Lab on a Chip*, **10**, 27–29.
28. Nagrath, S., Sequist, L.V., Maheswaran, S. *et al.* (2007) Isolation of rare circulating tumour cells in cancer patients by microchip technology. *Nature*, **450**, 1235–1239.
29. Stott, S.L., Hsu, C.H., Tsukrov, D.I. *et al.* (2010) Isolation of circulating tumor cells using a microvortex-generating herringbone-chip. *Proceedings of the National Academy of Sciences of the United States of America*, **107**, 18392–18397.
30. Sheng, W., Ogunwobi, O.O., Chen, T. *et al.* (2014) Capture, release and culture of circulating tumor cells from pancreatic cancer patients using an enhanced mixing chip. *Lab on a Chip*, **14**, 89–98.
31. Dong, Y., Skelley, A.M., Merdek, K.D. *et al.* (2013) Microfluidics and circulating tumor cells. *Journal of Molecular Diagnostics*, **15**, 149–157.
32. Huang, S.B., Wu, M.H., Lin, Y.H. *et al.* (2013) High-purity and label-free isolation of circulating tumor cells (CTCs) in a microfluidic platform by using optically-induced-dielectrophoretic (ODEP) force. *Lab on a Chip*, **13**, 1371–1383.
33. Ohnaga, T., Shimada, Y., Moriyama, M. *et al.* (2013) Polymeric microfluidic devices exhibiting sufficient capture of cancer cell line for isolation of circulating tumor cells. *Biomedical Microdevices*, **15**, 611–616.
34. Casavant, B.P., Mosher, R., Warrick, J.W. *et al.* (2013) A negative selection methodology using a microfluidic platform for the isolation and enumeration of circulating tumor cells. *Methods*, **64**, 137–143.
35. Vona, G., Sabile, A., Louha, M. *et al.* (2000) Isolation by size of epithelial tumor cells : a new method for the immunomorphological and molecular characterization of circulating tumor cells. *American Journal of Pathology*, **156**, 57–63.
36. Zheng, S., Lin, H., Liu, J.Q. *et al.* (2007) Membrane microfilter device for selective capture, electrolysis and genomic analysis of human circulating tumor cells. *Journal of Chromatography, A*, **1162**, 154–161.
37. Tan, S.J., Yobas, L., Lee, G.Y. *et al.* (2009) Microdevice for the isolation and enumeration of cancer cells from blood. *Biomedical Microdevices*, **11**, 883–892.
38. Tan, S.J., Lakshmi, R.L., Chen, P. *et al.* (2010) Versatile label free biochip for the detection of circulating tumor cells from peripheral blood in cancer patients. *Biosensors and Bioelectronics*, **26**, 1701–1705.
39. Lecharpentier, A., Vielh, P., Perez-Moreno, P. *et al.* (2011) Detection of circulating tumour cells with a hybrid (epithelial/mesenchymal) phenotype in patients with metastatic non-small cell lung cancer. *British Journal of Cancer*, **105**, 1338–1341.
40. Krivacic, R.T., Ladanyi, A., Curry, D.N. *et al.* (2004) A rare-cell detector for cancer. *Proceedings of the National Academy of Sciences of the United States of America*, **101**, 10501–10504.
41. Hsieh, H.B., Marrinucci, D., Bethel, K. *et al.* (2006) High speed detection of circulating tumor cells. *Biosensors and Bioelectronics*, **21**, 1893–1899.
42. Hoshino, K., Huang, Y.Y., Lane, N. *et al.* (2011) Microchip-based immunomagnetic detection of circulating tumor cells. *Lab on a Chip*, **11**, 3449–3457.

43. Huang, Y.Y., Hoshino, K., Chen, P. *et al.* (2013) Immunomagnetic nanoscreening of circulating tumor cells with a motion controlled microfluidic system. *Biomedical Microdevices*, **15**, 673–681.
44. Kojima, T., Hashimoto, Y., Watanabe, Y. *et al.* (2009) A simple biological imaging system for detecting viable human circulating tumor cells. *Journal of Clinical Investigation*, **119**, 3172–3181.
45. Shigeyasu, K., Tazawa, H., Hashimoto, Y. *et al.* (2014) Fluorescence virus-guided capturing system of human colorectal circulating tumour cells for non-invasive companion diagnostics. *Gut*, **64**, 627–635.
46. von Knebel Doeberitz, M. and Lacroix, J. (1999) Nucleic acid based techniques for the detection of rare cancer cells in clinical samples. *Cancer and Metastasis Reviews*, **18**, 43–64.
47. Keilholz, U., Willhauck, M., Rimoldi, D. *et al.* (1998) Reliability of reverse transcription-polymerase chain reaction (RT-PCR)-based assays for the detection of circulating tumour cells: a quality-assurance initiative of the EORTC Melanoma Cooperative Group. *European Journal of Cancer*, **34**, 750–753.
48. Van der Auwera, I., Peeters, D., Benoy, I.H. *et al.* (2010) Circulating tumour cell detection: a direct comparison between the CellSearch System, the AdnaTest and CK-19/mammaglobin RT-PCR in patients with metastatic breast cancer. *British Journal of Cancer*, **102**, 276–284.
49. Zieglschmid, V., Hollmann, C., and Bocher, O. (2005) Detection of disseminated tumor cells in peripheral blood. *Critical Reviews in Critical Laboratory Sciences*, **42**, 155–196.
50. Racila, E., Euhus, D., Weiss, A.J. *et al.* (1998) Detection and characterization of carcinoma cells in the blood. *Proceedings of the National Academy of Sciences of the United States of America*, **95**, 4589–4594.
51. Wang, S., Wang, H., Jiao, J. *et al.* (2009) Three-dimensional nanostructured substrates toward efficient capture of circulating tumor cells. *Angewandte Chemie International Edition in English*, **48**, 8970–8973.
52. Sekine, J., Luo, S.C., Wang, S. *et al.* (2011) Functionalized conducting polymer nanodots for enhanced cell capturing: the synergistic effect of capture agents and nanostructures. *Advanced Materials*, **23**, 4788–4792.
53. Zhang, N., Deng, Y., Tai, Q. *et al.* (2012) Electrospun TiO_2 nanofiber-based cell capture assay for detecting circulating tumor cells from colorectal and gastric cancer patients. *Advanced Materials*, **24**, 2756–2760.
54. Hsiao, Y.S., Luo, S.C., Hou, S. *et al.* (2014) 3D bioelectronic interface: capturing circulating tumor cells onto conducting polymer-based micro/nanorod arrays with chemical and topographical control. *Small*, **10**, 3012–3017.
55. Wang, S., Liu, K., Liu, J. *et al.* (2011) Highly efficient capture of circulating tumor cells by using nanostructured silicon substrates with integrated chaotic micromixers. *Angewandte Chemie International Edition in English*, **50**, 3084–3088.
56. Lu, Y.T., Zhao, L., Shen, Q. *et al.* (2013) NanoVelcro Chip for CTC enumeration in prostate cancer patients. *Methods*, **64**, 144–152.
57. Shen, Q., Xu, L., Zhao, L. *et al.* (2013) Specific capture and release of circulating tumor cells using aptamer-modified nanosubstrates. *Advanced Materials*, **25**, 2368–2373.
58. Hou, S., Zhao, L., Shen, Q. *et al.* (2013) Polymer nanofiber-embedded microchips for detection, isolation, and molecular analysis of single circulating melanoma cells. *Angewandte Chemie International Edition in English*, **52**, 3379–3383.

59. Zhao, L., Lu, Y.T., Li, F. *et al.* (2013) High-purity prostate circulating tumor cell isolation by a polymer nanofiber-embedded microchip for whole exome sequencing. *Advanced Materials*, **25**, 2897–2902.
60. Hou, S., Zhao, H., Zhao, L. *et al.* (2013) Capture and stimulated release of circulating tumor cells on polymer-grafted silicon nanostructures. *Advanced Materials*, **25**, 1547–1551.
61. Bettinger, C.J., Langer, R., and Borenstein, J.T. (2009) Engineering substrate topography at the micro- and nanoscale to control cell function. *Angewandte Chemie International Edition in English*, **48**, 5406–5415.
62. Liu, X. and Wang, S. (2014) Three-dimensional nano-biointerface as a new platform for guiding cell fate. *Chemical Society Reviews*, **43**, 2385–2401.
63. Saracino, G.A., Cigognini, D., Silva, D. *et al.* (2013) Nanomaterials design and tests for neural tissue engineering. *Chemical Society Reviews*, **42**, 225–262.
64. Fischer, K.E., Aleman, B.J., Tao, S.L. *et al.* (2009) Biomimetic nanowire coatings for next generation adhesive drug delivery systems. *Nano Letters*, **9**, 716–720.
65. Yoon, H.J., Kozminsky, M., and Nagrath, S. (2014) Emerging role of nanomaterials in circulating tumor cell isolation and analysis. *ACS Nano*, **8**, 1995–2017.
66. Lee, H., Jang, Y., Seo, J. *et al.* (2011) Nanoparticle-functionalized polymer platform for controlling metastatic cancer cell adhesion, shape, and motility. *ACS Nano*, **5**, 5444–5456.
67. Park, G.S., Kwon, H., Kwak, D.W. *et al.* (2012) Full surface embedding of gold clusters on silicon nanowires for efficient capture and photothermal therapy of circulating tumor cells. *Nano Letters*, **12**, 1638–1642.
68. Samant, M., Banerjee, S.S., Taneja, N. *et al.* (2014) Biophysical interactions of polyamidoamine dendrimer coordinated Fe3O4 nanoparticles with insulin. *Journal of Biomedical Nanotechnology*, **10**, 1286–1293.
69. Liu, X., Chen, L., Liu, H. *et al.* (2013) Bio-inspired soft polystyrene nanotube substrate for rapid and highly efficient breast cancer-cell capture. *NPG Asia Materials*, **5**, e63.
70. He, R., Zhao, L., Liu, Y. *et al.* (2013) Biocompatible TiO_2 nanoparticle-based cell immunoassay for circulating tumor cells capture and identification from cancer patients. *Biomedical Microdevices*, **15**, 617–626.
71. Zhang, P., Chen, L., Xu, T. *et al.* (2013) Programmable fractal nanostructured interfaces for specific recognition and electrochemical release of cancer cells. *Advanced Materials*, **25**, 3566–3570.
72. Yoon, H.J., Kim, T.H., Zhang, Z. *et al.* (2013) Sensitive capture of circulating tumour cells by functionalized graphene oxide nanosheets. *Nature Nanotechnology*, **8**, 735–741.
73. Huang, Q., Chen, B., He, R. *et al.* (2014) Capture and release of cancer cells based on sacrificeable transparent MnO nanospheres thin film. *Advanced Healthcare Materials*, **3**, 1420–1425.
74. Banerjee, S.S., Paul, D., Bhansali, S.G. *et al.* (2012) Enhancing surface interactions with colon cancer cells on a transferrin-conjugated 3D nanostructured substrate. *Small*, **8**, 1657–1663.
75. Meng, J., Liu, H., Liu, X. *et al.* (2014) Hierarchical biointerfaces assembled by leukocyte-inspired particles for specifically recognizing cancer cells. *Small*, **10**, 3735–3741.
76. Benson, K., Prasetyanto, E.A., Galla, H.-J., and Kehr, N.S. (2012) Self-assembled monolayers of bifunctional periodic mesoporous organosilicas for cell adhesion and cellular patterning. *Soft Matter*, **8**, 10845–10852.

77. Kumeria, T., Kurkuri, M.D., Diener, K.R. *et al.* (2012) Label-free reflectometric interference microchip biosensor based on nanoporous alumina for detection of circulating tumour cells. *Biosensors and Bioelectronics*, **35**, 167–173.
78. Stroock, A.D., Dertinger, S.K., Ajdari, A. *et al.* (2002) Chaotic mixer for microchannels. *Science*, **295**, 647–651.
79. Sun, J., Masterman-Smith, M.D., Graham, N.A. *et al.* (2010) A microfluidic platform for systems pathology: multiparameter single-cell signaling measurements of clinical brain tumor specimens. *Cancer Research*, **70**, 6128–6138.
80. Fang, X. and Tan, W. (2010) Aptamers generated from cell-SELEX for molecular medicine: a chemical biology approach. *Accounts of Chemical Research*, **43**, 48–57.
81. Ankeny, J.S., Hou, S., Lin, M. *et al.* (2014) Capture, isolation, and mutational analysis of single pancreatic circulating tumor cells using NanoVelcro technology. *Cancer Research*, **74**, (19 Suppl): Abstract nr 3070.
82. Gallagher, R.I., Blakely, S.R., Liotta, L.A., and Espina, V. (2012) Laser capture microdissection: Arcturus(XT) infrared capture and UV cutting methods. *Methods in Molecular Biology*, **823**, 157–178.
83. Pound, C.R., Partin, A.W., Eisenberger, M.A. *et al.* (1999) Natural history of progression after PSA elevation following radical prostatectomy. *JAMA*, **281**, 1591–1597.
84. Gerlinger, M., Rowan, A.J., Horswell, S. *et al.* (2012) Intratumor heterogeneity and branched evolution revealed by multiregion sequencing. *N Engl J Med.*, **366**, 883–892.
85. Bubendorf, L., Schopfer, A., Wagner, U. *et al.* (2000) Metastatic patterns of prostate cancer: an autopsy study of 1,589 patients. *Human Pathology*, **31**, 578–583.
86. Liu, L., Hou, S., Lin, M. *et al.* (2015) A pilot study of NIPS-24 using circulating fetal nucleated cells (CFNCs) isolated with Nanovelcro microchips. *Fertility and Sterility*, **103**, e17–e18.
87. Lu, Y.-T., Jiang, R., Tseng, H.-R. *et al.* (2014) Single-cell whole-genome sequencing verifies the surrogacy of circulating tumor cells for prostate cancer. *Cancer Research*, **74**, 938–938.
88. Okano, T., Yamada, N., Okuhara, M. *et al.* (1995) Mechanism of cell detachment from temperature-modulated, hydrophilic-hydrophobic polymer surfaces. *Biomaterials*, **16**, 297–303.
89. Ke, Z., Lin, M., Chen, J.F. *et al.* (2015) Programming thermoresponsiveness of NanoVelcro substrates enables effective purification of circulating tumor cells in lung cancer patients. *ACS Nano*, **9**, 62–70.
90. Lin, M., Chen, J.F., Lu, Y.T. *et al.* (2014) Nanostructure embedded microchips for detection, isolation, and characterization of circulating tumor cells. *Accounts of Chemical Research*, **47**, 2941–2950.

10

ACOUSTOPHORESIS IN TUMOR CELL ENRICHMENT

PER AUGUSTSSON

Department of Biomedical Engineering, Lund University, Lund, Sweden

CECILIA MAGNUSSON

Department of Translational Medicine, Lund University, Lund, Sweden

HANS LILJA

Department of Surgery, Memorial Sloan Kettering Cancer Center, New York, NY, USA

THOMAS LAURELL

Department of Biomedical Engineering, Lund University, Lund, Sweden

10.1 INTRODUCTION

This chapter outlines the opportunities and current developments of label-free separation of circulating tumor cell (CTC) from blood cells using acoustophoresis. After a brief introduction to microchannel acoustophoresis, where differences in acoustic properties and size are utilized to sort cells, two platforms are described that have been developed to enable subsequent evaluation of acoustophoresis in clinical settings.

10.1.1 Background

CTC enumeration and characterization of peripheral venous blood ("liquid biopsy") collected from patients with advanced cancer have recently been implemented in

Circulating Tumor Cells: Isolation and Analysis, First Edition. Edited by Z. Hugh Fan.

clinical practice as detection of ≥5 CTCs in 7.5 ml of blood from patients with metastatic colorectal, breast, and prostate cancer and was found to predict overall survival and response to therapy. This development has largely been based on the CellSearch™ assay, which is the only CTC assay to receive clearance for routine use by the Food and Drug Administration (FDA) in the United States and monitor patients with advanced stages of these malignancies (see Chapter 19 for details). CellSearch uses immunomagnetic beads to extract EpCAM-positive cells from blood, followed by $CD45^-$ and $cytokeratin^+$ (CK8, CK18, and CK19) fluorescence labeling combined with nuclear staining using DAPI to verify the identification of CTCs. However, these selection criteria do not capture events where tumor cells disseminated into circulation display a low-to-modest expression of EpCAM or have undergone epithelial-to-mesenchymal transition (EMT), with downregulation of the epithelial epitope expression [1,2]. Hence, there is a current unmet need for novel techniques to provide more efficient and unbiased CTC enrichment to enable comprehensive characterization of CTCs and contribute enhanced diagnostic performance and prognostic value and to support development of targeted therapies.

With the rapid development within the lab-on-a-chip community, novel microfluidic approaches have recently targeted the need for improved CTC separation from blood, where label-based techniques combined with active (externally applied force fields) or passive (hydrodynamic controlled principles) have shown promise [3]. These methods primarily use label-free separation technology and differences in mechanical properties to target the rare occurrence of CTCs in a vast background of billions of red blood cells (RBCs) and millions of nucleated white blood cells (WBCs). The discriminatory accuracy of these methods relies on physical properties such as size [4,5], density [6], deformability [7], and electrical properties [8]. From a clinical perspective, the current need is a label-free CTC separation technology that preferably enables isolated CTCs to be subject to comprehensive molecular profiling, subsequent expansion *in vitro* or *in vivo*, and drug resistance screening. In this context, microchannel acoustophoresis holds great promise mainly for enabling moderate-to-high throughput, being gentle to cells [9–11], and for separating cells based on multiple parameters [12–14].

In microchannel acoustophoresis, cells in suspension are subject to a volume-dependent force, which stems from scattering effects in an acoustic field. For resonant acoustic fields, this force causes the cells to migrate toward points of minimal pressure variation. Figure 10.1 shows a schematic of cells flowing through a channel in which an acoustic standing wave has been established perpendicular to the direction of the flow. From an initial position near the side walls, the cells migrate toward the central part of the channel under the influence of the acoustic field, so-called free-flow acoustophoresis [13]. The acoustic migration velocity is determined by the size of the cell and its density and compressibility relative to that of the suspending medium. The sideways displacement of a cell when arriving at the end of the channel is thus correlated with these acoustophysical properties, which enables cells of high acoustic migration velocity to be branched

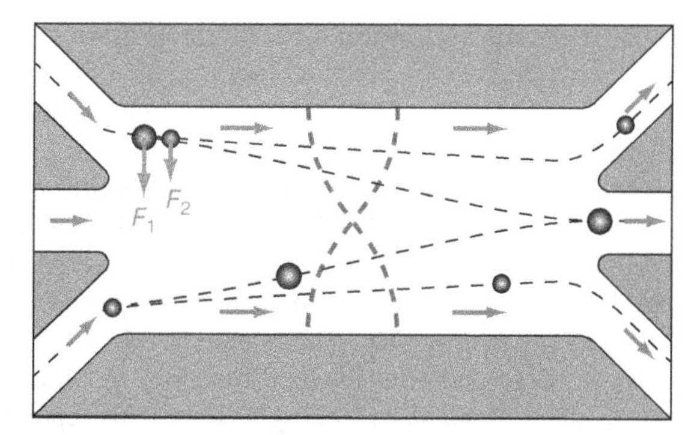

Figure 10.1 The fundamental separation mechanism in free-flow acoustophoresis. Larger objects will experience a stronger acoustophoretic force (F_1) and migrate faster to the channel center as compared to a smaller sized object (F_2) having the same mechanical properties.

off to a central outlet stream while cells of low migration velocity can be collected in the side branches of a trifurcation outlet.

10.1.2 System Specification

The acoustophoresis system described in this chapter was developed to target the clinical requirements for tumor cell separation in blood and yet offer additional benefits as compared to current clinical standard based on immunomagnetic EpCAM epitope-specific selection. A summary of the minimum requirements established for the system is presented in the following list.

1. Processing time: 1 h
2. Processed sample volume: 10 ml
3. Total recovery of target cells: ≥85%
4. Cells shall remain viable and preserve function.

The system should be able to process a 5–10 ml blood sample within an hour, which corresponds to a sample flow rate of 160 μl/min. The greatest challenge in CTC enrichment from whole blood is the extremely low abundance of CTCs. Using the established clinical detection limit of ≥5 CTC per 7.5 ml of blood in the CellSearch assay, the anticipated number of CTCs is less than 10 CTC in 7 ml of blood, which requires a single-cell resolution of detection in blood (i.e., detecting 1 CTC per milliliter in the background of 10^9 RBC and 10^6 WBCs). RBCs are commonly eliminated by targeted lysis or by gradient centrifugation.

When performing acoustophoretic separations, a limiting factor in terms of throughput is that the quality of separation starts to degenerate when the cell concentration exceeds a few million cells per milliliter (breakdown cell concentration). This is due to hydrodynamic interaction between cells that are close to each other [15].

To perform the acoustophoretic CTC separation, the breakdown cell concentration of the blood samples described below were preprocessed in an RBC lysis step, after which all nucleated cells were pelleted and resuspended in a volume corresponding to the initial blood volume.

A further requirement we established for the acoustophoretic separation system was a CTC recovery above 85% and WBC depletion of two to three orders of magnitude. This would yield a CTC frequency in the recovered fraction versus WBCs of close to 1:1000, which enables molecular biology tools to target the sample for molecular CTC profiling [16]. The system should ideally also deliver a CTC population that is viable and with retained function, enabling postseparation culturing. When performing molecular profiling or doing postseparation culturing, purity of 1:100 or higher is desirable.

10.2 FACTORS DETERMINING ACOUSTOPHORESIS CELL SEPARATION

The method for separating cancer cells from WBCs described in this chapter is based on acoustophoresis. All nucleated cells in suspension flow along a microchannel wherein the cells are exposed to an acoustic field. The individual cells experience a force that is perpendicular to the flow, which deflects the cells sideways at a magnitude that depends on the size and acoustic properties of the cell (Figure 10.2). If the properties of two cell types differ, they can be separated in the acoustophoresis chip at a performance that depends on the overlap in the distributions of these properties in the two populations. This section describes the acoustic field, the

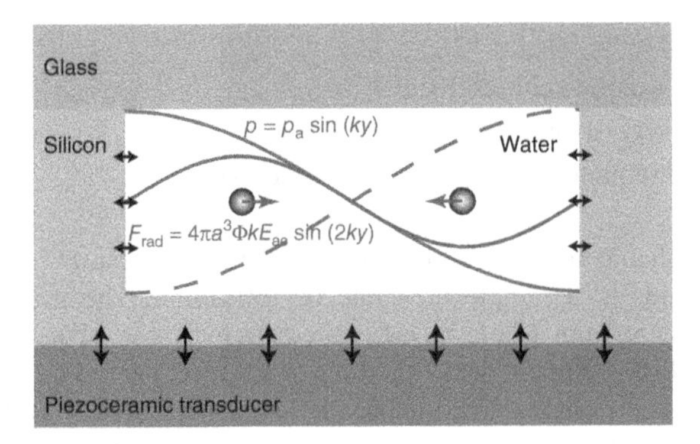

Figure 10.2 Schematic of the cross section, transverse to the flow of the acoustophoresis microchannel. Vibrations (black arrows) from the piezoceramic actuator are transmitted via the bulk material of the chip to the interior walls of the flow chamber. When the frequency of oscillation matches the resonance condition determined by the channel width and the sound velocity in the liquid, a resonance builds up (solid and dashed black lines). Cells suspended in the liquid experience a force (gray arrows and solid gray line) toward the center of the channel. Dimensions are not to scale.

acoustic radiation force on a cell, and the influence of the laminar flow on separation outcome.

10.2.1 Acoustic Field

In this work, an acoustic standing wave was set up, transversely oriented to the flow, along the whole length of a water-filled silicon and glass microchannel, by vibrating the chip using a piezoceramic actuator (Figure 10.1). For a standing wave, the relation between the speed of sound (c), frequency (f), and wavelength (λ) can be expressed as

$$c = \lambda \cdot f \tag{10.1}$$

By matching the frequency of oscillation, $f = 2\,\text{MHz}$, to the channel width, $w = \lambda/2 = 375\,\mu\text{m}$, a half wavelength resonance condition is fulfilled for liquids at the speed of sound $c = 1500\,\text{m/s}$. The time harmonic acoustic pressure field (p), which is plotted in Figure 10.2, can be expressed as

$$p = p_{\text{a}} \cos(ky) \exp(-i\omega t) \tag{10.2}$$

where p_{a} is the pressure amplitude, y is the position along the width of the channel, $k = 2\pi/\lambda$ is the wave vector, ω is the angular frequency, and t is time. Equation 10.2 describes a standing wave that has a pressure node in the center of the channel and where maximum pressure oscillations occur near each of the channel sidewalls such that when the pressure is maximal on one of the sides, the pressure is at its minimum at the opposing wall.

10.2.2 Acoustic Radiation Force

A cell submerged in a liquid that is exposed to a resonant acoustic field experiences an acoustic radiation force that causes it to move with respect to the surrounding liquid. The force stems from a local distortion of the acoustic field due to scattering from the cell. Momentum is transferred to the cell in this process. The cell experiences a time average net force directed along the pressure and velocity gradients of the acoustic field.

In this acoustic field, the acoustic radiation force (F_{rad}) on a particle of radius (a), relative density ($\tilde{\rho}$), and compressibility ($\tilde{k}$) with respect to the suspending liquid can be expressed as

$$F_{\text{rad}} = 4\pi a^3 \Phi k E_{\text{ac}} \sin(2ky) \tag{10.3a}$$

$$\Phi = \frac{1-\tilde{\kappa}}{3} + \frac{\tilde{\rho}-1}{2\tilde{\rho}+1} \tag{10.3b}$$

$$E_{\text{ac}} = \frac{p_{\text{a}}^2 \kappa_{\text{m}}}{4} \tag{10.3c}$$

where Φ is the acoustic contrast factor, E_{ac} is the acoustic energy density, and κ_{m} is the compressibility of the suspending medium.

Equation 10.3a shows that the acoustic contrast of a cell depends on its relative density and relative compressibility with respect to the suspending medium. The sign of the contrast must not necessarily be positive, but for the various blood cells or a cancer cell suspended in isotonic aqueous solution such as PBS, NaCl, or standard cell culture media, the acoustic contrast will be positive. The force will therefore be directed toward the pressure node at the center of the channel. A consequence of the contrast factor is that denser and less compressible cells will experience a stronger acoustic radiation force than lighter and more compressible cells do.

10.2.3 Trajectory of a Cell

When cells flow through the acoustophoresis channel, their trajectories, shown in Figure 10.3, are determined by the velocity distribution in the channel, the acoustic radiation force on the individual cells, and their initial positions upon entering the channel. The velocity component of a cell along the channel can be approximated to be the same as the liquid flow velocity while its velocity transverse to the flow depends on the acoustic contrast of the cell and the magnitude and shape of the acoustic field.

The motion of a cell exposed to an acoustic field is impeded by the friction at the cell–liquid interface, commonly referred to as Stokes' drag force. Assuming the cell to be spherical, this velocity can be expressed as

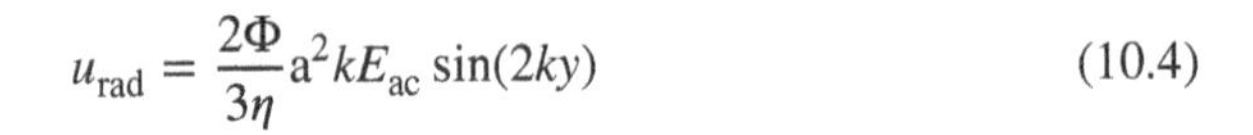

$$u_{\text{rad}} = \frac{2\Phi}{3\eta} a^2 k E_{\text{ac}} \sin(2ky) \tag{10.4}$$

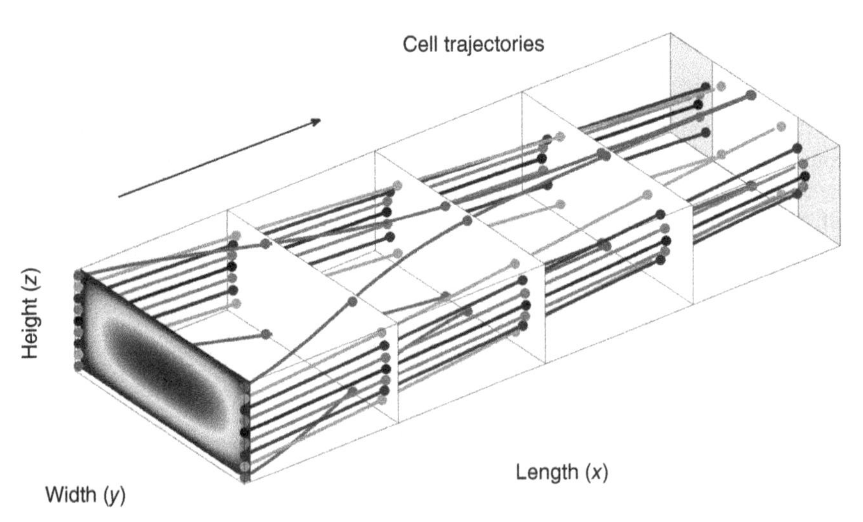

Figure 10.3 Simulated trajectories of identical particles undergoing acoustophoresis with different starting positions in height along the sidewalls. The rainbow-colored surface indicates the flow velocity distribution of the suspending liquid ranging from zero (blue) on the walls to high (red) in the central part of the flow. Yellow rectangles demark the regions of the flow that will exit through the side's outlet. (*See color plate section for the color representation of this figure.*)

where η is the dynamic viscosity of the liquid. As can be seen in Eq. 10.4, the size (a) of the cell has significant influence of its migration rate. It can therefore be anticipated that when separating cancer cells from WBCs, the outcome primarily depends on differences in size between the two populations.

The trajectory of a cell flowing through the channel can be altered either by changing the flow rate of the suspension or by changing the amplitude of the acoustic field. The acoustic energy density is a function of the second power of the voltage amplitude driving the oscillation of the piezoceramic actuator. Hence, the force exerted on the cells can be varied by altering the actuator voltage and thereby altering the cells' trajectories when flowing through the acoustic field. To achieve a higher throughput in the channel, the flow rate (Q) and the acoustic energy density E_{ac} can be increased so as to maintain a constant ratio E_{ac}/Q. This preserves the shape of the trajectories and thus leads to identical separation outcome. The throughput of a device can thus be increased up to a point where the heat dissipation in the piezoceramic actuator becomes substantial and the temperature in the channel exceeds 37 °C, causing damage to the cells.

10.2.4 The Microchannel Flow Profile

Flow of a liquid through a microchannel is highly laminar. This implies that any object submerged in the liquid will maintain its initial position in the cross section transverse to the flow, unless the acoustic field is active. Another feature of laminar flow is that the flow velocity is not uniform across the transverse cross section. The flow velocity is always zero at channel walls, floor, and ceiling and has a maximum in the very center of the transverse cross section. The velocity distribution has parabolic features near the boundaries, which is most pronounced across the smallest dimension, which in this case is the channel height.

The parabolic flow profile has the effect that the flow velocity of an object flowing through the acoustophoresis channel is a function of its position in the transverse cross section. This means that two identical objects starting out at the same position in the channel width dimension will follow different trajectories if their initial position in the height dimension differs. A cell flowing near the top or bottom of the channel has a relatively slow forward motion and will therefore move toward the equilibrium position, the centrally located pressure node, at a steeper trajectory than a cell that flows in the faster, mid-height region of the flow. Figure 10.3 shows simulated trajectories of identical spherical objects, taking into account the acoustic radiation force and the microchannel flow profile. The initial position in the height dimension clearly impacts the final position in width dimension at the end of the channel where the flow is to be divided into separate outlets. The deviation is most critical for particles near the top and bottom of the channel where the flow rate is low.

However, the low flow velocities near the top and bottom of the channel have the effect that only a small percentage of all processed particles will traverse those regions. The major part of the liquid flows in high flow velocity regions and will therefore experience similar retention times and separation conditions. This leads to an overall moderate dispersion. A fundamental condition is, however, that particles are introduced near either of the sidewalls of the channel leaving a central region of particle-free liquid to traverse.

10.3 ACOUSTOPHORESIS SYSTEM FOR SEPARATING CELLS

This section describes a platform for assessing acoustophoresis as a viable method for discriminating cancer cells from WBCs. In doing this, the requirement regarding processed sample volume was set aside while separation performance, sample flow rate, and cell viability were of main interest.

10.3.1 Acoustophoresis Chip

To achieve separation between cancer cells and blood cells, an acoustophoresis microfluidic chip was fabricated in silicon/glass, similar in design to what was first proposed by Petersson *et al.* [17] and later applied for cell cycle synchronization of mammalian cells by Thevoz *et al.* [18]. The chip, shown schematically in Figure 10.4, has a trifurcation inlet, designed such that a stream of cells in suspension, entering via a common inlet, is divided and laminated along both sides of an acoustophoresis microchannel while cell-free buffer solution, infused via a central inlet, occupies the central part of the channel. The outlet of the channel is constructed in the same way as the inlet, allowing the central fraction of the flow to be collected, whereas objects flowing in proximity to the channel walls are collected in a common side outlet.

10.3.2 Actuation of Ultrasound

The cell separation mechanism relies on acoustic forces directed perpendicular to the flow stream. The force stems from an alternating pressure field caused by an acoustic resonance condition being fulfilled along the whole length of the acoustophoresis channel. The source of vibrations is a piezoelectric actuator glued to the backside of the chip. The resonance frequency ($f = 2.0$ MHz) of the actuator was chosen such that the width of the channel (375 μm) corresponded to one half wavelength of sound in aqueous media, analogous to earlier work on microchip acoustophoresis [19, 20].

10.3.3 Flow System

The volume flow in the system was controlled by three 10 ml glass syringes, two of which were mounted so as to withdraw liquid from the outlets in a 1:1 ratio between central outlet flow and the combined side's outlet flow. The flow rates of these two syringes dictated the total flow rate in the main channel. The third syringe infused cell-free buffer solution in the central inlet to the channel at a volume flow rate corresponding to 7/8 of the total flow rate. Cell suspension was drawn into the side's inlet

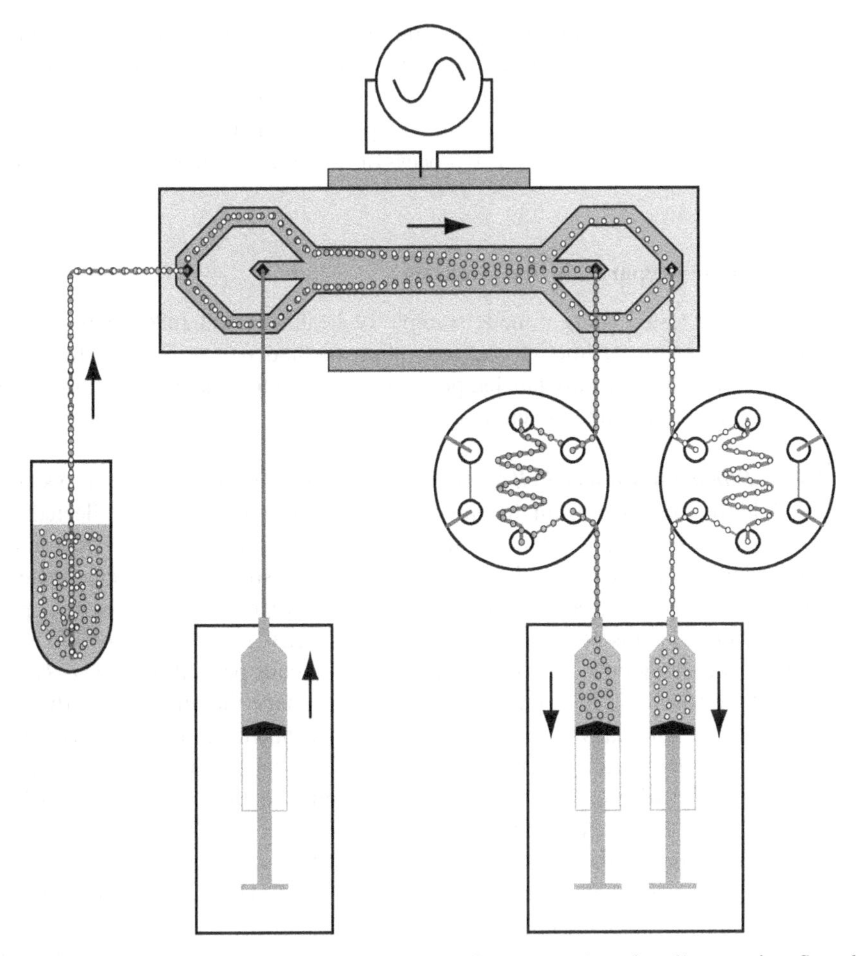

Figure 10.4 Schematic of the experimental setup for acoustophoretic cell separation. Sample was drawn from a test tube via a common side's inlet and laminated near the walls of the separation channel, while cell-free medium injected through the central inlet occupied the central part of the flow. Cells were subjected to ultrasound in the separation channel and moved toward the channel center at a rate determined by their acoustic properties and size. Two sample loops were used to collect cells from the side's and center outlet, respectively.

from the bottom of a test tube at a flow rate of 1/8 of the total flow rate. The flow ratio in the inlets determines the positioning of the cells upon entering the channel. A high central inlet flow rate results in cells entering closer to the channel sidewalls, which increases the width of the cell-free central region and leads to better separation performance.

Two six-port, two-way sample loops, each of volume 100 μl, were connected in series with the outlets to probe the separation performance while running the system.

Upon initiating the separation procedure, the loops are bypassed by the flow until the sound is activated and the flows have stabilized. The valves are then switched such that the outlet flows are routed through the loops. After switching the valves back, the loops are again bypassed and the loops, now containing cells, can be collected without disturbing the flow. The process of sampling the flow can thereafter be repeated.

10.3.4 Sample Preparation and Analysis

The device was tested using a model sample of healthy blood spiked with tumor cells cultured *in vitro*. Although the separation method *per se* is label free, the tumor cells were fluorescently labeled prior to separation to enable evaluation of the acoustophoretic separation efficiency.

10.3.4.1 Blood and Cancer Cell Preparation To optimize the separation system, we used blood obtained from healthy volunteers. The blood was collected in a vacutainer containing ethylenediaminetetraacetic acid (EDTA) or heparin as anticoagulant. The maximum cell concentration the acoustic microchip can process without decreased cell separation capacity is on the order of millions of cells per milliliter. Therefore, it is necessary to first remove the RBCs as the number of RBCs in undiluted whole blood is on the order of billions of cells per milliliter. The RBCs were lysed with isotonic lysis buffer that does not damage the nucleated WBCs or the cancer cells. Alternatively, a classic buffy coat centrifugation can be used for an initial blood fractionation, which results in a clear upper fraction of plasma, a middle buffy coat fraction of nucleated WBCs, tumor cells, and platelets, and a bottom fraction of RBCs. Another alternative is to use a Ficoll density centrifugation separation to minimize the RBCs contamination of the buffy coat. The WBCs are thereafter resuspended in a buffer containing serum and EDTA to prevent cells from aggregating and clogging the microchannel.

Tumor cell lines established from prostate cancer and breast cancer patients and obtained from ATCC (American Type Culture Collection) were cultured *in vitro* and spiked into blood from healthy volunteers at different concentrations to evaluate the discriminatory accuracy of the separation method. The cancer cells were grown in culture flasks and harvested prior to each experiment. The cells were counted and spiked into the blood samples from healthy volunteers. We used different cancer cell lines to comprehensively evaluate the performance of the acoustophoretic cell separation system.

10.3.4.2 Cell Labeling To enable postseparation cell identification, the cells may be fluorescently labeled with antibodies. This does not affect how the cells move in the acoustic standing wave field, so the cells can be labeled either before or after the acoustic separation. Depending on postseparation applications, the cells can either be fixed with, for example, paraformaldehyde, or left alive. Although fixed cells generate slightly better separation performance than viable cells [9], it is sometimes desirable with minimal pretreatment of the cells for subsequent characterizations.

10.3.5 Device Testing

Experiments were carried out to evaluate the performance of the device regarding separation quality, throughput, and sample composition. Three different prostate cancer cell lines (DU145, PC3, and LNCaP) were investigated with various settings of the acoustophoresis system.

10.3.5.1 Varying the Flow Rate The ability to discriminate tumor cells from WBCs was shown to be highly dependent on cell retention time in the chip. Retention time was assessed by varying the total flow rate in the chip while maintaining the same relative flow ratios in inlets and outlets, at constant piezoactuator driving voltage and frequency. The total flow rate here refers to the sum of the sample flow and the central cell-free medium.

For fixed cells, close to complete WBC depletion was achieved for a total flow rate of 800 µl/min while the recovery of DU 145 cells was low to modest. Lowering the flow rate to 320 µl/min leads to very high recovery of DU145 cells but also results in a lower WBC depletion, Figure 10.5a. The sample-processing rate was kept at 1/8 of the total flow rate and, thus, ranged from 40 to 100 µl/min. At a flow rate of 560 µl/min, the recovery of cancer cells was above 90% while depleting more than 90% of the WBCs.

The aforementioned sample-processing rates do not constitute an absolute upper limit of the capacity of the system. By increasing the amplitude of the piezoactuator drive voltage, the sample throughput can be raised to levels well above 10 ml/h while maintaining similar separation behavior as shown for lower flow rates.

In summary, this first demonstration of acoustophoresis to discriminate cancer cells from WBCs was fruitful. The recovery of cancer cells was high, and the sample flow rate corresponded to a processing time of approximately 1 h for a 10 ml sample. The depletion of WBCs was, despite being above 90%, not high enough to produce CTC purities of relevant levels in clinical samples. The major cause of the WBC contamination was found not to be overlapping acoustic properties or size but rather an effect of the microchannel laminar flow described in Section 10.2.4.

10.3.5.2 Cancer Cell Number Samples of decreasing number of the prostate cancer cell line DU145 spiked in RBC lysed blood were processed in the system. Results showed that cancer cell recovery was independent of the number of spiked cancer cells, Figure 10.5b. Not surprisingly, however, as the number of spiked cancer cells in the input sample decreases, so does the purity of the recovered cancer cells. In this particular implementation of the setup, it was not possible to detect extremely rare cells due to the relatively small volumes (100 µl) in the sample loops.

10.3.5.3 Cancer Cell Diversity In an attempt to address the expected diversity of CTCs in the peripheral blood of metastatic prostate cancer patients, two additional prostate cancer cell lines (LNCaP and PC3) were spiked in blood and processed in the acoustophoresis chip. All three different cancer cell lines exhibited similar separation behavior with tumor cell recovery and WBC depletion efficiencies each above

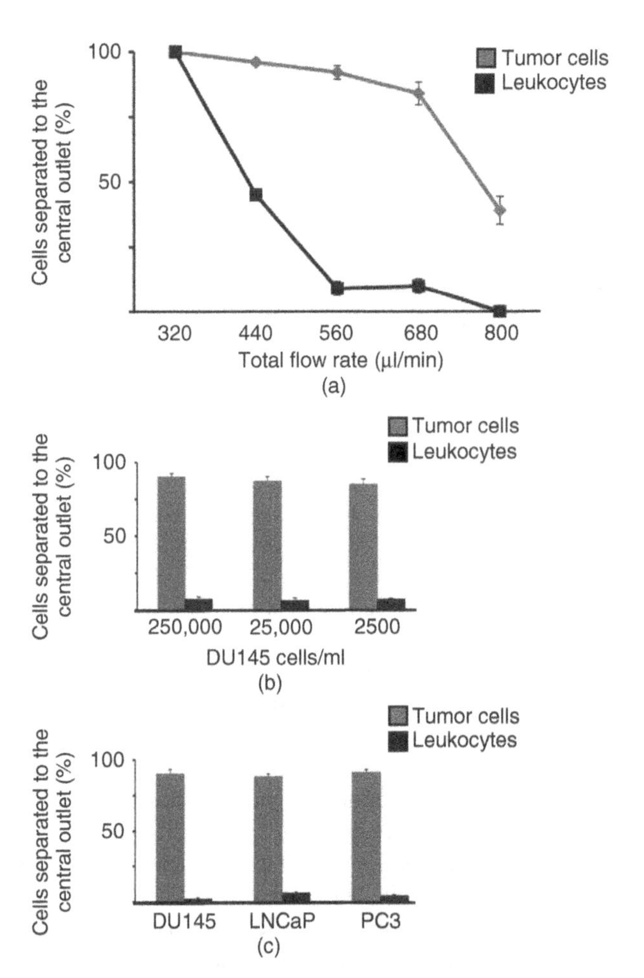

Figure 10.5 Analysis of the effects of retention time, cancer cell number, and diversity on acoustophoresis-mediated enrichment of cancer cells: Fluorescently labeled cultured prostate cancer cells were spiked into RBC lysed control blood (diluted 1:10). The cells were run through an acoustophoresis microfluidic chip in the presence of ultrasound. The cell content in the outlet fractions was analyzed by flow cytometry. The tumor cell capture efficiency and WBC depletion efficiency for acoustophoresis were determined for different total flow rates under constant voltage, for (a) fixed DU145 cells spiked in blood. (b) A dilution series with decreasing number of DU145 cells and constant level of WBCs was performed at a flow rate of 560 μl/min. (c) Acoustophoresis of three different prostate cancer cell lines (fixed in PFA) spiked in RBC lysed blood at the total flow rate of 560 μl/min. Data displayed as mean ± SD, $n = 5$.

90% for a net flow rate of 560 μl/min, Figure 10.5c. The result holds promise for acoustophoresis as a valuable tool for CTC extraction from patients' blood samples.

10.4 ACOUSTOPHORESIS PLATFORM FOR CLINICAL SAMPLE PROCESSING

After assessing the feasibility of using acoustophoresis to separate cancer cells from WBCs as outlined in Section 10.3, the acoustophoresis cell separation system was redesigned to better meet the criteria outlined in the introduction. The new platform was first presented in Augustsson *et al.* [9]. The main changes in design related to improvement of separation performance, enabling processing of large sample volumes, reducing the processing time, and improving reproducibility and long-term stability.

Three major changes were made compared to the previous design. First, an acoustophoresis cell prealignment channel was added to the separation channel to minimize dispersion of cells due to the laminar flow profile. Second, to improve short- and long-term reproducibility, the chip temperature was controlled using a temperature sensor and a Peltier element. Third, to accommodate larger sample volumes and, more importantly, to improve ease of use, the syringe-based flow system was changed to a pressure-driven flow system with flow sensor feedback to the pressure control system. Figure 10.6 shows a photo of the second-generation acoustophoretic CTC system.

10.4.1 The Acoustophoresis Chip

A schematic of the chip and flow system is shown in Figure 10.7. Cells enter the chip through a 10-mm-long prealignment channel. The width and height of this channel

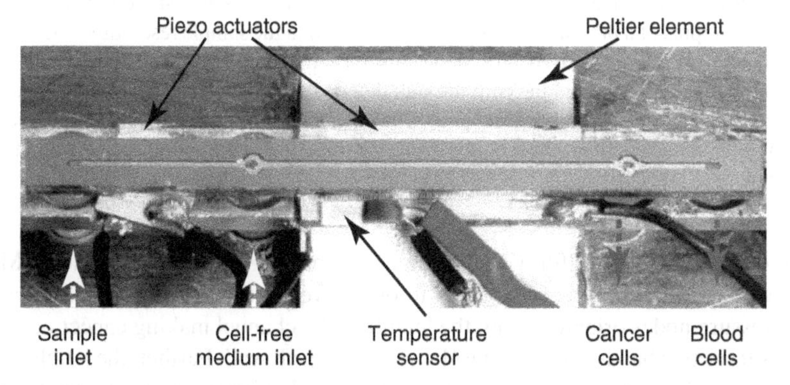

Figure 10.6 Photo of the assembled second-generation acoustophoresis cell separation chip. White arrows indicate inlets and outlets accessible from the backside of the base plate. Piezoceramic actuators resonant at 2 and 5 MHz were glued to the backside of the chip for controlling prealignment and separation, respectively. Augustsson *et al.* [9]. Adapted with permission of the American Chemical Society.

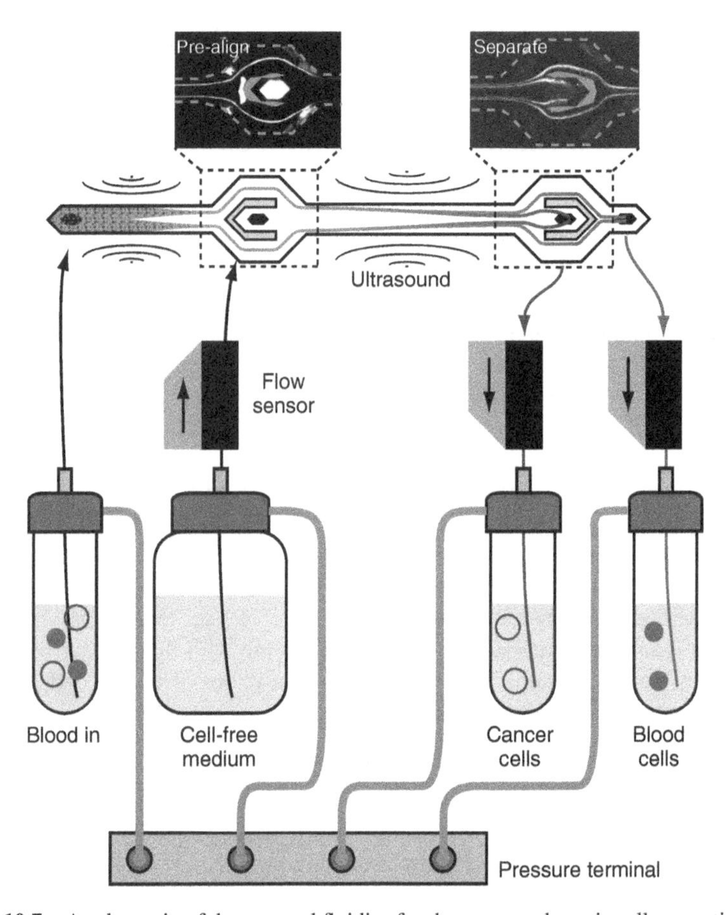

Figure 10.7 A schematic of the external fluidics for the acoustophoresis cell separation platform. The outputs of a pressure terminal were connected to the sample reservoirs, one for each inlet or outlet. When pressurized, an aqueous suspension of cells (filled and open circles) from the sample input tube entered the chip through the prealignment channel. The separated cells were collected in two test tubes. The pressures in each container were set using a feedback control loop, taking as input the readings from the three thermal flow sensors. Adapted from Deshmukh *et al.* [27].

were chosen so as to fulfill a resonance condition for sound of frequency 5 MHz. The width, $w = 300\,\mu m$, corresponded to one wavelength of sound, which rendered two pressure nodes oriented along the length of the channel making cancer cells and WBCs form two bands positioned $w/4$ away from each wall. Further, the height of the channel, $h = 150\,\mu m$, was chosen so as to support a half wavelength of sound, which drew cells away from the channel ceiling and floor to a pressure node, which was centered in height. After passing the prealignment channel, all cells were confined to two symmetrically equivalent locations in the cross section orthogonal to the flow and were thus entering the separation channel with identical starting conditions.

The separation channel was 375 μm wide and supported a half wavelength of sound of 2 MHz, similar to the first-generation design. The two bands of prealigned cells entered the separation channel on each side of an inlet of cell-free medium. The high relative flow rate in the central inlet pushed cells further toward the sidewalls upon entering the separation channel. This improved the resolution of the separation mechanism. In the end of the separation channel, the flow was divided into three flow streams. The central stream contained the cells of highest acoustophoretic mobility while the two side streams contained low mobility cells. These two streams were combined to a common side's outlet.

10.4.2 Flow System

In designing the system, care was taken to ensure that sample could be processed from one test tube to another with minimal user involvement. This is important in order to achieve good reproducibility in the system in a multi-user environment. Successful priming of a microfluidic system requires a strict sequence of events and precise timing in order to avoid bubbles from getting trapped in the flow path, which will affect the liquid flow and thereby the separation of cells. In the final manifestation of the acoustophoretic cell separation system, a pressure-driven flow approach was chosen.

10.4.2.1 Pressure-Driven Flow Pressure-driven flow was employed to transfer cells and liquids to and from the inlets and outlets of the chip. All liquids were kept in test tubes or flasks that could be pressurized. Tubing connected to the microfluidic chip extended to the bottom of each test tube such that all liquid would flow out of the tubes upon increased pressure. The flow could be controlled in all inlets and outlets of the system by adjusting the pressures in the test tubes.

The main benefits of pressure-driven flow relate mainly to the ease of use of the system in combination with providing oscillation free flow. The main disadvantage of pressure-driven flow is that when accurate flow rates are of essence, as in this setup, the flow must be monitored in a feedback control loop and the volume of the flow path is increased by the sensor dead volumes.

Pressure-driven flow systems are offered by a number of companies for a range of prices. Systems with integrated flow sensors are generally quite expensive and can have different performances in terms of how fast the flow can be regulated and whether each flow line is controlled individually or if a more elaborate system level control scheme is employed. For this application, however, the system time constants enabled a rather simple in-house developed control system.

10.4.2.2 Operation of the Flow System for Cell Separation Figure 10.7 shows schematically the flow system for acoustophoresis cell separation. The pressure control system was assembled in-house and based on an eight-port proportional pressure control valve terminal (VEMA, FESTO, Germany) with a pressure range of 0–1 bar and liquid thermal flow sensors (SLI-1000, Sensirion AG, Staefa ZH, Switzerland) having a maximum calibrated flow of 1 ml/min for water. Input and output samples are held in air-sealed pressurized test tubes in a spring-loaded sample tube rack. The pressures in the outlet containers are controlled in a closed feedback loop based on the readings from the flow sensors.

With a pressure-driven flow system, the priming of the system can be performed by removing the cell inlet tube and the two outlet tubes and applying an elevated pressure to the wash-fluid inlet container. The cell-free liquid then enters the central inlet at high flow rate and flows out through the open ends for a period of 15–30 s. To decontaminate or, in other ways, treat the interior of the system, any appropriately chosen liquid can be supplied to the central inlet reservoir prior to flushing/priming.

After the system is primed with running buffer, two empty test tubes are attached to the sample outlets and the sample containing the cells to be sorted is attached to the sample inlet. A spring-loaded horizontal bar pushes on the bottom of the test tubes to seal the tubes against O-rings in the test tube rack.

The user initiates the separation procedure in a computer-based user interface, waits until the whole sample has been withdrawn from the sample inlet tube, and thereafter stops the flow from the software. The test tubes are dismounted for further analysis, and the system is then flushed with clean liquid for 30 s before loading the next sample.

10.4.3 Temperature Control System

With increasing sample flow rates, higher voltages need to be applied to the piezoceramic actuator to focus cells in the separation channel. Higher amplitude leads to more heat being dissipated in the piezoceramic material and an elevated temperature of the chip.

System temperatures that are too high lead to cell damage and protein denaturation. Further, fluctuating temperatures lead to varying focusing performance. It has also recently been reported that acoustophoresis may be sensitive to temperature changes as small as 1–2 °C [21]. This means that even normal variations of the ambient temperature may cause a drift in the system's separation performance.

The redesign therefore included temperature control of the chip in order to accommodate for heat dissipation and variations in ambient temperature. The temperature was measured on the surface of the piezoceramic transducer close to the chip, and the measured signal was used in a feedback loop to control the voltage over a Peltier element glued to the base of the assembly.

10.4.4 Software Interface

The software was developed to facilitate the operating procedure of the instrument and to minimize the need for user training. Being a multiuser instrument in a translational research environment, the platform must be robust and produce reproducible output over time and hence the control software is of key importance.

In the software, the user can initiate priming of the system prior to loading cells in the sample inlet tube. The flow rates in the inlets and outlets to the chip can be configured, and the amplitude of the ultrasound can be set.

Once a run has been initiated, the user interface displays the measured flow rates to and from all vessels and the corresponding applied pressures. The user monitors the flow rates, observing any deviations from the set flow velocities, which are indicative of clogging or entrapment of air in the chip. All values are logged to a file, which allows for post separation analysis of the individual runs.

10.4.5 System Calibration using Microbeads

Samples containing 5 and 7 μm polystyrene beads were used to account for long-term drift of the platform. The distributions of microbeads in the outlets were analyzed for a range of piezoceramic actuator voltages. From this data, in combination with simulated trajectories of polystyrene beads, the acoustic intensity inside the separation channel can be determined. This renders a scaling factor for the actuator voltage, which can be used to maintain the same separation conditions for cells over long periods of time.

The bead samples also revealed the system's ability to discriminate objects based on size. It was possible to separate polystyrene beads with a diameter of 5 or 7 μm in a suspension with an initial purity of 50% of each bead size. In the two outlet fractions, each bead size was found to be of purity >99% when operating the microfluidic system in prealignment mode. Figure 10.8 shows a plot of the ratio of beads collected at the central outlet for each bead size. This diagram also demonstrates the importance of prealignment for a more distinct size separation. The instrument has

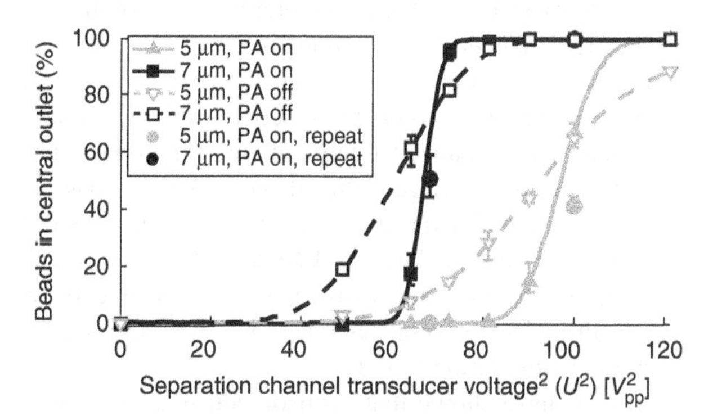

Figure 10.8 The plot shows that the larger (7 μm) beads appeared in the central outlet at lower voltage amplitudes than the smaller (5 μm) beads. Operating the microsystem with active acoustic prealignment (PA) of the particles before separation allowed a steeper transition slope from side's to central outlet. As a result, a vast majority of the larger beads could be collected in the central outlet, while the smaller beads exited through the side's outlet. Augustsson *et al.* [9]. Adapted with permission of the American Chemical Society.

excellent long-term stability indicated by the two repeated data points acquired on a separate occasion.

10.4.6 Cell Separations

To measure the performance of the pressure-driven acoustophoresis separation system, similar experiments were performed. Prealignment of cells and temperature stabilization of the system resulted in increased acoustic cell separation performance (Figure 10.9a), as previously shown with polystyrene beads. The temperature stabilization also allowed higher focusing energy levels, which opens the possibility of faster sample processing. Several different prostate cancer cell lines were used to address the expected diversity of CTCs in clinical samples. Figure 10.9b shows that the recovery of cancer cells at the central outlet increased with higher acoustic energy levels. However, the WBC contamination also increased with higher acoustic energy. An optimal energy level should be chosen dependent on application where either the purity or cell recovery is crucial for further cell analyses.

Up to a concentration of 3×10^6 cells/ml, the cell concentration in the microchannel is not a crucial parameter for the cell separation, as shown in Figure 10.9c. However, for exceedingly high concentrations, the hydrodynamic interactions of cells increase as the cell–cell distance diminishes. Hence, at higher cell concentrations, the separation performance between WBCs and CTCs degenerates.

10.5 UNPERTURBED CELL SURVIVAL AND PHENOTYPE AFTER MICROCHIP ACOUSTOPHORESIS

Questions regarding possible cell damage due to hydrodynamic shear, ultrasonic exposure, and possible exposure to high temperatures are commonly raised when reporting data on microchannel acoustophoresis processing of cells. It is vital for future clinical characterization of CTCs that the derived CTCs are viable and display an unperturbed phenotype to enable discovery of novel therapeutic targets and to provide means to personalized treatment. In response to this, Augustsson *et al.* [9] have shown that the damage inflicted on cells when passing through the acoustophoresis chip is minimal and does not seem to influence the cell viability for any of the cancer cell lines tested (DU145, LNCaP, VCaP, and PC3) as measured by Trypan blue staining. Neither did acoustophoresis affect cell proliferation or the ability to be recultured for several passages, nor was any impact on the PSA secretory function upon stimulation found.

Several other studies have shown that acoustic cell manipulation in microchannels does not impose any monitorable damage to cells, including proliferation assays and studies showing unaltered mitochondrial respiratory function of human thrombocytes and leukocytes [10]. Effects of ultrasonic trapping and in-chip culturing of yeast cells have been demonstrated [22], and acoustophoresis has been employed for blood component separation with no hemolysis of erythrocytes [23] or significant release of intracellular components [11]. Additionally, cell proliferation has been studied and

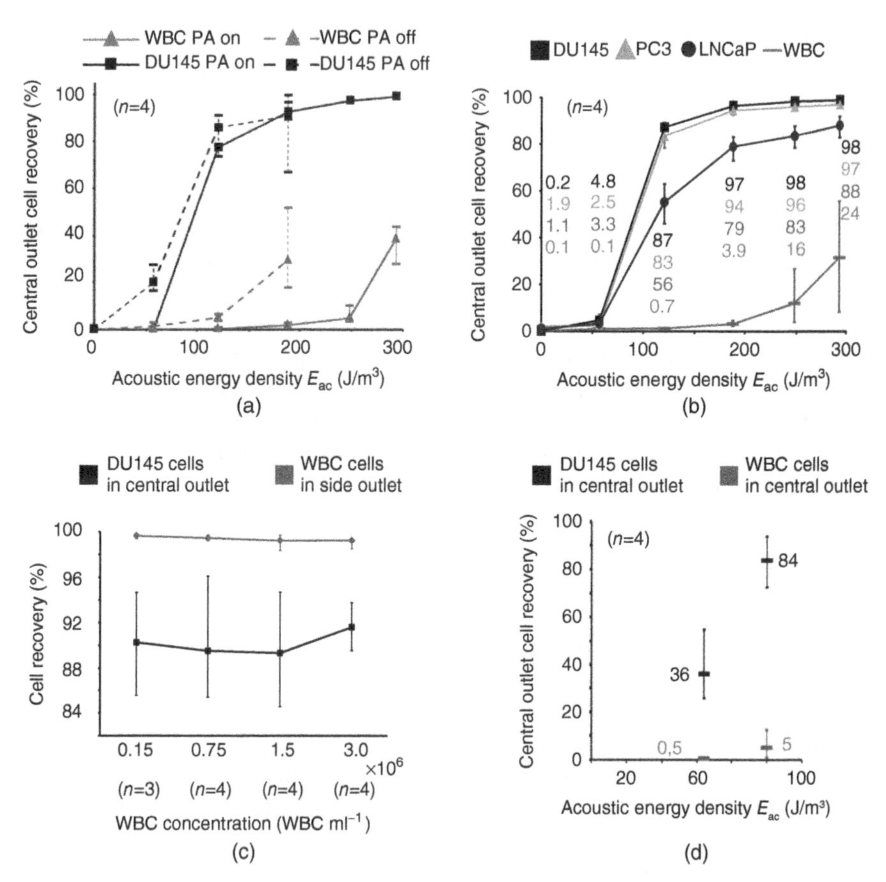

Figure 10.9 Samples of prostate cancer cells spiked in RBC lysed blood were run through a pressure-driven acoustophoretic microfluidic chip. The central outlet cancer cell recovery, that is, the relative amount of cells collected in the central outlet to the total amount of collected cells, was measured by flow cytometry. (a) The graph illustrates how prealignment of cells improves cell separation of PFA-fixed prostate cancer cells DU145 (black) and WBC (gray). Cells were processed through the microchip with acoustic prealignment turned on (solid) and off (dashed). (b) Three different PFA-fixed prostate cancer cell lines (DU145, PC3, and LNCaP) were separated from blood cells by acoustophoresis (prealignment on). (c) Acoustic separation of DU145 cells (2.5×10^5/ml) spiked in increasing concentrations of WBC with active cell prealignment. (d) Viable nonfixed DU145 cells spiked in blood were separated by acoustophoresis (prealignment on). All data is presented as mean, max, and min values, $n = 4$. Augustsson *et al.* [9]. Adapted with permission of the American Chemical Society.

revealed unperturbed growth rate after exposure to an acoustic standing wave field [11, 18, 24, 25].

The importance of collecting viable circulating cancer cells can be exemplified by prostate cancer. The molecular background of prostate tumor progression is not fully understood. To determine this, it is crucial for future cell analysis that the separation technology's intrinsic properties do not alter any physical appearances or functional properties of the processed cells. One way to conclude this is to monitor prostate-specific antigen (PSA) secretion from prostate cancer cells. PSA is a sensitive indicator for androgen receptor (AR) signaling activity in AR-dependent tumor cells. Most prostate cancer cells need androgen stimulation for growth and proliferation, and androgen deprivation therapy is the first line of treatment for most prostate cancer patients with disseminating disease. The treatment is effective in reducing tumor size and growth. However, it selects for androgen-independent cancer cells, generating tumors resistant to androgen therapy and progression into castration-resistant prostate cancer [26]. Molecular interrogation of isolated CTCs would provide valuable information on tumor progression and metastasis. Burguillos *et al.* show that acoustophoresis affects neither cell responsiveness nor the endogenous PSA secretion or PSA secretion induced by R1881 (a synthetic androgen), in prostate cancer cells. Acoustophoresis may therefore be a suitable technique for live CTC isolation from blood [10].

10.6 SUMMARY

Microchannel acoustophoresis has been evaluated for isolation of prostate and breast cancer cells from blood. Cancer cells are discriminated primarily based on their higher acoustophoretic mobility as compared to the WBCs. A recovery of 90% of spiked tumor cells is feasible while depleting WBCs 1000-fold. Viability, proliferation, or function of cells was not altered after acoustophoresis. An acoustophoresis cell separation instrument has been developed that can reproducibly isolate tumor cells from clinically relevant sample volumes in a multiuser nonengineering setting.

REFERENCES

1. Rhim, A.D., Mirek, E.T., Aiello, N.M. *et al.* (2012) Emt and dissemination precede pancreatic tumor formation. *Cell*, **148** (1–2), 349–361.
2. Polyak, K. and Weinberg, R.A. (2009) Transitions between epithelial and mesenchymal states: acquisition of malignant and stem cell traits. *Nature Reviews Cancer*, **9** (4), 265–273.
3. Nagrath, S., Sequist, L.V., Maheswaran, S. *et al.* (2007) Isolation of rare circulating tumour cells in cancer patients by microchip technology. *Nature*, **450** (7173), 1235–1239.
4. Karabacak, N.M., Spuhler, P.S., Fachin, F. *et al.* (2014) Microfluidic, marker-free isolation of circulating tumor cells from blood samples. *Nature Protocols*, **9** (3), 694–710.
5. Lee, M.G., Shin, J.H., Bae, C.Y. *et al.* (2013) Label-free cancer cell separation from human whole blood using inertial microfluidics at low shear stress. *Analytical Chemistry*, **85** (13), 6213–6218.
6. Rosenberg, R., Gertler, R., Friederichs, J. *et al.* (2002) Comparison of two density gradient centrifugation systems for the enrichment of disseminated tumor cells in blood. *Cytometry*, **49** (4), 150–158.

7. Mohamed, H., Murray, M., Turner, J.N., and Caggana, M. (2009) Isolation of tumor cells using size and deformation. *Journal of Chromatography. A*, **1216** (47), 8289–8295.
8. Gascoyne, P.R.C., Noshari, J., Anderson, T.J., and Becker, F.F. (2009) Isolation of rare cells from cell mixtures by dielectrophoresis. *Electrophoresis*, **30** (8), 1388–1398.
9. Augustsson, P., Magnusson, C., Nordin, M. *et al.* (2012) Microfluidic, label-free enrichment of prostate cancer cells in blood based on acoustophoresis. *Analytical Chemistry*, **84** (18), 7954–7962.
10. Burguillos, M.A., Magnusson, C., Nordin, M. *et al.* (2013) Microchannel acoustophoresis does not impact survival or function of microglia, leukocytes or tumor cells. *Plos One*, **8** (5), e64233.
11. Hultstrom, J., Manneberg, O., Dopf, K. *et al.* (2007) Proliferation and viability of adherent cells manipulated by standing-wave ultrasound in a microfluidic chip. *Ultrasound in Medicine and Biology*, **33** (1), 145–151.
12. Gupta, S., Feke, D.L., and Manas-Zloczower, I. (1995) Fractionation of mixed particulate solids according to compressibility using ultrasonic standing-wave fields. *Chemical Engineering Science*, **50** (20), 3275–3284.
13. Petersson, F., Aberg, L., Sward-Nilsson, A.M., and Laurell, T. (2007) Free flow acoustophoresis: microfluidic-based mode of particle and cell separation. *Analytical Chemistry*, **79** (14), 5117–5123.
14. Ding, X., Peng, Z., Lin, S.-C.S. *et al.* (2014) Cell separation using tilted-angle standing surface acoustic waves. *Proceedings of the National Academy of Sciences*, **111** (36), 12992–12997.
15. Mikkelsen, C. and Bruus, H. (2005) Microfluidic capturing-dynamics of paramagnetic bead suspensions. *Lab on a Chip*, **5** (11), 1293–1297.
16. Leversha, M.A., Han, J., Asgari, Z. *et al.* (2009) Fluorescence in situ hybridization analysis of circulating tumor cells in metastatic prostate cancer. *Clinical Cancer Research*, **15** (6), 2091–2097.
17. Petersson, F., Nilsson, A., Jonsson, H., and Laurell, T. (2005) Carrier medium exchange through ultrasonic particle switching in microfluidic channels. *Analytical Chemistry*, **77** (5), 1216–1221.
18. Thevoz, P., Adams, J.D., Shea, H. *et al.* (2010) Acoustophoretic synchronization of mammalian cells in microchannels. *Analytical Chemistry*, **82** (7), 3094–3098.
19. Hawkes, J.J.a. and Coakley, W.T. (2001) Force field particle filter, combining ultrasound standing waves and laminar flow. *Sensors and Actuators B: Chemical*, **75** (3), 213–222.
20. Nilsson, A., Petersson, F., Jonsson, H., and Laurell, T. (2004) Acoustic control of suspended particles in micro fluidic chips. *Lab on a Chip*, **4** (2), 131–135.
21. Augustsson, P., Barnkob, R., Wereley, S.T. *et al.* (2011) Automated and temperature-controlled micro-piv measurements enabling long-term-stable microchannel acoustophoresis characterization. *Lab on a Chip*, **11** (24), 4152–4164.
22. Evander, M., Johansson, L., Lilliehorn, T. *et al.* (2007) Noninvasive acoustic cell trapping in a microfluidic perfusion system for online bioassays. *Analytical Chemistry*, **79** (7), 2984–2991.
23. Jonsson, H., Nilsson, A., Petersson, F. *et al.* (2005) Particle separation using ultrasound can be used with human shed mediastinal blood. *Perfusion*, **20** (1), 39–43.
24. Pui, P.W., Trampler, F., Sonderhoff, S.A. *et al.* (1995) Batch and semicontinuous aggregation and sedimentation of hybridoma cells by acoustic resonance fields. *Biotechnology Progress*, **11** (2), 146–152.

25. Radel, S., McLoughlin, A.J., Gherardini, L. *et al.* (2000) Viability of yeast cells in well controlled propagating and standing ultrasonic plane waves. *Ultrasonics*, **38** (1–8), 633–637.
26. Javidan, J., Deitch, A.D., Shi, X.B., and de Vere White, R.W. (2005) The androgen receptor and mechanisms for androgen independence in prostate cancer. *Cancer Investigation*, **23** (6), 520–528.
27. Deshmukh, S., Brzozka, Z., Laurell, T., and Augustsson, P. (2014) Acoustic radiation forces at liquid interfaces impact the performance of acoustophoresis. *Lab on a Chip*, **14** (17), 3394–3400.
28. Augustsson, P. (2011) *On Microchannel Acoustophoresis – Experimental Considerations and Life Science Applications*, Lund University, Sweden.

11

PHOTOACOUSTIC FLOW CYTOMETRY FOR DETECTION AND CAPTURE OF CIRCULATING MELANOMA CELLS

JOHN A. VIATOR

Biomedical Engineering Program, Duquesne University, Pittsburgh, PA, USA

BENJAMIN S. GOLDSCHMIDT

Department of Bioengineering, University of Missouri, Columbia, MO, USA

KIRAN BHATTACHARYYA

Department of Biomedical Engineering, Northwestern University, Evanston, IL, USA

KYLE ROOD

Biodesign Program, One Hospital Drive, Columbia, MO, USA

11.1 INTRODUCTION

Melanoma is one of the leading causes of cancer death in the world, with ever-increasing incidence in some regions. In Australia, for example, the lifetime risk of having melanoma is 1 in 29 [1]. In the United States, the lifetime risk is 1 in 55 and has been doubling every decade or two for the past century [2, 3]. In contrast to other skin cancers, such as basal cell carcinoma, melanoma can spread aggressively throughout the body, entering into the *metastatic phase* of disease. During metastasis, melanoma cells shed from the primary tumor and spread through the blood and lymph systems, laying seeds for secondary tumors in distant organs,

Circulating Tumor Cells: Isolation and Analysis, First Edition. Edited by Z. Hugh Fan.

such as the liver, lungs, and brain, or even bone structures. It is in this phase that cancer becomes dangerous, as the spread to distant organs sets off a cascade of events that results in death. Among other things, metastatic spread of tumors interferes with normal organ function and draws nutrients away from the body to feed these growing cancerous masses.

In order to better manage therapy and to begin necessary treatment of metastatic disease, early detection is of primary importance. Detecting metastatic disease is currently performed using imaging modalities, such as computed tomography (CT) and magnetic resonance imaging (MRI). These methods detect macroscopic tumors, although in many cases, they may be too large or too invasive to be treated [4]. There have been many attempts to detect metastatic spread sooner. Cancer researchers have spent the past decade or so studying circulating tumor cells (CTCs) to this end. CTCs are those tumor cells that spread throughout the circulatory systems.

While researchers believe their presence promotes metastatic disease, little is actually known about the biology of CTCs and their precise relationship to disease state. In the context of tumor biology, there is a wide chasm between the understanding of primary tumors and metastatic spread, although many believe there is an epithelial–mesenchymal transition (EMT) that allows single cells to separate from a tumor and remain viable in the blood stream before spreading to other organs [5, 6]. Of course, beyond understanding the basic cancer biology of metastasis, detection and capture of CTCs may provide the key to diagnosing and, perhaps, halting metastasis.

Much of the biology of CTCs is unknown due to the difficulty in detecting and isolating these rare cells. Researchers have found that the incidence of CTCs among normal blood cells is on the order of one in a billion [7, 8]; thus, their detection and isolation constitute a major engineering challenge. Only after CTCs can be quickly, reliably, and inexpensively captured, can the fundamental questions about their biology be investigated. Of equal importance, monitoring CTCs may provide clinicians with diagnostic and prognostic information about their patients, possibly ushering a new era of therapy in which advanced cancer is fought cell by cell, rather than against large, macroscopic tumors, as is the present case.

The past decade has seen a vast increase in CTC detection research that includes a diverse set of technologies for detecting and capturing these rare cells. These technologies can primarily be categorized into five distinct groups: (i) molecular assays [9–13], (ii) immunohistochemical separation [14–17], (iii) microfluidics and filtration [18–20], (iv) flow cytometry [21–23], and (v) separation by dielectric or acoustic properties [24, 25]. Many of these technologies are developing rapidly as the need for finding CTCs is ever more relevant. We have developed a means for detecting circulating melanoma cells (CMCs), the CTCs for melanoma.

Our method is a type of flow cytometry that uses photoacoustics, or laser-induced ultrasound, rather than reflectance and fluorescence signals as in conventional systems. Photoacoustics has the advantage of optical selection of pigmented cells and using robust acoustic signals for detection.

11.1.1 Biomedical Photoacoustics

Photoacoustics is the phenomenon in which optical energy is converted into acoustic energy. There are several mechanisms to accomplish this conversion, including ablation, cavitation, radiation pressure, and thermoelastic expansion [26]. Most of the work in biomedical photoacoustics, that is, photoacoustics applied to medical research and application, uses the later mechanism. Thermoelastic expansion involves stress-confined absorption of laser energy into a medium, generally with less energy than that needed to create ablation or cavitation. The stress-confined pulses are often of nanosecond duration. This rapid deposition of photons results in rapid heating and subsequent expansion, leading to a pressure wave. This pressure disturbance is referred to as the *photoacoustic wave* and, in the linear regime, carries information about the optical absorption of the medium. This fact is one of the benefits of photoacoustic generation and detection in biological tissue, as the resulting acoustic wave can be analyzed for the optical properties. In addition to this advantage, photoacoustics allows the resulting acoustic wave to carry information about the biological tissue relatively unimpeded. Compared to optical signals that degrade within a millimeter or so due to the turbid nature of most tissue, acoustic waves propagate for possibly centimeters while maintaining useful information about the environment.

We have used photoacoustic generation to perform depth profiling of skin for vascular and pigmented lesions, exploiting the dual natured methods [27–31]. We also used this to perform depth profiling and imaging in burn injury [32–34]. These methods view the skin as a layered medium with different optical properties. In some cases, two or more laser wavelengths are used to differentiate absorption spectra of hemoglobin and melanin. For instance, by comparing photoacoustic responses from laser wavelengths of 532 and 650 nm, one can infer whether the source was from hemoglobin or from melanin because the response will be greater at 532 nm than 650 nm, while for melanin, the response will be similar. This discrimination scheme can be made more formal using classical or Bayesian methods [31, 33], thus enabling reconstruction of skin with respect to pigmented and vascular structures.

11.1.2 Photoacoustic Flow Cytometry

Later, we adapted our use of photoacoustics to find particles in flow. Specifically, we investigated the use of laser light to detect CMCs in blood of patients as a possible means for diagnosing metastatic disease [35, 36]. Our original setup included two laser wavelengths for discrimination between pigmented cells and red blood cells (Figure 11.1). Whole blood from melanoma patients was prepared using a standard protocol that resulted in extracting the buffy coat comprising leukocytes and any possible CMCs. This procedure, if done properly, should remove all red blood cells and thus obviate the need for two laser wavelengths because the only other optical absorbers of any significance would be pigmented cells. However, we were not sure that the blood separation procedure removed all red blood cells, so we continued

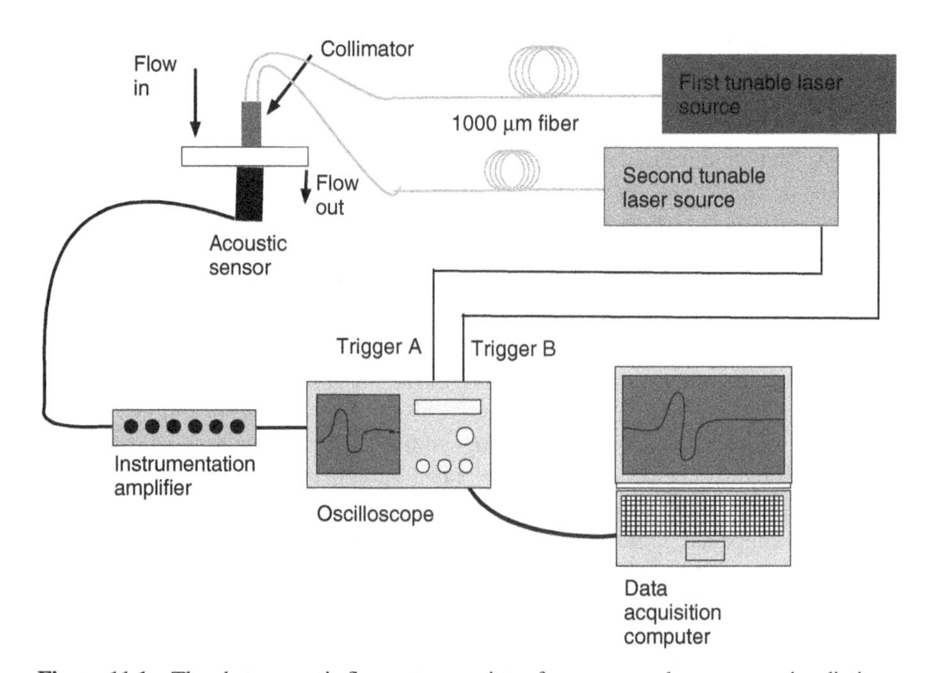

Figure 11.1 The photoacoustic flowmeter consists of one or more laser sources irradiating a flow chamber through which blood cells are circulated. Light-absorbing particles emit photoacoustic waves indicating their presence, while other cells pass silently. Cell throughput is high, as thousands or even millions of normal cells can pass through the laser beam at any time as they generate no acoustic signal.

performing hemoglobin and melanin classification using the two-wavelength method. Eventually, we were certain that the blood separation protocol was removing all sources of hemoglobin, so we now use only a single wavelength for exciting melanin within CMCs.

The apparatus consisted of the laser sources, usually tuned to green and red laser wavelengths, both coupled to 1000 μm, multimode optical fibers. The lasers were tunable systems consisting of Q-switched, frequency-tripled Nd:YAG lasers pumping optical parametric oscillators. The pulse duration was typically 5–10 ns with pulse repetition rates of 1–20 Hz, generally operating at the higher end of this range, allowing us to increase flow rate and reduce time to test the sample. These systems allowed wavelength tuning within most of the visible spectrum. In some cases, when using a single wavelength, we used frequency-doubled Nd:YAG systems operating at 532 nm.

We launched between 1 and 10 mJ of energy into the fibers, although robust photoacoustic waves can be generated in single melanoma cells at the lower end of this range. The fibers were directed to a glass chamber through which a suspension of the buffy coat was flowing. The chambers were between 1 and 3 mm in diameter. In contrast to conventional flow cytometry, where cells are polled in single file [37], in photoacoustic flow cytometry, multiple cells pass through the laser beam at any given time. In fact, because leukocytes have no appreciable optical absorbers, we flow thousands of such cells in the laser beam at any given time. Leukocytes do not generate photoacoustic waves and only scatter light somewhat. Only when an absorbing particle, such as a CMC, enters the beam a photoacoustic wave will be generated. This fact is one advantage of photoacoustic detection of CMCs. Since thousands of leukocytes can flow through the laser beam without confounding detection of CMCs, samples can be rapidly tested. In fact, we would typically have flow rates of 0.2 mm/s for 10 ml, giving a testing time of less than an hour.

The chamber was in-line with a flow system that began with a syringe pump containing the sample, flexible tubing from the pump to the chamber, followed by more flexible tubing to a receiver to contain the sample after laser irradiation. Typical photoacoustic waves arising from these tests are shown in Figure 11.2. Initially, we had used piezoelectric films made from polyvinylidene difluoride set in the glass chamber for acoustic sensing. These films were difficult to maintain electrical connection to and had a tendency to burn when the laser beam was misaligned. We subsequently investigated the use of an all-optical acoustic sensing method based on work by Paltauf *et al.* [38].

The optical sensor for photoacoustic waves generated in CMCs is shown schematically in Figure 11.3. In this setup, rather than a piezoelectric sensor, we used a probe laser beam from a HeNe laser reflected by a prism in contact with the flow chamber. The prism was in direct contact with the flowing medium. The reflectance of the HeNe beam was determined by the index of refraction of the medium and the index of the prism glass. In the presence of a compressive wave, such as a photoacoustic event, the index of refraction of the medium changes, thus changing the reflectance of the HeNe beam. The HeNe beam was reflected onto a high-speed photodetector through

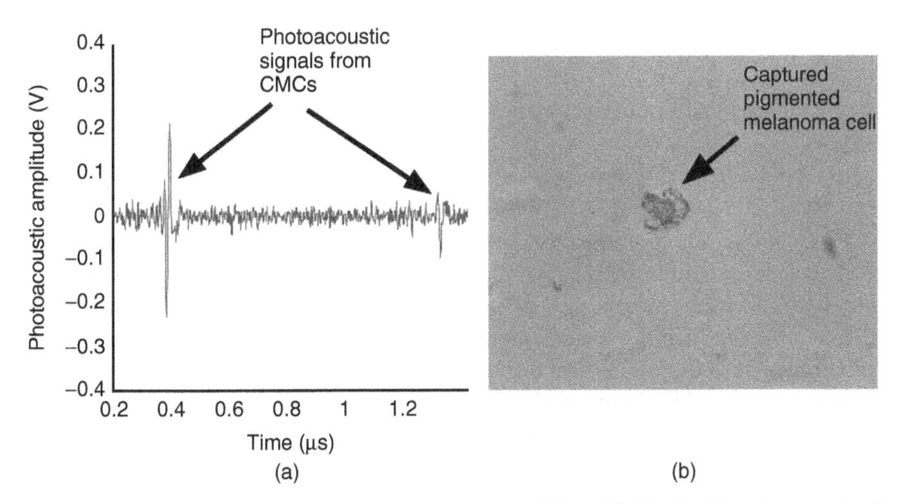

Figure 11.2 (a) Typical photoacoustic waves generated from CMCs. (b) The pigmented cell generating the photoacoustic wave.

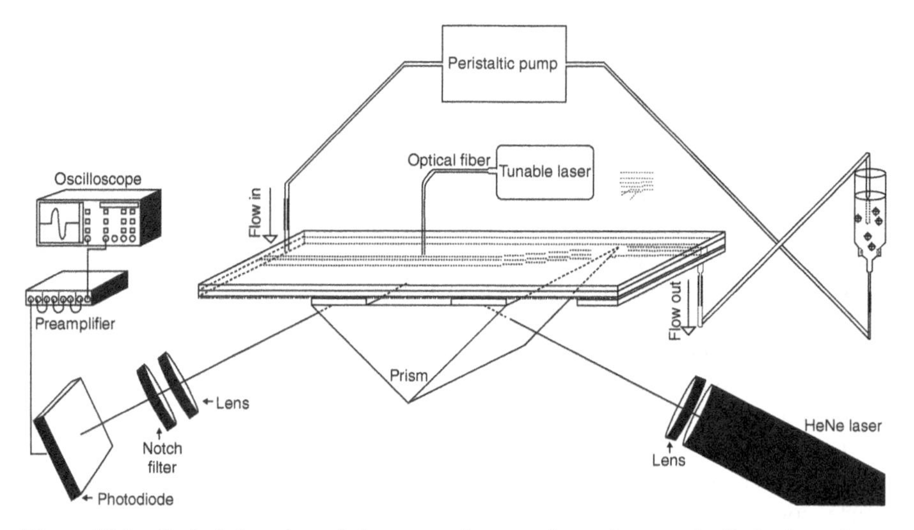

Figure 11.3 Optical detection of photoacoustic waves from pigmented cells is shown here. A HeNe laser provided a probe beam that reflected off of the detection chamber via a prism that was in contact with the flowing medium. The reflected light was detected by a photodiode. The change in reflectance of the probe beam as shown on an oscilloscope indicated any photoacoustic waves generated in melanoma cells.

a lens and notch filter. The output of the photodetector was sent to an oscilloscope after amplification. The photoacoustic wave was thus represented by the change in optical reflectance at the chamber. The setup along with a microcuvette ground from a glass slide is shown in Figure 11.4. The microcuvette contains cultured melanoma cells, as indicated by the brown dots. Although we were successful in detecting single cultured melanoma cells, the alignment of the system was difficult to maintain.

11.2 CURRENT METHODS FOR DETECTION AND CAPTURE OF CMCs

In order to enable repeatable, sensitive detection of CMCs, we moved to a focused, piezoelectric sensor. Additionally, we were interested in capturing the CMCs. Our original scheme simply suctioned a small volume of the effluent by determining the time it would take, given a flow rate, to travel from the detection chamber to the suction port. Our inability to track these particles was due to the nonuniform velocity of particles in the system, making our capture rate less than 25%. Our current apparatus avoids this timing problem by inducing two-phase flow in our system. That is, we separate our continuous flow of buffer into microdroplets separated by an immiscible fluid, such as air or oil. The current detection and capture process is described in this section.

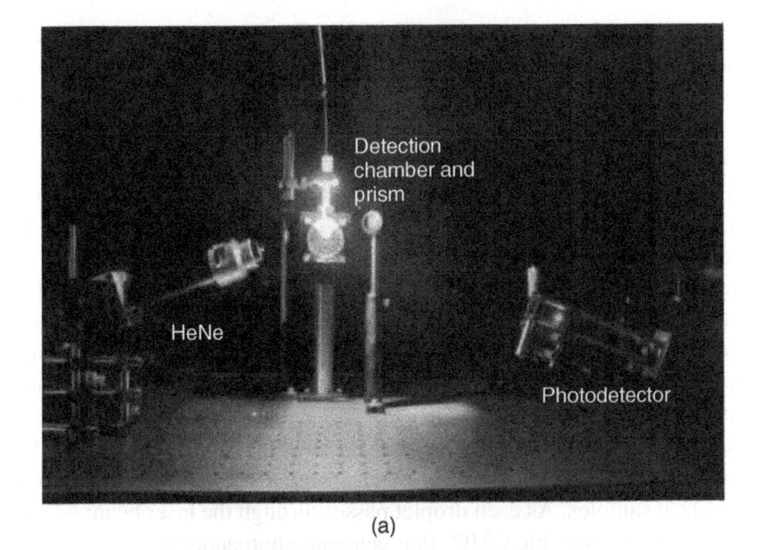

(a)

(b)

Figure 11.4 (a) The optical detection setup is shown here. (b) Pigmented melanoma cells in the microcuvette are shown here. (*See color plate section for the color representation of this figure.*)

11.2.1 Two-Phase Flow for Cell Capture

Two-phase flow is a phenomenon in microfluidics in which two immiscible fluids exist in a flow path. Two-phase flow occurs when these fluids are injected into a flow system while maintaining a capillary number less than 0.01 [39–41]. Steijn *et al.* [42] developed a model showing that only flow rates of the two fluids and the junction shape determined the droplet sizes. Using these principles, we were able to separate blood cell suspensions between air droplets.

Because the junction between the air and buffer was fixed, we only needed to adjust the fluid flow rate to modify droplet size and spacing. Ideally, we created buffer droplets approximately 1–10 μl. The exact size will be balanced by the ability to isolate CMCs in small volumes, while separating the original saline volume into a tractable number of droplets to process. As each droplet will take approximately 2–5 s to process, dividing a 1 ml sample into 10 μl droplets would require less than 20 min to process, allowing for equal-sized air bubbles in between. In addition to sequestering CMCs, droplet formation also creates a self-contained acoustic chamber, since photoacoustic waves will not be able to transmit across the interface of air and buffer, due to differences in acoustic impedance. Thus, there will be no cross talk from surrounding droplets, limiting the photoacoustic signals to the central bubble in the chamber. This effect will serve to improve our ability to isolate cells.

Figure 11.5 shows the photoacoustic flowmeter using two-phase flow [43, 44]. The separation of cell buffer into droplets essentially creates a conveyor belt of microliter-sized samples. As each droplet passes through the laser beam delivered by the optical fiber, any possible CMCs that generate photoacoustic waves are detected by the sensor. This droplet is diverted to a collection plate, while the droplets that did not generate photoacoustic waves are sent to waste. The collection plate can be later used for imaging or molecular assays.

11.2.2 Photoacoustic Flow Cytometer

Using the photoacoustic flow cytometer and two-phase flow, we tested 20 samples of human, stage IV, melanoma patients and 10 healthy human subjects for the presence of CMCs in 10 ml of whole blood. The photoacoustic flow cytometer comprised a nanosecond pulsed laser, a flow chamber coupled to an acoustic sensor, and two syringe pumps. We imaged the captured droplets for the presence of CMCs and captured some of them using micromanipulators. These cells can be kept for further analysis. The laser system was a Q-switched, frequency-doubled Nd:YAG laser (Continuum, San Jose, CA) with a pulse duration of 5 ns and a pulse repetition rate of 20 Hz. The energy delivered by the optical fiber was about 2 mJ. We couple the laser into a 1000 μm multimode optical fiber (Thorlabs, Newton, NJ) via a neutral density filter and focusing lens. The optical fiber directed light to the flow chamber, as shown in Figure 11.6. We used two syringe pumps (Braintree Scientific, Inc., Braintree, MA); one pump was used for the cell sample suspended in buffer and the other was used to introduce air for induction of two-phase flow. We captured droplets that generated photoacoustic waves directly from the flow chamber.

11.2.3 Blood Sample Preparation

Blood samples were either obtained as frozen buffy coats from melanoma patients or directly from whole blood. In the case of the frozen samples, we resuspended them

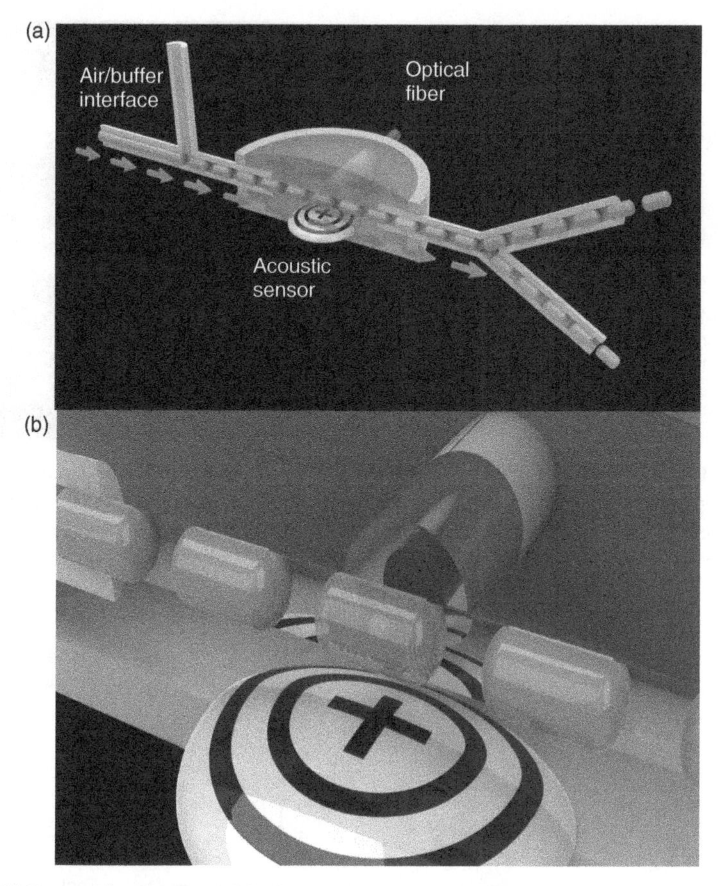

Figure 11.5 (a) The photoacoustic flowmeter separates continuous flow of blood cells with air bubbles. The resulting blood cell suspension droplets are irradiated by laser light. Droplets that contain CTCs generate photoacoustic waves that are sensed by an acoustic transducer. (b) The irradiated droplet contains a CMC and generates photoacoustic waves that are restricted to the droplet due to the acoustic impedance mismatch with the air separating it from other droplets. (*See color plate section for the color representation of this figure.*)

in a cell buffer. The separation procedure from whole blood is shown in Figure 11.7. The blood was centrifuged, from which we extracted the leukocytes, washed them with a buffer, and then introduced them to the photoacoustic system via the syringe pump. The process included introducing Histopaque 1077 into whole blood. This layer allowed separation of the red blood cells from the buffy coat during and after centrifugation. The extracted buffy coat is further centrifuged and mixed with red blood cell lysis buffer to ensure that all such cells are destroyed. This process ensures that no photoacoustic waves are produced by red blood cells, obviating the need for a two-wavelength classification scheme as mentioned earlier.

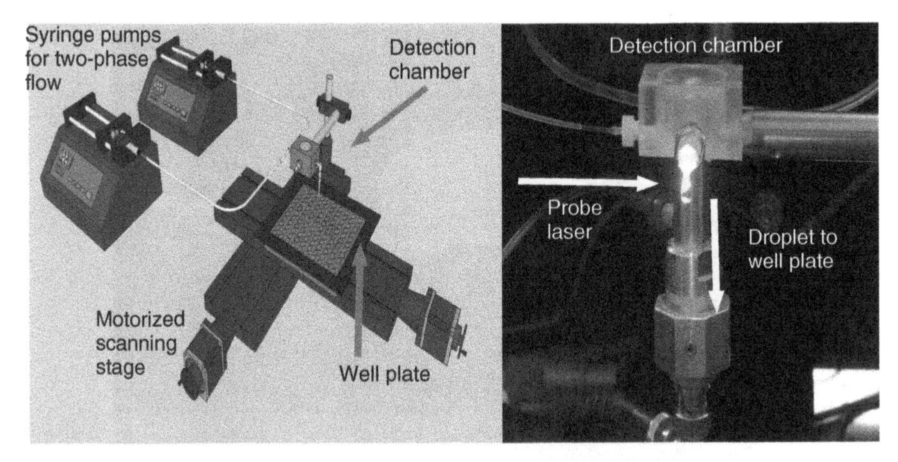

Figure 11.6 (a) The photoacoustic flowmeter uses a computer-controlled scanning stage that sends droplets into a 96-well plate. (b) The droplets are monitored by a blue laser probe beam. The probe beam checks when droplets separate from the flow chamber, triggering the computer controller to advance the well plate to a vacant well in preparation for the next droplet.

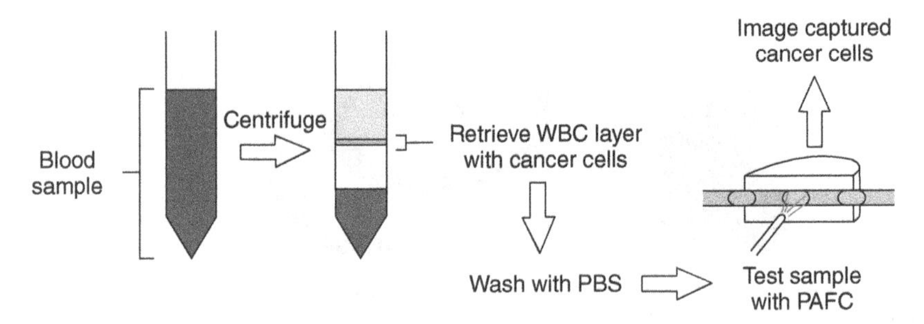

Figure 11.7 Samples are centrifuged in order to separate blood into its components. The buffy coat, containing leukocytes and possibly CMCs, is taken and processed in the photoacoustic flow cytometer (PAFC).

11.2.4 Capture Process

The captured droplets were placed onto microscope slides and then imaged using a fluorescent microscope (Zeiss, Jena, Germany). The microscope system had pipette micromanipulators for actual capture. We scanned this microliter volume for pigmented cells in bright field. Whenever such apparent pigmented cells were found, we imaged them under fluorescence, because melanin, under blue excitation, fluoresces in the green when treated with formaldehyde [45, 46]. We fixed the cell suspension with 4% formaldehyde after air-drying. The cells were excited at 410–460 nm and fluoresced in a broadband manner in the green through a 480–535 nm filter. Exposure time of the camera was set at 100 ms. No cells identified as leukocytes showed such fluorescence except for a minimal amount at 530 nm.

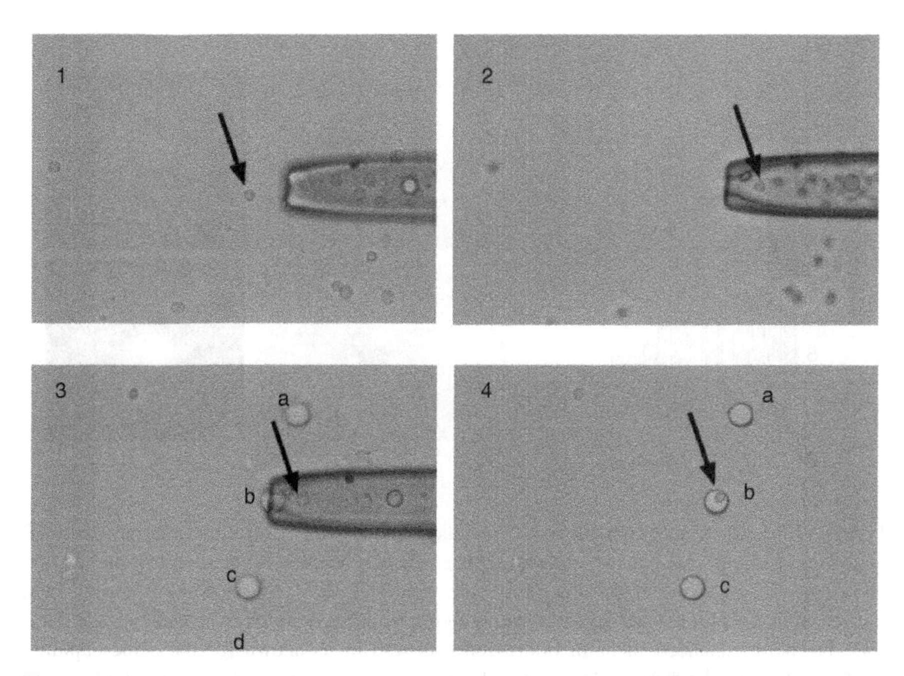

Figure 11.8 Captured droplets are deposited onto microscope slides for capture and isolation. (1) A micropipette is positioned near a melanoma cell, indicated by a black arrow. (2) The melanoma cell is captured within the micropipette. (3) The micropipette is moved near a group of cytophilic isolation wells. There are four wells labeled a, b, c, and d. (4) The melanoma cell is deposited in well b.

Once cells were identified as having fluorescence, we classified them as pigmented. We positioned a micropipette near the cell and suctioned them. The cell was then transferred to a cytophilic well in a ground glass slide, as shown in Figure 11.8.

11.2.5 Results of CMC Capture Study

The results of CMC detection are shown graphically in Figure 11.9. This bar graph shows only results of the 20 melanoma patients. There were no photoacoustics waves generated in samples from healthy human subjects, so no CMCs were detected in that group. All but two of the melanoma patients showed CMCs, with numbers of CMCs ranging from 0 to 25 for the 10 ml blood sample. The figure also shows examples of two captured CMCs in bright field and fluorescence. The bright field shows pigmented structures that show green fluorescence corresponding with the pigmented area.

11.3 DISCUSSION

Our results indicate that photoacoustic flow cytometry is a rapid, sensitive method for capturing CMCs from blood samples, while maintaining the cells in suspension until

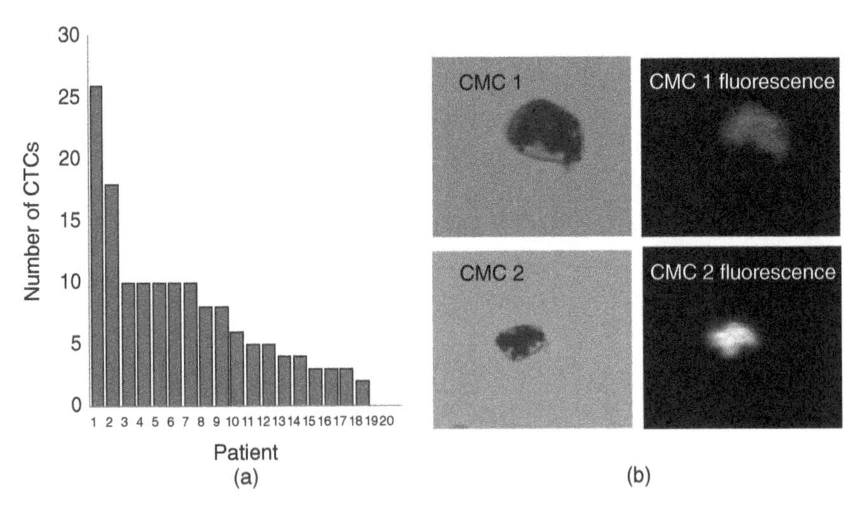

Figure 11.9 (a) A bar graph shows the results of processing 10 ml blood samples from 20 melanoma patients. All but two showed CMCs. Sixteen healthy human controls showed no photoacoustic waveforms, indicating they were free of CMCs (not shown). (b) Bright-field images of two captured CMCs are shown, with pigmented structures indicated by dark shadows. Formalin-enhanced melanin fluorescence at 490 nm is shown for the same captured CMCs, indicating the dark shadows are melanin.

the point of capture. This method has the potential to inform patient care, as well as to conduct studies on basic cancer biology on CTCs.

Although the study of CTCs has grown steadily over the past decade, the definition of a CTC is still evolving and generates controversy. A common definition involves immunohistochemistry and states that a CTC is positive for DAPI and cytokeratins and negative for CD-45, which is indicative of leukocytes [47, 48]. Some definitions include cells being positive for epithelial cell adhesion molecule (EpCAM), particularly for efforts in breast cancer diagnosis. For this work, we restricted our study to CMCs. Because melanin pigmentation is neither specific nor uniform and constant for melanoma, we will extend our definition beyond pigmentation to include the proteins MART-1 and SOX-10. The inclusion of these antigens in positively identifying CMCs will remove ambiguity when we conduct detection and capture studies. We performed such a study on cultured melanoma cells suspended among leukocytes derived from a healthy human subject, with results shown in Figure 11.10. This figure shows pigmented cells corresponding to MART-1 and DAPI, while leukocytes fluoresce for CD-45.

11.3.1 Extension to Nonpigmented CTCs

We have extended this work to nonpigmented CTCs using antibody labeling. While CMCs often have natural pigmentation, some melanomas are amelanotic. Most melanomas are highly pigmented, with estimates of amelanotic melanoma being less

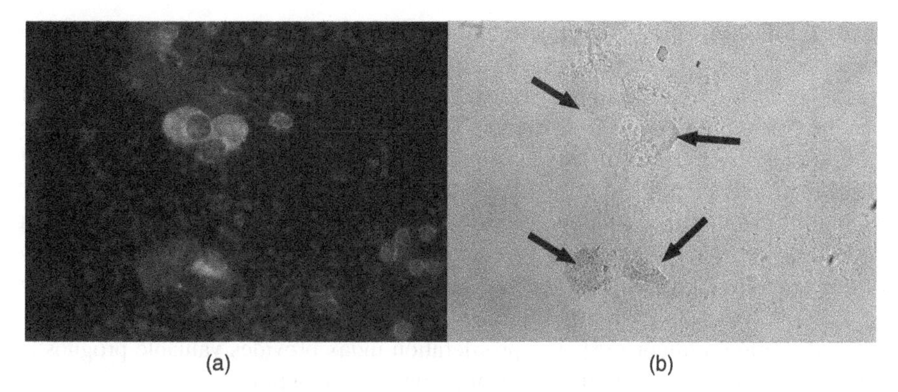

Figure 11.10 (a) Cultured melanoma cells are shown here after photoacoustic capture and immunohistochemical staining. Green fluorescence indicates MART-1 for melanoma, blue indicates DAPI, while red indicates CD-45 for leukocytes. (b) The same region under bright-field microscopy shows pigmented structures corresponding to the MART-1-positive cells. (*See color plate section for the color representation of this figure.*)

than 5% [49] or 1.8–8.1%, although this latter figure includes partially pigmented melanoma [50]. We wanted to extend this technique to those lightly pigmented or amelanotic CMCs. Additionally, if we can find some way to provide optical absorption to CMCs, we can extend this technique to a solid tumor with known surface chemistry. We were able to boost the melanin signature using antibody labeling with cultured melanoma cells [51]. In addition, we were successful in tagging dyed microspheres and nanoparticles to cultured prostate and breast cancer cells [52–54]. This method simply connects light-absorbing microspheres from the 300 nm to 2 μm diameter to CTCs using select monoclonal antibodies. Another technique exploits the plasmon resonance of gold nanoparticles in a similar vein. Both methods produce robust photoacoustic waves that should allow general detection of nonpigmented CTCs.

11.4 FUTURE WORK

We have shown that photoacoustic flow cytometry can be used to detect, count, and capture CMCs in human blood samples. We are working to improve enrichment and automate capture. In some cases, the presence of leukocytes in the captured droplet prevents proper use of assays, complicates imaging, and introduces superfluous material when attempting to study melanoma cells. By using a second- or third-pass method, it is possible to provide pure or nearly pure isolation of CMCs. The concept is simple. After the first pass, the captured droplet is resuspended in a larger volume of buffer. A second pass through the system will yield a purity that is two to three orders of magnitude greater, depending on the amount of buffer used in resuspension. If necessary, additional resuspensions and passes through the system are used for further purity. The question remains whether multiple passes may be too stressful on the CMCs. While overall temperature rise in the CMCs after laser irradiation is probably small, there may be peak temperatures in the melanin granules that may be damaging to the cell, disrupting the membrane, and resulting in loss of the CMCs. This area must be studied and modeled in order to have a better understanding of the thermodynamics and its relation to cell viability.

With the ability to detect and capture CMCs from blood samples of melanoma patients, we are planning to use these captured cells in assays to study various aspects of cancer biology. For instance, the Ki-67 antibody is highly specific for proliferating cells and is widely used in the clinical setting to measure the fraction of proliferating cells within a tumor [55]. The proliferation index provides valuable prognostic information that is often used to determine subsequent therapy.

We also will study the EMT, mentioned earlier. EMT is a complex phenotypic change that results in increased cell motility and invasiveness. Several transcription factors, including members of the Twist (Twist1 and Twist2) and Snail (Snail and Slug) families, have been identified as key regulators of the EMT in both normal developmental processes and epithelial-derived cancer cells [56]. We hypothesize that CMCs must have gone through an EMT-like transition in order to escape from either the primary tumor or secondary metastasis and enter the blood. Indeed, in CTCs isolated from breast cancer patients, Twist1 expression correlates with the level of CTCs. We will use reverse transcriptase polymerase chain reaction (RT-PCR) on captured CMCs to measure expression of Twist1, Twist2, Snail, and Slug to determine the status of EMT. Our goal is to use these factors as prognostic indicators in metastatic melanoma.

Finally, we have begun collecting data in a multiyear study that will check for CMCs in Stage II and III melanoma patients. While these patients have had no spread beyond the lymph nodes, they are all concerned about transitioning into Stage IV disease. We have set up a study to determine if the presence of CMCs in Stage II and III patient blood can predict later metastasis. If this is the case, this data will provide the motivation for monitoring early-stage cancer patients. The added benefit is that CMC detection may be accomplished well before overt metastasis, possibly enabling more successful treatment.

Research reported in this chapter was supported by the National Cancer Institute of the National Institutes of Health under award number NIH 1R01CA161367-01.

REFERENCES

1. Australian Institute of Health and Welfare (2014) *Australian Cancer Incidence and Mortality*, Australian Institute of Health and Welfare.
2. Lens, M. and Dawes, M. (2004PMID:14996086) Global perspectives of contemporary epidemiological trends of cutaneous malignant melanoma. *British Journal of Dermatology*, **150**, 179–185.
3. Surveillance, Epidemiology, and End Results (xxxx), http://seer.cancer.gov/statfacts/html/melan.html (accessed 26 November 2007).

4. Muller, V., Hayes, D., and Pantel, K. (2006) Tecent translational research: circulating tumor cells in breast cancer patients. *Breast Cancer Research*, **8**, 110.
5. Maheswaran, S. and Haber, D. (2010PMC2846729) Circulating tumor cells: a window into cancer biology and metastasis. *Current Opinion in Genetics and Development*, **20**, 96–99.
6. Bonnomet, A., Brysse, A., Tachsidis, A. *et al.* (2010) Epithelial-to-mesenchymal transitions and circulating tumor cells. *Journal of Mammary Gland Biology and Neoplasia*, **15**, 261–273.
7. Gage, T. and Fan, S. (2010) What goes around, comes around: a review of circulating tumor cells. *Analysis*, **11**, 18.
8. Yu, M., Stott, S., Toner, M. *et al.* (2011PMC3101098) Circulating tumor cells: approaches to isolation and characterization. *The Journal of Cell Biology*, **192**, 373–382.
9. Punnoose, E., Atwal, S., Spoerke, J. *et al.* (2010PMC2935889) Molecular biomarker analyses using circulating tumor cells. *PloS One*, **5**, e12517.
10. Koyanagi, K., O'Day, S., Gonzalex, R. *et al.* (2005PMC28564) Serial monitoring of circulating melanoma cells during neoadjuvent biochemotherapy for stage iii melanoma: outcome prediction in a multicenter trial. *Journal of Clinical Oncology*, **23**, 8057–8064.
11. Gogas, H., Kefala, G., Bafaloukus, D. *et al.* (2003PMC2376114) Prognostic significance of the sequential detection of circulating melanoma cells by rt-pcr in high-risk melanoma patients receiving adjuvant interferon. *British Journal of Cancer*, **88**, 981–982.
12. Pellegrino, D., Bellina, C., Manca, G. *et al.* (2000) Detection of melanoma cells in peripheral blood and sentinel lymph nodes by rt-pcr analysis: a comparative study with immunochemistry. *Tumori*, **86**, 336–338.
13. Berking, C., Schlupen, E., Schraeder, A. *et al.* (1999) Tumor markers in peripheral blood of patients with malignant melanoma: multimarker rt-pcr versus a luminoimmunometric assay for s-100. *Archives of Dermatological Research*, **291**, 479–484.
14. Riethdorf, S., Fritsche, H., Müller, V. *et al.* (2007) Detection of circulating tumor cells in peripheral blood of patients with metastatic breast cancer: a validation study of the CellSearch system. *Clinical Cancer Research*, **13**, 920.
15. Okegawa, T., Hayashi, K., Hara, H. *et al.* (2010) Immunomagnetic quantification of circulating tumor cells in patients with urothelial cancer. *International Journal of Urology*, **17**, 254–258.
16. Budd, G., Cristofanilli, M., Ellis, M. *et al.* (2006) Circulating tumor cells versus imaging predicting overall survival in metastatic breast cancer. *Clinical Cancer Research*, **12**, 6403.
17. Cristofanilli, M. (2006) Circulating tumor cells, disease progression, and survival in metastatic breast cancer. *Seminars in Oncology*, **33**, 9–14.
18. Vona, G., Sabile, A., Louha, M. *et al.* (2000) Isolation by size of epithelial tumor cells: a new method for the immunomorphological and molecular characterization of circulating tumor cells. *American Journal of Pathology*, **156**, 57.
19. Zheng, S., Lin, H., Liu, J. *et al.* (2007) Membrane microfilter device for selective capture, electrolysis and genomic analysis of human circulating tumor cells. *Journal of Chromatography. A*, **1162**, 154–161.
20. Nagrath, S., Sequist, L., Maheswaran, S. *et al.* (2007PMC3090667) Isolation of rare circulating tumour cells in cancer patients by microchip technology. *Nature*, **450**, 1235–1239.
21. Barlogie, B., Raber, M., Schumann, J. *et al.* (1983) Flow cytometry in clinical cancer research. *Cancer Research*, **43**, 3982.

22. Campana, D. and Coustan-Smith, E. (1999) Detection of minimal residual disease in acute leukemia by flow cytometry. *Cytometry. Part B, Clinical Cytometry*, **38**, 139–152.
23. Dworzak, M., Froschl, G., Printz, D. *et al.* (2002) Prognostic significance and modalities of flow cytometric minimal residual disease detection in childhood acute lymphoblastic leukemia. *Blood*, **99**, 1952.
24. Gascoyne, P., Noshari, J., Anderson, T., and Becker, F. (2009) Isolation of rare cells from cell mixtures by dielectrophoresis. *Electrophoresis*, **30**, 1388–1398.
25. Hu, X., Bessette, P., Qian, J. *et al.* (2005PMC1276091) Marker-specific sorting of rare cells using dielectrophoresis. *Proceedings of the National Academy of Sciences of the United States of America*, **102**, 15757–15761.
26. Sigrist, M. (1986) Laser generation of acoustic waves in liquids and gases. *Journal of Applied Physics*, **60**, R83.
27. Viator, J., Au, G., Paltauf, G. *et al.* (2002) Clinical testing of a photoacoustic probe for port wine stain depth determination. *Lasers in Surgery and Medicine*, **30**, 141–148.
28. Viator, J., Choi, B., Ambrose, M. *et al.* (2003) *In vivo* port wine stain depth determination using a photoacoustic probe. *Applied Optics*, **42**, 3215–3224.
29. Viator, J., Komadina, J., Svaasand, L. *et al.* (2004) A comparative study of photoacoustic and reflectance methods for determination of epidermal melanin content. *Journal of Investigative Dermatology*, **122**, 1432–1439.
30. Feldman, M., Swearingen, J., and Viator, J (2008). *The Use of Photoacoustics for Distinction Among Melanin, Hemoglobin, and Exogenous Pigment in Skin Lesions In Vivo.* International Investigative Dermatology 2008, Kyoto, Japan, volume 1186.
31. Swearingen, J., Holan, S., Feldman, M., and Viator, J. (2010) Photoacoustic discrimination of vascular and pigmented lesions using classical and Bayesian methods. *Journal of Biomedical Optics*, **15**, 016019.
32. Viator, J. and Jacques, S. (2005) Depth limitations for photoacoustic imaging of burn injury, *in vivo*. ASME Proceedings 2005.
33. Talbert, R., Holan, S., and Viator, J. (2007) Photoacoustic discrimination of viable and thermally coagulated blood using a two wavelength method for burn injury monitoring. *Physics in Medicine and Biology*, **52**, 1815–1829.
34. Holan, S. and Viator, J. (2008) Automated wavelet denoising of photoacoustic signals for circulating melanoma cell detection and burn image reconstruction. *Physics in Medicine and Biology*, **53**, N227.
35. Weight, R., Viator, J., Dale, P. *et al.* (2006) Photoacoustic detection of metastatic melanoma cells in the human circulatory system. *Optics Letters*, **31**, 2998–3000.
36. Weight, R., Dale, P. and Viator, J. (2009). Detection of circulating melanoma cells in human blood using photoacoustic flowmetry. Engineering in Medicine and Biology Society, 2009. EMBC 2009. Annual International Conference of the IEEE. IEEE 2009, pp. 106–109.
37. Givan, A. (2001) *Flow Cytometry: First Principles*, John Wiley & Sons.
38. Paltauf, G., Schmidt-Kloiber, H., and Guss, H. (1996) Light distribution measurements in absorbing materials by optical detection of laser-induced stress waves. *Applied Physics Letters*, **69**, 1526–1528.
39. Thorsen, T., Roberts, R., Arnold, F., and Quake, S. (2001) Dynamic pattern formation in a vesicle- generating microfluidic device. *Physical Review Letters*, **86**, 4163–4166.

40. Garstecki, P., Fuerstman, M., Stone, H., and Whitesides, G. (2006) Formation of droplets and bubbles in a microfluidic T-junction scaling and mechanism of break-up. *Lab on a Chip*, **6**, 437–446.
41. Martinez, M. and Udell, K. (2006) Axisymmetric creeping motion of drops through circular tubes. *Journal of Fluid Mechanics*, **210**, 565–591.
42. Steijn, V., Kleijn, C., and Kreutzer, M. (2010) Predictive model for the size of bubbles and droplets created in microfluidic T-junctions. *Lab on a Chip*, **10**, 2513–2518.
43. OBrien, C., Rood, K., Shramik, S. *et al.* (2011) Detection and isolation of circulating melanoma cells using photoacoustic flowmetry. *Journal of Visualized Experiments*, **25**, e3559.
44. OBrien, C., Rood, K., Bhattacharyya, K. *et al.* (2012) Capture of circulating tumor cells using photoacoustic flowmetry and two phase flow. *Journal of Biomedical Optics*, **17**, 061221.
45. Rost, F. and Polak, J. (1969) Fluorescence microscopy and microspectrofluorimetry of malignant melanomas, naevi and normal melanocytes. *Virchows Archiv*, **347**, 321–326.
46. Harke, A., Thakar, P., Deshmukh, S. *et al.* (1995) Study of formaldehyde-induced fluorescence in cutaneous melanomas and naevi. *Indian Journal of Dermatology, Venereology and Leprology*, **61**, 140.
47. Armakolas, A., Panteleakou, Z., Nezos, A. *et al.* (2010) Detection of the circulating tumor cells in cancer patients. *Future Oncology*, **6**, 1849–1856.
48. Paterlini-Brechot, P. and Benali, N. (2007) Circulating tumor cells (CTC) detection: clinical impact and future directions. *Cancer Letters*, **253**, 180–204.
49. Swetter, S. (2008) Malignant melanoma. www.emedicine.com/DERM/topic257.htm.
50. Wain, E., Stefanato, C., and Barlow, R. (2008) A clinicopathological surprise: amelanotic malignant melanoma. *Clinical and Experimental Dermatology*, **33**, 365–366.
51. McCormack, D., Bhattacharyya, K., Kannan, R. *et al.* (2011) Enhanced photoacoustic detection of melanoma cells using gold nanoparticles. *Lasers in Surgery and Medicine*, **43**, 333–338.
52. Nune, S., Chanda, N., Shukla, R. *et al.* (2009) Green nanotechnology from tea: phytochemicals in tea as building blocks for production of biocompatible gold nanoparticles. *Journal of Material Chemistry*, **19**, 2912–2920.
53. Viator, J., Gupta, S., Goldschmidt, B. *et al.* (2010) gold nanoparticle mediated detection of prostate cancer cells using photoacoustic flowmetry with optical reflectance. *Journal of Biomedical Nanotechnology*, **6**, 187–191.
54. Bhattacharyya, K., Goldschmidt, B., Hannink, M. *et al.* (2012) Gold nanoparticle mediated detection of circulating cancer cells. *Clinics in Laboratory Medicine*, **32**, 89–101.
55. Gerdes, J., Lemke, H., Baisch, H. *et al.* (1984) Cell cycle analysis of a cell proliferation-associated human nuclear antigen defined by the monoclonal antibody Ki-67. *The Journal of Immunology*, **133**, 1710.
56. Bonnomet, A., Brysse, A., Tachsidis, A. *et al.* (2010) Epithelial-to-mesenchymal transitions and circulating tumor cells. *Journal of Mammary Gland Biology and Neoplasia*, **15**, 261–273.

12

SELECTIN-MEDIATED TARGETING OF CIRCULATING TUMOR CELLS FOR ISOLATION AND TREATMENT

JOCELYN R. MARSHALL AND MICHAEL R. KING

Nancy E. and Peter C. Meinig School of Biomedical Engineering, Cornell University, Ithaca, NY, USA

12.1 INTRODUCTION

Circulating tumor cells (CTCs) have provided researchers with ample information about metastatic cancer. However, the exact mechanisms of metastasis are not definitively known. Mounting evidence indicates that CTCs utilize the same pathway through which immune cells are recruited to sites of inflammation. This pathway involves a variety of adhesion molecules, including those in the selectin family. In this chapter, we discuss how these adhesion molecules, namely E-selectin, can be used to recapitulate the early steps of CTC adhesion *in vitro* to achieve capture and characterization of CTCs from patients' blood samples. We also discuss how this interaction may be exploited to target and kill CTCs *in vivo*.

12.1.1 Selectin Adhesion

Cell–cell adhesion is necessary for the maintenance of tissue homeostasis as well as communication between cells. Because of their importance, cell adhesion molecules have been the focus of extensive research for decades. In the late 1980s, an important group of adhesion proteins, called selectins, were discovered on different cells in the vasculature. L-selectin was found on leukocyte membranes; P-selectin was found in the granules of platelets, although it was also discovered to be expressed

Circulating Tumor Cells: Isolation and Analysis, First Edition. Edited by Z. Hugh Fan.

by the vascular endothelium; E-selectin was shown to be expressed by inflamed endothelium. All three of these proteins facilitate immune cell interactions, including leukocyte recruitment to areas of inflammation [1–4]. The selectins were first discovered independently using antibodies that inhibited different adhesion interactions, but it was quickly discovered that the proteins were related and belonged to the same protein family. Not only are the selectins found within a 300 base pair section of chromosome 1 [5], but also they have very similar structures. Each selectin protein contains a lectin domain, an EGF (epidermal growth factor) repeat domain, a series of complement regulatory-like regions, and transmembrane and cytoplasmic domains. Where the proteins differ is in the number of complement regulatory-like regions; L-selectin has 2, P-selectin 9, while E-selectin has 6 [6].

Once the sequence and structure of the selectin molecules were discovered, focus transitioned to the function of the molecules. One of the main functions of these proteins is to mediate leukocyte recruitment to areas of inflammation, an important process for both the innate and adaptive immune systems [7–9]. It is known that this pathway has a number of discrete steps: tethering and rolling, activation, firm adhesion, and extravasation (Figure 12.1) [10]. The selectin family is important in mediating this process, in particular the tethering and rolling steps, while integrins have been shown to mediate the fast-to-slow rolling transition as well as firm adhesion [8, 9, 11]. Soon after their discovery, the selectins were shown to bind to sialylfucosylated ligands, sialyl Lewis x (sLe^x) and sialyl Lewis a (sLe^a). While these are the minimum binding structures for the selectins, these carbohydrate groups are usually found on a carrier backbone of either a glycoprotein or glycolipid [12, 13].

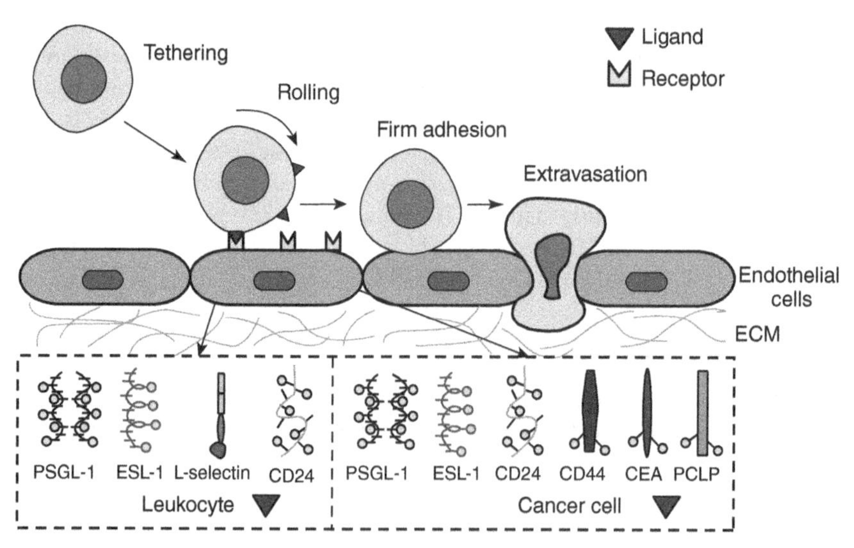

Figure 12.1 Leukocyte adhesion cascade. Geng *et al.* [10]. Reproduced with permission from Springer.

P-selectin glycoprotein ligand-1 (PSGL-1) is one of the most extensively studied selectin ligands, and it is one of the few ligands that has been shown to bind to all three selectins [14].

The selectins, namely E-selectin, are also important at mediating the hematogeneous metastasis of many types of cancer. It is believed that cancer cells use a similar mechanism to leukocyte adhesion to migrate to distant sites in the body. This phenomenon was first studied in detail for colon cancer cells [15–21], but since then different types of cancer cells have been shown to adhere to E-selectin including, but not limited to, prostate [22–24], breast [25–30], lung [31], and pancreatic cancer cells [32–35]. The importance of these adhesive interactions in metastasis was solidified further when it was discovered that the expression of sLe^x and sLe^a, two important selectin ligands, correlated with survival rate and metastasis [36–41]. Moreover, reagents blocking certain glycosylation processes have been shown to inhibit metastasis in several *in vivo* studies [42].

Cancer metastasis differs from leukocyte recruitment to areas of inflammation in some key ways. First, E-selectin is the major selectin implicated in cancer metastasis, while leukocyte recruitment uses a combination of interactions involving all three selectins. Colon, gastric, and lung cancer cell lines have been shown to bind to E-selectin but not to P-selectin [17, 43]. However, P- and L-selectin have been shown to influence cancer metastasis *in vivo*, as P-/L-deficiency in mice inhibits metastasis [11, 13], although it may be that these selectins influence the adhesion of platelets and immune cells that enable cancer adhesion to endothelium rather than adhering to the cancer cells directly [11, 13]. Secondly, sLe^a is found on a number of different cancer types in abundance, whereas leukocytes express low to no sLe^a. Takada *et al.* [44] showed that some cancer cell lines rely solely on sLe^a for adhesion to human umbilical vein endothelial cells (HUVECs), while others rely heavily on sLe^x. Also, in a colon cancer xenograft study, sLe^a-expressing cells were shown to generate larger tumors when compared to their nonexpressing counterparts [41]. The implication is that while cancer cells and leukocytes share a number of selectin ligands, such as PSGL-1, some cancers express novel ligands, which could serve as a target for anticancer therapies. One example is hematopoietic cell E-selectin/L-selectin ligand (HCELL), a CD44 glycoform, which on cancer cells is expressed on O-glycans, rather than N-glycans as on hematopoietic progenitor cells, thus demonstrating that HCELL is expressed as the variant form of CD44 in cancer [45].

Cancer-related inflammation has been the subject of much research, especially relating to metastasis. E-selectin is generally only expressed on cytokine-stimulated endothelial cells. Some cancers have been shown to secrete inflammatory cytokines such as Il-1β. The Hakomori lab showed that while not all cancer cells have the ability to secrete these cytokines, the interaction of cancer cells with platelets could induce this secretory ability, thereby promoting inflammation [46]. Moreover, tumors are infiltrated by a variety of immune cells, which may contribute to cancer-related inflammation [47]. It should also be noted that E-selectin is constitutively expressed on the skin and bone marrow endothelium [48].

12.1.2 Circulating Tumor Cells

CTCs are epithelial cells that have dissociated from the primary tumor and entered the blood stream. CTCs were first discovered in 1869 in the blood of a cancer patient during an autopsy [49]. However, CTCs did not become a major subject of cancer research until two decades ago. Most work has centered around the detection and isolation of CTCs from blood samples of cancer patients, although in recent years, research has expanded to the identification of important mutations present in CTCs, EMT (epithelial–mesenchymal transition) status of CTCs, and the correlation of CTC count with drug efficacy [50–59]. CTCs have been identified in the blood of patients with breast [50, 53, 60–65], lung [66], pancreatic [67], prostate [52, 68, 69], head and neck [51], colorectal [70–72], renal [73, 74], gastric [75, 76], and bladder cancers [77, 78], among others [79–82]. Many studies have shown in both animal models and human studies that CTC count correlates with prognosis, suggesting that a higher CTC count is generally indicative of a worse survival rate and greater disease progression [51, 61, 63, 83–88].

There is no single definitive approach for isolating or identifying CTCs from patients' blood. The problem with isolating CTCs is their rareness in the blood, with CTC counts on the order of 1 per billion blood cells [89]. CTCs can be enriched by density centrifugation, which removes the majority of red blood cells, neutrophils, platelets, and plasma. However, the remaining leukocytes still greatly outnumber the CTCs; therefore, density centrifugation is usually used in conjunction with another isolation technique. Once CTCs are isolated, their identity must be confirmed. This is usually done by polymerase chain reaction (PCR) or fluorescence microscopy [90]. A cell is considered a CTC if it meets the following criteria: a misshapen nucleus, positive for epithelial cell adhesion molecule (EpCAM) or cytokeratin (CK) expression, and negative for CD45, a leukocyte marker [91]. However, recent work has shown that CTCs are heterogeneous and one identification method may not work to identify all CTC subtypes [71, 82]. Only one isolation technique, CellSearch (see Chapter 19 for details), has been Food and Drug Administration (FDA) approved for the enumeration of patient CTC counts in prostate and breast cancer. CellSearch uses anti-EpCAM antibody-coated magnetic beads to capture CTCs from blood samples and then a strong magnet is used to precipitate the aggregates out of solution. The cells attached to the beads are then stained for CK, CD45, and DAPI, a nuclear stain. Samples are processed through a specialized detector that detects and enumerates positively identified, intact CTCs [90]. Other isolation techniques have been developed using similar approaches. For example, the Chalmers lab (see Chapter 14 for details), among others, uses magnetic bead technology to isolate CTCs via negative selection. In their process, magnetic beads are coated with anti-CD45 antibodies that capture contaminating leukocytes. This process allows for the identification of both EpCAM-positive and EpCAM-negative CTCs [51]. Nagrath *et al.* [92] created a microfluidic "CTC-chip" that applies anti-EpCAM antibodies to microfabricated posts. A blood sample is applied to the chip, where fluid flow produces collisions between the CTCs and the posts. The same group extended this idea by designing a chip without posts but with a herringbone pattern etched onto a glass substrate.

Anti-EpCAM antibody was then applied to the whole chip. This allows for a greater capture area, and the herringbone etching increases the turbidity of the flow, thereby increasing CTCs collisions with the chip surface [93].

It is believed that CTCs facilitate metastasis by hijacking the process that leukocytes use when responding to inflammation signals. This involves the adhesion of cells to the vascular endothelium mediated by adhesion molecules such as E-selectin. Our group has employed this concept to develop a physiologic mimic microtube device that uses E-selectin adhesion to recruit CTCs to the surface and facilitate firm adhesion via antibody binding [81]. In the next section, the development and use of this technology are further detailed.

12.2 CTC CAPTURE BY E-SELECTIN

E-selectin's use as a capture agent started about a decade ago when the Liepmann lab developed a cell separation device that uses E-selectin to concentrate cell solutions. In this microfluidic device, microfabricated posts are coated in recombinant human E-selectin (rhES) and cell solution is pumped through, allowing cells to be captured on the surface of each post. Two post geometries were tested: square pillars and slender, offset pillars. The offset pillars were shown to capture more cells initially, but the square pillars enriched the cell solution better due to slower rolling speeds and less cell detachment. This device was not tested for CTC capture specifically, but in theory could be applied for that purpose [94]. In addition, Kim *et al.* [95] developed a biomimetic chip to capture blood cells. Rather than target one selectin interaction, they coated their chip with E-selectin, ICAM-1, and VCAM for the purpose of targeting all three selectins. However, they found that E-selectin was necessary for capture [95].

CTCs from patients with prostate cancer were only recently shown to adhere to microtubes functionalized with rhES [96], but work to create a capture device using immobilized selectin started years earlier. Much of the work, in fact, was pioneered using a hematopoietic stem cell (HSPC) model. In this device, microrenathane tubes coated in P-selectin-enriched HSPC samples sevenfold with an average purity of about 30% [97]. While this process represented a major advance, further work showed that by roughening the surface area of the tube with nanomaterials, one could greatly increase the efficiency of the device. For example, coating the tube with colloidal silica prior to protein absorption increased cell capture [98]. Capture was further increased with halloysite coatings; halloysite is naturally occurring aluminosilicate nanoparticles that range from 500 to 1200 nm in length and 40–200 nm in diameter. Combined hallyosite and P-selectin surfaces captured both KG1a and COLO205 cells at increased rates compared to tubes with P-selectin alone [99]. This technology was subsequently adapted for CTC isolation.

To capture CTCs, a length of microrenathane (MRE) tubing is first coated with poly-L-lysine to endow the tube with a positively charged surface that will facilitate the adsorption of the halloysite nanotubes. The halloysite coating is allowed to cure overnight. Protein G is then added to the tube followed by a mixture of rhES and anti-EpCAM antibody. Protein G binds to the F_c region of both the rhES and anti-EpCAM antibody allowing for consistent molecular orientation within the tube interior [100]. The technology was optimized and characterized using a leukemia cell line prior to testing with patient samples. This approach typically yields a capture efficiency of 50% and CTC purities of >50% [81].

Primary buffy coat samples were processed through tubes without halloysite ("smooth tubes"), with halloysite-coated tubes, and using CellSearch. For each method, half a tube of blood (3.75 ml) was processed and the resulting normalized cell counts were calculated and compared (Figure 12.2). For nearly all cancer blood

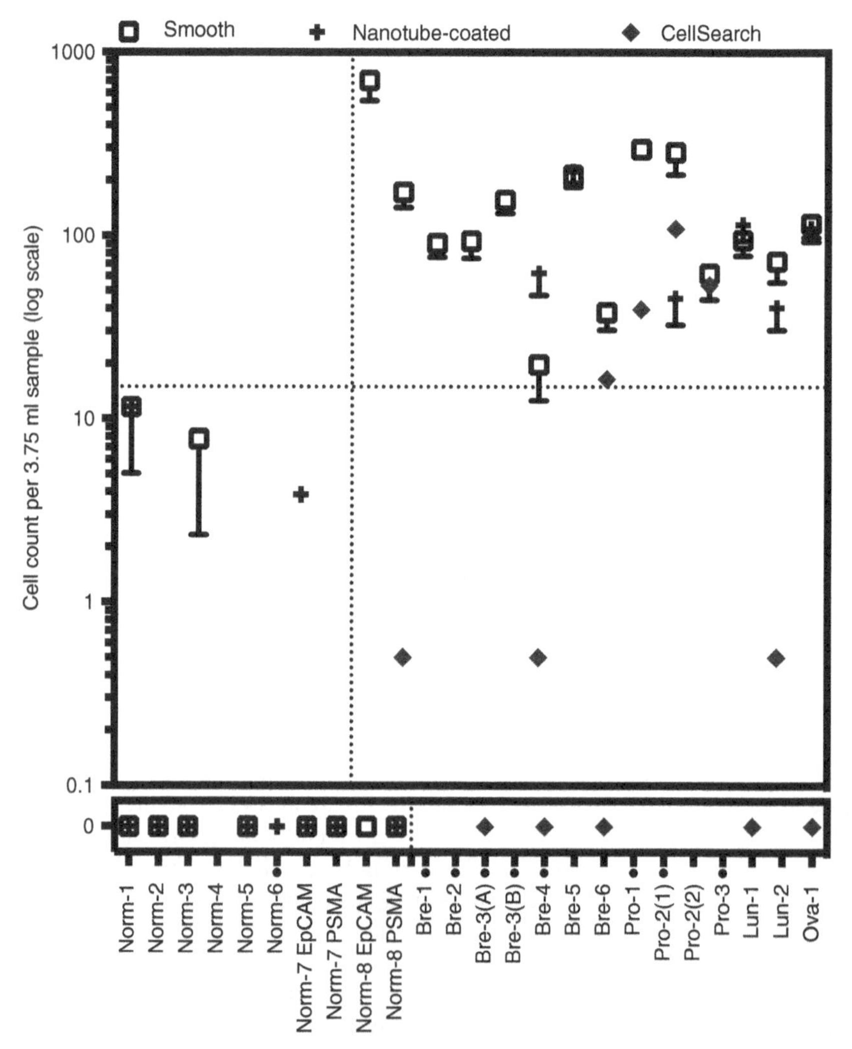

Figure 12.2 Number of CTCs captured from the blood of cancer patients, along with the results of samples collected from healthy patients. Smooth tube device, halloysite-coated device, and CellSearch methods were compared in parallel experiments, data from [81]. Hughes *et al.* [101]. Reproduced with permission from Elsevier.

samples, both the smooth tubes and halloysite-coated tubes captured more cells than CellSearch. Halloysite-coated tubes show generally higher CTC purities than the smooth tubes, due to the inability of leukocytes to spread and firmly attach on halloysite [81].

12.3 APPLICATIONS FOR E-SELECTIN IN CANCER DIAGNOSIS AND TREATMENT

While CTC counts have primarily been used as a marker of disease progression, CTCs hold much potential for targeted, personalized therapy development. In fact, in the study that led to FDA approval of CellSearch, CTC counts were correlated not only with disease progression but also with response to treatment. Patients with >4 CTCs per sample were noted to have responded poorly to treatment [102]. Two other studies in prostate cancer also used CTC count to track drug efficacy [103, 104]. In the study by Reid *et al.* [103], 63% of patients showed a CTC count decrease by half after treatment.

12.3.1 E-Selectin Capture for Drug Efficacy Testing

Our lab has exploited knowledge of E-selectin adhesion to cancer cells to explore the potential for a broadly applicable diagnostic test for the development of personalized treatment regimes. We hypothesized that by employing the same device as described in the previous section, one could use CTC samples to test for drug sensitivity prior to administration. CTCs represent an advantageous medium for drug testing because one CTC sample can contain multiple different subsets of cancer cells believed to reflect the heterogeneous nature of the primary tumor [105] and because of their demonstrated correlation with metastasis, which accounts for 90% of all cancer-related deaths [106].

In previous work, a simple procedure to test for drug efficacy using CTC count was created. Each blood sample was split into multiple aliquots, leaving one aliquot as an untreated reference control. Blood samples were treated with chemotherapeutic drugs and incubated overnight. The drug concentrations used in the study were chosen based on the peak plasma concentration determined by pharmacokinetic studies. After incubation, CTCs were isolated and enumerated. A reduction in CTC count, when compared to the reference control, was concluded to indicate drug sensitivity [101].

The method was first characterized using a cell line model that consisted of spiking a known number of prostate or breast cancer cells into the blood of a healthy donor. All of the cell lines tested were found to be sensitive to each drug in media and in blood, and the trend in cell count reduction followed the same pattern for both conditions. Following characterization, this method was tested on the blood from seven cancer patients (three breast, two prostate, one renal, and one colon). Each patient's response to the drugs varied, with three patients responding to both drugs, three were sensitive to one drug, and one showed no sensitivity. More detailed results of the primary sample experiments can be seen in Figure 12.3 [101].

There are several advantages to this system. First, it is a straightforward procedure that can be performed on standard laboratory equipment. Second, the method can be used with any CTC isolation device, for instance, CellSearch or CTC-Chip. Moreover, the technique supports the testing of intravenous drugs or combinations of therapeutics [101].

12.3.2 E-Selectin for use in Targeted Cancer Therapy

The fact that E-selectin can be used to recruit CTCs to surfaces suggests that it may also be useful in targeted therapeutics. Previous attempts have been made to create cancer drugs that exploit E-selectin adhesion including drugs targeting glycosylation, selectin ligands, but none of these formulations have reached the clinic [13, 42]. Unlike previous technologies, we have explored the use of E-selectin as a targeting agent and employed a secondary protein, TNF-related apoptosis-inducing

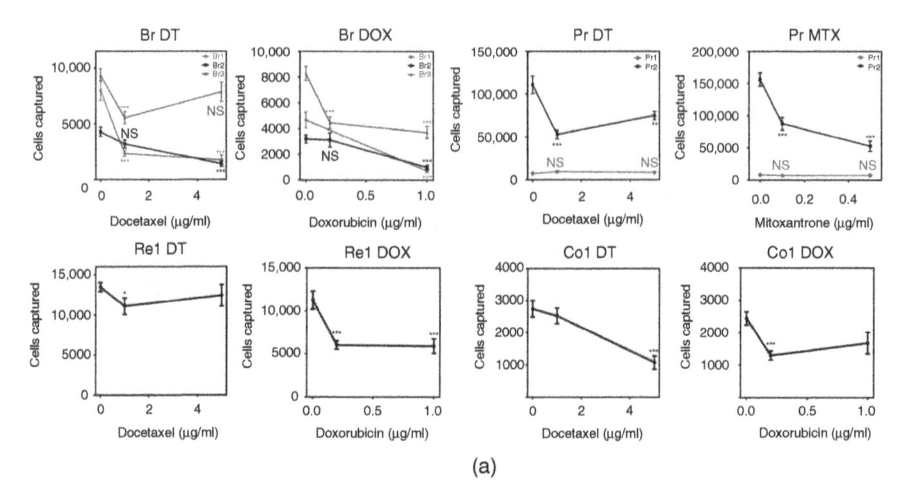

Figure 12.3 Patient samples were collected from three breast cancer patients (Br1, Br2, and Br3), two prostate cancer patients (Pr1 and Pr2), one renal cancer patient, and one colon cancer patient (Re1 and Co1, respectively). Each tube of whole blood was split into three aliquots and treated with vehicle control, 20% peak plasma concentration (PPC), or 100% PPC of the appropriate drug. (a) CTC counts of patient samples treated with chemotherapeutic drugs. (b) Example micrographs of two patient samples. Patients showed varied responses to the drugs as shown in the line graphs. Pr2 shows significant response to both drugs, which can be seen in the top set of fluorescent images as a reduction in the number of cancer cells in the treatment images compared the control. Re1 shows no sensitivity to docetaxel, but is sensitive doxorubicin, which is observed in the lower set of fluorescent images as no reduction in green cells and a reduction in the number of cancer cells, respectively. Larger cells typically represent cancer cells. For more detail refer to original color format (blue: DAPI, green: EpCAM, red: CD45). DT, docetaxel; DOX, doxorubicin; MTX, mitoxantrone. Error bars represent standard error of the mean. $*p<0.05$, $**p<0.01$, $***p<0.001$; scale bar = 50 μm. Hughes *et al.* [101]. Reproduced with permission from Elsevier. (*See color plate section for the color representation of this figure.*)

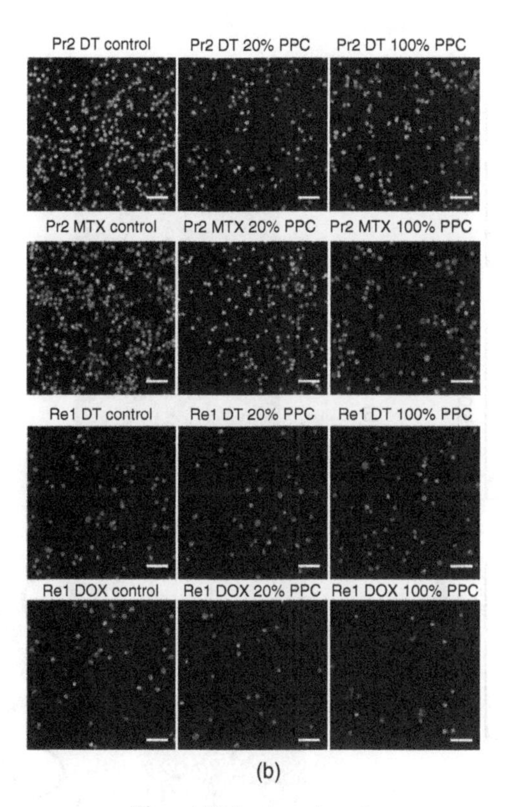

Figure 12.3 (*continued*)

ligand (TRAIL) to induce apoptosis in CTCs [107]. TRAIL is a protein expressed by cells of the immune system, and the receptors for TRAIL, termed death receptors, have been shown to be upregulated on the surface of many types of cancer cells [108]. By employing the combination of TRAIL and E-selectin, this therapeutic increases its targeting of cancer cells and decreases its cytotoxicity toward healthy cells.

Huang and King showed that selectin molecules have the ability to target cells by using P-selectin-coated liposomes to deliver siRNA to HL60 cells. Liposomes were absorbed onto the surface of MRE tubes and HL60 cells rolled across the surface, and the encapsulated siRNA was successfully delivered to the cells to induce a knockdown of the elastase gene (ELA2) to less than half of the original expression level [109]. E-selectin has also served as a successful targeting protein for liposomes. Rh-ES was conjugated to the surface of liposomes encapsulating doxorubicin and used to treat cancer cells *in vitro*. E-selectin was successful in targeting cancer cells as shown by liposomes adhering to nearly 100% of cells. Under static conditions, the targeted liposomes experienced a similar kill rate to normal liposomal doxorubicin, but killed fewer cells than soluble doxorubicin. To study the effect of shear on the targeting and killing of cancer cells by E-selectin functionalized liposomes, they were adsorbed onto the surface of MRE tubes and cells were perfused through the tubes.

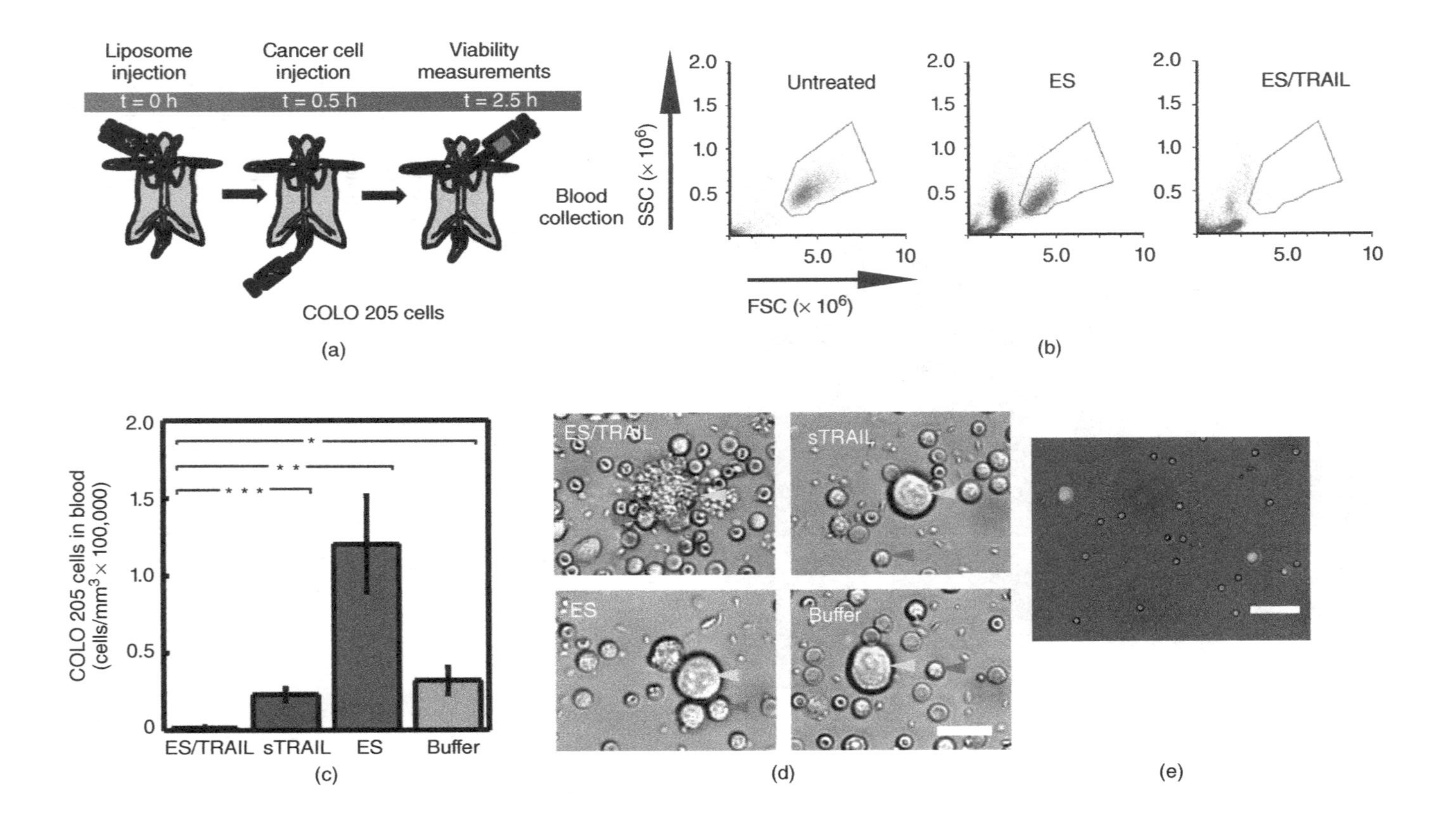
Liposome injection
Cancer cell injection
Viability measurements
t = 0 h
t = 0.5 h
t = 2.5 h
Blood collection
COLO 205 cells
(a)
SSC (× 10^6)
FSC (× 10^6)
2.0
1.5
1.0
0.5
5.0
10
Untreated
ES
ES/TRAIL
(b)
COLO 205 cells in blood (cells/mm^3 × 100,000)
0
0.5
1.0
1.5
2.0
*
* *
* * *
ES/TRAIL
sTRAIL
ES
Buffer
(c)
ES/TRAIL
sTRAIL
ES
Buffer
(d)
(e)

Under dynamic conditions, E-selectin functionalized liposomes killed significantly more cells than identical liposomes containing no doxorubicin, although the kill rate was less than that of soluble doxorubicin. When a halloysite coating was added to the MRE tubing before liposome adsorption, however, the number of apoptotic cells increased greatly and the observed difference in the kill rates of the functionalized liposomes and soluble doxorubicin was lower [110].

In a study by Rana *et al.*, the dual power of TRAIL and E-selectin was observed. MRE tubing was coated with a combination of rhES and TRAIL and COLO205 cells were flowed through the device for 1–2 h. While 1 h of flow did not produce significant apoptosis, after 2 h, less than 40% of COLO-205 cells remained viable. Pretreatment with 1 mM aspirin significantly increased the effectiveness of the surface, bringing the kill rate of 1 h of rolling to that of the 2 h treatment without aspirin [111].

The work culminated in the development of E-selectin/TRAIL-functionalized liposomes that target CTCs within the bloodstream. Dual-functionalized liposomes were shown to kill significantly more cancer cells than naked liposomes, E-selectin-only liposomes, or TRAIL-only liposomes. The liposomes also bound to human leukocytes *in vitro* without causing apoptosis. The cancer cell killing effect was greatly enhanced when cells were in the whole blood environment, with less than 5% of cells remaining viable. It was shown in this study that the liposomes are capable of not only killing cancer cells directly but more importantly also attach to leukocytes and remain available to bind to cancer cells and induce death. This dual action enhances the utility of the therapy by increasing the circulation time of the therapy. One limitation of previous TRAIL therapies has been its short circulation time (15–30 min) and rapid clearance. Interestingly, E-selectin liposomes without therapeutic were found to cause CTCs to be retained in the circulation *in vivo* rather than adhering to the vascular endothelium by competitive inhibition. This was shown in a mouse model where COLO205 cells were injected via the tail vein, followed by liposome therapy, control liposomes, buffer, or soluble TRAIL. The CTCs were collected via cardiac puncture and enumerated after 2–3 h in culture. Mice treated with E-selectin/TRAIL liposomes had the fewest remaining CTCs after treatment

Figure 12.4 (a) Schematic of procedure for *in vivo* liposome experiments. (b) Flow cytometry of COLO205 cells *in vitro* (L), recovered from cardiac puncture of ES-liposome-treated mice (c), recovered from ES/TRAIL-liposome-treated mice (R). SSL, side-scattered light; FSC, forward-scattered light. (c) Number of viable cells recovered from blood of mice compared by treatments. $n = 3$ for all samples. Bars represent the mean $\pm$ SD in each treatment group. $*p < 0.01$, $**p < 0.001$, $***p < 0.0001$ (one-way ANOVA with Tukey posttest). (d) Representative micrographs of cells recovered from mouse blood. Scale bar = 20 μm. (e) Leukocytes functionalized with fluorescent ES/TRAIL liposomes recovered during cardiac puncture. Scale bar = 50 μm. ES/TRAIL liposomes bound to leukocytes in the circulation of mice and successfully killed COLO205 cells in the circulation of mice. Mitchell *et al.* [107]. Reproduced with permission from the Proceedings of the National Academy of Sciences. (*See color plate section for the color representation of this figure.*)

(Figure 12.4). In addition, the ES/TRAIL liposome treatment decreased the number of cells that lodged in the lungs of the mice. This study shows the importance and potential for E-selectin as a CTC-targeting molecule [107].

12.4 CONCLUSIONS

Ever since the selectin family of adhesion molecules was discovered in the late 1980s, research has looked at the importance of this molecule in immune and cancer interactions. Recent work has revealed new ways that E-selectin adhesion can be used to detect and kill CTCs (Figure 12.5). It has been demonstrated that E-selectin, in combination with other specific cancer antibodies, can be used to capture CTCs from the blood of patients to better characterize a patient's prognosis and also to potentially test living tumor cells for the most effective treatment option. Going

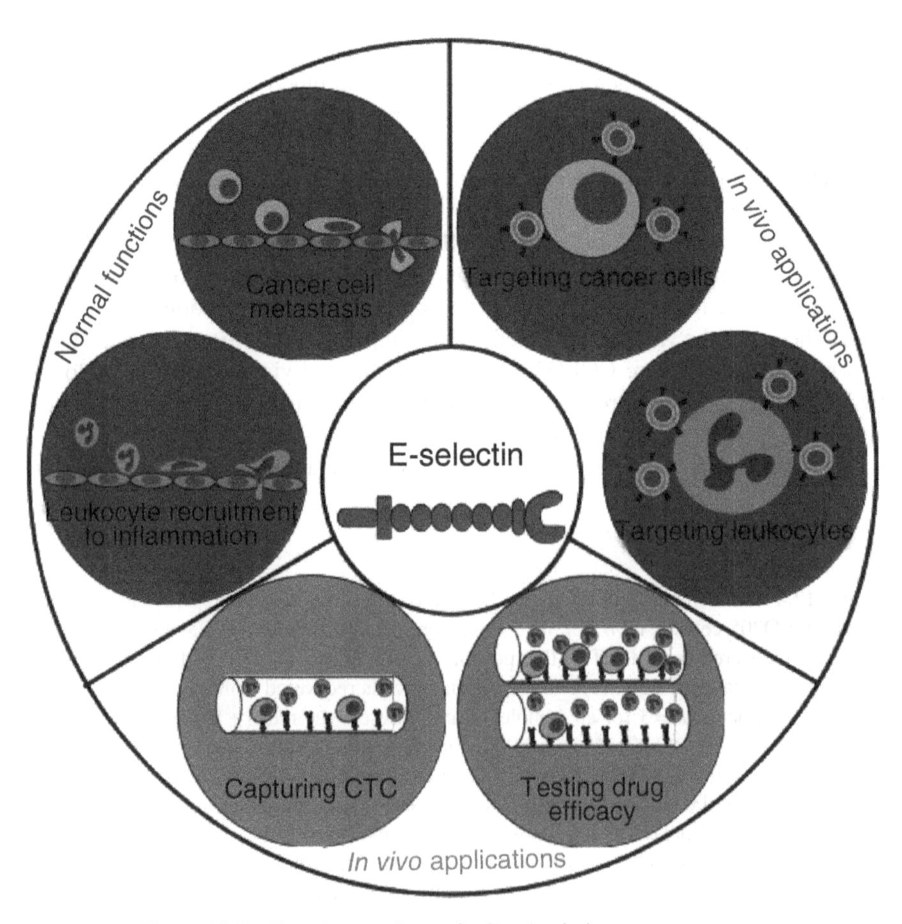

Figure 12.5 Functions and uses for E-selectin in cancer treatment.

further, E-selectin has been used as a targeting molecule for cancer treatment. By combining the targeting of E-selectin with the killing and targeting of TRAIL, CTCs in mice were killed in the bloodstream and the number of embedded cells in the lungs was decreased significantly. E-selectin has great potential in the realm of cancer treatment, and its full utility is just now being fully appreciated.

REFERENCES

1. Siegelman, M.H., van de Rijn, M., and Weissman, I.L. (1989) Mouse lymph node homing receptor cDNA clone encodes a glycoprotein revealing tandem interaction domains. *Science*, **243**, 1165–1172.
2. Lasky, L.A., Singer, M.S., Yednock, T.A. *et al.* (1989) Cloning of a lymphocyte homing receptor reveals a lectin domain. *Cell*, **56**, 1045–1055.
3. Johnston, G.I., Cook, R.G., and McEver, R.P. (1989) Cloning of GMP-140, a granule membrane protein of platelets and endothelium: sequence similarity to proteins involved in cell adhesion and inflammation. *Cell*, **56**, 1033–1044.
4. Bevilacqua, M.P., Stengelin, S., Gimbrone, M.A., and Seed, B. (1989) Endothelial leukocyte adhesion molecule 1: an inducible receptor for neutrophils related to complement regulatory proteins and lectins. *Science*, **243**, 1160–1165.
5. Watson, M.L., Kingsmore, S.F., Johnston, G.I. *et al.* (1990) Genomic organization of the selectin family of leukocyte adhesion molecules on human and mouse chromosome 1. *Journal of Experimental Medicine*, **172**, 263–272.
6. Bevilacqua, M.P. and Nelson, R.M. (1993) Selectins. *Journal of Clinical Investigation*, **91**, 379–387.
7. Toothill, V.J., Van Mourik, J.A. *et al.* (1990) Characterization of the enhanced adhesion of neutrophil leukocytes to thrombin-stimulated endothelial cells. *Journal of Immunology*, **145**, 283–291.
8. Lawrence, M.B. and Springer, T.A. (1991) Leukocytes roll on a selectin at physiologic flow rates: distinction from and prerequisite for adhesion through integrins. *Cell*, **65**, 859–873.
9. Vestweber, D. and Blanks, J.E. (1999) Mechanisms that regulate the function of the selectins and their ligands. *Physiological Reviews*, **79**, 181–213.
10. Geng, Y., Marshall, J.R., and King, M.R. (2012) Glycomechanics of the metastatic cascade: tumor cell-endothelial cell interactions in the circulation. *Annals of Biomedical Engineering*, **40**, 790–805.
11. Barthel, S.R., Gavino, J.D., Descheny, L., and Dimitroff, C.J. (2007) Targeting selectins and selectin ligands in inflammation and cancer. *Expert Opinion on Therapeutic Targets*, **11**, 1473–1491.
12. Crocker, P.R. and Feizi, T. (1996) Carbohydrate recognition systems: functional triads in cell-cell interactions. *Current Opinion in Structural Biology*, **6**, 679–691.
13. Läubli, H. and Borsig, L. (2010) Selectins promote tumor metastasis. *Seminars in Cancer Biology*, **20**, 169–177.
14. Moore, K.L. (1998) Structure and function of P-selectin glycoprotein ligand-1. *Leukemia & Lymphoma*, **29**, 1–15.

15. Lauri, D., Needham, L., Martin-Padura, I., and Dejana, E. (1991) Tumor cell adhesion to endothelial cells: endothelial leukocyte adhesion molecule-1 as an inducible adhesive receptor specific for colon carcinoma cells. *Journal of National of Cancer Institute*, **83**, 1321–1324.
16. Rice, G.E. and Bevilacqua, M.P. (1989) An inducible endothelial cell surface glycoprotein mediates melanoma adhesion. *Science*, **246**, 1303–1306.
17. Mannori, G., Crottet, P., Cecconi, O. *et al.* (1995) Differential colon cancer cell adhesion to E-, P-, and L-selectin: role of mucin-type glycoproteins. *Cancer Research*, **55**, 4425–4431.
18. Burdick, M.M., McCaffery, J.M., Kim, Y.S. *et al.* (2003) Colon carcinoma cell glycolipids, integrins, and other glycoproteins mediate adhesion to HUVECs under flow. *American Journal of Physiology - Cell Physiology*, **284**, C977–C987.
19. Tremblay, P.-L., Auger, F.A., and Huot, J. (2006) Regulation of transendothelial migration of colon cancer cells by E-selectin-mediated activation of p38 and ERK MAP kinases. *Oncogene*, **25**, 6563–6573.
20. Laferrière, J., Houle, F., and Huot, J. (2004) Adhesion of HT-29 colon carcinoma cells to endothelial cells requires sequential events involving E-selectin and integrin beta4. *Clinical and Experimental Metastasis*, **21**, 257–264.
21. Sawada, R., Tsuboi, S., and Fukuda, M. (1994) Differential E-selectin-dependent adhesion efficiency in sublines of a human colon cancer exhibiting distinct metastatic potentials. *Journal of Biological Chemistry*, **269**, 1425–1431.
22. Dimitroff, C.J., Lechpammer, M., Long-Woodward, D., and Kutok, J.L. (2004) Rolling of human bone-metastatic prostate tumor cells on human bone marrow endothelium under shear flow is mediated by E-selectin. *Cancer Research*, **64**, 5261–5269.
23. Barthel, S.R., Wiese, G.K., Cho, J. *et al.* (2009) Alpha 1,3 fucosyltransferases are master regulators of prostate cancer cell trafficking. *Proceedings of National Academy of Science of the United States of America*, **106**, 19491–19496.
24. Li, J., Guillebon, A.D., Hsu, J. *et al.* (2013) Human fucosyltransferase 6 enables prostate cancer metastasis to bone. *British Journal of Cancer*, **109**, 3014–3022.
25. Narita, T., Kawasaki-Kimura, N., Matsuura, N. *et al.* (1996) Adhesion of human breast cancer cells to vascular endothelium mediated by sialyl lewis/E-selectin. *Breast Cancer*, **3**, 19–23.
26. Tözeren, A., Kleinman, H.K., Grant, D.S. *et al.* (1995) E-selectin-mediated dynamic interactions of breast- and colon-cancer cells with endothelial-cell monolayers. *International Journal of Cancer*, **60**, 426–431.
27. Shirure, V.S., Henson, K.A., Schnaar, R.L. *et al.* (2011) Gangliosides expressed on breast cancer cells are E-selectin ligands. *Biochemical and Biophysical Research Communications*, **406**, 423–429.
28. Shirure, V.S., Reynolds, N.M., and Burdick, M.M. (2012) Mac-2 binding protein is a novel E-selectin ligand expressed by breast cancer cells. *PloS One*, **7**, e44529.
29. Chandrasekaran, S., Geng, Y., DeLouise, L.A., and King, M.R. (2012) Effect of homotypic and heterotypic interaction in 3D on the E-selectin mediated adhesive properties of breast cancer cell lines. *Biomaterials*, **33**, 9037–9048.
30. Geng, Y., Yeh, K., Takatani, T., and King, M.R. (2012) Three to tango: MUC1 as a ligand for both E-selectin and ICAM-1 in the breast cancer metastatic cascade. *Frontiers in Oncology*, **2**, 76.

31. Richter, U., Schröder, C., Wicklein, D. *et al.* (2011) Adhesion of small cell lung cancer cells to E- and P-selectin under physiological flow conditions: implications for metastasis formation. *Histochemistry and Cell Biology*, **135**, 499–512.
32. ten Kate, M., Hofland, L.J., van Koetsveld, P.M. *et al.* (2006) Pro-inflammatory cytokines affect pancreatic carcinoma cell. Endothelial cell interactions. *Journal of the Pancreas*, **7**, 454–464.
33. Nozawa, F., Hirota, M., Okabe, A. *et al.* (2000) Tumor necrosis factor alpha acts on cultured human vascular endothelial cells to increase the adhesion of pancreatic cancer cells. *Pancreas*, **21**, 392–398.
34. Takada, M., Yamamoto, M., Hasegawa, Y., and Saitoh, Y. (1995) Endothelial leukocyte adhesion molecule-1-mediated vasoinvasion of human pancreatic adenocarcinoma. *Journal of Surgical Research*, **59**, 653–657.
35. Iwai, K., Ishikura, H., Kaji, M. *et al.* (1993) Importance of E-selectin (ELAM-1) and sialyl Lewis(a) in the adhesion of pancreatic carcinoma cells to activated endothelium. *International Journal of Cancer*, **54**, 972–977.
36. Numahata, K., Satoh, M., Handa, K. *et al.* (2002) Sialosyl-Le(x) expression defines invasive and metastatic properties of bladder carcinoma. *Cancer*, **94**, 673–685.
37. Hoff, S.D., Matsushita, Y., Ota, D.M. *et al.* (1989) Increased expression of sialyl-dimeric LeX antigen in liver metastases of human colorectal carcinoma. *Cancer Research*, **49**, 6883–6888.
38. Nakamori, S., Kameyama, M., Imaoka, S. *et al.* (1993) Increased expression of sialyl Lewis x antigen correlates with poor survival in patients with colorectal carcinoma: clinicopathological and immunohistochemical study. *Cancer Research*, **53**, 3632–3637.
39. Nakayama, T., Watanabe, M., Katsumata, T. *et al.* (1995) Expression of sialyl Lewis(a) as a new prognostic factor for patients with advanced colorectal carcinoma. *Cancer*, **75**, 2051–2056.
40. Kannagi, R., Izawa, M., Koike, T. *et al.* (2004) Carbohydrate-mediated cell adhesion in cancer metastasis and angiogenesis. *Cancer Science*, **95**, 377–384.
41. Terraneo, L., Avagliano, L., Caretti, A. *et al.* (2013) Expression of carbohydrate-antigen sialyl-Lewis a on colon cancer cells promotes xenograft growth and angiogenesis in nude mice. *International Journal of Biochemistry and Cell Biology*, **45**, 2796–2800.
42. Hakomori, S. (1996) Tumor malignancy defined by aberrant glycosylation and sphingo(glyco)lipid metabolism. *Cancer Research*, **56**, 5309–53018.
43. Handa, K., White, T., Ito, K. *et al.* (1995) P-selectin-dependent adhesion of human cancer-cells - requirement for coexpression of a psgl-1-like core protein and the glycosylation process for sialosyl-le(x) or sialosyl-le(a). *International Journal of Oncology*, **6**, 773–781.
44. Takada, A., Ohmori, K., Yoneda, T. *et al.* (1993) Contribution of carbohydrate antigens sialyl Lewis A and sialyl Lewis X to adhesion of human cancer cells to vascular endothelium. *Cancer Research*, **53**, 354–361.
45. Hanley, W.D., Burdick, M.M., Konstantopoulos, K., and Sackstein, R. (2005) CD44 on LS174T colon carcinoma cells possesses E-selectin ligand activity. *Cancer Research*, **65**, 5812–5817.
46. Hakomori, S. (1994) Novel endothelial cell activation factor(s) released from activated platelets which induce E-selectin expression and tumor cell adhesion to endothelial cells: a preliminary note. *Biochemical and Biophysical Research Communications*, **203**, 1605–1613.

47. Hanahan, D. and Weinberg, R.A. (2011) Hallmarks of cancer: the next generation. *Cell*, **144**, 646–674.
48. Sipkins, D.A., Wei, X., Wu, J.W. *et al.* (2005) In vivo imaging of specialized bone marrow endothelial microdomains for tumour engraftment. *Nature*, **435**, 969–973.
49. Ashworth, T.R. (1869) A case of cancer in which cells similar to those in the tumors were seen in the blood after death. *Australian Medical Journal*, **14**, 146–149.
50. Armstrong, A.J., Marengo, M.S., Oltean, S. *et al.* (2011) Circulating tumor cells from patients with advanced prostate and breast cancer display both epithelial and mesenchymal markers. *Molecular Cancer Research*, **9**, 997–1007.
51. Balasubramanian, P., Lang, J.C., Jatana, K.R. *et al.* (2012) Multiparameter analysis, including EMT markers, on negatively enriched blood samples from patients with squamous cell carcinoma of the head and neck. *PloS One*, **7**, e42048.
52. Chen, C.-L., Mahalingam, D., Osmulski, P. *et al.* (2012) Single-cell analysis of circulating tumor cells identifies cumulative expression patterns of EMT-related genes in metastatic prostate cancer. *The Prostate*, **73**, 813–826.
53. Kallergi, G., Papadaki, M.a., Politaki, E. *et al.* (2011) Epithelial to mesenchymal transition markers expressed in circulating tumour cells of early and metastatic breast cancer patients. *Breast Cancer Research*, **13**, R59.
54. Rhim, A.D., Mirek, E.T., Aiello, N.M. *et al.* (2012) EMT and dissemination precede pancreatic tumor formation. *Cell*, **148**, 349–361.
55. Yu, M., Bardia, A., Wittner, B.S. *et al.* (2013) Circulating breast tumor cells exhibit dynamic changes in epithelial and mesenchymal composition. *Science*, **339**, 580–584.
56. Zhang, L., Ridgway, L.D., Wetzel, M.D. *et al.* (2013) The identification and characterization of breast cancer CTCs competent for brain metastasis. *Science Translational Medicine*, **5**, 180ra48.
57. Lu, C.-Y., Tsai, H.-L., Uen, Y.-H. *et al.* (2013) Circulating tumor cells as a surrogate marker for determining clinical outcome to mFOLFOX chemotherapy in patients with stage III colon cancer. *British Journal of Cancer*, **108**, 791–797.
58. Markiewicz, A., Książkiewicz, M., Wełnicka-Jaśkiewicz, M. *et al.* (2014) Mesenchymal phenotype of CTC-enriched blood fraction and lymph node metastasis formation potential. *PloS One*, **9**, e93901.
59. Rossi, E., Fassan, M., Aieta, M. *et al.* (2012) Dynamic changes of live/apoptotic circulating tumour cells as predictive marker of response to sunitinib in metastatic renal cancer. *British Journal of Cancer*, **107**, 1286–1294.
60. Aktas, B., Müller, V., Tewes, M. *et al.* (2011) Comparison of estrogen and progesterone receptor status of circulating tumor cells and the primary tumor in metastatic breast cancer patients. *Gynecologic Oncology*, **122**, 356–360.
61. Bidard, F.-C., Hajage, D., Bachelot, T. *et al.* (2012) Assessment of circulating tumor cells and serum markers for progression-free survival prediction in metastatic breast cancer: a prospective observational study. *Breast Cancer Research*, **14**, R29.
62. Fehm, T., Hoffmann, O., Aktas, B. *et al.* (2009) Detection and characterization of circulating tumor cells in blood of primary breast cancer patients by RT-PCR and comparison to status of bone marrow disseminated cells. *Breast Cancer Research*, **11**, R59.
63. Green, T.L., Cruse, J.M., and Lewis, R.E. (2013) Circulating tumor cells (CTCs) from metastatic breast cancer patients linked to decreased immune function and response to treatment. *Experimental and Molecular Pathology*, **95**, 174–179.

64. Lee, H.J., Oh, J.H., Oh, J.M. *et al.* (2013) Efficient isolation and accurate in situ analysis of circulating tumor cells using detachable beads and a high-pore-density filter. *Angewandte Chemie, International Edition*, **52**, 8337–8340.
65. Jaeger, B.A.S., Jueckstock, J., Andergassen, U. *et al.* (2014) Evaluation of two different analytical methods for circulating tumor cell detection in peripheral blood of patients with primary breast cancer. *BioMed Research International*, **2014**, 491459.
66. Igawa, S., Gohda, K., Fukui, T. *et al.* (2014) Circulating tumor cells as a prognostic factor in patients with small cell lung cancer. *Oncology Letters*, **7**, 1469–1473.
67. Thege, F.I., Lannin, T.B., Saha, T.N. *et al.* (2014) Microfluidic immunocapture of circulating pancreatic cells using parallel EpCAM and MUC1 capture: characterization, optimization and downstream analysis. *Lab on Chip*, **14**, 1775–1784.
68. M.J.M. Magbanua, E. V. Sosa, J.H. Scott *et al.*, Isolation and genomic analysis of circulating tumor cells from castration resistant metastatic prostate cancer, *BMC Cancer* 2012, **12**, 78.
69. Gleghorn, J.P., Pratt, E.D., Denning, D. *et al.* (2010) Capture of circulating tumor cells from whole blood of prostate cancer patients using geometrically enhanced differential immunocapture (GEDI) and a prostate-specific antibody. *Lab on Chip*, **10**, 27–29.
70. Barbazán, J., Alonso-Alconada, L., Muinelo-Romay, L. *et al.* (2012) Molecular characterization of circulating tumor cells in human metastatic colorectal cancer. *PloS One*, **7**, e40476.
71. Denève, E., Riethdorf, S., Ramos, J. *et al.* (2013) Capture of viable circulating tumor cells in the liver of colorectal cancer patients. *Clinical Chemistry*, **59**, 1–9.
72. Iinuma, H., Okinaga, K., Adachi, M. *et al.* (2000) Detection of tumor cells in blood using CD45 magnetic cell separation followed by nested mutant allele-specific amplification of p53 and K-ras genes in patients with colorectal cancer. *International Journal of Cancer*, **89**, 337–344.
73. Gradilone, A., Iacovelli, R., Cortesi, E. *et al.* (2011) Circulating tumor cells and "suspicious objects" evaluated through CellSearch® in metastatic renal cell carcinoma. *Anticancer Research*, **31**, 4219–4221.
74. McKiernan, J.M., Buttyan, R., Bander, N.H. *et al.* (1999) The detection of renal carcinoma cells in the peripheral blood with an enhanced reverse transcriptase-polymerase chain reaction assay for MN/CA9. *Cancer*, **86**, 492–497.
75. Wang, S., Zheng, G., Cheng, B. *et al.* (2014) Circulating tumor cells (CTCs) detected by RT-PCR and its prognostic role in gastric cancer: a meta-analysis of published literature. *PloS One*, **9**, e99259.
76. Pituch-Noworolska, A., Wieckiewicz, J., Krzeszowiak, A. *et al.* (1998) Evaluation of circulating tumour cells expressing CD44 variants in the blood of gastric cancer patients by flow cytometry. *Anticancer Research*, **18**, 3747–3752.
77. Gazzaniga, P., Gradilone, A., de Berardinis, E. *et al.* (2012) Prognostic value of circulating tumor cells in nonmuscle invasive bladder cancer: a CellSearch analysis. *Annals of Oncology*, **23**, 2352–2356.
78. Okegawa, T., Hayashi, K., Hara, H. *et al.* (2010) Immunomagnetic quantification of circulating tumor cells in patients with urothelial cancer. *International Journal of Urology*, **17**, 254–258.
79. Joshi, P., Jacobs, B., Derakhshan, A. *et al.* (2014) Enrichment of circulating melanoma cells (CMCs) using negative selection from patients with metastatic melanoma. *Oncotarget*, **5**, 2450–2461.

80. Rao, C.G., Chianese, D., Doyle, G.V. *et al.* (2005) Expression of epithelial cell adhesion molecule in carcinoma cells present in blood and primary and metastatic tumors. *International Journal of Oncology*, **27**, 49–57.
81. Hughes, A.D., Mattison, J., Western, L.T. *et al.* (2012) Microtube device for selectin-mediated capture of viable circulating tumor cells from blood. *Clinical Chemistry*, **58**, 846–853.
82. Grover, P.K., Cummins, A.G., Price, T.J., and Hardingham, J.E. (2014) Circulating tumour cells: the evolving concept and the inadequacy of their enrichment by EpCAM-based methodology for basic and clinical cancer research. *Annals in Oncology*, **25**, 1–11.
83. Baccelli, I., Schneeweiss, A., Riethdorf, S. *et al.* (2013) Identification of a population of blood circulating tumor cells from breast cancer patients that initiates metastasis in a xenograft assay. *Nature Biotechnology*, **31**, 539–544.
84. Fan, Z.C., Yan, J., Da Liu, G. *et al.* (2012) In vivo flow cytometry visualizes the effects of tumor resection on metastasis by real-time monitoring of rare circulating cancer cells. *Cancer Research*, **72**, 2683–2691.
85. Smerage, J.B., Budd, G.T. *et al.* (2013) Monitoring apoptosis and Bcl-2 on circulating tumor cells in patients with metastatic breast cancer. *Molecular Oncology*, **7**, 680–692.
86. Sun, Y.-F., Xu, Y., Yang, X.-R. *et al.* (2013) Circulating stem cell-like epithelial cell adhesion molecule-positive tumor cells indicate poor prognosis of hepatocellular carcinoma after curative resection. *Hepatology*, **57**, 1458–1468.
87. Naito, T., Tanaka, F., and Ono, A. (2012) Prognostic impact of circulating tumor cells in patients. *Journal of Thoracic Oncology*, **7**, 1–8.
88. Zhao, L., Li, P., Li, F. *et al.* (2013) The prognostic value of circulating tumor cells lacking cytokeratins in metastatic breast cancer patients. *Journal of Cancer Research Therapy*, **9**, 29–37.
89. Hughes, A.D. and King, M.R. (2012) Nanobiotechnology for the capture and manipulation of circulating tumor cells. *WIREs Nanotechnology Nanbiotechnology*, **4**, 291–309.
90. Diamond, E., Lee, G.Y., Akhtar, N.H. *et al.* (2012) Isolation and characterization of circulating tumor cells in prostate cancer. *Frontiers in Oncology*, **2**, 131.
91. Allard, W.J., Matera, J., Miller, M.C. *et al.* (2004) Tumor cells circulate in the peripheral blood of all major carcinomas but not in healthy subjects or patients with nonmalignant diseases. *Clinical Cancer Research*, **10**, 6897–6904.
92. Nagrath, S., Sequist, L.V., Maheswaran, S. *et al.* (2007) Isolation of rare circulating tumour cells in cancer patients by microchip technology. *Nature*, **450**, 1235–1239.
93. Stott, S.L., Hsu, C.-H., Tsukrov, D.I. *et al.* (2010) Isolation of circulating tumor cells using a microvortex-generating herringbone-chip. *Proceedings of National Academy of Sciences of United States of America*, **107**, 18392–18397.
94. Chang, W.C., Lee, L.P., and Liepmann, D. (2005) Biomimetic technique for adhesion-based collection and separation of cells in a microfluidic channel. *Lab on Chip*, **5**, 64–73.
95. Kim, S.K., Moon, W.K., Park, J.Y., and Jung, H. (2012) Inflammatory mimetic microfluidic chip by immobilization of cell adhesion molecules for T cell adhesion. *The Analyst*, **137**, 4062–4068.
96. Gakhar, G., Navarro, V.N., Jurish, M. *et al.* (2013) Circulating tumor cells from prostate cancer patients interact with E-selectin under physiologic blood flow. *PloS One*, **8**, e85143.

97. Wojciechowski, J.C., Narasipura, S.D., Charles, N. *et al.* (2008) Capture and enrichment of CD34-positive haematopoietic stem and progenitor cells from blood circulation using P-selectin in an implantable device. *British Journal of Haematology*, **140**, 673–681.
98. Han, W., Allio, B.A., Foster, D.G., and King, M.R. (2010) Nanoparticle coatings for enhanced capture of flowing cells in microtubes. *ACS Nano*, **4**, 174–180.
99. Hughes, A.D. and King, M.R. (2010) Use of naturally occurring halloysite nanotubes for enhanced capture of flowing cells. *Langmuir*, **26**, 12155–12164.
100. Hughes, A.D., Mattison, J., Powderly, J.D. *et al.* (2012) Rapid isolation of viable circulating tumor cells from patient blood samples. *Journal of Visualized Experiments*, **64**, e4248.
101. Hughes, A.D., Marshall, J.R., Keller, E. *et al.* (2013) Differential drug responses of circulating tumor cells within patient blood. *Cancer Letters*, **352**, 28–35.
102. Scher, H.I., Jia, X., de Bono, J.S. *et al.* (2009) Circulating tumour cells as prognostic markers in progressive, castration-resistant prostate cancer: a reanalysis of IMMC38 trial data. *Lancet*, **10**, 233–239.
103. Reid, A.H.M., Attard, G., Danila, D.C. *et al.* (2010) Significant and sustained antitumor activity in post-docetaxel, castration-resistant prostate cancer with the CYP17 inhibitor abiraterone acetate. *Journal of Clinical Oncology*, **28**, 1489–1495.
104. Danila, D.C., Morris, M.J., de Bono, J.S. *et al.* (2010) Phase II multicenter study of abiraterone acetate plus prednisone therapy in patients with docetaxel-treated castration-resistant prostate cancer. *Journal of Clinical Oncology*, **28**, 1496–1501.
105. Davnall, F., Yip, C.S.P., Ljungqvist, G. *et al.* (2012) Assessment of tumor heterogeneity: an emerging imaging tool for clinical practice? *Insights Imaging*, **3**, 573–589.
106. Longley, D.B. and Johnston, P.G. (2005) Molecular mechanisms of drug resistance. *Journal of Pathology*, **205**, 275–292.
107. Mitchell, M.J., Wayne, E., Rana, K. *et al.* (2014) TRAIL-coated leukocytes that kill cancer cells in the circulation. *Proceedings of National Academy of Sciences of United States of America*, **111**, 930–935.
108. Kim, M. and Seol, D. (nd) (2005) Chapter 7: Death signaling and therapeutic applications of TRAIL. *Aids*, **2**, 133–148.
109. Huang, Z. and King, M.R. (2009) An immobilized nanoparticle-based platform for efficient gene knockdown of targeted cells in the circulation. *Gene Therapy*, **16**, 1271–1282.
110. Mitchell, M.J., Chen, C.S., Ponmudi, V. *et al.* (2012) E-selectin liposomal and nanotube-targeted delivery of doxorubicin to circulating tumor cells. *Journal of Controlled Release*, **160**, 609–617.
111. Rana, K., Reinhart-King, C.A., and King, M.R. (2012) Inducing apoptosis in rolling cancer cells: a combined therapy with aspirin and immobilized TRAIL and E-selectin. *Molecular Pharmaceutics*, **9**, 2219–2227.

13

APTAMER-ENABLED TUMOR CELL ISOLATION

JINLING ZHANG

Department of Mechanical and Aerospace Engineering, University of Florida, Gainesville, FL, USA

Z. HUGH FAN

Department of Mechanical and Aerospace Engineering, University of Florida, Gainesville, FL, USA; J. Crayton Pruitt Family Department of Biomedical Engineering, University of Florida, Gainesville, FL, USA; Department of Chemistry, University of Florida, Gainesville, FL, USA

13.1 INTRODUCTION

The detection and enumeration of circulating tumor cells (CTCs) are expected to play an important role in cancer-related research as well as in clinical practice including cancer diagnosis, therapy monitoring, and drug development. CTCs are tumor cells in peripheral blood that escape from the tumor site in a primary organ and enter the circulatory system. Some of these CTCs migrate to the secondary site, completing the metastases process [1, 2]. As a result, CTC detection represents one of "liquid biopsy" methods, and it has a potential to help cancer diagnostics, complementing the current practice of tissue biopsies for guiding cancer therapy [3–5].

However, CTC isolation is a technical challenge due to the fact that their number is very low in comparison with the healthy cells in peripheral blood [6–9]. Antibody-based methods such as CellSearch® (see Chapter 19 for the details) and other immunoidentification approaches are useful for isolating CTCs from complex cellular fluids with high efficiency, sensitivity, and throughput [10–13].

Circulating Tumor Cells: Isolation and Analysis, First Edition. Edited by Z. Hugh Fan.

However, antibody-based strategies are limited by the availability and specificity of the antibodies, biomarkers that have been clinically validated, and the difficulty of releasing captured cells for downstream analysis [14–16].

Aptamers are single-stranded nucleic acid molecules, either RNA or DNA [14, 15, 17–19]. The molecular weight of an aptamer is about 8–15 kDa, which is much smaller than antibodies. Aptamers, especially DNA aptamers, possess a number of advantages over antibodies, including stability over a long period, resistance to harsh conditions such as extreme pH and detergents, high heat resistance (i.e., not denatured), and programmable sequences [15, 20]. Furthermore, aptamers can be developed by selecting against a binding target, ranging from a small molecule to a protein, even to a cell without knowing particular target molecules. Finally, aptamers can be reproducibly synthesized and can be modified to meet different requirement. Therefore, engineered nucleic acid aptamers could become an alternative to antibodies for cancer diagnosis and prognosis [21–23].

13.2 APTAMERS AND THEIR BIOMEDICAL APPLICATIONS

13.2.1 Identification of Aptamers

Aptamers are generated by a technique called the systematic evolution of ligands by exponential enrichment (SELEX) process [24]. The development of this *in vitro* selection and amplification technique has allowed the discovery of specific nucleic acid aptamers that can bind to a variety of target molecules with high affinity and specificity. The selected oligonucleotides are called "aptamer," which is derived from the Latin word "aptus" meaning "fitting." The molecules that bind with aptamers are called "apatopes," which is in analogous to "epitope" used in antibody terminology [15]. However, "epitope" or "target" is frequently used in the aptamer literature. As explained next, the SELEX process is an iterative process that screens a very large combinatorial library of oligonucleotides through *in vitro* selection and amplification [24]. For aptamers used for cell capture and enrichment, cell-SELEX would be employed as follows.

As shown in Figure 13.1, the cell-SELEX procedure usually begins with a random library of 10^{13}–10^{15} DNA or RNA sequences obtained from combinatorial chemical synthesis. The library goes through an iterative process with positive selections and negative/counterselections to obtain sequences with specific binding with the target cells. Sequence amplification process such as polymerase chain reaction (PCR) boosts the population of those oligonucleotides with specific binding with the target cells in the library. Structural analogs and the sequences without specific binding are eluted or discarded. Due to their unique secondary or tertiary structures, aptamers can bind with their target cells with high affinity, which is often defined by dissociation constants (K_d). The K_d values of aptamers are often in the nanomolar to picomolar range, indicating that their binding affinity is comparable to antibodies [25]. In addition to biological cells and bacteria, aptamers have been selected for a broad range of targets including small molecules and viruses [26].

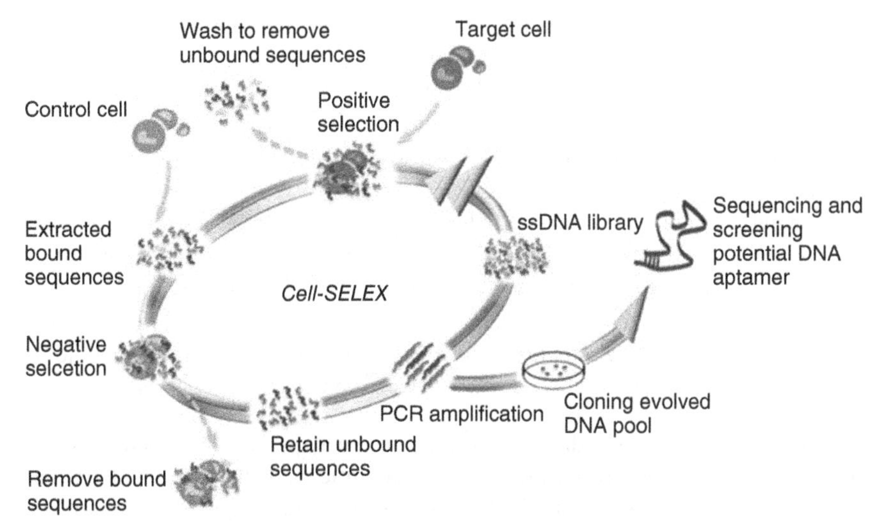

Figure 13.1 Schematic of the cell-SELEX process. (from "ssDNA library" at the right, counterclockwise) A library of single-stranded DNA (ssDNA) is created and then incubated with target cells. Unbound DNAs are washed away, and the bound DNAs are retained with cells, followed by thermal denaturation to release them. The eluted DNAs are subjected to control cells and those bound DNAs are removed with the cells. The remaining DNAs are specific to the target cells, but not to control cells. They are amplified by PCR, followed by many rounds of the same selection/amplification cycles. The final PCR product is sequenced, and the aptamer's DNA sequence is then identified. Sefah *et al.* [25], pp. 1169–1185. Nature Protocols.

13.2.2 Aptamers versus Antibodies

Aptamers generally possess strong specificity. Aptamers can discriminate targets on the basis of subtle structural differences such as the presence or absence of a methyl [27] or a hydroxyl group [28]. In comparison to antibodies, aptamers have the following advantages [29]:

- Aptamers are developed via an *in vitro* process, whereas antibodies are generally produced *in vivo*, such as in animals. Therefore, aptamers are not animal dependent.
- Aptamers can be selected in nonphysiological conditions and they are suitable for *in vitro* diagnostics.
- Aptamers can be developed for toxins or molecules that do not elicit good immune responses. In contrast, antibodies are hard to generate for them because (i) toxins are not well tolerated by animals or (ii) some molecules are inherently not immunogenic and do not produce antibodies in animals.
- Aptamers are chemically synthesized and thus can be reproducibly produced with high purity. There is little variation among batches. In contrast, the properties of the same antibody vary from one manufacturer to the other and are somewhat different between two batches even from the same manufacturer.

- Aptamers can be easily modified with a reporter molecule such as fluorescein or a functional molecule such as biotin for subsequent conjugation reactions. Although antibodies can also be modified, multiple molecules are likely attached in various locations of an antibody. In contrast, these molecules can be attached to aptamers at a precise location during the chemical synthesis.
- The binding affinity of aptamers can be taken away by nucleases or complementary sequence. As a result, when they are used for CTC capture, aptamers offer the possibility to release CTCs captured. In contrast, cells captured by antibodies are difficult to be released for downstream analysis.
- Aptamers can be denatured at a high temperature, and the denaturation process is reversible. They are stable at ambient temperature. In contrast, antibodies are irreversibly denatured; thus, they must be stored at a freezing temperature.

On the other hand, antibodies have advantages over aptamers. Antibodies are well known and generally accepted by research communities and other users. Additionally, many commercially available antibodies have been validated by the work reported in the literature, whereas this is generally not the case for many aptamers.

13.2.3 List of Aptamers

Based on cell-SELEX [30], a panel of aptamers has been developed by Dr. Weihong Tan and his coworkers, as well as other research groups, targeting different types of cancer cells, including leukemia, liver cancer, small-cell lung cancer, and colorectal cancer. Table 13.1 lists the sequences of these aptamers as well as their target molecules.

13.3 APTAMER-BASED TUMOR CELL ISOLATION

13.3.1 Aptamers with Microfluidics

Microfluidics technology has been applied for CTC detection as early as in 2007 [10]. The use of aptamer-based microfluidic devices for efficient capture of tumor cells is discussed next. Microfluidics offers significant advantages over macroscale methods (see Chapter 2 for the details) because more efficient interactions with tumor cells can take place in microchannels due to shorter diffusion distance, with minimal damage to cells at low shear forces.

13.3.1.1 Device Designs A number of microfluidic devices that incorporate aptamers for isolating CTCs have been reported [14, 21–23, 44]. Different designs of microstructures inside microfluidic devices for CTC capture have been reported [10, 11, 21]. As shown in Figure 13.2, we have developed aptamer-immobilized microfluidic devices for tumor cell isolation. The devices were in the same size as a microscope slide (3 in. × 1 in.), consisting of eight parallel channels to achieve high sample throughput. In these channels, different microstructures have been fabricated to enhance cell capture. Figure 13.2a showed a herringbone structure,

TABLE 13.1 List of Aptamers with Their Names, Targets, and Sequences

Aptamer Name	Target Name	Sequence
Sgc8	CCRF-CEM cells (PTK7)	5′-ATC TAA CTG CTG CGC CGC CGG GAA AAT ACT GTA CGG TTA GA-3′ [31]
TD05	Ramos cells (human Burkitt's lymphoma)	5′-AAC ACC GTG GAG GAT AGT TCG GTG GCT GTT CAG GGT CTC CTC CCG GTG-3′ [32]
Sgd5	Toledo cells (human diffuse large cell lymphoma)	5′-ATA CCA GCT TAT TCA ATT ATC GTG GGT CAC AGC AGC GGT TGT GAG GAA GAA AGG CGG ATA ACA GAT AAT AAG ATAGTAAGTGCAATCT-3′ [33]
KH1C12	HL60 cells (acute myeloid leukemia)	5′-ATC CAG AGT GAC GCA GCA TGC CCT AGT TAC TAC TAC TCT TTT TAG CAA ACG CCC TCG CTT TGG ACA CGG TGG CTT AGT-3′ [34]
TLS11a	LH86 (human hepatoma cells)	5′-ACAGCA TCCCCATGTGAACAATCG-CATTGTGATTGTTACGGT TTC CGC CTC ATG GAC GTG CTG-3′ [35]
PP3	Hemagglutinin (HA, also known as A56R)	5′-ATC CAG AGT GAC GCA GCA CGA GCC AGA CAT CTC ACA CCT GTT GCA TAT ACA TTT TGC ATG GAC ACG GTG GCT TAG T-3′ [36]
TV02	Vaccinia-infected A549 cells	5′-ATC GTCTGCTCCGTCCAATACCTG-CATATACACTTTGCATGT GGT TTG GTG TGA GGT CGT GC-3′ [37]
HCH07	NCI-H23 (adenocarcinoma)	5′-TAC CAG TGC GAT GCT CAG GCC GAT GTC AAC TTT TTC TAA CTC ACT GGT TTT GCC TGA CGC ATT CGG TTG AC-3′ [38]
KDED2a-3	DLD-1 cells (colorectal cancer cells)	5′-TGC CCG CGA AAA CTG CTA TTA CGT GTG AGA GGA AAG ATC ACG CGG GTT CGT GGA CAC GG-3′ [39]
aptTOV1	TOV-21G cells (ovarian cancer cells)	5′-ATC CAG AGT GAC GCA GCA GAT CTG TGT AGG ATC GCA GTG TAG TGG ACA TTT GAT ACG ACT GGC TCG ACA CGG TGG CTT A-3′ [40]

TABLE 13.1 (*Continued*)

Aptamer Name	Target Name	Sequence
KMF2-1a	Breast cancer cells	5′-AGG CGG CAG TGT CAG AGT GAA TAG GGG ATG TAC AGG TCT GCA CCC ACT CGA GGA GTG ACT GAG CGA CGA AGA CCC C-3′ [41]
EJ2	H23 cells (lung adenocarcinoma cells)	5′-AGT GGT CGA ACT ACA CAT CCT TGA ACT GCG GAA TTA TCT AC-3′ [42]
CSC01	Prostate cancer cells	5′-ACC TTG GCT GTC GTG TTG TAG GTG GTT TGC TGC GGT GGG CTC AAG AAG AAA GCG CAA AGT CAG TGG TCA GAG CGT-3′ [43]
SYL3C	EpCAM	5′-CAC TAC AGA GGT TGC GTC TGT CCC ACG TTG TCA TGG GGG GTT GGC CTG-3′ [44]
Anti-EGFR aptamer	EGFR	5′-GGC GCU CCG ACC UUA GUC UCU GUG CCG CUA UAA UGC ACG GAU UUA AUC GCC GUA GAA AAG CAU GUC AAA GCC GGA ACC GUG UAG CAC AGC AGAGAAUUAAAUGCCCG-CCAUGACCAG-3′ [45]
Anti-PSMA aptamer	PSMA (prostate-specific membrane antigen)	5′-ACCAAGACCUGACUUC-UAACUAAGUCUACGUUCC-3′ [46]

which contains 50-μm-deep herringbone grooves. The grooves are staggered, and they disrupt streamlines of a fluid flow, inducing chaotic mixing and maximizing interactions between cells and capture agents on channel surfaces [47]. This device was made from an upper polydimethylsiloxane (PDMS) layer and a glass slide substrate. PDMS substrates were fabricated using lithography with SU-8 as a photoresist [48]. In a micropillar array device shown in Figure 13.2b, the distance between the micropillars is 60 μm and each pillar's diameter is 90 μm, which was optimized for cell interaction with microchannels [49]. Also, we have explored a glass device, which consists of elliptical pillar arrays inside the microchannels as shown in Figure 13.2c [21]. The dimension of the elliptical pillars is 30 μm (major axis) × 15 μm (minor axis) × 32 μm (height), with an interpillar distance of 80 μm (center to center) and an 80 μm shift after every three rows in the direction of the minor axis. Through photolithography, patterns on a photomask can be transferred to glass substrates coated with chromium and photoresist layers. The glass substrate was then chemically etched using a mixture of HF, HNO_3, and H_2O while the etching time determined the channel depth. The channel depth is pretty reproducible if the

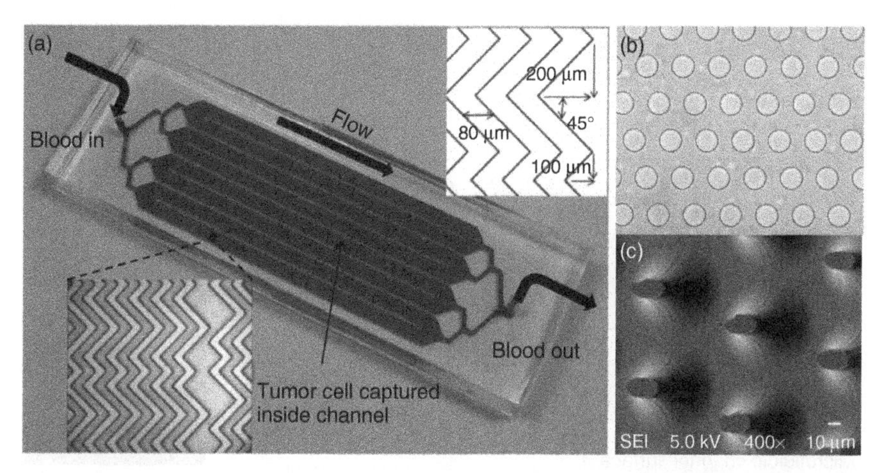

Figure 13.2 (a) Picture of a 3″ × 1″ microfluidic device, consisting of eight parallel channels with single inlet and outlet. Insert: micrograph of herringbone grooves inside channels. (b) Microstructure of circular micropillars, and (c) elliptical micropillars, inside a channel of microfluidic device. Sheng *et al.* [21], pp. 4199–4206. Analytical Chemistry; Sheng *et al.* [47], pp. 7067–7076. ACS Nano.

same solution is used to calibrate a test piece on the same day. The glass substrate was then sealed with a 5-mm-thick PDMS sheet.

13.3.1.2 Surface Functionalization The surfaces of microchannels were prepared using an avidin–biotin reaction [21]. As shown in Figure 13.3a, a two-step surface modification scheme was used. First, after being washed with ethanol and phosphate-buffered saline (PBS), the channels were incubated with avidin, which physically adsorbed onto the glass surfaces. The second step was to introduce biotinylated sgc8 aptamer with a polythymine (10-T) linker, which conjugated with avidin on the surfaces of microchannels and micropillars. Afterward, the devices were ready for cell capture as discussed next.

13.3.1.3 Tumor Cell Isolation As shown in Figure 13.3b and c, we used human leukemia cells (CCRF-CEM) as target cells and isolated them using sgc8 aptamers [21]. Immediately before the capture experiments, both target cells (CEM) and control cells (Ramos, Burkitt's lymphoma cells) were rinsed with the washing buffer and resuspended at appropriate concentrations. They were mixed at either 1:1 or a variation up to 1:1000 of target cells to control cells. The cell mixture was introduced into the device as discussed next.

13.3.1.4 Instrument Setup The cell mixture or a blood sample was introduced into the microfluidic device using a syringe pump. At the end of the experiment, the device was imaged under a microscope via an automatic stage. Images were then analyzed using Image J and cell numbers were counted. Note that cells were verified by

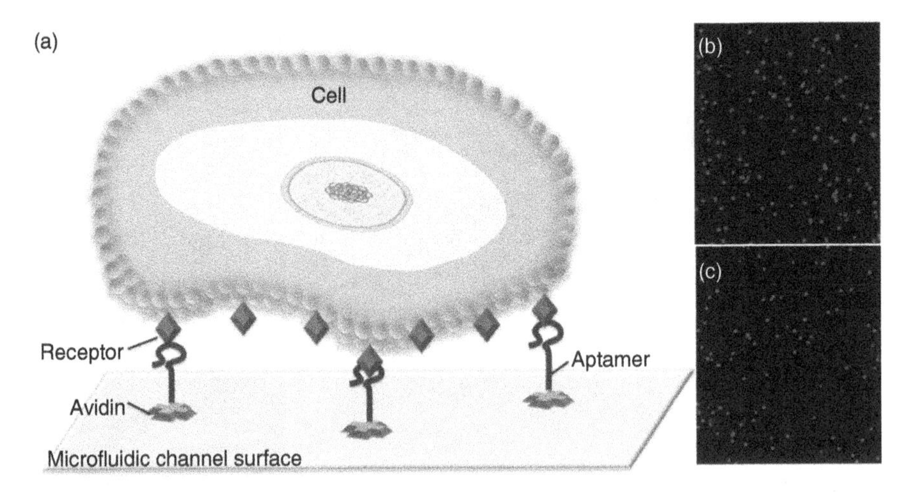

Figure 13.3 (a) Schematic of surface modification and cell capture. (b) Representative image of a couple of target cells stained in red among a large number of control cells stained in blue. (c) Image of the same cell mixture after the capture experiment, showing significant enrichment of the target cells. Sheng *et al.* [21], pp. 4199–4206. Analytical Chemistry. (*See color plate section for the color representation of this figure.*)

comparing fluorescent images with transmission images; only those with appropriate cell morphology in the transmission images were counted.

13.3.2 Aptamers with Nanoparticles

After aptamers were demonstrated for capturing cancer cells, nanoparticle–aptamer conjugates have been studied for enhancing the cell capture efficiency. We have developed a multivalent binding system using gold nanoparticles (AuNPs) as a vehicle to assemble multiple DNA aptamers [47]. In this work, AuNP is conjugated with about 95 aptamers via self-assembly, leading to multivalent binding effects and resulting in enhanced cell capture efficiency. Capture efficiency was measured by dividing the number of the target cells captured on the device by the number of total target cells introduced into the device. The enhanced binding affinity afforded by the AuNP–aptamers likely resulted from higher aptamer density on channel surfaces than using aptamers alone, as schematically illustrated in Figure 13.4. Moreover, AuNP–aptamers maintained high capture efficiency with increased flow rate, which leads to higher sample throughput in the device.

A related scheme used a label-free DNA–silver nanoclusters (DNA-AgNCs) for detecting tumor cells [50]. In this effort, two tailored DNA sequences were involved. The first DNA sequence consisted of an aptamer at the 3′-end, a guanine-rich DNA sequence, and a recognition probe at the 5′-end. This sequence was designed to be in a hairpin-shaped structure. The second DNA sequence contained a sequence for AgNCs-templated synthesis and a sequence complementary to the recognition

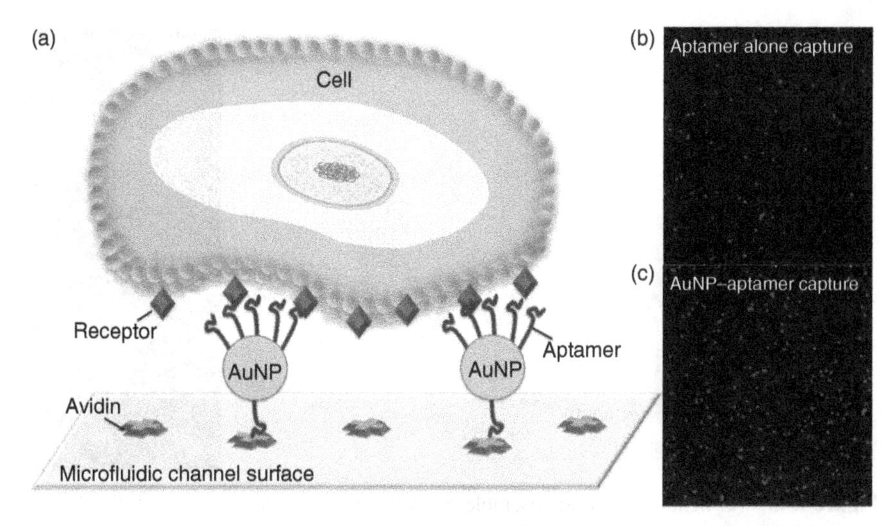

Figure 13.4 (a) Schematic of cell capture using AuNP–aptamers that were immobilized onto channel surfaces. Since multiple aptamers on the AuNP can simultaneously bind with a few receptors on the cell surface, cell capture efficiency would be enhanced. (b and c) Comparison in the capture efficiency of target cells in red between (b) with aptamer alone and (c) using AuNP–aptamers. Sheng *et al.* [47], pp. 7067–7076. ACS Nano. (*See color plate section for the color representation of this figure.*)

probe. This sequence served as the signal probe. When the aptamer was interacting with a receptor, the conformation change of the aptamer disrupted the hairpin structure, allowing hybridization between the recognition probe in the first DNA and the complementary sequence in the second DNA. After hybridization, AgNCs was in close proximity to guanine-rich DNA, leading to enhanced fluorescence intensity. The scheme was incorporated in the experiments, in which aptamers recognized the target cells and the event could be detected without fluorescence labeling.

Another platform has been developed that uses aptamer-conjugated upconversion nanoparticles (UCNPs) for tumor cell capture [51]. The isolated cells are enriched in a magnetic field due to their attachment with magnetic nanoparticles. The magnetic separation enables cell isolation and the autofluorescence of upconversion luminescence imaging enhances the detection sensitivity, both of which lead to sensitive detection of rare tumor cells.

13.3.3 Aptamers with Innovative Schemes

Many researchers have focused on engineering aptamers on functionalized surface to realize enhanced cell capture and isolation. We have developed an ensemble of aptamers and antibodies that is able to enhance cell capture efficiency [49]. Due to their differences in the size, the ensemble effectively captures cells by binding to cell surface markers in a cooperative manner as shown in Figure 13.5.

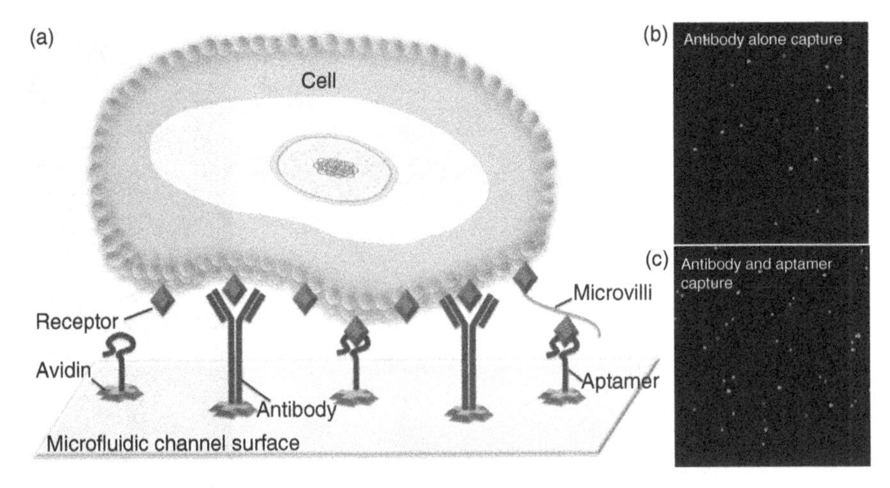

Figure 13.5 (a) Schematic of an ensemble of antibodies and aptamers immobilized on the channel surfaces and the multivalent interactions of one cell with the ensemble. The drawing is not to scale. (b and c) Comparison in the capture efficiency of target cells in green between (b) with antibody alone and (c) using antibody–aptamer ensemble. Zhang *et al.* [49], pp. 6722–6725. Chemical Communications. (*See color plate section for the color representation of this figure.*)

Aptamers have also been developed to target cancer cells via extra cellular membrane proteins, including prostate-specific membrane antigen (PSMA), human epidermal growth factor receptor 3 (HER-3), and MUC1 [52]. A method using anti-EGFR (epidermal growth factor receptor) aptamers to capture hGBM cells was developed in a microfluidic device, allowing for the isolating of EGFR-overexpressed tumor cells [53]. In addition, anti-EGFR aptamers were immobilized on the glass beads in an array, followed by subsequent release of cells with antisense molecules that hybridized with cell-bound aptamers [16].

13.3.4 Aptamers for CTC Isolation

As discussed earlier, Cell-SELEX could be employed to generate aptamers with specific binding with target CTCs. It is possible to find one panel of aptamers, each of which binds one specific antigen on the surface of a tumor cell. As a result, the panel could produce a pattern to differentiate cell types such as epithelial CTCs and mesenchymal CTCs [17]. In other words, aptamers could play a role in analyzing the intrinsic heterogeneity of CTCs.

One example of using structured aptamers has been reported by Zhao *et al.* [54] for isolating of CTCs. They reported a microfluidic device immobilized with DNA aptamers, which were prepared by using rolling circle amplification. Because each aptamer molecule consists of multiple binding sites, it was able to capture each cancer cell with improved efficiency due to enhanced avidity. A three-dimensional network of DNA aptamers could be formed in the microfluidic channel, enhancing cell capture efficiency even under a high flow rate [55].

Soper's group [14] has developed an aptamer-based platform for the isolation of prostate-specific CTCs. These tumor cells overexpress PSMA; thus, anti-PSMA aptamers can be used to selectively isolate tumor cells from clinical samples. To demonstrate the application, anti-PSMA aptamers was used to modify the surfaces of a capture bed in a microfluidic device, which was fabricated in a way described in Chapter 4.

In addition, a series of DNA aptamers against cancer biomarker protein EpCAM (epithelial cell adhesion molecule) has been generated [44]. By screening against EpCAM, aptamers with specific binding were identified. These aptamers have been validated using a number of cancer cells expressing EpCAM, and the authors found that the aptamers do not bind to EpCAM-negative cells.

13.4 CONCLUSION AND OUTLOOK

In this chapter, we introduced aptamers, explained their selection process, pointed out their differences from antibodies, and listed a number of aptamers for a variety of cancer cells. In addition, we showed the use of aptamers for tumor cell isolation, including their integration with microfluidics, immobilization on nanoparticles, and assembly with antibodies. We illustrated a couple of examples of using aptamers for CTC isolation in clinical samples.

As we argued earlier, a panel of aptamers has the potential to distinguish different types of cancer cells such as metastatic stages due to a pattern (i.e., fingerprint) generated with different level of various antigens on the cell surfaces. As an alternative to antibodies, aptamers could be developed in future diagnostic products. They have a potential role to play in cancer diagnosis and prognosis, either with or without antibodies.

REFERENCES

1. Chaffer, C.L. and Weinberg, R.A. (2011) A perspective on cancer cell metastasis. *Science*, **331**, 1559–1564.
2. Hanahan, D. and Weinberg, R.A. (2011) Hallmarks of cancer: the next generation. *Cell*, **144**, 646–674.
3. Friedlander, T.W., Premasekharan, G., and Paris, P.L. (2014) Looking back, to the future of circulating tumor cells. *Pharmacology & Therapeutics*, **142**, 271–280.
4. Allard, W.J., Matera, J., Miller, M.C. *et al.* (2004) Tumor cells circulate in the peripheral blood of all major carcinomas but not in healthy subjects or patients with nonmalignant diseases. *Clinical Cancer Research*, **10**, 6897–6904.
5. Cristofanilli, M., Budd, G.T., Ellis, M.J. *et al.* (2004) Circulating tumor cells, disease progression, and survival in metastatic breast cancer. *New England Journal of Medicine*, **351**, 781–791.

6. Helo, P., Cronin, A.M., Danila, D.C. *et al.* (2009) Circulating prostate tumor cells detected by reverse transcription-PCR in men with localized or castration-refractory prostate cancer: concordance with CellSearch assay and association with bone metastases and with survival. *Clinical Chemistry*, **55**, 765–773.
7. Guo, J., Yao, F., Lou, Y. *et al.* (2007) Detecting carcinoma cells in peripheral blood of patients with hepatocellular carcinoma by immunomagnetic beads and rt-PCR. *Journal of Clinical Gastroenterology*, **41**, 783–788.
8. Riethdorf, S., Fritsche, H., Muller, V. *et al.* (2007) Detection of circulating tumor cells in peripheral blood of patients with metastatic breast cancer: a validation study of the CellSearch system. *Clinical Cancer Research*, **13**, 920–928.
9. Dharmasiri, U., Witek, M.A., Adams, A.A., and Soper, S.A. (2010) Microsystems for the capture of low-abundance cells. *Annual Review of Analytical Chemistry*, **3**, 409–431.
10. Nagrath, S., Sequist, L.V., Maheswaran, S. *et al.* (2007) Isolation of rare circulating tumour cells in cancer patients by microchip technology. *Nature*, **450**, 1235–U1210.
11. Stott, S.L., Hsu, C.H., Tsukrov, D.I. *et al.* (2010) Isolation of circulating tumor cells using a microvortex-generating herringbone-chip. *Proceedings of the National Academy of Sciences of the United States of America*, **107**, 18392–18397.
12. Wang, S., Liu, K., Liu, J. *et al.* (2011) Highly efficient capture of circulating tumor cells by using nanostructured silicon substrates with integrated chaotic micromixers. *Angewandte Chemie, International Edition*, **50**, 3084–3088.
13. Dharmasiri, U., Njoroge, S.K., Witek, M.A. *et al.* (2011) High-throughput selection, enumeration, electrokinetic manipulation, and molecular profiling of low-abundance circulating tumor cells using a microfluidic system. *Analytical Chemistry*, **83**, 2301–2309.
14. Dharmasiri, U., Balamurugan, S., Adams, A.A. *et al.* (2009) Highly efficient capture and enumeration of low abundance prostate cancer cells using prostate-specific membrane antigen aptamers immobilized to a polymeric microfluidic device. *Electrophoresis*, **30**, 3289–3300.
15. Bunka, D.H.J. and Stockley, P.G. (2006) Aptamers come of age - at last. *Nature Reviews Microbiology*, **4**, 588–596.
16. Wan, Y., Liu, Y.L., Allen, P.B. *et al.* (2012) Capture, isolation and release of cancer cells with aptamer-functionalized glass bead array. *Lab on a Chip*, **12**, 4693–4701.
17. Cerchia, L. and de Franciscis, V. (2010) Targeting cancer cells with nucleic acid aptamers. *Trends in Biotechnology*, **28**, 517–525.
18. Xu, Y.H., Yang, X.R., and Wang, E.K. (2010) Review: Aptamers in microfluidic chips. *Analytica Chimica Acta*, **683**, 12–20.
19. Xiong, X.L., Liu, H.P., Zhao, Z.L. *et al.* (2013) DNA aptamer-mediated cell targeting. *Angewandte Chemie International Edition*, **52**, 1472–1476.
20. Tang, Z.W., Mallikaratchy, P., Yang, R.H. *et al.* (2008) Aptamer switch probe based on intramolecular displacement. *Journal of the American Chemical Society*, **130**, 11268–11269.
21. Sheng, W.A., Chen, T., Kamath, R. *et al.* (2012) Aptamer-enabled efficient isolation of cancer cells from whole blood using a microfluidic device. *Analytical Chemistry*, **84**, 4199–4206.
22. Xu, Y., Phillips, J.A., Yan, J.L. *et al.* (2009) Aptamer-based microfluidic device for enrichment, sorting, and detection of multiple cancer cells. *Analytical Chemistry*, **81**, 7436–7442.

23. Phillips, J.A., Xu, Y., Xia, Z. *et al.* (2009) Enrichment of cancer cells using aptamers immobilized on a microfluidic channel. *Analytical Chemistry*, **81**, 1033–1039.
24. Tuerk, C. and Gold, L. (1990) Systematic evolution of ligands by exponential enrichment: RNA ligands to bacteriophage T4 DNA polymerase. *Science*, **249**, 505–510.
25. Sefah, K., Shangguan, D., Xiong, X.L. *et al.* (2010) Development of DNA aptamers using Cell-SELEX. *Nature Protocols*, **5**, 1169–1185.
26. Famulok, M., Hartig, J.S., and Mayer, G. (2007) Functional aptamers and aptazymes in biotechnology, diagnostics, and therapy. *Chemical Reviews*, **107**, 3715–3743.
27. Jenison, R.D., Gill, S.C., Pardi, A., and Polisky, B. (1994) High-resolution molecular discrimination by RNA. *Science*, **263**, 1425–1429.
28. Manniron, C., Di Nardo, A., Fruscoloni, P., and Tocchini-Valentini, G.P. (1997) In vitro selection of dopamine RNA ligands. *Biochemistry*, **36**, 9726–9734.
29. Jayasena, S.D. (1999) Aptamers: An emerging class of molecules that rival antibodies in diagnostics. *Clinical Chemistry*, **45**, 1628–1650.
30. Shangguan, D., Li, Y., Tang, Z. *et al.* (2006) Aptamers evolved from live cells as effective molecular probes for cancer study. *Proceedings of the National Academy of Sciences of the United States of America*, **103**, 11838–11843.
31. Shangguan, D., Cao, Z.H., Meng, L. *et al.* (2008) Cell-specific aptamer probes for membrane protein elucidation in cancer cells. *Journal of Proteome Research*, **7**, 2133–2139.
32. Mallikaratchy, P., Tang, Z.W., Kwame, S. *et al.* (2007) Aptamer directly evolved from live cells recognizes membrane bound immunoglobin heavy mu chain in Burkitt's lymphoma cells. *Molecular and Cellular Proteomics*, **6**, 2230–2238.
33. Shangguan, D.H., Cao, Z.H.C., Li, Y., and Tan, W.H. (2007) Aptamers evolved from cultured cancer cells reveal molecular differences of cancer cells in patient samples. *Clinical Chemistry*, **53**, 1153–1155.
34. Sefah, K., Tang, Z.W., Shangguan, D.H. *et al.* (2009) Molecular recognition of acute myeloid leukemia using aptamers. *Leukemia*, **23**, 235–244.
35. Meng, L., Yang, L., Zhao, X.X. *et al.* (2012) Targeted delivery of chemotherapy agents using a liver cancer-specific aptamer. *PLOS One*, **7**, e33434.
36. Parekh, P., Tang, Z.W., Turner, P.C. *et al.* (2010) Aptamers recognizing glycosylated hemagglutinin expressed on the surface of vaccinia virus-infected cells. *Analytical Chemistry*, **82**, 8642–8649.
37. Tang, Z.W., Parekh, P., Turner, P. *et al.* (2009) Generating aptamers for recognition of virus-infected cells. *Clinical Chemistry*, **55**, 813–822.
38. Chen, H.W., Medley, C.D., Sefah, K. *et al.* (2008) Molecular recognition of small-cell lung cancer cells using aptamers. *ChemMedChem*, **3**, 991–1001.
39. Sefah, K., Meng, L., Lopez-Colon, D. *et al.* (2010) DNA aptamers as molecular probes for colorectal cancer study. *PLOS One*, **5**, e14269.
40. Van Simaeys, D., Lopez-Colon, D., Sefah, K. *et al.* (2010) Study of the molecular recognition of aptamers selected through ovarian cancer cell-SELEX. *PLOS One*, **5**, e13770.
41. Zhang, K.J., Sefah, K., Tang, L.L. *et al.* (2012) A novel aptamer developed for breast cancer cell internalization. *ChemMedChem*, **7**, 79–84.
42. Jimenez, E., Sefah, K., Lopez-Colon, D. *et al.* (2012) Generation of lung adenocarcinoma DNA aptamers for cancer studies. *PLOS One*, **7**, e46222.
43. Sefah, K., Bae, K.M., Phillips, J.A. *et al.* (2013) Cell-based selection provides novel molecular probes for cancer stem cells. *International Journal of Cancer*, **132**, 2578–2588.

44. Song, Y.L., Zhu, Z., An, Y. *et al.* (2013) Selection of DNA aptamers against epithelial cell adhesion molecule for cancer cell imaging and circulating tumor cell capture. *Analytical Chemistry*, **85**, 4141–4149.

45. Wan, Y.A., Kim, Y.T., Li, N. *et al.* (2010) Surface-immobilized aptamers for cancer cell isolation and microscopic cytology. *Cancer Research*, **70**, 9371–9380.

46. McNamara, J.O., Andrechek, E.R., Wang, Y. *et al.* (2006) Cell type-specific delivery of siRNAs with aptamer-siRNA chimeras. *Nature Biotechnology*, **24**, 1005–1015.

47. Sheng, W.A., Chen, T., Tan, W.H., and Fan, Z.H. (2013) Multivalent DNA nanospheres for enhanced capture of cancer cells in microfluidic devices. *ACS Nano*, **7**, 7067–7076.

48. Sia, S.K. and Whitesides, G.M. (2003) Microfluidic devices fabricated in poly(dimethylsiloxane) for biological studies. *Electrophoresis*, **24**, 3563–3576.

49. Zhang, J., Sheng, W., and Fan, Z.H. (2014) An ensemble of aptamers and antibodies for multivalent capture of cancer cells. *Chemical Communications*, **50**, 6722–6725.

50. Yin, J.J., He, X.X., Wang, K.M. *et al.* (2013) Label-free and turn-on aptamer strategy for cancer cells detection based on a DNA-silver nanocluster fluorescence upon recognition-induced hybridization. *Analytical Chemistry*, **85**, 12011–12019.

51. Shuai Fang, C.W., Xiang, J., Cheng, L. *et al.* (2014) Aptamer-conjugated upconversion nanoprobes assisted by magnetic separation for effective isolation and sensitive detection of circulating tumor cells. *Nano Research*, **7**, 1327–1336.

52. Vorhies, J.S. and Nemunaitis, J.J. (2007) Nucleic acid aptamers for targeting of shRNA-based cancer therapeutics. *Biologics*, **1**, 367–376.

53. Wan, Y., Tan, J., Asghar, W. *et al.* (2011) Velocity effect on aptamer-based circulating tumor cell isolation in microfluidic devices. *Journal of Physical Chemistry B*, **115**, 13891–13896.

54. Zhao, W., Ali, M.M., Brook, M.A., and Li, Y. (2008) Rolling circle amplification: applications in nanotechnology and biodetection with functional nucleic acids. *Angewandte Chemie International Edition in English*, **47**, 6330–6337.

55. Zhao, W., Cui, C.H., Bose, S. *et al.* (2012) Bioinspired multivalent DNA network for capture and release of cells. *Proceedings of the National Academy of Sciences of the United States of America*, **109**, 19626–19631.

14

DEPLETION OF NORMAL CELLS FOR CTC ENRICHMENT

JEFFREY J. CHALMERS

William G. Lowrie Department of Chemical and Biomolecular Engineering, The Ohio State University, Columbus, OH, USA

MARYAM B. LUSTBERG

Stefanie Spielman Comprehensive Breast Center, Wexner Medical Center, The Ohio State University, Columbus, OH, USA

CLAYTON DEIGHAN, KYOUNG-JOO JENNY PARK, YONGQI WU, AND PETER AMAYA

William G. Lowrie Department of Chemical and Biomolecular Engineering, The Ohio State University, Columbus, OH, USA

14.1 INTRODUCTION

In contrast to the various approaches to isolate circulating tumor cells (CTCs) through a positive selection approach such as targeting specific cell surface markers or specific cell size or morphology, an alternative approach is to remove as many normal blood cells as possible and then attempt to identify and/or separate the remaining cells through a positive targeting of specific cell characteristics.

As with all separation/isolation methodologies that manipulate blood samples, there exist specific advantages and disadvantages. Advantages of a methodology that includes a depletion of normal cells prior to CTC identification (i.e., negative selection step) include the following:

Circulating Tumor Cells: Isolation and Analysis, First Edition. Edited by Z. Hugh Fan.

1. No bias with respect to what marker/characteristic defines a CTC (except that a CTC does not have characteristics used to define a normal cell!).
2. Not limited to the requirement that a specific level of surface marker expression is needed to achieve a high-performance separation. Most cell surface marker expression on a population of cells is not "binary"; it exhibits a Gaussian or more complex range of expression distribution. If a cell marker is specifically targeted, how many of these cell surface markers need to be present on a CTC before it can be separated?
3. The final cell suspension will be significantly smaller, which reduces the amount of subsequent targeting reagents used.
4. The removal of the normal cells can be achieved rapidly.

Disadvantages of a negative depletion methodology include the following:

1. Further analysis is required of the typically not pure, enriched sample because there are typically still contaminating normal cells.
2. The potential addition of extra steps, as opposed to a single-step positive selection, introduces the strong possibility of nonspecific loss of CTCs (i.e., a general rule to achieve the highest recovery is to use the absolute minimum number of processing steps).

14.2 ESTIMATES OF NUMBER AND TYPE OF CELLS IN BLOOD

As is well known, human blood contains a high number of a large variety of cell types. The generally accepted number of red blood cells, RBCs, range from 4 to 6×10^9 per milliliter of blood, platelets on the order of 1.5 to 4.4×10^8 per milliliter, and nucleated cells (typically considered white blood cells) range from 4 to 11×10^6 per milliliter of blood. Within this nucleated cell population are a large number of subtypes, some of which probably have not been fully characterized. Further, these numbers, both the broad groups listed, as well as the various subtypes of nucleated cells, can have significant ranges in concentration based on a patient's medical condition. Figure 14.1 is presented to further assist in putting all of these cells in a

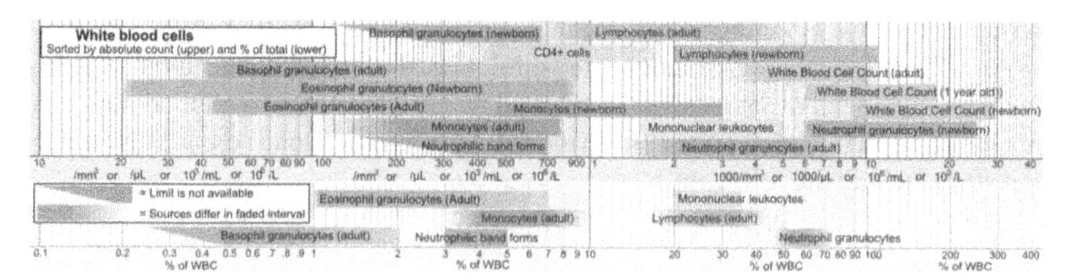

Figure 14.1 Range of concentration of various types of blood cells. Häggström (2014), DOI:10.15347/wjm/2014.008.

broader context. As a point of comparison, for 140 metastatic breast cancer patient samples analyzed in our laboratory, we counted an average RBC concentration of $4.4 \pm 1.3 \times 10^9$ per milliliter, and the average number of nucleated cells was $6.4 \pm 2.9 \times 10^6$ per milliliter.

14.3 SUMMARY OF EXAMPLES OF NEGATIVE DEPLETION

If one assumes that the traditional definition of a CTC is a cell of epithelial origin with no hematopoietic linage, the typical approach to deplete normal cells would be to remove the RBCs and all of the normal blood cells, including the various types presented in Figure 14.1. In the following sections, we discuss approaches to remove RBCs and various normal nucleated cells. In the end, we discuss other potential cell surface markers beyond traditional CD45 to remove other cells of hematopoietic origin.

14.3.1 Removal of RBCs

Based on a literature and conference presentation survey, there are three routinely used, and two proposed, methodologies to remove RBCs from blood. The three most common methodologies are the following: (i) density gradient, (ii) RBC lysis, and (iii) a combination of density gradient with immunologically based approaches. Two more recent, predominately research scale approaches, are the following: (iv) deterministic lateral displacement, DLD, and (v) magnetic separation based on the intrinsic magnetic moment of deoxygenated or met-hemoglobin form of RBCs.

In a direct comparison between Ficoll–Hypaque density–gradient separation and RBC lysis process, Lara *et al.* [1] demonstrated that the recovery of spiked breast cancer cells (MCF-7 cell line) into human blood was a function of the hematocrit, with a maximum recovery of 73% at a hematocrit of 22%. In contrast, MCF-7 recovery from direct spiking into normal, undiluted human blood using an RBC lysis approach was 89%. Because targeting CD45 on the normal cells was the intended next step in the Lara *et al.* [1] study, a flow cytometry study was conducted on the remaining nucleated population after either the Ficoll–Hypaque density–gradient separation or the RBC lysis process. By gating on the leukocyte population, (forward and side scatter location), it was observed that 99% versus 96% of the leukocyte could bind CD45, after RBC lysis or Ficoll–Hypaque density–gradient separation, respectively. These two groups of tests convinced the authors of that work that while not optimum, RBC lysis was superior to Ficoll–Hypaque density–gradient separation.

RosetteSep™ from Stem Cell Technologies provides a technology in which a density gradient and an immunologically based separation are combined [2]. Using highly purified monoclonal antibodies, bispecific tetrameric antibody complexes are formed; one end targets the cell surface antigens on human hematopoietic cells, such as CD2, CD16, CD19, CD36, CD38, CD45, and CD66b, while the other end binds to glycophorin A on erythrocytes [2]. Subsequently, these hematopoietic cells are cross-linked to multiple erythrocytes and removed by Ficoll–Hypaque density–gradient separation [2].

Unfortunately, we do not know of any head-to-head comparison studies to demonstrate the overall performance of this approach to the more traditional density gradient or RBC lysis approaches.

It has long been known, both theoretically and experimentally, that deoxygenated RBCs, or the met form of hemoglobin created chemically, impart a magnetic moment on RBCs [3]. However, this low magnetic moment, both in absolute terms and in relative terms to oxygenated RBCs, limited the practical applications. However, the availability of low-cost, high-powered, permanent magnets and computational capabilities allow the exploration of the possibility of RBC removal based on this intrinsic magnetic moment. Moore *et al.* [4] demonstrated that it is possible to remove 95% of the RBCs in 1 μl of whole, human blood in 1 min. This was accomplished with a specially designed permanent magnetic assembly generating a quadrupole field with a maximum value of 1.68 T and nearly constant radial field gradient of 1.75 T/mm. The purpose of this instrument was originally for the separation of mature RBCs from hematopoietic progenitors; hence, the recovery of rare cancer cells has not yet been tested.

A more developed and experimentally demonstrated new method to remove RBC without significant external manipulation is DLD [5]. Karabacak *et al.* [6] applied the concept of DLD to a complete microfluidic device (iChip), which not only removes the RBCs with DLD but also immunomagnetically labels and subsequently removes the normal nucleated cells. While they report high recovery of spiked cancer cells (>90%), with a high level of depletion of normal nucleated cells, the RBC depletion level is unclear [6].

14.3.2 Removal of Normal Nucleated Cells

As presented with RBC removal, a number of approaches have been used to remove normal nucleated cells, and most involve the use of magnetic forces through immunomagnetic labeling. The various instrumental approaches, and performance, of immunomagnetic cell separation systems can be divided into the actual methodology (i.e., complete flow through the instrument of the labeled and unlabeled cells or some sort of catch and subsequent release of the magnetically labeled cells) and the sensitivity and selectivity of the immunomagnetic labeling itself.

Catch, with a subsequent release, is the most common methodology for magnetic cell separation technology. Commercial examples include CellSearch (see Chapter 19 for details), Miltenyi Biotec family of separators, MACS™, and Stem Cell Technologies, RoboSep™. Examples of flow through separation of magnetically labeled and unlabeled cells into different exit streams based on the degree of magnetic labeling include the dipole separator, the quadrupole magnetic separator (QMS), and the latest version of the iChip technology [6–8]. Using the QMS system, Lara *et al.* [1] demonstrated that the QMS system, while able to operate in a complete flow through methodology, is able to provide a superior negative depletion of cells when operated in a catch-and-release mode of operation under specific operating conditions.

Unlike other catch-and-release methodologies, this optimized protocol with the QMS system used specific types of magnetic labeling, the very high magnetic energy gradient in the QMS (on the order of 300 T/m), and a minimum and maximum of hydrodynamic forces that significantly reduce nonspecific binding of cells to the surfaces in contact with the cells within the assembly, yet not too high to remove immunomagnetically labeled cells from the capture region. The magnetic depletion is accomplished as the cell suspension flows through an open annular channel (as opposed to a column packed with steel balls such as a MACS column), which is surrounded by the quadrupole magnet. The magnetically labeled cells are retained on the inner surface of the outer column wall (the column that is in contact with the surrounding quadrupole magnets on its outer surface), Figure 14.2. This open annular channel approach (with respect to the flow of the cell suspension) reduces or prevents clumping and the subsequent nonspecific capture of cells. This magnetic pressure created to hold the magnetically labeled cells against the annulus wall is sufficient to allow a relatively rapid fluid flow through the annulus, with corresponding shear stress, which reduces nonspecific binding. As has been presented previously, the recovery of spiked cancer cells into buffy coats can range from as low as 40% to as high as 90%, depending on the cancer cell lines used [9].

14.4 TYPES OF CELLS OBSERVED AFTER DEPLETION OF NORMAL CELLS

Using this methodology of CTC enrichment on studies conducted on patients with breast cancer and with squamous cell carcinoma of the head and neck, a large number of different combination of markers have been investigated/observed [10–14]. To assist in summarizing these combinations, a matrix is presented in Figure 14.3. For all of the analyses represented in this matrix, a fundamental criterion was that the cells have an intact nucleus, which was identified using DAPI. Therefore, it only left three other markers to be stained using a traditional four-color fluorescent staining approach. All of these various combinations and potential combinations indicate the need for a more systematic and higher number of signals per cell approach.

14.5 INCOMPLETE DEPLETION OF NORMAL CELLS

While it sounds rational that targeting a ubiquitous marker, such as CD45, to remove hematopoietic cells would be effective, the performance of targeting such a single marker varies significantly between normal human blood and metastatic patients. It can also vary significantly within a subset of specific types of metastatic patients [10]. A number of reasons can be suggested for this nonideal performance, including (i) significant variation in the absolute number of CD45 receptors per cell, which implies that a minimum number of receptors needs to be immunomagnetically labeled for the separation to be effective; (ii) overall effectiveness of the antibody targeting the CD45 receptor (see Reference [9] for a further discussion); (iii) the potential of platelets to

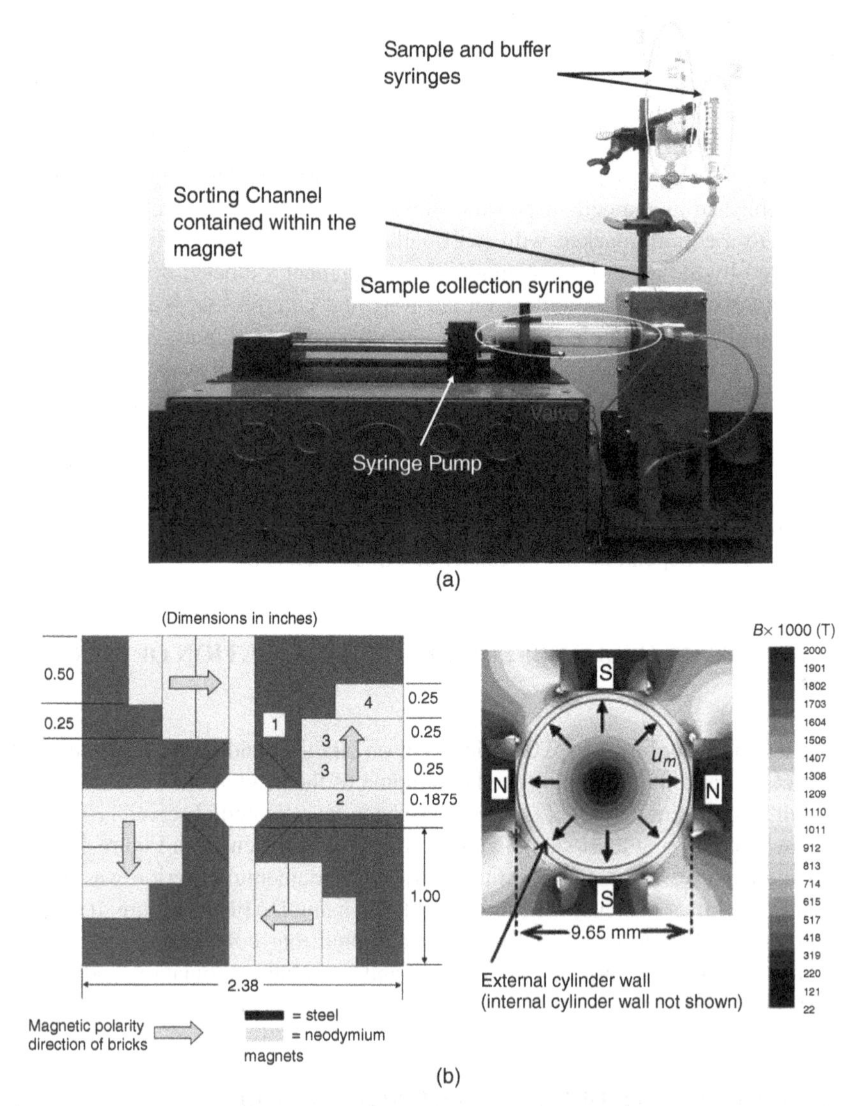

Figure 14.2 (a) A photograph of the QMS system, while (b) presents a diagram of the quadrupole magnet on the left-hand side and a computer-generated field map on the RHS.

"cloak" the CD45 receptor; and (iv) the to-be-proved potential that the CD45 receptor can vary from cell type to cell type.

Figure 14.4 presents flow cytometry analysis (forward scatter, side scatter, and CD45 expression) of representative examples of normal blood (a), and a metastatic breast cancer patient blood (b), prior to and after immunomagnetic depletion using the QMS system presented previously. What is obvious in the dot plots of both blood types is the significantly higher number of presumably granulocytic cells (cells with

Patient	CD45	CK	Vimentin	N-Cadherin	EGFR	Her2	EpCAM	SMO	GLI1	Gamma-H2AX
BC	-	Y	Y	Y	-	-	-	-	-	-
BC	-	Y	Y	-	Y	-	-	-	-	-
BC	-	Y	Y	-	-	Y	-	-	-	-
BC	-	Y	Y	-	-	-	Y	-	-	-
BC	-	Y	-	-	-	-	Y	-	-	Y
BC	-	Y	-	-	-	-	N	-	-	Y
BC		Y	-	-	-	-	-	Y	-	-
BC	N	Y	Y	-	-	-	-	-	-	-
BC	N	Y	-	-	-	-	Y	-	-	-
BC	N	Y	-	-	-	Y	-	-	-	-
BC	N	Y	-	-	-	-	-	-	-	Y
BC	N	N	Y	-	-	-	-	-	-	-
BC	N	N	-	-	-	-	-	-	-	Y
BC	Y	Y	Y	-	-	-	-	-	-	-
BC	Y	Y	-	-	Y	-	-	-	-	-
BC	Y	Y	-	-	-	-	Y	-	-	-
BC	Y	-	Y	-	-	-	Y	-	-	-
BC	Y	Y	-	-	-	-	-	-	-	Y
BC	Y	N	Y	-	-	-	-	-	-	
BC	Y	N	-	-	-	Y	-	-	-	-
BC	Y	N	-	-	-	-	Y	-	-	-
BC	Y	N	-	-	-	-	-	-	-	Y
SCCHN	-	Y	Y	Y	-	-	-	-	-	-
SCCHN	-	Y	Y	-	Y	-	-	-	-	-
SCCHN	-	N	Y	Y		-	-	-	-	-
SCCHN	-	N	Y	-	Y	-	-	-	-	-
SCCHN	-	Y	Y	-		-	-	-	-	-
SCCHN	-	Y	Y	-	-	-	Y	-	-	-
SCCHN	-	Y	Y	-	-	-	N	-	-	-
SCCHN	-	Y	N	-	-	-	Y	-	-	-
SCCHN	Y	Y	Y	-	-	-	-	-	-	-
SCCNH	Y	Y	N	-	-	-	-	-	-	-

Figure 14.3 A matrix of the various combination of staining patterns on peripheral blood cells from breast cancer, BC, and squamous cell carcinoma of the head and neck, SCCHN, patients after an enrichment through depletion of normal cells.

a high level of side scatter) in the metastatic patient samples, both prior to and after magnetic depletion. On both the RHS plots of side scatter versus CD45 expression and the histograms in the middle plots, the average CD45 expression level in the depleted fraction is significantly lower than the predepletion samples. It should also be noted that the overall level of depletion of CD45 positive cells, which was determined by independent nucleated cell counts, was similar in both patient samples.

Granulocytes are a subset of leukocytes found in normal peripheral blood derived from myeloid progenitor cells. These cells are so named because they contain cytoplasmic granules. In addition, the granules found in these cells cause laser light directed at them to be scattered at a right angle. Consequently, viable granulocytes typically have a higher side scatter signal than the agranulocytes.

Neutrophils are the most abundant granulocyte and therefore have the most prominent effect on the quality of leukocyte depletion among the granulocytic cells in the peripheral blood.

One method of improving the depletion of these hematopoietic cells is to increase the number of immunomagnetically bound receptors by targeting markers other than CD45. Flow cytometry analysis of the depleted sample using several known leukocyte

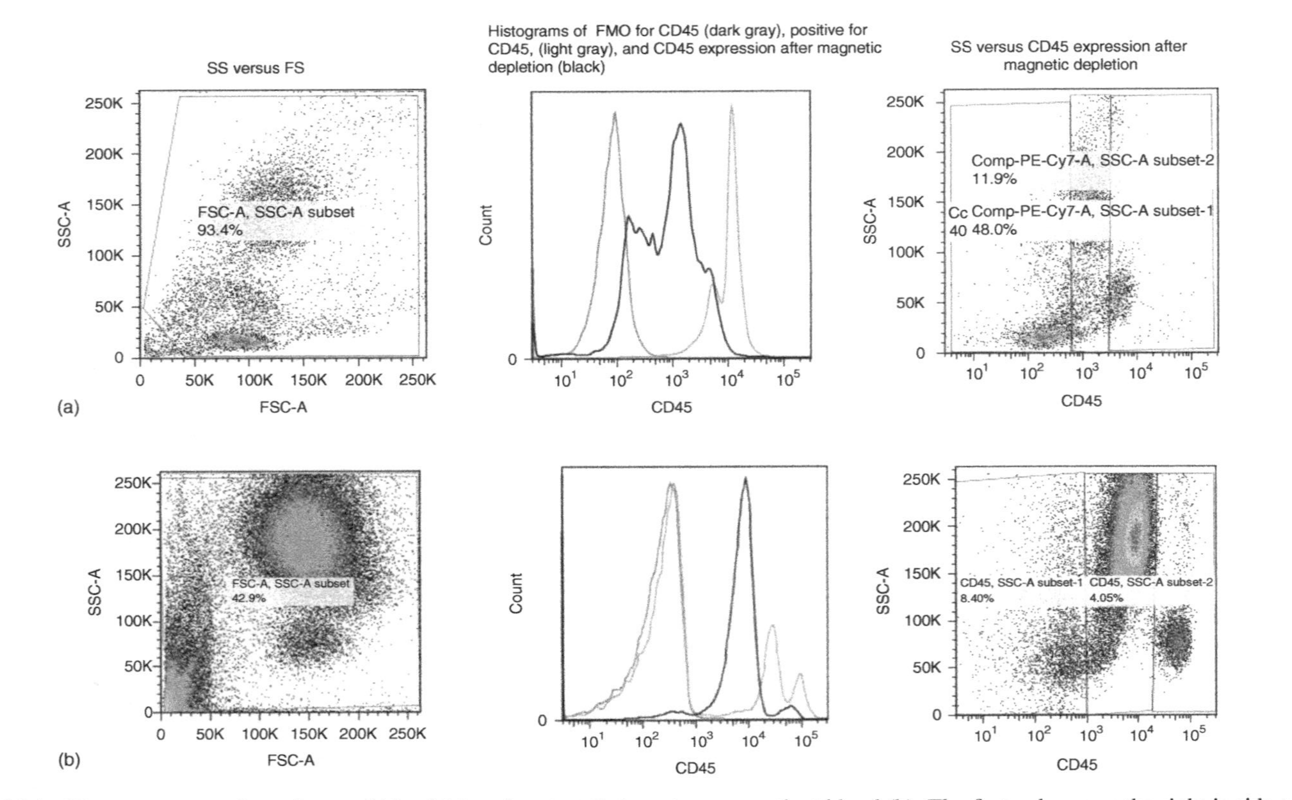

Figure 14.4 Flow cytometry plots of normal blood (a) and metastatic breast cancer patient blood (b). The first column on the right is side scatter, SSC, versus forward scatter, FSC, (i.e., a measure of granularity versus size), the middle plots are histograms of unstained (red), and CD45 stained, prior to, blue, and after magnetic depletion, black. The last plots on the right are side scatter versus CD45 expression of the samples after depletion targeting CD45. The depletion on both samples was on the order of 2.5 $\log_{10}$. (*See color plate section for the color representation of this figure.*)

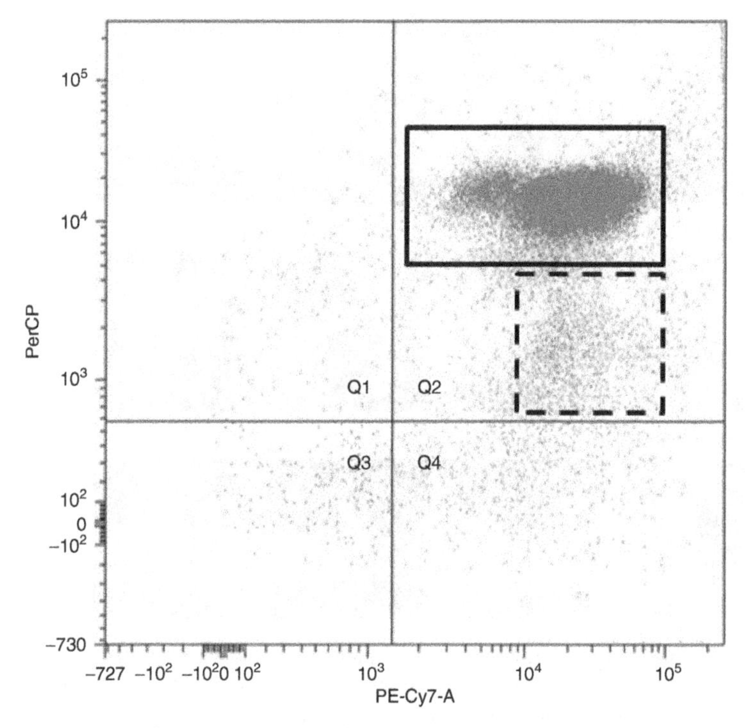

Figure 14.5 Typical flow cytometry data from peripheral blood from a metastatic breast cancer patient after immunomagnetic enrichment. Cells that were not magnetically captured were stained with calcein to distinguish viable cells from dead cells and other events. The cells were also stained with anti-CD45 PE/Cy7 and anti-CD3, CD14, CD16, and CD19 PerCP antibodies. The largest portion of cells after magnetic enrichment are granulocytes, most of which are neutrophils ($CD45^+$, CD16hi, solid box). A smaller fraction of remaining cells are lymphocytes ($CD45^+$, $CD3^+$ or $CD19^+$, dashed box). The ratio of granulocytes to lymphocytes is higher after magnetic enrichment than before magnetic enrichment, suggesting that magnetic enrichment captures more lymphocytes than granulocytes.

antigens including CD3, CD14, CD16, CD19 revealed that the weak and moderate CD45 granulocytes that were not captured by the magnetic column expressed CD16 at a high level (Figure 14.5). CD16 is a receptor that binds to the Fc portion of IgG antibodies. Circulating IgG antibodies coat pathogens such as bacteria, allowing the bacteria to easily bind to immune cells during phagocytosis. Hence, the CD16 receptor is expressed on the surface of phagocytic immune cells: mature neutrophils, natural killer cells, monocytes, and macrophages.

It is expected that immunomagnetically labeling the CD16 receptor will improve the depletion of normal leukocytes in the blood. Immunomagnetic depletion of CD16 positive cells may have the added benefit of removing a population of cells that is CK+CD45+ when visualized with fluorescence microscopy. It is plausible that phagocytic cells in metastatic cancer patients "clean up" apoptotic CTCs as well as

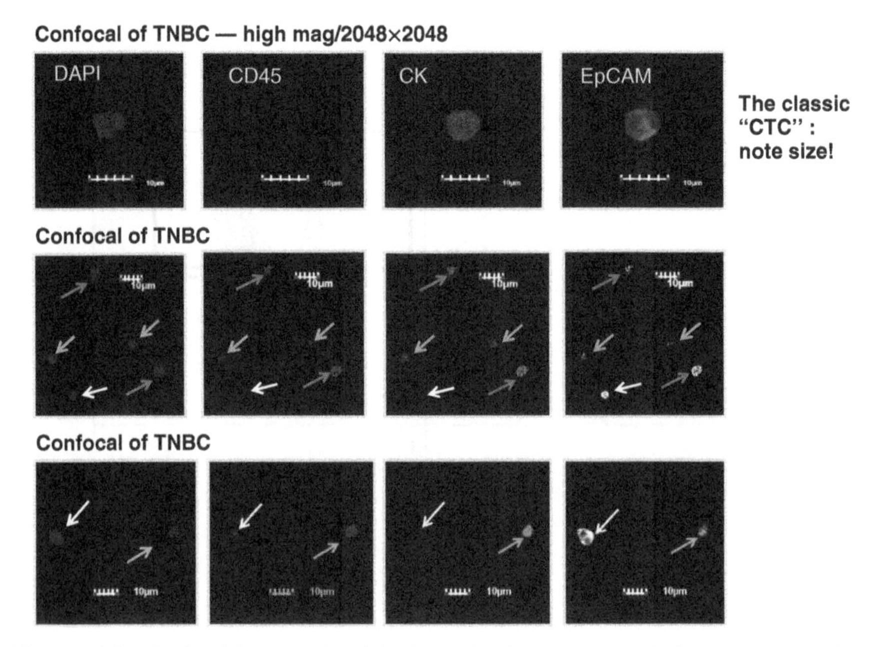

Figure 14.6 Confocal images of enriched samples from a metastatic breast cancer patient. The columns of images, from left to right, correspond to cells stained for DAPI, CD45, cytokeratins, and EpCAM. The color arrows correspond to cells highly positive for all markers, green, weak for CD45, positive of other markers, pink, and cells positive for EpCAM, negative for CK and CD45, white. (*See color plate section for the color representation of this figure.*)

endothelia that have sloughed off leaky vasculature. These cells can be troublesome in phenotypic analysis because they contain cytokeratins, can be relatively large among blood cells, and can be weak for CD45, making them prone to be a "false-positive CTC." Figure 14.6 presents confocal images of traditional CTC as well as cells that are positive and weakly positive cells for CD45 and CK and/or EpCAM. In addition to CD16, other granulocyte markers such as CD66b have been used in other negative selection technology (third-generation iChip and Stem Cell Technologies CTC enrichment cocktail) to improve negative depletion performance.

14.6 CONCLUSION

As it is becoming increasingly clear, CTCs are heterogeneous with respect to specific marker expression. Further, it has been long known that weakly positive CD45 cells, with many of the CTC markers, are also present. In fact, it was recently shown that many of these weakly CD45-positive cells that express CTC markers also express CD68 and macrophage markers in metastatic breast cancer patients, adding further evidence of these cells having an association of the solid tumor [14]. These results, along with others, suggest that negative depletion of normal cells, combined with a systematic analysis of the remaining cells, allows further basic information on the type of cells associate with a patient's cancer. Through this type of systematic analysis, either further surface markers can be added to the depletion step to improve the final purity, or, with the further refinement of specific combination of markers on CTCs, this information can be used to develop more focused positive selection technology for diagnostic studies.

REFERENCES

1. Lara, O., Tong, X., Zborowski, M., and Chalmers, J.J. (2004) Enrichment of rare cancer cells through depletion of normal cells using density and flow-through, immunomagnetic cell separation. *Experimental Hematology*, **32**, 891–904.
2. Bland, K.I. and Copeland, E.M. (2009) *The Breast: Comprehensive Management of Benign and Malignant Diseases*, 4th edn, Saunders Elsevier, Philadelphia.
3. Pauling, L. and Coryell, C.D. (1936) The magnetic properties and structure of hemoglobin, oxyhemoglobin and carbonmonoxyhemoglobin. *Proceedings of the National academy of Sciences of the United States of America*, **22** (4), 210–216.
4. Moore, L.R., Williams, P.S., Nehl, F. *et al.* (2014) Feasibility study of red blood cell debulking by magnetic field-flow fractionation with step-programmed flow. *Analytical and Bioanalytical Chemistry*, **406** (6), 1661–1670.
5. Huang, L.R., Cox, E.C., Austin, R.H., and Sturm, J.C. (2004) Continuous particle separation through deterministic lateral displacement. *Science*, **304** (5673), 987–990.
6. Karabacak, N.M., Spuhler, P.S., Fachin, F. *et al.* (2014) Microfluidic, marker-free isolation of circulating tumor cells from blood samples. *Nature Protocols*, **9**, 694–710.
7. Moore, L.R., Zborowski, M., Sun, L., and Chalmers, J.J. (1998) Lymphocyte fractionation using immunomagnetic colloid and dipole magnet flow cell sorter. *Journal of Biochemical and Biophysical Method*, **37**, 11–33.
8. Williams, P.S., Zborowski, M., and Chalmers, J.J. (1999) Flow rate for the quadrupole magnetic cell sorter. *Analytical Chemistry*, **71**, 3799–3807.
9. Wu, Y., Deighan, C.J., Miller, B.L. *et al.* (2013) Isolation and analysis of rare cells in the blood of cancer patients using a negative depletion methodology. *Methods*, **64**, 169–182.
10. Yang, L., Lang, J.C., Balasubramanian, P. *et al.* (2009) Optimization of an enrichment process for circulating tumor cells from the blood of head and neck cancer patients through depletion of normal cells. *Biotechnology and Bioengineering*, **102** (2), 521–534.
11. Balasubramanian, P., Yang, L., Lang, J.C. *et al.* (2009) Confocal images of circulating tumor cells obtained using a methodology and technology that removes normal cells. *Molecular Pharmaceutics*, **6** (5), 1402–1408.
12. Balasubramanian, P., Lang, J.C., Jatana, K.R. *et al.* (2012) Multiparameter analysis, including EMT markers, on negatively enriched blood samples from patients with squamous cell carcinoma of the head and neck. *PLoS One*, **7** (7), e42048.
13. Garcia-Villa, A., Balasubramanian, P., Miller, B.L. *et al.* (2012) Assessment of γ-H2AX levels in circulating tumor cells from patients receiving chemotherapy. *Frontiers in Oncology*, **2**, 128.
14. Lustberg, M.B., Balasubramanian, P., Miller, B. *et al.* (2014) Heterogeneous atypical cell populations are present in blood of metastatic breast cancer patients. *Breast Cancer Research*, **16** (2), R23.

PART III

POST-ISOLATION ANALYSIS AND CLINICAL TRANSLATION

15

TUMOR HETEROGENEITY AND SINGLE-CELL ANALYSIS OF CTCs

Evelyn K. Sigal and Stefanie S. Jeffrey

Department of Surgery, Stanford University School of Medicine, Stanford, CA, USA

15.1 INTRODUCTION

Cancer cells generally carry oncogenic mutations, alterations in tumor-suppressor genes, chromosomal changes, and/or epigenetic changes that facilitate unchecked cell growth and division and/or evasion of cell death [1]. With genetic instability – including acquired point mutations; defective mismatch, nucleotide, or base excision repair; or chromosome segregation errors – additional genetic variations may be acquired and passed on to progeny [2–4]. Moreover, during the evolution of malignant tumors, cells are exposed to various microenvironmental changes associated with inflammatory and immune responses, increasing degrees of hypoxia within the tissue and space and architectural biophysical constraints involving the basement membrane or physical compartments during ongoing proliferation [5–7]. Exposure of the organism to certain external factors, for example, smoke, or even chemo- or radiotherapeutic treatments, can cause death of some tumor cells, whereas others may survive and continue to proliferate, evolving by subclonal selection [4, 5, 8]. Thus, the tumor macro- and microenvironment play an important role that leads to a selection of cell clones with advantageous phenotypes.

CTCs represent an important link between the primary tumor and the occurrence of distant metastasis; originating from potentially diverse and surviving subclones of the primary tumor, these are cells that can enter the bloodstream and disseminate. Some will have the capability to extravasate (leave the bloodstream and invade

Circulating Tumor Cells: Isolation and Analysis, First Edition. Edited by Z. Hugh Fan.

into other tissues), seed, and grow in distant sites, leading to the development of metastases [9]. To this point, Baccelli *et al.* have shown that a subpopulation of CTCs isolated from primary luminal breast cancer patients have "metastasis-initiating" capabilities, which can be identified by their expression of EpCAM (epithelial cell adhesion molecule), CD44 (a stem cell marker and bone-homing receptor), CD47 (inhibits phagocytosis, also referred to as a "don't eat me signal" to immune cells [10]), and MET (a hepatocyte growth factor receptor, thought to activate migration and invasion in different cancers) [11].

The number of CTCs circulating in a cancer patient's bloodstream has been shown to be of prognostic significance for colorectal cancer (>3 CTCs per 7.5 ml of blood) [12] and for breast and prostate cancers (>5 CTCs per 7.5 ml of blood) when using the FDA-approved CellSearch® device [13–16]. Beyond enumeration, the analysis and characterization of CTCs are crucial to the understanding of tumor biology and the development of therapies that can eliminate the cells capable of promoting metastatic growth and disease recurrence [6]. Genomic and transcriptional analyses may also elucidate the means by which cells survive the shear stress of blood flow and escape immune defense mechanisms [17, 18]. Yet, CTCs circulate in the blood together with approximately 5 billion red blood cells and 5–10 million white blood cells per milliliter. Thus, analyses of CTCs require separation from other blood cells that can confound molecular studies [19]. Multiple separation techniques are discussed in earlier chapters of this book.

An important consideration is that molecular analyses of a group of cells pooled together can mask specific characteristics of a subset of single cells or even single clusters of a few adherent cells that may be capable of spreading, lodging, and growing in distant sites. Therefore, once CTCs have been separated from contaminating blood cells using one of many possible enrichment methods, single-cell analysis of CTCs (or tiny cancer cell clusters) becomes an increasingly important strategy. In this section, we focus on tumor and CTC heterogeneity, existing profiling methods following CTC enrichment, and implications and perspectives for the use of CTCs in clinical cancer care management.

15.2 TUMOR HETEROGENEITY

It has been known for a long time that tumors consist of malignant cells that show a wide range of genetic and phenotypic variation, referred to as tumor heterogeneity [2]. These variations can be found between different tumor entities; between patients with the same tumor type, known as intertumor heterogeneity; and as subclonal diversity within one tumor, known as intratumor heterogeneity [3, 20, 21]. Over a century ago, in the late 1800s, pathologists first described structural intratumor heterogeneity, and later, in the 1900s, this was further demonstrated by the introduction of new staining techniques for chromosomes [22].

In 2012, a landmark paper showed evidence of the widespread intratumor heterogeneity in patients with renal cell carcinoma using next-generation sequencing of multiple core biopsy samples: Gerlinger *et al.* demonstrated that samples from different parts of the same tumor from a single patient showed different genomic profiles. Furthermore, they also showed "ubiquitous mutations" present in all tumor parts, "shared mutations" present in several tumor portions, as well as "private mutations," which were present in only one specific piece of the tumor. They suggested that these private mutations represent ongoing regional clonal evolution and demonstrated that taking a biopsy from a single region of a tumor does not reveal all mutations that may be present at different sites in that tumor or from regional and distant metastases related to that tumor [23, 24]. In a different study that removed clusters of tumor cells with laser capture microdissection, it was shown that breast cancer cell clusters from three closely approximated portions of tumor within a small sample area on a microscope slide were either wild-type or contained mutations in two different exons of the *PIK3CA* (phosphatidylinositol-4,5-biphosphate 3-kinase, catalytic subunit alpha) gene. This reflects microheterogeneity and, again, the subclonal diversity resulting during tumor evolution [25]. Similarly, underlying intratumor heterogeneity has recently been proven to exist in several other cancer types, including colorectal cancer, oropharyngeal cancer, and lung cancer [26–29].

In a recently published and well-designed study, the fact that individual tumor cells are not only phenotypically but also functionally heterogeneous was proved. The authors propagated tumor cells from 10 colorectal cancer patients in cell cultures and transduced them with lentiviral vectors encoding green fluorescent protein. After xenotransplantation of these tumor cells into immunodeficient mice, they analyzed single-cell-derived tumor transplants by DNA copy number alteration profiling and sequencing. They were able to show that the xenografts predominantly maintained the mutational patterns and copy number alterations that were originally found in the cancer patient's tumor. They were also able to define cell populations according to distinct behavioral pattern during serial passaging, which they classified into five types: type I or persistent clones that were present in all serial transplants; type II or short-term clones that did not persist up to the final passage, showing less tumor-propagating ability; type III or transient clones that were only detected in the first passage and then disappeared in the following recipients; type IV or resting clones that were not detectable in the first recipients, but could be detected in later passages, and therefore were thought to represent dormant or slowly proliferating cells; type V or fluctuating clones, which were first detectable, then became undetectable in the following passage, but interestingly reappeared in later passages. Next, the authors analyzed the response of those cell clones to Oxaliplatin, a routinely used drug in colorectal cancer treatment, which revealed that actively proliferating cell clones were killed, while the dormant or resting type IV clones proliferated later causing tumor reinitiation after therapy [30]. This suggests that drug resistance is not solely caused by genetic mutation and emphasizes the importance of single-cell analysis to further understand and identify the cells that can escape the effects of common drug therapies and are capable of driving tumor growth after chemotherapy. Those are the cells that should be specifically targeted.

As clinicians aim for patient-specific treatments, intratumor heterogeneity as well as temporal changes in existing subclones represent major problems and emphasize the need for diagnostic procedures other than relying on a primary tumor biopsy for decision-making concerning targeted therapy. Tumor tissues or metastases are often not easily accessible for biopsies, as they are frequently present in internal organs and may be near nerves, blood vessels, or other structures that potentially could be harmed during serial biopsy. The procedures to obtain a biopsy are usually invasive, so that the risk of resulting complications obviously needs to be taken into account. Because a tissue biopsy only provides a very small sample with a relatively small number of cells, tumor cells could be missed, leading to a false-negative result, or that a definite diagnosis cannot be made. Further, a single biopsy often does not reflect the complexity of the mutations present in a cancer patient, which might lead to a tumor sampling bias. Therefore, taking serial blood biopsies is far more feasible, especially in advanced stages of cancer. CTCs that result from detachment of different parts of the primary and metastatic tumor have been shown to be heterogeneous and represent different subclones existent in a cancer patient [19, 31]. Therefore, taking simple blood draws (liquid biopsies) for isolation and characterization of those rare cells may help guide therapy more closely and efficiently throughout the clinical course of disease [32, 33].

15.3 SINGLE-CELL ANALYSIS OF CTCs AND CTC HETEROGENEITY

Single-cell analysis of CTCs is a challenge because each individual cell carries picogram (pg) quantities of RNA and DNA, requiring highly sensitive analytical methods. Nevertheless, characterization of CTCs offers a great opportunity to better identify the cells capable of surviving dissemination from the primary tumor site, extravasation, and ultimately formation of metastasis. It is hypothesized that CTCs in a cancer patient's blood derive from different tumor cell subpopulations, as well as from different metastatic sites. Thus, it is expected that they more accurately represent existing heterogeneity than could be revealed by a single tissue biopsy.

Originally, cytological methods such as fluorescence *in situ* hybridization (FISH) and other immunohistochemical methods allowed the analysis of genomic rearrangements of single tumor cells [22]. Later, during the 1980s, the emergence of the polymerase chain reaction (PCR) made it possible to amplify and sequence DNA from single cells. Recent advances in next-generation sequencing and single-nucleus sequencing have enabled far better identification and characterization of tumor cells and continue to be improved, leading to increased overall throughput [34]. Single-cell analysis methods can be separated into two large groups:

1. Gene expression analysis, which investigates changes at the RNA level.
2. Mutational analysis (including base substitutions, insertions and deletions – indels – of bases, structural rearrangements, and DNA copy number changes [35]), which helps elucidate changes at the DNA level.

Coupling new CTC enrichment devices with novel single-cell sequencing approaches will likely provide better insight into distinct characteristics of CTCs and also tumor biology.

15.4 GENE EXPRESSION ANALYSIS

As RNA degradation can occur rather quickly, RNA is more difficult to preserve than DNA [36]. Therefore, after isolation and identification of CTCs, it is preferable that these cells be preserved live without fixation or staining procedures. Subsequently, because a single cell only contains picogram quantities of RNA, the RNA may be first converted to complementary DNA (cDNA) or cDNA preamplified for specific targets and then amplified [37]. Following that, quantitative real-time polymerase chain reaction (qRT-PCR) or high-throughput microfluidic qRT-PCR or deep sequencing technologies can be utilized [38, 39].

We were one of the first groups to perform high-dimensional single-cell gene expression analysis rather than analyzing pooled cells. For isolation of live CTCs from whole blood, we used the MagSweeper, an immunomagnetic enrichment device that isolates highly purified, live tumor cells with high specificity. Briefly, whole blood was labeled with magnetic beads coated with anti-EpCAM antibodies (although any cell surface antigen could be targeted). Subsequently, the blood sample was processed on the MagSweeper, which consists of magnetic rods sheathed in a plastic selected for its lack of nonspecific cell binding. The magnetic rods sweep through the whole-blood sample in a preset pattern and with a specific velocity so that labeled, EpCAM-expressing CTCs attach to the magnets. The movements of the magnetic rods produce a controlled shear force, with release of nonspecifically bound blood cells. The device goes through two rounds of capture, wash, and release to optimize purity and minimize leukocyte contamination [40]. In our analysis, following enrichment, single CTCs or other cancer cells were individually picked by micromanipulation. After targeted preamplification of each cell to cDNA, gene expression analysis was performed using a microfluidic technique for real-time quantitative PCR gene expression analysis (BioMark Real-Time PCR system, Fluidigm Corporation), which enabled the simultaneous measurement of up to 96 genes in a single cell. Only cells that (i) expressed three reference genes (*ACTB*, *GAPDH*, and *UBB*); (ii) showed expression of one of the epithelial markers *KRT 7*, *8*, *18* and/or *19*; and (iii) showed no expression of the leukocyte marker *CD45* were identified as CTCs and further analyzed. We were able to characterize single CTCs from patients with breast cancer and showed that CTCs are highly heterogeneous, even those isolated out of the same blood draw of one patient [19].

Additionally, we were able to demonstrate that most of the CTCs we isolated showed a triple-negative phenotype (i.e., they were lacking expression of estrogen receptor (ER), progesterone receptor (PR), or human epidermal growth factor receptor 2 (HER2)), which did not always correspond to the phenotype of the primary tumor. This could explain why therapies that target ER, PR, or HER2 may not be effective in some patients. Patient CTCs showed strong expression of genes associated with metastasis, such as *NPTN*, *S100A4*, and *S100A9*. Also, genes associated with epithelial-to-mesenchymal transition (EMT), which is thought to be an important step that CTCs undergo prior to extravasation and that increases cell invasiveness and the development of metastases, such as *VIM*, *TGFβ1*, *ZEB2*, *FOXC1*, and *CXCR4*, showed elevated expression. Further, by also analyzing single breast cancer cells from cell lines or primary cultures, we were able to show that breast cancer cell lines commonly used for drug testing were not reflective of the gene expression profiles of the CTCs nor the heterogeneity that exists in patient blood draw, bringing into question their reliability for drug testing experiments [19]. Interestingly, another study that isolated CTCs from breast cancer patients using the Herringbone-Chip, a microfluidic device coated with antibodies, confirmed that the majority of the enriched CTCs expressed EMT-genes such as *TGFβ* and *FOXC1* [41].

Using the MagSweeper, our group was also able to isolate CTCs in a metastatic tumor xenograft model: for induction of breast cancer, MDA-MB-231 cells were injected into the second mammary fat pads of NOD/SCID (nonobese diabetic-severe combined immunodeficiency) mice, and 55 days postinjection, the animals were euthanized, the blood was collected, and the CTCs were isolated. We were able to grow the CTCs in culture. Further, we demonstrated that hypoxic CTCs showed a different phenotype than the cells from which they were derived. They showed higher expression of anoxia-induced factors ATF3, ATF4, and its target gene *ASNS* (ATF4-mediated asparagine synthetase), which are associated with survival, cell proliferation, and aggression. The CTCs grew more aggressive *in vitro* than their parental cell line, which was revealed by higher clonogenicity. Reimplantation of the cultured CTCs into mice resulted in markedly pronounced tumorigenicity and metastatic potential *in vivo* [42]. This suggests that CTCs that derive from rapidly proliferating, hypoxic tumor areas, might comprise the populations that separate from the tumor, extravasate, and eventually settle in other organ sites to form metastases.

Recent technical advances have made it possible to perform CTC single-cell profiling by RNA sequencing. A new study by Lohr *et al.* introduced a method to perform RNA sequencing on single CTCs from prostate cancer patients. After CTC enrichment using the MagSweeper, CTCs were isolated using the CellCelector, a fully automated cell-picking device. The picked single cells were automatically deposited into arrays of subnanoliter wells. The cells were then imaged, and EpCAM-positive/CD45-negative CTCs were collected for whole-genome amplification (WGA). The CTCs were subsequently subjected to low-pass whole-genome sequencing. This was done to evaluate the level of amplification bias, which facilitated the exact identification of well-covered targets in whole-exome sequencing (i.e., sequencing of only the part of the genome formed by exons, which remains in the mature messenger RNA after removal of the introns). Whole-exome sequencing then enabled the identification and accurate assessment of somatic single-nucleotide variants (SSNVs). When the CTCs collected out of the blood of one prostate cancer patient were sequenced and compared to the detected SSNVs in one of his metastases, they found that CTCs carried mutations that were found both in the primary tumor and in the metastasis but that they clustered markedly more closely to the metastasis rather than the primary [43]. In another study, a single-cell mRNA-sequencing protocol with low technical variation and high reproducibility was developed. CTCs isolated from a patient suffering from recurrent metastatic melanoma using the MagSweeper were analyzed and compared to primary melanocytes, melanoma cell lines, and white blood cells. They found that the gene expression profiles from the melanoma CTCs were similar to the ones found in melanoma cell lines and primary melanocytes but clearly different from the ones of white blood cells [44].

In another study, CTCs were isolated from the blood of metastatic colorectal cancer patients using immunomagnetic enrichment, and after RNA extraction and WGA, gene expression analysis was performed. This characterization showed that those CTCs express genes that provide them with adhesive, migratory, invasive, and cell survival capabilities [45]. Further, using the same methods, this group developed a panel of six genes for the accurate detection of CTCs in stage IV colorectal cancer patients, meaning patients with distant metastases, including GAPDH (as housekeeping gene), VIL1 (as specific marker for intestinal origin of CTCs), TIMP1, and CLU (as specific colorectal cancer genes) as well as LOXL3 and ZEB2 (as EMT markers, which are important during metastatic dissemination [46]). Other studies have characterized CTCs isolated from prostate cancer, using mRNA sequencing or a microfluidics-based quantitative real-time PCR array [47, 48].

These studies support the assumption that analyzing CTCs can help understand the underlying biological mechanisms by which CTCs survive, extravasate, and escape the immune system. Also, taking blood ("liquid biopsies") to isolate and analyze CTCs potentially is a better alternative to taking multiple and sequential tissue biopsies, as the mutations of both the primary cancer and the metastases may be revealed.

15.5 MUTATIONAL ANALYSIS

WGA has existed since around the 2000s, after the discovery of two DNA polymerases that showed high processivity for DNA synthesis [34]. Subsequently, the combination of WGA and array-comparative genomic hybridization (aCGH) has been used in several studies to analyze copy number variations of single CTCs [49, 50].

Another possibility for mutational analysis is next-generation sequencing, which enables the calculation of absolute copy numbers from single cells [34]. It has to be taken into account that DNA sequencing of only protein-coding genes, the exons of the genome, covers only ~1% of the human genome [35]. Therefore, whole-genome sequencing provides far more information but involves the major drawback of being very costly [34].

In a recent study, CTCs that were isolated using the CellSearch system (currently, the only FDA-approved CTC enrichment device) from patients with metastasized colorectal cancer were subjected to WGA, after which massive parallel sequencing of a panel of 68 genes associated with colorectal cancer was performed. Additionally, aCGH was used to analyze copy number changes of the CTCs. This analysis was done not only on CTCs but also on the corresponding primary tumors and liver metastasis. They found that certain mutations in driver genes (genes that are thought to be responsible for tumor progression) for colorectal cancer (e.g., *APC*, *KRAS*) found in the primary tumor and the metastasis could also be identified in the CTCs but that the CTCs additionally revealed the existence of differing mutations. Additional deep sequencing analysis of the primary and liver metastasis was performed to determine whether the mutations found uniquely in CTCs originated from a subclone in the primary tumor or represented new mutations responsible for metastases. This led to the discovery that the majority of the mutations that were at first thought to be exclusively present in CTCs could also be detected in the primary and metastatic tumor tissue at subclonal levels [49].

To allow the accurate identification of structural copy number changes in single CTCs, another group enhanced the aCGH method to allow the detection of signals as low as 0.1 Mb. This enabled the analysis of chromosomal changes throughout the course of disease and therapy and especially may help in identifying the development of genomic alterations leading to chemotherapeutic resistance of CTCs [51].

In another study on single CTCs from colorectal cancer patients isolated using the CellSearch system, mutational analysis of the genes *KRAS*, *BRAF*, and *PIK3CA* revealed heterogeneity of CTCs from the same patient. This might explain why the therapeutic response to EGFR inhibitors, which are often administered to patients with *KRAS* wild-type (nonmutated) tumors, is variable [52].

These studies emphasize the importance of analyzing CTCs and implementing them as biomarkers for cancer patients, as they can give an insight into the heterogeneous population existent in the primary tumors as well as mutations present in metastases. Further, they can pave the way to the personalized, targeted therapy we aim for, in order to kill the cells responsible for disease progression with the right therapeutic agent, rather than diminishing only certain populations, leaving other subclones to proliferate and lead to relapse and/or further development of metastases.

Another report also analyzed CTCs from colorectal cancer patients isolated using the CellSearch device. They performed comprehensive genomic analysis of the isolated CTCs and the corresponding primary tumor using array CGH, mutational profiling, as well as analysis of microsatellite instability (MSI) and also gene expression analysis by qRT-PCR. Similarly to the previously mentioned study, they confirmed that mutation and MSI analysis of the CTCs showed some similarities but also some differences compared to the primary tumor. A very interesting finding by transcriptomic (gene expression) analysis was that CTCs circulate in a dormant state, as most tested genes were downregulated. Further, CTCs seem to possess immune-escape mechanisms by overexpressing CD47, which is a protein that inhibits the cytotoxic activity of activated immune cells [10, 50]. This explains how CTCs can survive the harsh environment in the blood stream. The finding is also in line with another study mentioned before, which showed that CTCs expressing CD47, among other cell-surface markers (EpCAM, CD44, and MET), are not only capable of escaping the immune system but also of forming metastases [11].

A study on CTCs from lung cancer patients examining copy number and single-nucleotide variations using WGA followed by next-generation sequencing demonstrated that apart from being heterogeneous, several tumor-related genes in the patients CTCs, which are also involved in drug resistance (e.g. EGFR), were frequently mutated not only in the CTCs but also in the primary tumor and the metastasis. Interestingly, they also found that different patients with adenocarcinomas had similar copy number variation patterns, which were markedly different than those observed in CTCs isolated from patients with small-cell lung cancer [53]. They suggest that copy number variations at specific gene loci in CTCs could possibly change gene expression patterns of different pathways and thus provide those cells with an advantage for metastasis.

15.6 CONCLUSION: CLINICAL IMPLICATIONS AND FUTURE PERSPECTIVES

A tumor tissue biopsy has many drawbacks, being an invasive method that does not reveal the genomic landscape of the tumor in a patient. It only gives a snapshot of the existing mutation and gene expression profile. It is known that tumors evolve over time, and hence, the genomic makeup constantly changes. Therefore, taking a "liquid biopsy" – a simple blood draw – at various different time points throughout the clinical course of disease is far more feasible and has great potential to help guide personalized therapy for every single patient: by isolation, enumeration, and characterization of CTCs before and after surgery, as well as before, during, and after chemotherapy, the response to therapy can potentially be monitored and clonal changes in the genetic or mutational profile can be detected earlier. This would enable clinicians to adjust the therapeutic regimen, eventually combine other drugs to the therapy, which also may lower the required dose and thus adverse effects, or change the drug therapy completely according to the detected mutations.

Even though cancer genome sequencing is still a rather new field, its rapid development has already influenced the clinical care of cancer patients. With the knowledge of driver mutations that lead to the development of cancer, new small-molecule inhibitors are being developed. For example, EGFR kinase inhibitors are used to treat cancer types with *EGFR* mutations (e.g., in lung cancer patients), and BRAF inhibitors are implemented in the treatment of cancers with *BRAF* mutations (e.g., in melanoma patients) [54].

Apart from helping therapeutic guidance, CTCs also have the potential to prove being a much better and more accurate biomarker than the conventional ones (e.g., CEA for colorectal cancer). Further, by continued analysis of CTCs from patients suffering from different tumor entities, we will learn more and more about the mechanisms of metastasis development, which also may lead to the development of new therapeutic options.

Nevertheless, before the analysis of CTCs can be used routinely in the clinic, multiplexing technologies that enable rapid profiling of a large number of single cells at the same time and with a reasonable price need to be developed [34]. Further, both the isolation and the detection and analysis methods of CTCs need to be standardized before they can be adapted for widespread clinical use.

REFERENCES

1. Marusyk, A., Almendro, V., and Polyak, K. (2012) Intra-tumour heterogeneity: a looking glass for cancer? *Nature Reviews Cancer*, **12**, 323–334.
2. Nowell, P.C. (1976) The clonal evolution of tumor cell populations. *Science*, **194**, 23–28.
3. Burrell, R.A., McGranahan, N., Bartek, J., and Swanton, C. (2013) The causes and consequences of genetic heterogeneity in cancer evolution. *Nature*, **501**, 338–345.
4. Burrell, R.A. and Swanton, C. (2014) The evolution of the unstable cancer genome. *Current Opinion in Genetics and Development*, **24**, 61–67.
5. Greaves, M. and Maley, C.C. (2012) Clonal evolution in cancer. *Nature*, **481**, 306–313.
6. Kidess, E. and Jeffrey, S.S. (2013) Circulating tumor cells versus tumor-derived cell-free DNA: rivals or partners in cancer care in the era of single-cell analysis? *Genome Medicine*, **5**, 70.
7. Gupta, G.P. and Massague, J. (2006) Cancer metastasis: building a framework. *Cell*, **127**, 679–695.
8. Jamal-Hanjani, M., Thanopoulou, E., Peggs, K.S. *et al.* (2013) Tumour heterogeneity and immune-modulation. *Current Opinion in Pharmacology*, **13**, 497–503.
9. Bednarz-Knoll, N., Alix-Panabieres, C., and Pantel, K. (2012) Plasticity of disseminating cancer cells in patients with epithelial malignancies. *Cancer and Metastasis Reviews*, **31**, 673–687.
10. Willingham, S.B., Volkmer, J.P., Gentles, A.J. *et al.* (2012) The CD47-signal regulatory protein alpha (SIRPa) interaction is a therapeutic target for human solid tumors. *Proceedings of National Academy of Sciences of the United States of America*, **109**, 6662–6667.
11. Baccelli, I., Schneeweiss, A., Riethdorf, S. *et al.* (2013) Identification of a population of blood circulating tumor cells from breast cancer patients that initiates metastasis in a xenograft assay. *Nature Biotechnology*, **31**, 539–544.
12. Cohen, S.J., Punt, C.J., Iannotti, N. *et al.* (2008) Relationship of circulating tumor cells to tumor response, progression-free survival, and overall survival in patients with metastatic colorectal cancer. *Journal of Clinical Oncology*, **26**, 3213–3221.
13. Cristofanilli, M., Budd, G.T., Ellis, M.J. *et al.* (2004) Circulating tumor cells, disease progression, and survival in metastatic breast cancer. *New England Journal of Medicine*, **351**, 781–791.
14. Giuliano, M., Giordano, A., Jackson, S. *et al.* (2011) Circulating tumor cells as prognostic and predictive markers in metastatic breast cancer patients receiving first-line systemic treatment. *Breast Cancer Research*, **13**, R67.
15. Giuliano, M., Giordano, A., Jackson, S. *et al.* (2014) Circulating tumor cells as early predictors of metastatic spread in breast cancer patients with limited metastatic dissemination. *Breast Cancer Research*, **16**, 440.

16. de Bono, J.S., Scher, H.I., Montgomery, R.B. *et al.* (2008) Circulating tumor cells predict survival benefit from treatment in metastatic castration-resistant prostate cancer. *Clinical Cancer Research*, **14**, 6302–6309.
17. Li, J. and King, M.R. (2012) Adhesion receptors as therapeutic targets for circulating tumor cells. *Frontiers in Oncology*, **2**, 79.
18. Smith, H.A. and Kang, Y. (2013) The metastasis-promoting roles of tumor-associated immune cells. *Journal of Molecular Medicine (Berlin)*, **91**, 411–429.
19. Powell, A.A., Talasaz, A.H., Zhang, H. *et al.* (2012) Single cell profiling of circulating tumor cells: transcriptional heterogeneity and diversity from breast cancer cell lines. *PloS One*, **7**, e33788.
20. Allison, K.H. and Sledge, G.W. Jr. (2014) Heterogeneity and cancer. *Oncology (Williston Park)*, **28**, 772–778.
21. Murugaesu, N., Chew, S.K., and Swanton, C. (2013) Adapting clinical paradigms to the challenges of cancer clonal evolution. *American Journal of Pathology*, **182**, 1962–1971.
22. Navin, N.E. (2014) Tumor evolution in response to chemotherapy: phenotype versus genotype. *Cell Reports*, **6**, 417–419.
23. Gerlinger, M., Rowan, A.J., Horswell, S. *et al.* (2012) Intratumor heterogeneity and branched evolution revealed by multiregion sequencing. *New England Journal of Medicine*, **366**, 883–892.
24. Gerlinger, M., Horswell, S., Larkin, J. *et al.* (2014) Genomic architecture and evolution of clear cell renal cell carcinomas defined by multiregion sequencing. *Nature Genetics*, **46**, 225–233.
25. Dupont Jensen, J., Laenkholm, A.V., Knoop, A. *et al.* (2011) PIK3CA mutations may be discordant between primary and corresponding metastatic disease in breast cancer. *Clinical Cancer Research*, **17**, 667–677.
26. Kosmidou, V., Oikonomou, E., Vlassi, M. *et al.* (2014) Tumor heterogeneity revealed by KRAS, BRAF, and PIK3CA pyrosequencing: KRAS and PIK3CA intratumor mutation profile differences and their therapeutic implications. *Human Mutation*, **35**, 329–340.
27. Zhang, X.C., Xu, C., Mitchell, R.M. *et al.* (2013) Tumor evolution and intratumor heterogeneity of an oropharyngeal squamous cell carcinoma revealed by whole-genome sequencing. *Neoplasia*, **15**, 1371–1378.
28. Tomonaga, N., Nakamura, Y., Yamaguchi, H. *et al.* (2013) Analysis of intratumor heterogeneity of EGFR mutations in mixed type lung adenocarcinoma. *Clinical Lung Cancer*, **14**, 521–526.
29. Jakobsen, J.N. and Sorensen, J.B. (2012) Intratumor heterogeneity and chemotherapy-induced changes in EGFR status in non-small cell lung cancer. *Cancer Chemotherapy and Pharmacology*, **69**, 289–299.
30. Kreso, A., O'Brien, C.A., van Galen, P. *et al.* (2013) Variable clonal repopulation dynamics influence chemotherapy response in colorectal cancer. *Science*, **339**, 543–548.
31. Deng, G., Krishnakumar, S., Powell, A.A. *et al.* (2014) Single cell mutational analysis of PIK3CA in circulating tumor cells and metastases in breast cancer reveals heterogeneity, discordance, and mutation persistence in cultured disseminated tumor cells from bone marrow. *BMC Cancer*, **14**, 456.
32. Kin, C., Kidess, E., Poultsides, G.A. *et al.* (2013) Colorectal cancer diagnostics: biomarkers, cell-free DNA, circulating tumor cells and defining heterogeneous populations by single-cell analysis. *Expert Review of Molecular Diagnostics*, **13**, 581–599.

33. Pantel, K. and Alix-Panabieres, C. (2013) Real-time liquid biopsy in cancer patients: fact or fiction? *Cancer Research*, **73**, 6384–6388.
34. Navin, N. and Hicks, J. (2011) Future medical applications of single-cell sequencing in cancer. *Genome Medicine*, **3**, 31.
35. Stratton, M.R. (2011) Exploring the genomes of cancer cells: progress and promise. *Science*, **331**, 1553–1558.
36. Krebs, M.G., Metcalf, R.L., Carter, L. *et al.* (2014) Molecular analysis of circulating tumour cells-biology and biomarkers. *Nature Reviews. Clinical Oncology*, **11**, 129–144.
37. Klein, C.A., Seidl, S., Petat-Dutter, K. *et al.* (2002) Combined transcriptome and genome analysis of single micrometastatic cells. *Nature Biotechnology*, **20**, 387–392.
38. Nolan, T., Hands, R.E., and Bustin, S.A. (2006) Quantification of mRNA using real-time RT-PCR. *Nature Protocols*, **1**, 1559–1582.
39. Rodriguez-Gonzalez, F.G., Mustafa, D.A., Mostert, B., and Sieuwerts, A.M. (2013) The challenge of gene expression profiling in heterogeneous clinical samples. *Methods*, **59**, 47–58.
40. Talasaz, A.H., Powell, A.A., Huber, D.E. *et al.* (2009) Isolating highly enriched populations of circulating epithelial cells and other rare cells from blood using a magnetic sweeper device. *Proceedings of the National Academy of Sciences of the United States of America*, **106**, 3970–3975.
41. Yu, M., Bardia, A., Wittner, B.S. *et al.* (2013) Circulating breast tumor cells exhibit dynamic changes in epithelial and mesenchymal composition. *Science*, **339**, 580–584.
42. Ameri, K., Luong, R., Zhang, H. *et al.* (2010) Circulating tumour cells demonstrate an altered response to hypoxia and an aggressive phenotype. *British Journal of Cancer*, **102**, 561–569.
43. Lohr, J.G., Adalsteinsson, V.A., Cibulskis, K. *et al.* (2014) Whole-exome sequencing of circulating tumor cells provides a window into metastatic prostate cancer. *Nature Biotechnology*, **32**, 479–484.
44. Ramskold, D., Luo, S., Wang, Y.C. *et al.* (2012) Full-length mRNA-Seq from single-cell levels of RNA and individual circulating tumor cells. *Nature Biotechnology*, **30**, 777–782.
45. Barbazan, J., Alonso-Alconada, L., Muinelo-Romay, L. *et al.* (2012) Molecular characterization of circulating tumor cells in human metastatic colorectal cancer. *PloS One*, **7**, e40476.
46. Barbazan, J., Muinelo-Romay, L., Vieito, M. *et al.* (2014) A multimarker panel for circulating tumor cells detection predicts patient outcome and therapy response in metastatic colorectal cancer. *International Journal of Cancer*, **135**, 2633–2643.
47. Cann, G.M., Gulzar, Z.G., Cooper, S. *et al.* (2012) mRNA-Seq of single prostate cancer circulating tumor cells reveals recapitulation of gene expression and pathways found in prostate cancer. *PloS One*, **7**, e49144.
48. Chen, C.L., Mahalingam, D., Osmulski, P. *et al.* (2013) Single-cell analysis of circulating tumor cells identifies cumulative expression patterns of EMT-related genes in metastatic prostate cancer. *Prostate*, **73**, 813–826.
49. Heitzer, E., Auer, M., Gasch, C. *et al.* (2013) Complex tumor genomes inferred from single circulating tumor cells by array-CGH and next-generation sequencing. *Cancer Research*, **73**, 2965–2975.
50. Steinert, G., Scholch, S., Niemietz, T. *et al.* (2014) Immune escape and survival mechanisms in circulating tumor cells of colorectal cancer. *Cancer Research*, **74**, 1694–1704.

51. Czyz, Z.T., Hoffmann, M., Schlimok, G. *et al.* (2014) Reliable single cell array CGH for clinical samples. *PloS One*, **9**, e85907.
52. Gasch, C., Bauernhofer, T., Pichler, M. *et al.* (2013) Heterogeneity of epidermal growth factor receptor status and mutations of KRAS/PIK3CA in circulating tumor cells of patients with colorectal cancer. *Clinical Chemistry*, **59**, 252–260.
53. Ni, X., Zhuo, M., Su, Z. *et al.* (2013) Reproducible copy number variation patterns among single circulating tumor cells of lung cancer patients. *Proceedings of the National Academy of Sciences of the United States of America*, **110**, 21083–21088.
54. Vogelstein, B., Papadopoulos, N., Velculescu, V.E. *et al.* (2013) Cancer genome landscapes. *Science*, **339**, 1546–1558.

16

SINGLE-CELL MOLECULAR PROFILES AND BIOPHYSICAL ASSESSMENT OF CIRCULATING TUMOR CELLS

Devalingam Mahalingam

Department of Medicine, University of Texas Health Science Center, San Antonio, TX, USA

Pawel Osmulski and Chiou-Miin Wang

Molecular Medicine, University of Texas Health Science Center at San Antonio, San Antonio, TX, USA

Aaron M. Horning

Integrated Biomedical Science Graduate Program, University of Texas Health Science Center, San Antonio, TX, USA

Anna D. Louie

School of Medicine, University of Nevada, Reno, NV, USA

Chun-Lin Lin, Maria E. Gaczynska, and Chun-Liang Chen

Molecular Medicine, University of Texas Health Science Center at San Antonio, San Antonio, TX, USA

16.1 INTRODUCTION

Circulating tumor cells (CTCs) are rare cells shed from primary tumors into the blood stream. They subsequently colonize in different organs to form the secondary tumors during metastasis (Figure 16.1) [1]. It has been estimated that CTCs are as rare as

Circulating Tumor Cells: Isolation and Analysis, First Edition. Edited by Z. Hugh Fan.

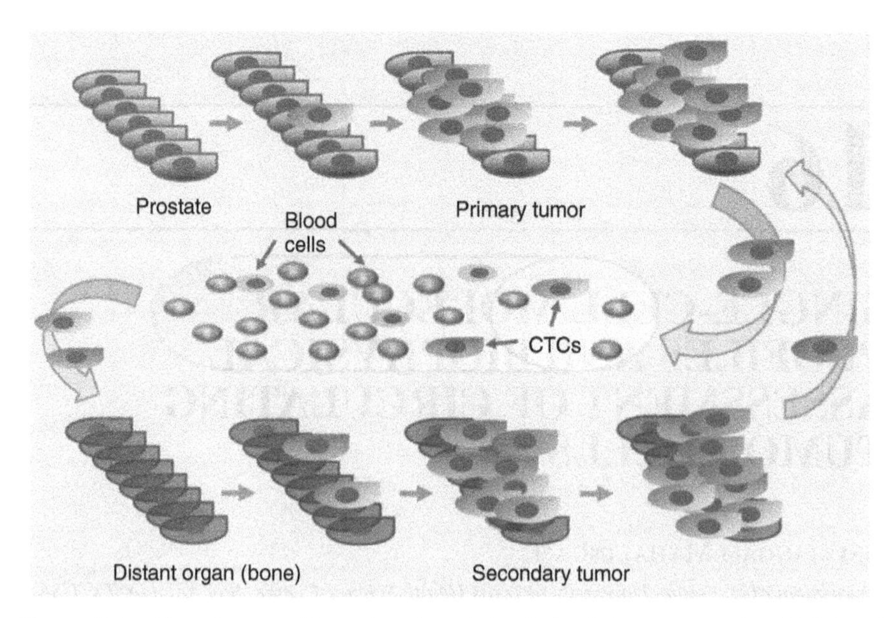

Figure 16.1 The schematic illustrating the origin of circulating tumor cells (CTCs) and metastasis. Primary tumors shed CTCs into the blood stream and subsequently colonize in the distant organs form secondary tumors and metastasis. The metastatic CTCs may home in to the primary tumors. (*See color plate section for the color representation of this figure.*)

1 in 10 billion blood cells. Enumeration of CTCs using immunomagnetic affinity has shown that a higher load of CTCs in blood associates with a poor prognosis and lower overall survival in several metastatic cancers [2–7]. Consequently, CTCs have emerged as an important "liquid biopsy" biomarker for detection and prognosis of metastatic cancers [6, 8]. Our current understanding of CTC properties and functions is still limited due to long-standing difficulties in their identification and isolation. Fortunately, some of these problems are being solved by the newest nanoscale and microfluidic innovations, including methods with promising clinical potential [1, 9]. It is anticipated that the recent advance in CTC studies will shed new light on the metastasis mechanism and lead to the discovery of clinical biomarkers and new therapeutic approaches for metastatic cancers.

The rare CTCs are very attractive subjects for one of the hottest experimental approaches in recent years: the single-cell studies. Single-cell sequencing was declared the Method of the Year 2013 by the Editorial Board of Nature Methods, predicting significant impact of the method for years to come [10]. As genetic, biochemical, and biophysical single-cell analyses reveal, mammalian cells are more heterogeneous than they appeared in bulk studies [11–23] and distinct subpopulations exist in seemingly homogeneous groups of cells, such as tumors. Not surprisingly, molecular profiles and biophysical features of single tumor cells are drawing much attention because the heterogeneity holds critical clues for understanding tumor physiology and cancer progression [6, 24–27]. The heterogeneity of cancer cells is increased in genomic and epigenetic modifications that result in phenotypic shifts in transcriptional diversities and in functional changes as cancer undergoes progress and clonal expansion [11, 14, 28–36]. Although stochastic outburst noise may occur at the single cells, high-throughput qRT-PCR (quantitative reverse transcription polymerase chain reaction) and transcriptome studies have overcome technical hurdles and allowed us to appreciate the evolution of cancer and generate novel molecular signatures and fine-tune tumor classifications [37–39]. Meanwhile, more evidence shows that cancer cells are also subject to cytoskeletal remodeling, alterations in intracellular transport, and redistribution of cell membrane adhesion molecules. These changes manifest in distinct mechanical and in general biophysical phenotype of cancer cells [40, 41]. Structure–function studies point out that the phenotypic changes in cell elasticity, deformability, or adhesion and the acquisition of cancerous activities, replication, and metastatic motility are closely related. Therefore, modifications of cytoarchitectural properties and the resulting changes in single-cell biophysical phenotype are worth attention as new potential biomarkers for cancer detection, stratification, and prognosis [40, 42]. In light of previous studies, gene transcript quantitation [43] and biophysical properties of CTCs should help explain the mechanism of CTC formation and provide potential biomarkers for detection of metastasis and risk stratification [6, 9, 40, 44–48]. Single-cell studies of CTCs are still limited and at an exploratory stage. Therefore, in this chapter, we focus on two major single-cell technology breakthroughs: molecular profiling and biophysical assessment of tumors cells and their potential applications in the translational study of CTCs. An alternative single-cell analysis approach using dieletrophoresis (DEP) is presented in Chapter 18.

16.2 METHODS

16.2.1 Single-Cell Molecular Profiling

Single-cell molecular profiling is powerful for identifying transcription factors, oncogene and tumor-suppressor gene signatures in rare malignant stem cells, and tumor initiation cells [25, 49, 50]. The upregulation and downregulation of master regulators and pathways are associated with cancer pathological status and have proved useful for cancer diagnosis and prognosis [51]. The transcriptional patterns of tumor cells shaped by the regulation of the cell–cell and cell–environment interactions are heterogeneous [52–55]. However, the extent of heterogeneity in gene expression has not been fully appreciated because it has been masked in the traditional gene expression profiling with bulk cells [56–59]. Particularly, the transcriptional signals of rare cells with clinical significance, such as cancer stem cells or initiation cells, become obscured by the majority background and may only be identifiable when single-cell approaches are applied. Several technical challenges, particularly the small amount of starting material of a single cell (~10–20 pg RNA), had limited the feasibility of single-cell molecular profiling. However, next we describe some recent technical advances improving the efficiency and reliability of single-cell molecular profiling.

16.2.1.1 High-Throughput Single-Cell qRT-PCR using Microfluidic BioMark^{TM} HD Semiquantitative RT-PCR has several advantages that can accommodate a wide range of target concentrations with high reproducibility and sensitivity [60–62]. Traditional qRT-PCR could investigate the expression of one gene or a few genes (multiplex) in a cell at a time. However, the advanced BioMark^{TM} HD can perform high-throughput single-cell real-time qRT-PCR simultaneously and analyze the expression of 48 genes in 48 cells or 98 genes in 98 cells based on single-cell RNA in reaction volumes as small as a few nanoliters [27, 63]. The target sequences are preamplified in low cycles from limited RNA templates (mRNA or miRNA) using multiplexed primers and followed by real-time qPCR. If one-third of single-cell lyses is used for preamplification, the maximal coverage can extend to ~294 genes (three 96 × 96 chips) and easily accommodate the integrative study of a major pathway or several small pathways of interest in one cell at a lower cost. The following qPCR can be performed in either Taqman assay or EvaGreen Green based PCR. The technique involves multiplex preamplifications, so the design of primers up to 96 pairs should be further verified to avoid nonspecific amplification if EvaGreen Green PCR is applied.

16.2.1.2 Single-Cell Transcriptome Analysis using Gene Expression Microarray Microarray technology has provided a view of the whole transcriptional profile of cancer cells for a decade [64]. It allows us to investigate the transcriptomes of tumors in a novel and comprehensive way that leads to molecular cancer classifications and prognosis. Significant clinical subtypes of tumors have been identified in patients that seemed to have similar tumors based on clinicopathological criteria. Early technology of DNA microarray or serial analysis of gene expression (SAGE) provided a comprehensive view of the expression of thousands of genes in a cancer cell population [65]. The difference between DNA microarray and SAGE is that SAGE has even more thorough coverage with the additional transcripts not annotated.

Single-cell transcriptomes have been studied in different tissues and cancer [25, 66, 67]. Microarray capability is dependent on whether the cDNA amplification truly reflects original RNA signals. Several strategies have been developed to improve cDNA amplification. In the case of linear amplification, T7 antisense RNA amplification and microarray were used for neural cell type classification, and as a result, several subpopulations of pyramid neurons were found in rat hippocampus CA1 [66, 68, 69]. To prevent the loss of RNA during purification, total RNA from single-cell lysates was directly used for cDNA synthesis without purification steps and subsequently for random PCR amplification using Super SMART PCR cDNA Synthesis Kit (Takara Clontech) [67] or WT-Ovation One-Direct Amplification System (Nugene) [25]. Several ug cRNA could be generated in subsequent amplifications for microarray hybridization. Linear amplification and random PCR amplification methods seem complementary to each other in amplification quality, but the former has better correlation with classic amplification, whereas the latter has higher 3′/5′ ratio and cDNA yield [70]. A hybrid strategy to combine the advantages of both amplification tactics produced more reliable, representative amplified cDNA [71–73]. The detection rates of probes on microarrays are dependent on the initial inputs of total RNA. There was a 25% decrease in probes detected in single-cell microarray as compared to 10-cell pool microarray [25].

There are several caveats of the expression microarray technology, such as cross-hybridization, hybridization [74, 75], unmapped transcripts, splicing forms, dye-based issues, and low and high expressions of transcript. Although some studies show that a strong congruence is found between microarray intensities and RNA-sequencing (RNA-Seq) read counts, RNA-Seq seems superior to the gene expression microarray method in single-cell transcriptome analysis [76].

16.2.1.3 Single-Cell RNA-Seq RNA-Seq has revolutionized the way we understand transcriptomes of mammalian cells [77, 78]. RNA-Seq outperforms microarray technology in many ways [77, 79]. With unprecedented depth and accuracy, RNA-Seq has revealed thousands of new species of unmapped transcripts and different spliced forms, miRNA, and nonencoding RNA in mammalian cells [79, 80]. RNA-Seq can map and quantify transcriptomes in mammalian cells at a scale spanning five orders of magnification [81]. A tremendous amount of transcriptomic data from high-throughput RNA-Seq has broadened our understanding of gene expression and regulation in mammalian cells in response to the external stimuli and internal network interactions.

With a powerful resolution, single-cell RNA-Seq is promising to dissect the heterogeneity in transcriptomes of rare cells, stem cells, and cancer cells [19, 21, 26, 28, 82–86]. The methods for whole transcriptome amplification in single-cell RNA-Seq are similar to protocols for the single-cell microarray except modifications in primers and adaptor ligation steps to accommodate NGS [83, 87–90]. SMART-Seq [26, 91], CEL-Seq [88], 3′-end poly(A) tailing [83], or an amplification-free low-quantity digital gene expression (LQ-DGE) [24, 92] has been proposed for different next-generation sequencing (NGS) platforms from Illumina, SOLiD, and Helico Biosciences [82, 87, 88]. In spite of such great progress, some substantial obstacles remain. Sampling noise and global cell-to-cell variation, two major technical variables, affect sequencing efficiency and complicate subsequent data analysis when minute input of single-cell RNA is applied [93]. The technical noise is even augmented for low copy transcripts. Single-cell RNA-Seq reproducibility ranges between 57% and 65% [94]. Several controls, such as Eternal RNA Control Consortium spiked-ins (ERCC) and unique molecular identifiers, as well as quantitative statistical software programs, have been developed to abate technical noise and identify meaningful biological variations [95–98]. Additionally, multiplex single-cell RNA-Seq could dramatically reduce the overall costs and make the technology more affordable [84, 89, 99].

16.2.2 Probing Cellular Biophysical Properties of Single Cells

16.2.2.1 The Instruments for Biophysical Assessment of Single Cells Nanomechanical and nanochemical assessment of cancer cells can be carried out using cutting-edge technologies and devices such as atomic force microscopy (AFM), creep cytoindentation apparatus (CCA), magnetic twisting cytometry (MTC), optical

Raman tweezers, and micropipette aspiration, among many other experimental techniques. With biophysical instruments, cellular elasticity (softness), adhesion, deformability, motility, and distribution of cell surface receptors could be mapped and probed at the single-cell level. These biophysical properties can identify the types and characterize the physiological status of cancer cells, as well as evaluate the extent of responses of single cells to microenvironmental influence. They are also surely complementary to molecular profiling in single-cell cancer research and as potential biomarkers for cancer detection and prognosis.

Micropipette aspiration uses fluid pressure directed on cells to record deformability of the cells [100, 101]. The pressure gradient is generated by an adjustable fluid reservoir and applied to the surface of the cells with a glass micropipette under the control of a pneumatic micromanipulator. The changes in fluid pressure are detected with a transducer and shown on the screen by a multiplexer. Cell responses to the pressures as deformation and aspiration are recorded through a charge-coupled device camera. The method is used to study not only whole cell but also nuclear mechanics [102].

The CCA can measure viscoelastic properties of an adherent cell by the displacement of a probe attached to a cantilever pressing on the cell with constant force [103, 104]. The instrument is sensitive enough to detect cellular indentions as low as 0.1 μm using the order of 1 nN forces.

Optical Raman tweezers technique offers the combination of optical collections of laser light with Raman spectroscopy and spectroscopic analysis of single cells [105, 106]. The back- and forward-scattering spectral configurations of cells allow discrimination of different cells, including normal and cancer cells at the single-cell level.

MTC [107] is a technique suitable for cell micromechanics analysis on the single cells. The technique is based on the measurement of the rotation of ligand-coated magnetic microbeads bound to membrane receptors using a magnetometer after stress or stimuli applied to the cells. MTC is able to analyze cellular and subcellular mechanics and apply localized forces at the cell surface and in the cytoplasm [108].

Among these, force-distance-based AFM is relatively versatile and measures cell morphometric as well as biomechanical parameters such as elasticity (softness), deformation, and adhesion (Figure 16.2a) [48, 109, 110]. The advanced AFM instruments are capable of processing simultaneously collected pseudo 3D images of multiple types of mechanical and chemical data on individual cells. Imaging with AFM is performed with a micrometer size sharp tip that can detect a set of morphological, mechanical, and chemical properties of a cell. An important advantage of AFM stems from the fact that live, unfixed cells are imaged with the nanometer lateral and vertical resolution [41]. AFM can deliver maps of subcellular properties and record intracellular events. Therefore, it has been used to detect morphological features upon pharmacological induction with drugs [48, 109], multiparametric imaging [41], formation of lipid rafts in human breast cancer cells [109, 111], apoptosis process recording [112], and membrane roughness [113].

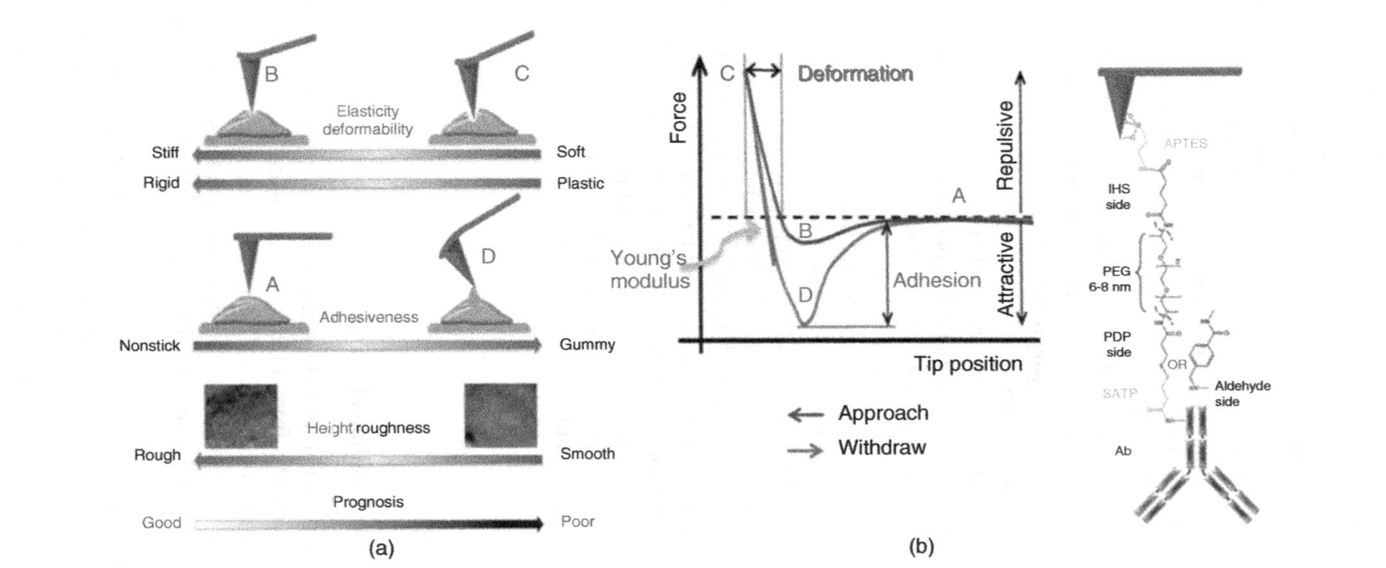

Figure 16.2 Components of CTCs' mechanical phenotype and their predicted association with cancer prognosis. (a, left), Stiffer and more rigid cells (dark gray) are less indented by an AFM tip than softer metastatic cells (light gray). More adhesive cells pull the tip back stronger. Smoother topography of a cell membrane is found in cancerous cells. (Right) A force plot cycle with the corresponding tip–cell interactions. Elasticity is calculated from a slope of the withdraw trace (the Young's modulus, Pa), adhesion from a depth of a negative dip (nN), and deformation from a depth of tip indentation (nm). (b) Functionalization of an amine activated tip with an Ab molecule attached via a PEG18 linker equipped with –SH or $–NH_2$ specific cross-linkers.

16.2.2.2 The Biophysical Parameters of Single Cells Independent from the instrument used, the major biophysical parameters of cancer cells usually include nanomechanical and nanochemical evaluations of mammalian cells including elasticity, deformability, adhesion, and the distribution of biomacromolecules, such as membrane receptors [45, 47, 114–116] (Figure 16.2). The figure illustrates how nanomechanical and nanochemical parameters of single cells are measured. Alterations in cellular biomechanical pathways that can be connected with the disease status [44] or induced by biochemical factors, pathogens, and drug treatments result in changes in cell morphology and mechanical status [44]. Reciprocally, changes in mechanical properties of cells can generate altered cell function in gene expression, viability and proliferation, and locomotion and mobility that, in turn, reflect disease status such as cancer cell metastasis or clogging of microvasculature.

When defining nanomechanical parameters in an AFM platform, several distinct models are used depending on the tip geometry and level of complication of its interactions with a probed object. The most popular in cell applications are the Young and Sneddon models assuming spherical and cone tip geometries, respectively. Based on the so-called force plots that show the dependence of a tip–object distance on the resulting force exercised on the object, several parameters such as elasticity, deformability, and adhesion can be calculated. Elasticity describes the capability of a cell to resist deformation—that is, the ability to return fully to its original shape after removal of the deforming force. It is measured in Pascal as the Young modulus (YM) [117]. Soft cells exhibit low YM, whereas stiff cells show high YM. Deformability, which is measured in units of length, is defined as the maximal indention of a cell induced by the AFM tip without breaking the continuum of a cell surface [118]. Adhesion (unit: Newton) measures the stickiness of the cell surface and physically expresses the force needed to lift the tip from a cell surface [116, 119–121]. The other useful parameters, viscoelastic analysis and stress–strain analysis, were also described previously [122, 123].

16.3 CTC APPLICATIONS

16.3.1 Single-cell Molecular Profiling of CTCs

Molecular profiling of CTCs is still at its early stage, and some groups investigated the transcriptional landscape of CTCs from different cancers using high-throughput qRT-PCR and transcriptomic analysis [24, 25, 27, 63, 124]. The characterization of gene expression of CTCs has identified several important signatures of regulatory networks and biomarkers for CTCs derived from breast, pancreatic, prostate cancers, and melanoma.

A microfluidic qRT-PCR study on prostate cancer CTCs showed that EMT-related genes are overexpressed in castration-resistant (CR) CTCs as compared to castration-sensitive (CS) CTCs (Figure 16.3a) [63]. Accumulative evidence indicates that epithelial-to-mesenchymal transition (EMT) and the formation of CTCs and metastasis are closely associated [125, 126]. Although biochemical recurrence

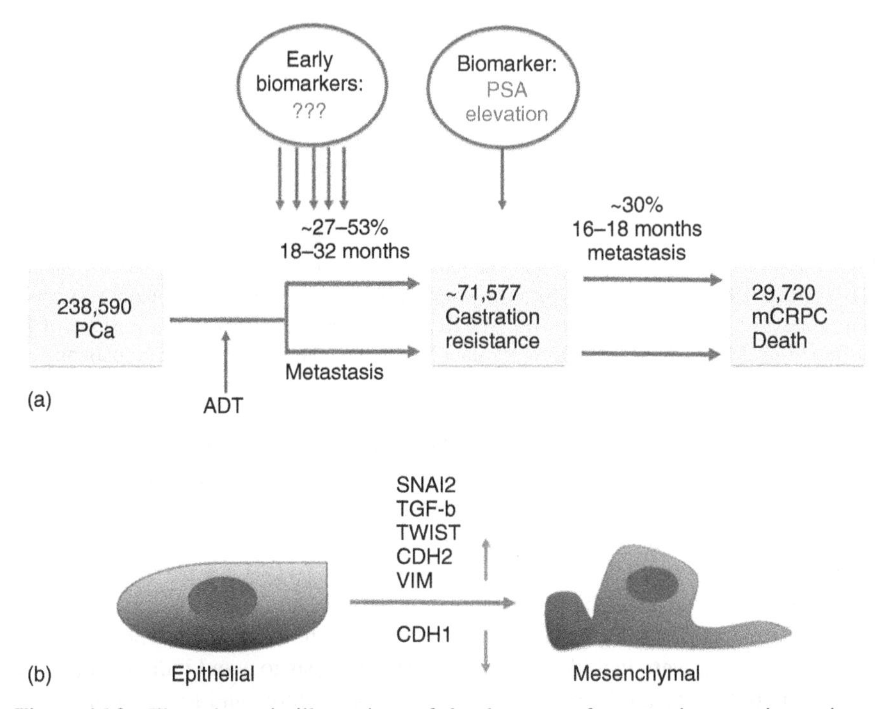

Figure 16.3 The schematic illustrations of development of metastatic castration-resistant prostate cancer (mCRPC) and epithelial-to-mesenchymal transition (EMT). (a) The schematic flow of mCRPC development. (b) During EMT, cells undergo dramatic molecular and phenotypic changes. Some of EMT-related genes are upregulated and some are downregulated. The morphological modifications of the cells are apparent externally. These lead to a hypothesis that EMT molecular profiles and biophysical modifications are potential biomarkers for mCRPC. ADT, androgen deprivation therapy.

with prostate-specific antigen (PSA) elevation is a standard clinical marker, fatal metastatic castration-resistant prostate cancer (mCRPC) still does not have biomarkers for early detection (Figure 16.3a). The majority of prostate cancer patients (~70%) under castration therapy have slowly growing disease that may not need any treatment at all. How to distinguish these fatal and indolent prostate cancer patients remains unknown even with all combinations of current biomarkers.

Several lines of evidence suggest that EMT-related genes expressed in CTCs could be potential biomarkers for mCRPC (Figure 16.3b) because high CTC counts and their mesenchymal stem-likeness are linked to drug resistance and the worst prognosis for prostate cancer [4, 50, 126, 127]. From eight patients (Figure 16.4), 38 prostate-derived CTCs were isolated using the combined microfiltration–microfluidics system (CM2S). The cells were subjected to high-throughput microfluidic expression profiling of 96 genes, including 56 EMT-related genes, some stem cell markers, prostate markers, and drug target genes (Figure 16.5). In spite of high heterogeneity of gene expression among the CTCs, the

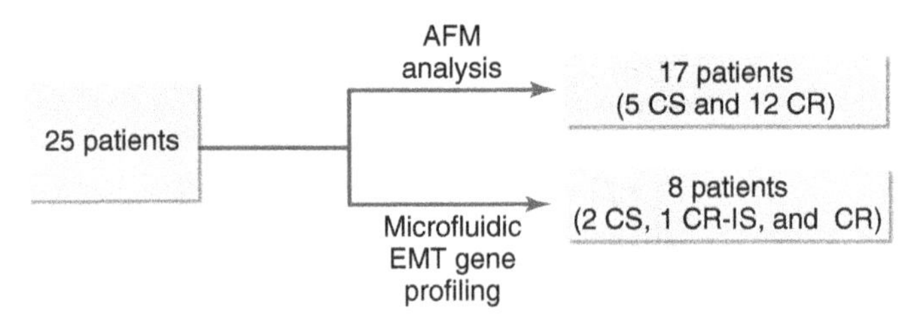

Figure 16.4 The prostate cancer patient recruitment and pathological status of patients. CTCs from 17 patients and 8 patients were subject to AFM nanomechanical parameter assessment and microfluidic qRT-PCR, respectively. CS: castration-sensitive; CR: castration-resistant; CR-IS: castration-resistant but immunotherapy-sensitive; Pt: patient.

increased accumulative expression of EMT-related genes was present in CR CTCs (Figure 16.6a). Some specific genes, Sonic Hedgehog, WNT, TGF-β, PTPRN2, and ALDH1 were upregulated in CR CTCs as well (Figure 16.6b).

Transcriptional analysis of a breast cancer patient cohort ($n = 35$) was carried out using the same microfluidic qRT-PCR [27]. A total of 105 CTCs were retrieved using a MagSweeper system from blood samples of 15 patients with primary breast cancer and 21 patients with metastatic disease. No more than five CTC expression data from each patient were included for downstream analysis to avoid individual patient bias. Although there are heterogeneous expression patterns and high variability in 87 genes analyzed, 31 of the genes were consistently expressed in at least 15% of the CTCs analyzed and involved in epithelial, EMT, metastasis, PI3K, apoptosis, cell proliferation, and metabolism pathways. Unsupervised hierarchical clustering analysis grouped the CTCs into two major clusters I and II. CTCs of 8 patients (23%) were present in both clusters, whereas 5 cases were only represented in cluster I and 22 cases are only found in cluster II.

Another study on 10 disseminated tumor cells (DTCs) from two patients identified two upregulated gene sets (TRKA and RAC pathways) using single-cell expression array and gene set enrichment analysis [25]. They are involved in prostate cancer growth and cell motility. The top 10 prostate marker genes were in the top 75th percentile gene list implicating prostate origin of DTCs. The single-cell DTC expression array data appeared robust and showed high correlation with 5-cell or 10-cell pools data with Pearson coefficients 0.875 and 0.783, respectively, among 10,000 randomly selected paired genes.

Cann *et al.* [124] used single-cell RNA-Seq to profile transcriptomes of CTCs from prostate cancer patients. CTCs were isolated from patient blood using Magsweeper technology. Bioanalyzer evaluation indicated that the cDNA amplification products reflected a wide range of CTC RNA quality that was good (21%), intermediate (37%), or poor (42%). RNA reverse transcription was performed using the SMARTer Ultra Low Input RNA (Clontech) system. Complementary DNA was amplified using the Advantage 2 PCR kit (Clontech) for 18–25 cycles before constructing an Illumina compatible DNA sequencing library using the Nextera

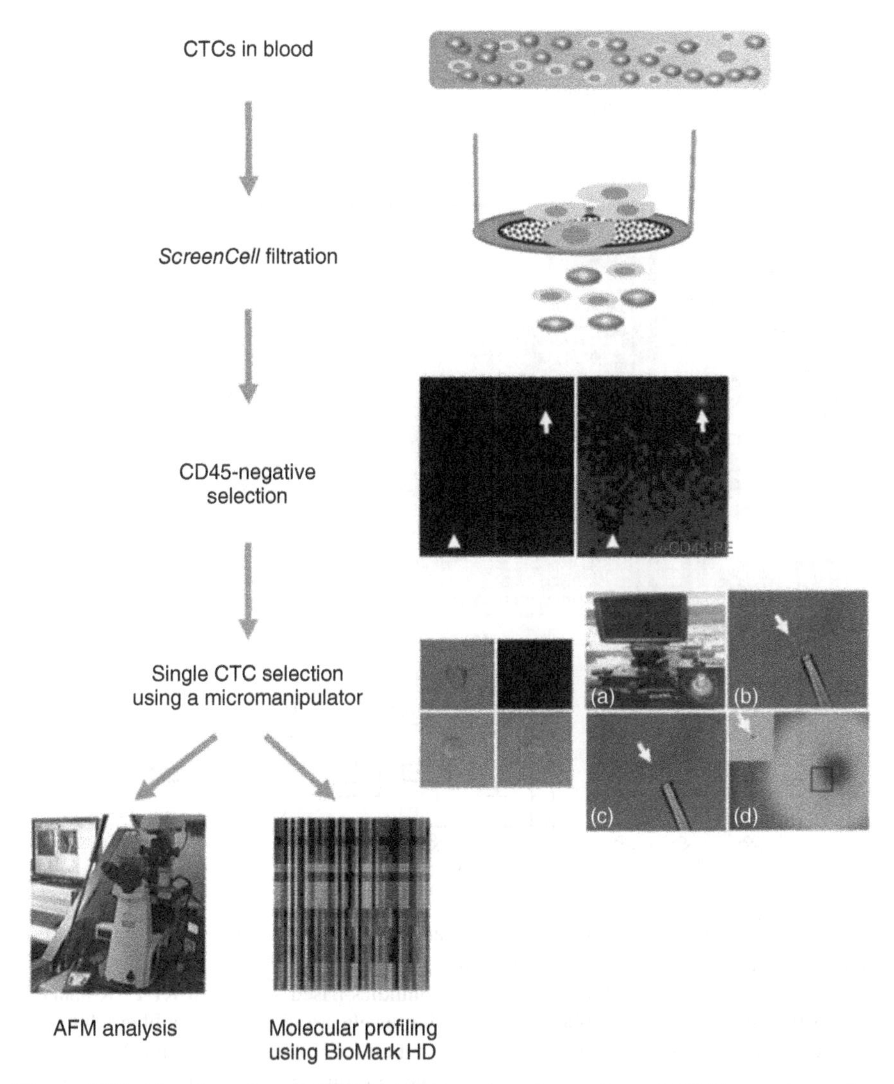

Figure 16.5 A schematic of CTC isolation and analyses. In anti-CD45 negative selection, representative microscopic photos show a CTC (white arrow head) was negative for anti-CD45-PE staining, whereas a blood cell (white arrow) positive. In single CTC selection, the left panel shows four representative CTCs and the right panel illustrates the single cell isolation using a micromanipulator and an Evos *fl* microscope. (a) An Evos *fl* microscope and a micromanipulator (inset). (b) A pipette tip pointing to a cell (white arrow) selected using a micromanipulator. (c) The single cell was aspirated into the pipette tip from the place it was previously located at (white arrow). (d) The selected single cell was placed on a Petri dish. A higher magnification of the single cell (black rectangle) is shown in the inset. Chen *et al.* [63]. Adapted with permission from John Wiley and Sons, Inc.

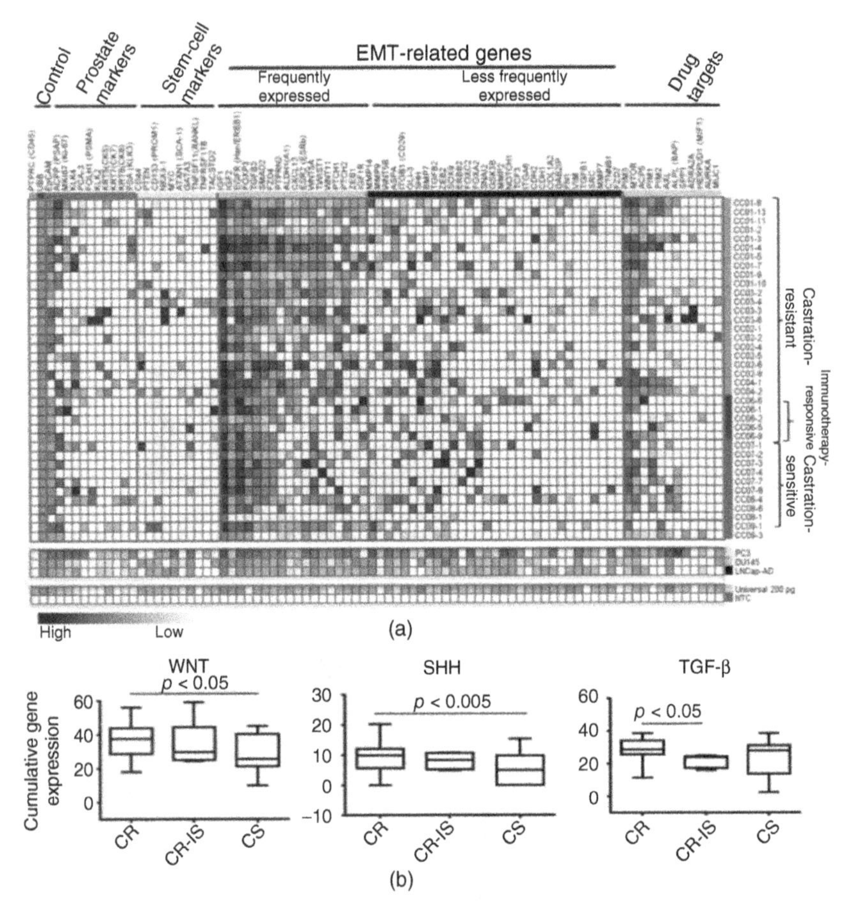

Figure 16.6 Elevated expression of EMT-related genes in CTC castration-resistant prostate cancer patients. (a) Single-cell expression profiles of EMT-related and control genes among CTCs. RNA from CTCs was subjected to microfluidics-based single-cell RT-PCR analysis using a BioMark device (in triplicate). Expression levels are displayed in a blue-white gradient. Gene symbols and gene groups were labeled on the top and individual CTC numbers and patient groups on the right. EMT-related genes are further divided into two groups: the frequently expressed group and the less frequently expressed group. (b) Elevated expression of EMT-related genes and signaling pathways in CTCs from castration-resistant patients. Cumulative gene expressions of WNT, SHH, and TGF-β signaling pathways. Data were analyzed using one-way ANOVA and unpaired Student's t-test. A p-value of <0.05 is considered as statistically significant. CR: castration-resistant; CR-IS: castration-resistant and immunotherapy-sensitive; CS: castration-sensitive. Chen *et al.* [63]. Adapted with permission from John Wiley and Sons, Inc. (*See color plate section for the color representation of this figure.*)

DNA Sample Prep Kit (Illumina) and 12 cycles of PCR. Twenty CTCs from 4 patients had 2362 ± 865 RefSeq transcripts with ≥10 fragments per kilobase of exon per million fragments mapped (FPKM), compared to LNCaP controls having 4622 ± 136 RefSeq transcripts with ≥10 FPKM. The androgen receptor (AR), KLK3 (PSA), and TMPRSS2 expressed higher in prostate CTCs compared to normal prostate samples. Ingenuity Pathway Analysis (IPA) showed that overexpressed pathways and gene sets were related to cancer and cell cycle. Additionally, nine upregulated genes were specific to prostate CTCs: AR, TK1, PLK1, MAGEA1, MAGEC1, MAGEC2, CTAGB1, BIRC5, and TOP2A.

Analysis of melanoma CTCs' transcriptomes using SMART-Seq revealed 289 upregulated genes and 436 downregulated genes as compared to melanoma primary cells and established cell lines [26]. The upregulated genes are involved in cancer and cell cycle regulation, whereas the downregulated genes are the regulators of cell death and MHC I genes. Compared to previous methods, SMART-Seq offers improved overall read coverage, nearly 40% coverage of 5′ ends, more detailed alternative splicing forms, and identification of single-nucleotide polymorphisms [26]. Nine melanoma CTC-specific genes that are highly expressed on cell membrane but not lymphocytes or melanomas could serve as potential melanoma CTC biomarkers.

Pancreatic CTCs from 12 patients and 4 healthy controls were isolated using anti-EpCAM HbCTC-Chips and subjected to low-quantity RNA-based single-molecule sequencing analysis [128]. In fact, 5 of 11 patients with CTC counts showed higher expression of non-canonical Wnt signaling pathway and 8 displayed upregulated Wnt signaling pathway, compared to 4 healthy patients and 1 patient with a low CTC count. With the same techniques, transcriptomes of breast cancer patients' CTCs were also investigated for the signature pathways [24]. Three gene signatures were found to be upregulated in breast cancer CTCs. They are a gene signature for bone relapse, an enriched group of 170 genes caught in predominant mesenchymal time point, and an enriched 717 genes group associated with EMT signature.

16.3.2 Analysis of Nanomechanical Phenotypes of Single CTCs

Unlike tumor cells, the nanomechanical analysis of CTCs is just beginning. During metastasis, EMT is crucial in the tumor cell morphogenesis (Figure 16.3b) [50]. A previous single-cell study indicates that upregulated expression of EMT-related genes in CTCs suggests more aggressive metastatic prostate cancer [63]. EMT-related nanomechanical attributes of CTCs could be potential biomarkers for mCRPC as well (Figure 16.3b). Topographical assessment using AFM has differentiated nonmetastatic mesothelial cells and metastatic nonsmall lung, breast, and pancreatic malignant cells [40, 109].

To determine the nanomechanical phenotype of prostate CTCs, an AFM was equipped with the PeakForce quantitative nanomechanical (PF-QNM) capabilities that collect multiple parameters of cells in a single scan with high resolution and accuracy. Prostate CTCs were as soft as cultured PCa cells as examined by conventional force AFM [63]. CTCs from 12 CR patients and 5 CS cases were subjected to microfiltration and further assessed. Retained CTCs on the filters were subject to AFM analysis and nanomechanical parameter data collection (Figure 16.4) [45]. Nanomechanical parameters were calculated with Nanoscope Analysis software v.4.1 using retrace images. The measurements of the elastic modulus followed the rules published by Sokolov *et al.* [129], assuming a high heterogeneity of cell surface properties (brush and rigidity). Additionally, we included adhesion forces in the analysis. Calculations were performed based on the Sneddon model that approximates the mechanics of conical tip interactions with an object. Cell mapping included elasticity, deformation, and adhesion, in the same scan. The nanomechanical phenotypes of CR CTCs were about three times softer, three times easier to deform, and seven times more adhesive than CS CTCs. Each parameter alone was not able to separate CR CTCs from CS CTCs. However, nonsupervised hierarchical clustering and principle component analysis showed that three parameters combined were able to differentiate CR patient group from CS patient group [45].

16.4 CONCLUSIONS

Single-cell transcriptional profiling and biophysical assessment of CTCs have emerged as important tools to dissect CTCs and mechanisms underlying their formation. Several technical advances have made these single-cell studies feasible. Some major strides were made in target amplification, PCR amplification, and linear RNA amplification from very limited starting materials of single CTCs. Microfluidic high-throughput qRT-PCR, gene expression microarray, and RNA-Seq have provided platforms for single-cell gene expression profiling. The technical noise from single-cell stochastic expression and cell-to-cell variability have been identified and significantly reduced for subsequent data analysis.

The massive amount of transcriptomic data has revolutionized our understanding and revealed many master regulatory networks and expression signatures in CTCs derived from different types of cancer. Moreover, several cutting-edge nanomechanical and biophysical devices are able to assess cellular phenotype of CTCs with high sensitivity and accuracy. The understanding of transcriptomes and microarchitectures of single CTCs showed promise in early detection of metastatic malignancy and drug resistance useful for disease prognosis.

Single-cell gene expression profiling and biophysical assessment of CTC still have some technical limits that need to be overcome in future. The preservation of RNA status and quality is crucial for single-cell transcriptomic profiling. Because the gene expression of CTCs is susceptible to environmental stimuli, CTC purification processes and duration should be simplified and shortened with minimal environmental changes introduced. Improvements in automated high-throughput CTC isolation and several newly developed devices for CTC isolation have addressed these issues and are discussed in other chapters in this book. The quantity of CTCs is another limiting factor for single-cell gene expression analysis and biophysical assessment. Single-cell analyses can only reach statistical significance with certain sizes of samples. How many CTCs are required may vary with different cancer types. To increase the sample sizes of CTCs, a larger clinical cohort, multiple blood sample collections from a patient, or longitudinal studies may be helpful for single-cell analyses.

The current technologies for CTC biophysical assessment are facing several challenges as well. Although the simultaneous mapping of several nanomechanical parameters in AFM functions has shortened the assessment process, AFM can analyze only a single cell at a time, and it takes ~15 min to collect the data. A high-throughput parallel platform or a high-speed device for biophysical mapping might be desirable for measuring multiple single cells at the same time and reducing the time needed for the process. Additionally, for biophysical assessment of CTCs, the newly isolated live CTCs are required to attach on a surface that should be sturdy enough for cellular mechanical manipulations by different types of detection devices. The optimal surface medium for biophysical assessment of CTCs remains to be determined.

Although some hurdles are expected for the improvement of technologies, the single-cell transcriptome and biophysical analyses have taken a step closer to understanding the CTC formation and its relationship to the metastatic pathological status. With the aids of these two approaches, novel biomarkers for metastatic tumors may be developed for personalized medicine with great potential for early detection, risk stratification, and therapeutic intervention.

REFERENCES

1. Maheswaran, S. and Haber, D.A. (2010) Circulating tumor cells: a window into cancer biology and metastasis. *Current Opinion in Genetics and Development*, **20**, 96–99.
2. Cristofanilli, M., Hayes, D.F., Budd, G.T. *et al.* (2005) Circulating tumor cells: a novel prognostic factor for newly diagnosed metastatic breast cancer. *Journal of Clinical Oncology*, **23**, 1420–1430.
3. Cristofanilli, M. (2006) Circulating tumor cells, disease progression, and survival in metastatic breast cancer. *Seminar in Oncology*, **33**, S9–14.
4. Danila, D.C., Heller, G., Gignac, G.A. *et al.* (2007) Circulating tumor cell number and prognosis in progressive castration-resistant prostate cancer. *Clinical Cancer Research*, **13**, 7053–7058.
5. Shaffer, D.R., Leversha, M.A., Danila, D.C. *et al.* (2007) Circulating tumor cell analysis in patients with progressive castration-resistant prostate cancer. *Clinical Cancer Research*, **13**, 2023–2029.
6. Danila, D.C., Fleisher, M., and Scher, H.I. (2011) Circulating tumor cells as biomarkers in prostate cancer. *Clinical Cancer Research*, **17**, 3903–3912.
7. Nagrath, S., Sequist, L.V., Maheswaran, S. *et al.* (2007) Isolation of rare circulating tumour cells in cancer patients by microchip technology. *Nature*, **450**, 1235–1239.
8. Danila, D.C., Pantel, K., Fleisher, M., and Scher, H.I. (2011) Circulating tumors cells as biomarkers: progress toward biomarker qualification. *Cancer Journal*, **17**, 438–450.
9. Yu, M., Stott, S., Toner, M. *et al.* (2011) Circulating tumor cells: approaches to isolation and characterization. *Journal of Cell Biology*, **192**, 373–382.
10. Eberwine, J., Sul, J.Y., Bartfai, T., and Kim, J. (2014) The promise of single-cell sequencing. *Nature Methods*, **11**, 25–27.

11. Meacham, C.E. and Morrison, S.J. (2013) Tumour heterogeneity and cancer cell plasticity. *Nature*, **501**, 328–337.
12. Kantlehner, M., Kirchner, R., Hartmann, P. *et al.* (2011) A high-throughput DNA methylation analysis of a single cell. *Nucleic Acids Research*, **39**, e44.
13. Bakstad, D., Adamson, A., Spiller, D.G., and White, M.R. (2012) Quantitative measurement of single cell dynamics. *Current Opinion in Biotechnology*, **23**, 103–109.
14. Hou, Y., Song, L., Zhu, P. *et al.* (2012) Single-cell exome sequencing and monoclonal evolution of a JAK2-negative myeloproliferative neoplasm. *Cell*, **148**, 873–885.
15. Li, Y., Guo, X., Xue, Q. *et al.* (2013) Single cell gene profiling revealed heterogeneity of paracrine effects of bone marrow cells in mouse infarcted hearts. *PloS One*, **8**, e68270.
16. Lovett, M. (2013) The applications of single-cell genomics. *Human Molecular Genetics*, **22**, R22–26.
17. Shalek, A.K., Satija, R., Adiconis, X. *et al.* (2013) Single-cell transcriptomics reveals bimodality in expression and splicing in immune cells. *Nature*, **498**, 236–240.
18. Jaitin, D.A., Kenigsberg, E., Keren-Shaul, H. *et al.* (2014) Massively parallel single-cell RNA-seq for marker-free decomposition of tissues into cell types. *Science*, **343**, 776–779.
19. Liu, N., Liu, L., and Pan, X. (2014) Single-cell analysis of the transcriptome and its application in the characterization of stem cells and early embryos. *Cellular and Molecular Life Sciences*, **71**, 2707–2715.
20. Macaulay, I.C. and Voet, T. (2014) Single cell genomics: advances and future perspectives. *PLoS Genetics*, **10**, e1004126.
21. Treutlein, B., Brownfield, D.G., Wu, A.R. *et al.* (2014) Reconstructing lineage hierarchies of the distal lung epithelium using single-cell RNA-seq. *Nature*, **509**, 371–375.
22. Xue, Z., Huang, K., Cai, C. *et al.* (2013) Genetic programs in human and mouse early embryos revealed by single-cell RNA sequencing. *Nature*, **500**, 593–597.
23. Zong, C., Lu, S., Chapman, A.R., and Xie, X.S. (1622-1626) Genome-wide detection of single-nucleotide and copy-number variations of a single human cell. *Science*, **2012**, 338.
24. Yu, M., Bardia, A., Wittner, B.S. *et al.* (2013) Circulating breast tumor cells exhibit dynamic changes in epithelial and mesenchymal composition. *Science*, **339**, 580–584.
25. Welty, C.J., Coleman, I., Coleman, R. *et al.* (2013) Single cell transcriptomic analysis of prostate cancer cells. *BMC Molecular Biology*, **14**, 6.
26. Ramskold, D., Luo, S., Wang, Y.C. *et al.* (2012) Full-length mRNA-Seq from single-cell levels of RNA and individual circulating tumor cells. *Nature Biotechnology*, **30**, 777–782.
27. Powell, A.A., Talasaz, A.H., Zhang, H. *et al.* (2012) Single cell profiling of circulating tumor cells: transcriptional heterogeneity and diversity from breast cancer cell lines. *PloS One*, **7**, e33788.
28. Patel, A.P., Tirosh, I., Trombetta, J.J. *et al.* (2014) Single-cell RNA-seq highlights intratumoral heterogeneity in primary glioblastoma. *Science*, **344**, 1396–1401.
29. Moller, E.K., Kumar, P., Voet, T. *et al.* (2013) Next-generation sequencing of disseminated tumor cells. *Frontiers in Oncology*, **3**, 320.
30. Almendro, V., Marusyk, A., and Polyak, K. (2013) Cellular heterogeneity and molecular evolution in cancer. *Annual Review of Pathology*, **8**, 277–302.
31. Hannemann, J., Meyer-Staeckling, S., Kemming, D. *et al.* (2011) Quantitative high-resolution genomic analysis of single cancer cells. *PloS One*, **6**, e26362.
32. Gupta, P.B., Fillmore, C.M., Jiang, G. *et al.* (2011) Stochastic state transitions give rise to phenotypic equilibrium in populations of cancer cells. *Cell*, **146**, 633–644.

33. Dalerba, P., Kalisky, T., Sahoo, D. *et al.* (2011) Single-cell dissection of transcriptional heterogeneity in human colon tumors. *Nature Biotechnology*, **29**, 1120–1127.
34. Xu, X., Hou, Y., Yin, X. *et al.* (2012) Single-cell exome sequencing reveals single-nucleotide mutation characteristics of a kidney tumor. *Cell*, **148**, 886–895.
35. Shibata, D. (2012) Cancer. Heterogeneity and tumor history. *Science*, **336**, 304–305.
36. Marusyk, A. and Polyak, K. (2013) Cancer. Cancer cell phenotypes, in fifty shades of grey. *Science*, **339**, 528–529.
37. Tang, F., Barbacioru, C., Nordman, E. *et al.* (2010) RNA-Seq analysis to capture the transcriptome landscape of a single cell. *Nature Protocols*, **5**, 516–535.
38. Costa, V., Aprile, M., Esposito, R., and Ciccodicola, A. (2013) RNA-Seq and human complex diseases: recent accomplishments and future perspectives. *European Journal of Human Genetics*, **21**, 134–142.
39. Mitra, S., Das, S., and Chakrabarti, J. (2013) Systems biology of cancer biomarker detection. *Cancer Biomarkers*, **13**, 201–213.
40. Cross, S.E., Jin, Y.S., Rao, J., and Gimzewski, J.K. (2007) Nanomechanical analysis of cells from cancer patients. *Nature Nanotechnology*, **2**, 780–783.
41. Dufrene, Y.F., Martinez-Martin, D., Medalsy, I. *et al.* (2013) Multiparametric imaging of biological systems by force-distance curve-based AFM. *Nature Methods*, **10**, 847–854.
42. Lekka, M., Laidler, P., Gil, D. *et al.* (1999) Elasticity of normal and cancerous human bladder cells studied by scanning force microscopy. *European Biophysics Journal*, **28**, 312–316.
43. Mego, M., Mani, S.A., and Cristofanilli, M. (2010) Molecular mechanisms of metastasis in breast cancer--clinical applications. *Nature Reviews. Clinical Oncology*, **7**, 693–701.
44. Suresh, S. (2007) Biomechanics and biophysics of cancer cells. *Acta Biomaterialia*, **3**, 413–438.
45. Osmulski, P., Mahalingam, D., Gaczynska, M.E. *et al.* (2014) Nanomechanical biomarkers of single circulating tumor cells for detection of castration resistant prostate cancer. *Prostate*, **74**, 1297–1307. doi: 10.1002/pros.22846
46. Bastatas, L., Martinez-Marin, D., Matthews, J. *et al.* (2012) AFM nano-mechanics and calcium dynamics of prostate cancer cells with distinct metastatic potential. *Biochimica et Biophysica Acta*, **1820**, 1111–1120.
47. Plodinec, M., Loparic, M., Monnier, C.A. *et al.* (2012) The nanomechanical signature of breast cancer. *Nature Nanotechnology*, **7**, 757–765.
48. Pillet, F., Chopinet, L., Formosa, C., and Dague, E. (2014) Atomic force microscopy and pharmacology: from microbiology to cancerology. *Biochimica et Biophysica Acta*, **1840**, 1028–1050.
49. Navin, N. and Hicks, J. (2011) Future medical applications of single-cell sequencing in cancer. *Genome Medicine*, **3**, 31.
50. Krebs, M.G., Metcalf, R.L., Carter, L. *et al.* (2014) Molecular analysis of circulating tumour cells-biology and biomarkers. *Nature Reviews. Clinical Oncology*, **11**, 129–144.
51. Stahlberg, A., Kubista, M., and Aman, P. (2011) Single-cell gene-expression profiling and its potential diagnostic applications. *Expert Review of Molecular Diagnostics*, **11**, 735–740.
52. Warren, L., Bryder, D., Weissman, I.L., and Quake, S.R. (2006) Transcription factor profiling in individual hematopoietic progenitors by digital RT-PCR. *Proceedings of the National Academy of Sciences of the United States of America*, **103**, 17807–17812.

53. Graf, T. and Stadtfeld, M. (2008) Heterogeneity of embryonic and adult stem cells. *Cell Stem Cell*, **3**, 480–483.

54. Ragone, G., Van De Bor, V., Sorrentino, S. *et al.* (2003) Transcriptional regulation of glial cell specification. *Developmental Biology*, **255**, 138–150.

55. Sigalotti, L., Fratta, E., Coral, S. *et al.* (2004) Intratumor heterogeneity of cancer/testis antigens expression in human cutaneous melanoma is methylation-regulated and functionally reverted by 5-aza-2′-deoxycytidine. *Cancer Research*, **64**, 9167–9171.

56. Morris, J., Singh, J.M., and Eberwine, J.H. (2011) Transcriptome analysis of single cells. *Journal of Visualized Experiments*, **25**, 2634.

57. Wang, L. and Janes, K.A. (2013) Stochastic profiling of transcriptional regulatory heterogeneities in tissues, tumors and cultured cells. *Nature Protocols*, **8**, 282–301.

58. Spiller, D.G., Wood, C.D., Rand, D.A., and White, M.R. (2010) Measurement of single-cell dynamics. *Nature*, **465**, 736–745.

59. Janes, K.A., Wang, C.C., Holmberg, K.J. *et al.* (2010) Identifying single-cell molecular programs by stochastic profiling. *Nature Methods*, **7**, 311–317.

60. Golding, I., Paulsson, J., Zawilski, S.M., and Cox, E.C. (2005) Real-time kinetics of gene activity in individual bacteria. *Cell*, **123**, 1025–1036.

61. Norrman, K., Strombeck, A., Semb, H., and Stahlberg, A. (2013) Distinct gene expression signatures in human embryonic stem cells differentiated towards definitive endoderm at single-cell level. *Methods*, **59**, 59–70.

62. Bengtsson, M., Stahlberg, A., Rorsman, P., and Kubista, M. (2005) Gene expression profiling in single cells from the pancreatic islets of Langerhans reveals lognormal distribution of mRNA levels. *Genome Research*, **15**, 1388–1392.

63. Chen, C.-L., Mahalingam, D., Osmulski, P. *et al.* (2013) Single-cell analysis of circulating tumor cells identifies cumulative expression patterns of EMT-related genes in metastatic prostate cancer. *Prostate*, **73**, 813–826.

64. Bucca, G., Carruba, G., Saetta, A. *et al.* (2004) Gene expression profiling of human cancers. *Annals of the New York Academy of Sciences*, **1028**, 28–37.

65. Polyak, K. and Riggins, G.J. (2001) Gene discovery using the serial analysis of gene expression technique: implications for cancer research. *Journal of Clinical Oncology*, **19**, 2948–2958.

66. Kamme, F., Salunga, R., Yu, J. *et al.* (2003) Single-cell microarray analysis in hippocampus CA1: demonstration and validation of cellular heterogeneity. *Journal of Neuroscience*, **23**, 3607–3615.

67. Esumi, S., Wu, S.X., Yanagawa, Y. *et al.* (2008) Method for single-cell microarray analysis and application to gene-expression profiling of GABAergic neuron progenitors. *Neuroscience Research*, **60**, 439–451.

68. Eberwine, J., Yeh, H., Miyashiro, K. *et al.* (1992) Analysis of gene expression in single live neurons. *Proceedings of the National Academy of Sciences of the United States of America*, **89**, 3010–3014.

69. Van Gelder, R.N., von Zastrow, M.E., Yool, A. *et al.* (1990) Amplified RNA synthesized from limited quantities of heterogeneous cDNA. *Proceedings of the National Academy of Sciences of the United States of America*, **87**, 1663–1667.

70. Klur, S., Toy, K., Williams, M.P., and Certa, U. (2004) Evaluation of procedures for amplification of small-size samples for hybridization on microarrays. *Genomics*, **83**, 508–517.

71. Kurimoto, K., Yabuta, Y., Ohinata, Y., and Saitou, M. (2007) Global single-cell cDNA amplification to provide a template for representative high-density oligonucleotide microarray analysis. *Nature Protocols*, **2**, 739–752.
72. Kurimoto, K. and Saitou, M. (2011) A global single-cell cDNA amplification method for quantitative microarray analysis. *Methods in Molecular Biology*, **687**, 91–111.
73. Kurimoto, K., Yabuta, Y., Ohinata, Y. *et al.* (2006) An improved single-cell cDNA amplification method for efficient high-density oligonucleotide microarray analysis. *Nucleic Acids Research*, **34**, e42.
74. Casneuf, T., Van de Peer, Y., and Huber, W. (2007) In situ analysis of cross-hybridisation on microarrays and the inference of expression correlation. *BMC Bioinformatics*, **8**, 461.
75. Eklund, A.C., Turner, L.R., Chen, P. *et al.* (2006) Replacing cRNA targets with cDNA reduces microarray cross-hybridization. *Nature Biotechnology*, **24**, 1071–1073.
76. Malone, J.H. and Oliver, B. (2011) Microarrays, deep sequencing and the true measure of the transcriptome. *BMC Biology*, **9**, 34.
77. Wang, Z., Gerstein, M., and Snyder, M. (2009) RNA-Seq: a revolutionary tool for transcriptomics. *Nature Reviews Genetics*, **10**, 57–63.
78. Cloonan, N. and Grimmond, S.M. (2008) Transcriptome content and dynamics at single-nucleotide resolution. *Genome Biology*, **9**, 234.
79. Marioni, J.C., Mason, C.E., Mane, S.M. *et al.* (2008) RNA-seq: an assessment of technical reproducibility and comparison with gene expression arrays. *Genome Research*, **18**, 1509–1517.
80. Wang, E.T., Sandberg, R., Luo, S. *et al.* (2008) Alternative isoform regulation in human tissue transcriptomes. *Nature*, **456**, 470–476.
81. Mortazavi, A., Williams, B.A., McCue, K. *et al.* (2008) Mapping and quantifying mammalian transcriptomes by RNA-Seq. *Nature Methods*, **5**, 621–628.
82. Lao, K.Q., Tang, F., Barbacioru, C. *et al.* (2009) mRNA-sequencing whole transcriptome analysis of a single cell on the SOLiD system. *Journal of Biomolecular Techniques*, **20**, 266–271.
83. Tang, F., Barbacioru, C., Wang, Y. *et al.* (2009) mRNA-Seq whole-transcriptome analysis of a single cell. *Nature Methods*, **6**, 377–382.
84. Islam, S., Kjallquist, U., Moliner, A. *et al.* (2011) Characterization of the single-cell transcriptional landscape by highly multiplex RNA-seq. *Genome Research*, **21**, 1160–1167.
85. Henley, B.M., Williams, B.A., Srinivasan, R. *et al.* (2013) Transcriptional regulation by nicotine in dopaminergic neurons. *Biochemical Pharmacology*, **86**, 1074–1083.
86. Yan, L., Yang, M., Guo, H. *et al.* (2013) Single-cell RNA-Seq profiling of human preimplantation embryos and embryonic stem cells. *Nature Structural and Molecular Biology*, **20**, 1131–1139.
87. Tang, F., Barbacioru, C., Bao, S. *et al.* (2010) Tracing the derivation of embryonic stem cells from the inner cell mass by single-cell RNA-Seq analysis. *Cell Stem Cell*, **6**, 468–478.
88. Hashimshony, T., Wagner, F., Sher, N., and Yanai, I. (2012) CEL-Seq: single-cell RNA-Seq by multiplexed linear amplification. *Cell Reports*, **2**, 666–673.
89. Picelli, S., Faridani, O.R., Bjorklund, A.K. *et al.* (2014) Full-length RNA-seq from single cells using Smart-seq2. *Nature Protocols*, **9**, 171–181.

90. Marinov, G.K., Williams, B.A., McCue, K. *et al.* (2014) From single-cell to cell-pool transcriptomes: stochasticity in gene expression and RNA splicing. *Genome Research*, **24**, 496–510.
91. Picelli, S., Bjorklund, A.K., Faridani, O.R. *et al.* (2013) Smart-seq2 for sensitive full-length transcriptome profiling in single cells. *Nature Methods*, **10**, 1096–1098.
92. Ozsolak, F., Ting, D.T., Wittner, B.S. *et al.* (2010) Amplification-free digital gene expression profiling from minute cell quantities. *Nature Methods*, **7**, 619–621.
93. Grun, D., Kester, L., and van Oudenaarden, A. (2014) Validation of noise models for single-cell transcriptomics. *Nature Methods*, **11**, 637–640.
94. Wu, A.R., Neff, N.F., Kalisky, T. *et al.* (2014) Quantitative assessment of single-cell RNA-sequencing methods. *Nature Methods*, **11**, 41–46.
95. Brennecke, P., Anders, S., Kim, J.K. *et al.* (2013) Accounting for technical noise in single-cell RNA-seq experiments. *Nature Methods*, **10**, 1093–1095.
96. Baker, S.C., Bauer, S.R., Beyer, R.P. *et al.* (2005) The external RNA controls consortium: a progress report. *Nature Methods*, **2**, 731–734.
97. Kivioja, T., Vaharautio, A., Karlsson, K. *et al.* (2012) Counting absolute numbers of molecules using unique molecular identifiers. *Nature Methods*, **9**, 72–74.
98. Shiroguchi, K., Jia, T.Z., Sims, P.A., and Xie, X.S. (2012) Digital RNA sequencing minimizes sequence-dependent bias and amplification noise with optimized single-molecule barcodes. *Proceedings of the National Academy of Sciences of the United States of America*, **109**, 1347–1352.
99. Fox-Walsh, K., Davis-Turak, J., Zhou, Y. *et al.* (2011) A multiplex RNA-seq strategy to profile poly(A+) RNA: application to analysis of transcription response and 3′ end formation. *Genomics*, **98**, 266–271.
100. Chen, C., Tambe, D.T., Deng, L., and Yang, L. (2013) Biomechanical properties and mechanobiology of the articular chondrocyte. *American Journal of Physiology. Cell Physiology*, **305**, C1202–1208.
101. Trickey, W.R., Lee, G.M., and Guilak, F. (2000) Viscoelastic properties of chondrocytes from normal and osteoarthritic human cartilage. *Journal of Orthopaedic Research*, **18**, 891–898.
102. Hochmuth, R.M. (2000) Micropipette aspiration of living cells. *Journal of Biomechanics*, **33**, 15–22.
103. Koay, E.J., Shieh, A.C., and Athanasiou, K.A. (2003) Creep indentation of single cells. *Journal of Biomechanical Engineering*, **125**, 334–341.
104. Jiang, C.C., Chiang, H., Liao, C.J. *et al.* (2007) Repair of porcine articular cartilage defect with a biphasic osteochondral composite. *Journal of Orthopaedic Research*, **25**, 1277–1290.
105. Harvey, T.J., Faria, E.C., Henderson, A. *et al.* (2008) Spectral discrimination of live prostate and bladder cancer cell lines using Raman optical tweezers. *Journal of Biomedial Optics*, **13**, 064004.
106. Faria, E.C. and Gardner, P. (2012) Analysis of single eukaryotic cells using Raman Tweezers. *Methods in Molecular Biology*, **853**, 151–167.
107. Wang, N., Butler, J.P., and Ingber, D.E. (1993) Mechanotransduction across the cell surface and through the cytoskeleton. *Science*, **260**, 1124–1127.

108. Guck, J., Schinkinger, S., Lincoln, B. *et al.* (2005) Optical deformability as an inherent cell marker for testing malignant transformation and metastatic competence. *Biophysical Journal*, **88**, 3689–3698.

109. Cross, S.E., Jin, Y.S., Tondre, J. *et al.* (2008) AFM-based analysis of human metastatic cancer cells. *Nanotechnology*, **19**, 384003.

110. Lekka, M. and Laidler, P. (2009) Applicability of AFM in cancer detection. *Nature Nanotechnology*, **4**, 72–73.

111. Orsini, F., Cremona, A., Arosio, P. *et al.* (2012) Atomic force microscopy imaging of lipid rafts of human breast cancer cells. *Biochimica et Biophysica Acta*, **1818**, 2943–2949.

112. Kim, K.S., Cho, C.H., Park, E.K. *et al.* (2012) AFM-detected apoptotic changes in morphology and biophysical property caused by paclitaxel in Ishikawa and HeLa cells. *PloS One*, **7**, e30066.

113. Wang, D.C., Chen, K.Y., Tsai, C.H. *et al.* (2011) AFM membrane roughness as a probe to identify oxidative stress-induced cellular apoptosis. *Journal of Biomechanics*, **44**, 2790–2794.

114. Lekka, M., Pogoda, K., Gostek, J. *et al.* (2012) Cancer cell recognition – mechanical phenotype. *Micron*, **43**, 1259–1266.

115. Rodriguez, M.L., McGarry, P.J., and Sniadecki, N.J. (2013) Review on cell mechanics: experimental and modeling approaches. *Applied Mechanics Reviews*, **65**, 060801–060801.

116. Stroka, K.M. and Konstantopoulos, K. (2014) Physical biology in cancer. 4. Physical cues guide tumor cell adhesion and migration. *American Journal of Physiology. Cell Physiology*, **306**, C98–C109. doi: 110.1152/ajpcell.00289.02013

117. Guo, Q., Xia, Y., Sandig, M., and Yang, J. (2012) Characterization of cell elasticity correlated with cell morphology by atomic force microscope. *Journal of Biomechanics*, **45**, 304–309.

118. Hoffman, B.D. and Crocker, J.C. (2009) Cell mechanics: dissecting the physical responses of cells to force. *Annual Review of Biomedical Engineering*, **11**, 259–288.

119. Puech, P.H., Poole, K., Knebel, D., and Muller, D.J. (2006) A new technical approach to quantify cell-cell adhesion forces by AFM. *Ultramicroscopy*, **106**, 637–644.

120. Wirtz, D., Konstantopoulos, K., and Searson, P.C. (2011) The physics of cancer: the role of physical interactions and mechanical forces in metastasis. *Nature Reviews Cancer*, **11**, 512–522.

121. Sariisik, E., Docheva, D., Padula, D. *et al.* (2013) Probing the interaction forces of prostate cancer cells with collagen I and bone marrow derived stem cells on the single cell level. *PloS One*, **8**, e57706.

122. Kohles, S.S., Liang, Y., and Saha, A.K. (2011) Volumetric stress-strain analysis of optohydrodynamically suspended biological cells. *Journal of Biomechanical Engineering*, **133**, 011004.

123. Mahaffy, R.E., Park, S., Gerde, E. *et al.* (2004) Quantitative analysis of the viscoelastic properties of thin regions of fibroblasts using atomic force microscopy. *Biophysical Journal*, **86**, 1777–1793.

124. Cann, G.M., Gulzar, Z.G., Cooper, S. *et al.* (2012) mRNA-Seq of single prostate cancer circulating tumor cells reveals recapitulation of gene expression and pathways found in prostate cancer. *PloS One*, **7**, e49144.

125. Kalluri, R. (2009) EMT: when epithelial cells decide to become mesenchymal-like cells. *Journal of Clinical Investigation*, **119**, 1417–1419.
126. Kalluri, R. and Weinberg, R.A. (2009) The basics of epithelial-mesenchymal transition. *Journal of Clinical Investigation*, **119**, 1420–1428.
127. Nauseef, J.T. and Henry, M.D. (2011) Epithelial-to-mesenchymal transition in prostate cancer: paradigm or puzzle? *Nature Reviews. Urology*, **8**, 428–439.
128. Yu, M., Ting, D.T., Stott, S.L. *et al.* (2012) RNA sequencing of pancreatic circulating tumour cells implicates WNT signalling in metastasis. *Nature*, **487**, 510–513.
129. Sokolov, I., Dokukin, M.E., and Guz, N.V. (2013) Method for quantitative measurements of the elastic modulus of biological cells in AFM indentation experiments. *Methods*, **60**, 202–213.

17

DIRECTING CIRCULATING TUMOR CELL TECHNOLOGIES INTO CLINICAL PRACTICE

Benjamin P. Casavant

Department of Biomedical Engineering, University of Wisconsin, Madison, WI, USA

David Kosoff

Department of Medicine, University of Wisconsin, Madison, WI, USA

Joshua M. Lang

Department of Medicine, University of Wisconsin Carbone Cancer Center, Madison, WI, USA

17.1 INTRODUCTION

Given the many therapeutic advances in medical oncology over the past decade, the need to tailor treatment recommendations to each individual has never been greater. However, clinical decision-making in oncology is often limited to radiographic assessments of total tumor burden and toxicity-based assessments for a given therapy. Oncologists weigh these factors in an "analog decision-making process" to determine if a patient is experiencing sufficient clinical benefit to continue on their current therapy. This rough assessment of the tumor burden in patients with metastatic tumor lesions is limited and often inaccurate. This was made evident in a study wherein biopsies of progressing tumor masses in patients with adenocarcinoma of the lung identified a new small-cell cancer phenotype during treatment with an epidermal growth factor receptor (EGFR) antagonist [1]. This limited biopsy study was among the first to identify an unappreciated extent of cellular diversity in

Circulating Tumor Cells: Isolation and Analysis, First Edition. Edited by Z. Hugh Fan.

patients with metastatic cancer. A critical finding in this study was that targeting the next treatment option (chemotherapy) to this new phenotype led to radiographic improvement but, on repeat biopsy, the original adenocarcinoma regained a foothold and returned as the dominant histologic pattern. This transition suggests a complex series of interactions imposed by the selective/evolutionary pressure of these markedly different anticancer therapies (chemotherapy vs EGFR targeted therapy). This diversity is not limited to lung cancer and is now being appreciated across nearly all malignancies, including prostate, renal, and breast cancer. However, this personalization of cancer therapies requires biomarkers that (i) predict therapeutic benefit, (ii) identify emerging mechanisms of resistance, and (iii) tailor subsequent treatment strategies to continually evolving tumors. Easy access to tumor samples, without the need for painful and expensive tumor biopsies, would revolutionize patient care and cancer research.

The great hope in circulating tumor cell (CTC) research lies in the potential for these rare cells to be an accessible "liquid biopsy" that permits frequent, minimally invasive sampling of tumor cells for the same molecular assays performed on traditional biopsies. The relevance of CTCs is supported by the belief that these rare cells are shed into peripheral circulation from primary and metastatic tumor sites (Figure 17.1) and may contribute to the development of metastatic tumor deposits [2, 3]. While CTCs have been noted as far back as the mid-1800s [4], interest in these cells has increased dramatically in recent years with technologic advances that have improved CTC detection, isolation, capture, and characterization [5]. The translation of these technologies and molecular assays as biomarkers for clinical care requires extensive testing and validation in defined contexts of use (COU) [6]. Thus, understanding biomarker categories and determining clinically relevant COUs for these biomarkers are critical prior to entering any development phase (Table 17.1).

17.2 DEFINING BIOMARKERS

Broadly speaking, a biomarker is defined as "a characteristic that is objectively measured and evaluated as an indicator of normal biologic processes, pathogenic processes, or biologic responses to a therapeutic intervention" [7]. Biomarkers are categorized based on clinical utility as diagnostic, prognostic, predictive, pharmacodynamic, or as a surrogate for clinical endpoints. In the field of oncology, biomarkers are an integral part of clinical practice. Tumor staging, imaging modalities such as computed tomography (CT) and positron emission tomography (PET) scans, secreted proteins in the blood (CEA; prostate-specific antigen, PSA; CA 19-9), chromosomal analysis (fluorescence *in situ* hybridization or FISH, cytogenetics), and Eastern Cooperative Oncology Group (ECOG) status are all biomarkers used to assist in the treatment of malignant disease. While the use of biomarkers has clearly helped to improve cancer survival and overall morbidity, there are significant limitations to the use of current biomarkers in clinical practice. Most blood biomarkers are organ-specific rather than disease-specific and therefore may not be completely representative of disease status. This has been demonstrated with the

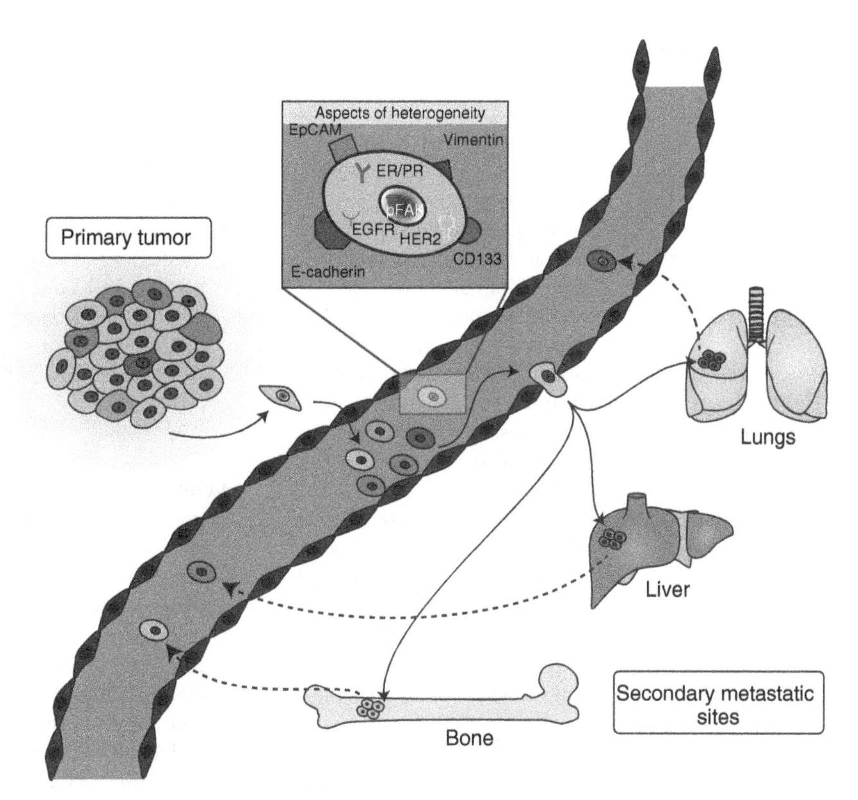

Figure 17.1 *Cancer metastasis and definition of Circulating Tumor Cells (CTCs).* Tumor cells from primary tumors are shed into the circulation and travel to other organs, where they exit the circulatory system and grow into new tumor foci. This process is known as metastasis. Tumor cells from metastatic sites are also shed into the circulation through the same process. While in circulation, these tumor cells are referred to as CTCs. Inset: CTCs express various molecular markers depending on the tissue of origin and properties of the cancer cell. These markers can be used to identify and isolate the tumor cell.

use of PSA in prostate cancer by Newling *et al.* in 2009 [8]. Additionally, many of the most reliable biomarkers require biopsy [9], which, due to cost and patient morbidity, is typically only performed at diagnosis. Over time, unfortunately, the natural progression of the disease, as well as selective pressure exerted by treatment, can produce a cancer with a very different genetic and molecular profile than the one identified at biopsy [10, 11]. Also, in the setting of metastatic disease, there is significant disease heterogeneity among the varied sites.

17.2.1 Prognostic CTC Biomarkers

Most of the research thus far has been focused on the use of CTCs for prognosis. Bono *et al.* showed that the number of CTCs in the blood can be used as a prognosticator. When analyzing the number of CTCs per 7.5 ml of blood from those

TABLE 17.1 Biomarker Classification

Type	Definition	Example
Diagnostic	Identifies the presence of a malignancy and may establish the site of origin	Histopathologic analyses of tissue biopsies, t(8:21) and inv(16) in AML
Prognostic	Baseline patient or disease characteristic that identifies degrees of risk for disease recurrence or progression. A prognostic biomarker informs about the natural history of the disorder in that particular patient regardless of a therapeutic intervention	ECOG performance status, Tumor staging and grading, FLT3 mutation in AML, Gleason score in prostate cancer
Predictive	Baseline characteristic that categorizes patients by their likelihood for response to a particular treatment. A predictive biomarker is used to identify whether a given patient may experience a favorable or an unfavorable clinical outcome	EGFR mutation in non-small cell lung cancer and sensitivity to erlotinib/gefitinib, KRAS mutation and insensitivity to cetuximab in colorectal cancer, ER/PR positivity in breast cancer and sensitivity to endocrine therapies
Pharmacodynamic (or activity)	Dynamic assessment that shows that a biological response has occurred in a patient after having received a therapeutic intervention. A pharmacodynamic biomarker may be treatment-specific or more broadly informative of disease response	CT scan, FDG PET scan
Surrogate endpoint	Biomarker intended to substitute for a clinical efficacy endpoint in therapeutic decision-making	Progression free survival (PFS)

Source: Adapted from Qualification Process for Drug Development Tools.
Abbreviations: AML, Acute Myeloid Leukemia; ECOG, Eastern Cooperative Oncology Group; FLT3, fms Related Tyrosine Kinase 3; EGFR, Epidermal Growth Factor Receptor; KRAS, Kirsten Rat Sarcoma; ER, Estrogen Receptor; PR, Progesterone Receptor; CT, Computed tomography; FDG, Fluorodeoxyglucose; PET, Positron Emission Tomography.

with prostate cancer, less than five CTCs were associated with improved overall survival (11.5 vs 21.7 months). Additionally, transition from ≥5 CTCs/7.5 ml blood before treatment to <5 CTCs/7.5 ml of blood after treatment is also associated with improved survival (6.8 vs 21.3 months) and transition from <5 CTCs/7.5 ml to ≥5 CTCs/7.5 ml posttreatment with poorer survival (9.3 vs >26 months) [12]. Similar prognostic utility has also been demonstrated in breast cancer [13], colon cancer [14], and melanoma [15], among others. In prostate cancer, CTC number has been shown to be a stronger prognostic factor than PSA [12, 16, 17].

17.2.2 Predictive CTC Biomarkers

Perhaps the most significant use of CTCs would be their ability to guide clinicians as they select therapy strategies throughout the course of treatment. Beyond merely forecasting outcomes, CTC enumeration has also been shown to predict response to therapy. Multiple studies have demonstrated that CTC number was predictive of treatment response, including androgen responsiveness in hormone-sensitive metastatic prostate cancer [18] and biochemotherapy in melanoma [19]. In metastatic breast cancer, CTC enumeration was even shown to be more predictive of treatment response than conventional imaging [14].

Still more promising is the potential to use CTC analysis to predict the most potent and focused therapy. Dramatic increase in knowledge of cancer at the molecular level has led to more targeted therapy. Imatinib and trastuzumab have dramatically improved treatment of chronic myeloid leukemia (CML) and breast cancer, respectively, by targeting molecular mechanisms underlying malignant expansion. These selective treatments have proven to be more efficacious, more selective, and less toxic than traditional chemotherapy, which uses cytotoxic drugs based on broad tumor characteristics such as origin, stage, and grade. The difficulty is selecting the patients most likely to benefit from a specific targeted therapy. CTCs provide a convenient approach to sample and profile tumor cells throughout therapy. As cancers are heterogeneous over time as well as space, the need to profile a cancer at diagnosis is likely to be equally as important as subsequent intervals. Drug development could also benefit from the predictive ability of CTCs. Only 1 in 20 drugs gains regulatory approval, of which 30% fail due to inadequate therapeutic activity. CTCs could provide molecular profile at time of study, accelerating drug development by selecting those most likely to respond [20].

17.2.3 Pharmacodynamic CTC Biomarkers

Drug development is both costly and timely. Cost of anticancer drug development can reach billions of dollars, and production typically takes 8–10 years from discovery to registration [21]. The prospect of CTC assessment of treatment response in real time offers the possibility of assessing a drug's efficacy early on in a trial and could predict early success or failure [22], preventing wasted time and money spent on inefficacious drugs. This would also apply to the clinic where patient response by CTC analysis could predict treatment failure faster and more accurately than current modalities leading to quicker adjustment of therapy. Additionally, clinical trials are currently designed to assess maximally tolerated dose and then assess efficacy. It is becoming clear, however, that the maximum tolerated dose may be higher than dose for maximum effect [23]. CTCs could help select optimal dose for Phase II studies [24], allowing for increased approval of drugs with better toxicity profiles. Continuous sampling of tumor cells also allows for monitoring of different pharmacodynamic targets such as membrane-bound growth factor receptors, DNA mutation, apoptosis, signal transduction cascade, and tumor aggressiveness [24]. The accessibility of CTCs for repeated sampling paired with blood collection for pharmacokinetic studies is certainly possible. However, it is yet to be determined if repeated sampling of this nature is reliable for such studies.

17.2.4 Diagnostic CTC Biomarkers

The challenge in using CTC analyses for the diagnosis of cancer lies in the rarity of the event and the importance of specificity. For example, atypical cells can often be identified in bone marrow aspirates but do not qualify as diagnostic for malignancy. Thresholds for each assay must be benchmarked against the current gold standard tests, tumor biopsies. From a clinical perspective, it is unclear if the performance characteristics of CTC analyses (on only a few cells) will ever eclipse the accuracy and specificity of tumor biopsies. There is a small amount data available on the use of CTCs for diagnostic purposes in various cancers [25–27]. Although results show potential, the data thus far is not strong. Further studies are needed.

17.2.5 Surrogate CTC Biomarkers

The potential for a blood test to act as a surrogate for a clinical endpoint such as overall survival would revolutionize the field of drug development. Currently, radiographic progression-free survival is the major surrogate marker in clinical use. However, this endpoint is not always accepted by the FDA in the registration strategies for new therapeutic agents. Can CTCs fill this gap? Data has been presented at the American Society of Clinical Oncology annual meeting suggesting enumeration of CTCs in men with prostate cancer treated with abiraterone met surrogacy endpoints. However, this finding must be repeated in multiple Phase III trials with different therapeutic agents to meet these purposes. Understanding the transition of these various biomarkers through clinical trials requires an understanding of the interplay of the clinical needs and the technological development. The technological development is more straightforward than the complexities that exist in clinical trials but are necessary in the creation of translated technologies that can advance biomarker research.

17.3 THE TECHNOLOGY

CTCs have the potential to be applied toward many classes of biomarker. However, the technology to isolate and analyze CTCs becomes critically important to using these rare but accessible cells. When speaking of CTC technologies, there exists an intersection between technological approaches to isolating these cells and increasingly precise methods to analyze them. Because the two aspects of the technology are difficult in their own right (isolating 1 in 1 billion cells, analyzing material from low numbers of total cells), there exists an interplay between the isolation and analysis that creates a diverse landscape of technologies that specialize in one or the other, but rarely a combination of both. Further, understanding the quality needs of a system that combines two precise and delicate technologies can be difficult. To begin a discussion on the technological advances of clinical applications of CTCs, a logical starting point is with the only currently FDA-approved test that uses CTCs, the Veridex CellSearch platform.

17.3.1 Translated Technologies

As reviewed many times previously [28–32] and detailed in Chapter 19, the Veridex CellSearch platform is currently the only FDA-approved CTC detection and analysis platform. From a technological perspective, it is very easy to list the problems with the CellSearch platform. To list a few, the cells isolated by CellSearch are reliant entirely on EpCAM positivity, excluding cells that have undergone epithelial–mesenchymal transition (EMT), which may be among the more important cells according to certain theories of metastasis [33–35]; the plasticity of CTCs may cause EpCAM expression to be up- or downregulated at the time of isolation [5]; CellSearch may be missing large numbers of CTCs [36, 37]; there may exist a bias based on the isolation methodology and the "missed" cells analyzed by the platform [24]. However easy it is to evaluate the flaws of the Veridex CellSearch platform, two important facts remain: (i) it is to date the only FDA-approved CTC analysis platform and (ii) the information being collected by the CellSearch platform continues to be proven relevant. Just recently, studies have demonstrated the potential of CellSearch results to be used as a surrogate biomarker [38, 39]. The question that remains is simple: what allows the Veridex to – nearly 10 years after its introduction as a CTC detection system – be the only FDA-approved device to use CTCs as a biomarker? To answer this question, we must take a closer look at what it means to create a translational technology.

17.4 TRANSLATING TECHNOLOGY

The process of creating a translational technology is complex and involves many factors. These factors, or the "stakeholders," include any party that is involved in the process of creating and applying the technology. For CTC technologies, these stakeholders (with some simplifications) include patients, clinicians, clinical labs, innovators, manufacturers, investors, and regulatory bodies (in this case, the FDA). Looking at technologies that are successful in academic settings and some that have companies that are built around them, CTC technologies are remarkably good at being innovative [40–45] and bringing a strong clinical need story with the developed technology. This is no surprise considering the way that academics is structured and rewarded – success is earned by innovation and collaboration, pushing the envelope for higher CTC capture quantities [46–48] or more precision on the analytical side [49, 50], all while pushing clinically relevant targets [51, 52]. One of the great benefits of developing innovative technologies comes with the patents associated with those technologies, allowing innovators and investors to build value on defensible technology and secure their application early.

One of the main stakeholders in the development of CTC technologies is the FDA. To understand the approaches to the FDA that may be overlooked by CTC technologies, we must understand the interest and concerns of the FDA. According to a draft FDA guidance document regarding the qualification of drug development tools [53], when using a device to measure a biomarker, the FDA will be concerned with two main features of the technology and will deal with them separately: (i) the evaluation of the device for its ability to "reliably and accurately measure the biomarker" and (ii) qualification of a clinical biomarker for "specific interpretation and application in drug development and regulatory review." In other words, the FDA is concerned with (i) the quality of the systems regarding the engineering design of the technology and (ii) the relevance of the biomarker in a clinical setting (Figure 17.2).

17.4.1 The Technology Side

Focusing on the FDA suggestion for device evaluations, the ability to "reliably and accurately measure the biomarker" focuses on two very direct engineering tasks. To address the term "reliability," the requirement focuses on the ability of the device to reproducibly create a measurement of the biomarker within a given context of use and reliably report errors or misreads appropriately as they arise. Taken this way, the FDA is asking technologies to demonstrate their repeatability in the collection and

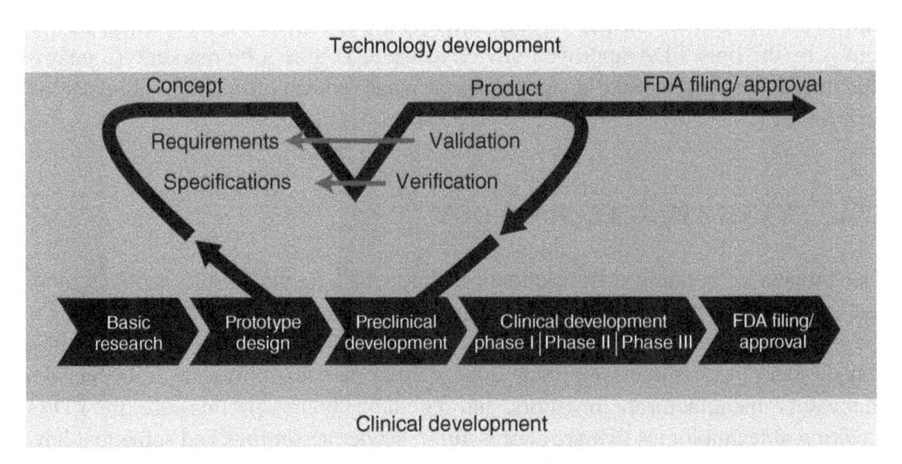

Figure 17.2 *Development of biomarker technologies for FDA approval.* Creating biomarker technology for FDA approval requires that the device/assay accurately and reliably measure a clinically useful biomarker. This process involves conceptualizing and designing the technology followed by various phases of study in clinical trials to demonstrate clinical utility *(bottom half)*. Before a technology can be studied in clinical trials, there must be rigorous testing to validate that a product does what it is proposed to do (requirements) within certain preset standards (specifications) *(top half)*.

analysis of CTCs. In this light, many CTC technologies can already be seen as having difficulty, given the high levels of precision required for the biomarker analysis to return a correct result. For example, batch variation in antibodies exists and can result in highly variable readouts. Securing the animal from which the batch originated is one way to ensure that this variability is avoided; however, as the animals are living creatures, there is always risk of the animal dying. Another example of variability exists with the high-precision fluorescent imaging equipment often used for CTC analysis, so having a documented control system that accounts for bulb type or bulb life may be important factors to control the reproducibility of the CTC analysis.

The accuracy guideline of the FDA is focused on a different aspect of quality systems, specifically the internal controls and reference standards that can be used to evaluate and maintain the accuracy of the system. Unfortunately, as the field is young, there are not yet internationally approved reference standards for many other biomarkers, but reference standards are important for the application of the technology within labs that run the CTC test. Reference standards typically exist as contrived samples that have a set amount of the biomarker of interest contained in them. Running them through the machine to evaluate the biomarker readout allows you to assess the machine for proper function and allows labs to run the test remotely. Internal controls are run typically within the sample to ensure that the quality of the sample is maintained through the instrument run. Examples of internal controls that are used often include using housekeeping genes in PCR or looking for hemolysis within plasma samples, as an illustration of the breadth of internal controls. In general, internal controls and reference standards are ways to ensure that samples are run properly and need to be designed and have a system in place to report the outcomes to ensure that a technology maintains its accuracy.

These two aspects of the FDA device approval guidance are typical systems engineering setups but are not implemented widely in academic settings, where devices are being developed to use the latest analytical method or use newly developed techniques to improve assay workflows. For typical publications in the engineering field, basic quality controls are performed; however, maintaining the reproducibility of a test in academia is all but forgotten as technologies are rewarded for performing 100 tests, not performing 100 tests on 100 samples. Keeping in mind the high levels of quality required for FDA approval of a technology will help guide CTC technologies during the development stage, allowing the technologies to transition well out of academia.

In typical product-development cycles, the creation of product requirements and product specifications are the beginning of the product realization process. The requirements of a product tell generally what the product is set out to do (e.g., the test is to quantify protein levels within CTCs). The specifications are set to determine the specific engineering standards that the product will meet (e.g., the test will quantify CTCs in reference samples within x%). The establishment of product requirements and specifications are to give specific goals for the device to meet during internal development of the technology and allow the FDA to observe the internal requirements and assess if the technology is able to use the requirements and specifications to perform a test to the standards needed in the clinic. The product realization process is a widely adopted method to achieve the reproducibility and reliability needed for FDA approval of medical devices and is important to understand and use during the translation of CTC technologies.

17.4.2 The Clinic Side

The second focus of the FDA lies on the clinical side regarding the utility of the biomarker. In parallel with addressing the fairly straightforward engineering concerns of creating a robust and reproducible reporting system, the biomarker that is being reported needs to build clinical relevance or be qualified. Qualification of biomarkers begins with a test system that is robust and well-characterized [54]. This allows the results to be trusted as consistent and allows quantification of the biomarker to be related back to clinical outcomes. This process can take a long time but begins by introducing a biomarker as a new "exploratory" biomarker. As the biomarker is evaluated using more patients and multicenter trials, the documentation for the biomarker moves it from "probable valid" to "known valid" (if appropriate) – a complex process that requires peer review and acceptance of the biomarker in addition to the costly and lengthy process of clinical trials. Importantly, the foundation of these processes is tied directly to proving the utility of the biomarker through patient accrual. In this context, the utility of CTCs as a biomarker is tied to the types of assay outputs and clinical team with which the CTC technology is paired. As discussed earlier, this is what enables CTCs to be used for so many different classes of biomarkers.

Tied to the clinical side of CTC evaluation, Clinical Laboratory Improvement Amendments (CLIA) certified labs provide a necessary testing environment for using CTC technologies. CLIA certified labs have achieved laboratory testing standards that allow these labs to process human samples for diagnostic purposes and include a wide network of clinical testing labs that have obtained the certification. Creating CTC technologies that can be easily integrated into the CLIA lab workflow simplifies multisite trials of the CTC platform during the qualification process and clarifies the internal control and reference standard needs of the technology for widespread use. Leveraging the experience and rigor of CLIA labs lends well to the qualification process, as the test will be performed at well-documented high standards. A challenge put forth to the developers of CTC technologies is to create systems that can be inserted seamlessly in the CLIA lab, as this will help in translating the technology.

17.5 CONCLUSIONS

The exciting promise of CTCs should not overshadow the need for significant advancement in both knowledge and technology necessary to unlock their potential. First and foremost, there is no single definition of a CTC and no single CTC biomarker [5]. Numerous assays are in use or under development, which employ various strategies and targets. These assays reveal unique applications of CTCs toward improving patient care; however, each technology needs to be taken through the appropriate technical validation and clinical qualification. As technologies are translated from the laboratory setting into the clinic, the utility of CTCs can be more vigorously tested and expanded toward use as a liquid biopsy for a variety of disease states. Importantly, the excitement that is created by the multitude of applications for CTC isolation and analysis should be harnessed and directed to advance the technologies, but the oft mundane and time-consuming processes of verification, validation, and qualifications are necessary foundations to offer clinically relevant guidance to direct patient care.

REFERENCES

1. Lecharpentier, A., Vielh, P., Perez-Moreno, P. *et al.* (2011) Detection of circulating tumour cells with a hybrid (epithelial/mesenchymal) phenotype in patients with metastatic non-small cell lung cancer. *British Journal of Cancer*, **105**, 1338–1341.
2. Müller, V., Stahmann, N., Riethdorf, S. *et al.* (2005) Circulating tumor cells in breast cancer: correlation to bone marrow micrometastases, heterogeneous response to systemic therapy and low proliferative activity. *Clinical Cancer Research*, **11**, 3678–3685.
3. Kim, M.-Y., Oskarsson, T., Acharyya, S. *et al.* (2009) Tumor self-seeding by circulating cancer cells. *Cell*, **139**, 1315–1326.
4. Ashworth, T.R. (1869) A case of cancer in which cells similar to those in the tumours were seen in the blood after death. *Australian Medical Journal*, **4**, 146–147.
5. Danila, D.C., Pantel, K., Fleisher, M., and Scher, H.I. (2011) Circulating tumors cells as biomarkers: progress toward biomarker qualification. *Cancer Journal*, **17**, 438–450.
6. Food Drug Administration Center for Drug Evaluation and Research (2011) Guidance for Industry E16 Biomarkers Related to Drug or Biotechnology Product Development: Context, Structure, and Format of Qualification Submissions. FDA Maryland.
7. Lesko, L.J. and Atkinson, A.J. (2001) Use of biomarkers and surrogate endpoints in drug development and regulatory decision making: criteria, validation, strategies. *Annual Review of Pharmacology and Toxicology*, **41**, 347–366.
8. Newling, D.W.W. (2009) Issues with the use of prostate-specific antigen as a surrogate end point in hormone-resistant prostate cancer. *European Urology Supplements*, **8**, 13–19.
9. King, J.D., Casavant, B.P., and Lang, J.M. (2013) Rapid translation of circulating tumor cell biomarkers into clinical practice: technology development, clinical needs and regulatory requirements. *Lab on a Chip*, **14**, 24–31.
10. Maheswaran, S., Sequist, L.V., Nagrath, S. *et al.* (2008) Detection of mutations in EGFR in circulating lung-cancer cells. *New England Journal of Medicine*, **359**, 366–377.
11. Holzbeierlein, J., Lal, P., LaTulippe, E. *et al.* (2004) Gene expression analysis of human prostate carcinoma during hormonal therapy identifies androgen-responsive genes and mechanisms of therapy resistance. *American Journal of Pathology*, **164**, 217–227.
12. De Bono, J.S., Scher, H.I., Montgomery, R.B. *et al.* (2008) Circulating tumor cells predict survival benefit from treatment in metastatic castration-resistant prostate cancer. *Clinical Cancer Research*, **14**, 6302–6309.
13. Cristofanilli, M., Budd, G.T., Ellis, M.J. *et al.* (2011) Circulating tumor cells, disease progression, and survival in metastatic breast cancer. *New England Journal of Medicine*, **351**, 781–791.

14. Cohen, S.J., Punt, C.J.A., Iannotti, N. *et al.* (2008) Relationship of circulating tumor cells to tumor response, progression-free survival, and overall survival in patients with metastatic colorectal cancer. *Journal of Clinical Oncology: Official Journal of the American Society of Clinical Oncology*, **26**, 3213–3221.

15. Khoja, L., Lorigan, P., Zhou, C. *et al.* (2013) Biomarker utility of circulating tumor cells in metastatic cutaneous melanoma. *Journal of Investigative Dermatology*, **133**, 1582–1590.

16. Scher, H.I., Jia, X., de Bono, J.S. *et al.* (2009) Circulating tumour cells as prognostic markers in progressive, castration-resistant prostate cancer: a reanalysis of IMMC38 trial data. *Lancet Oncology*, **10**, 233–239.

17. Halabi, S., Small, E.J., Kantoff, P.W. *et al.* (2003) Prognostic model for predicting survival in men with hormone-refractory metastatic prostate cancer. *Journal of Clinical Oncology: Official Journal of the American Society of Clinical Oncology*, **21**, 1232–1237.

18. Okegawa, T., Nutahara, K., and Higashihara, E. (2008) Immunomagnetic quantification of circulating tumor cells as a prognostic factor of androgen deprivation responsiveness in patients with hormone naive metastatic prostate cancer. *Journal of Urology*, **180**, 1342–1347.

19. Koyanagi, K., O'Day, S.J., Boasberg, P. *et al.* (2010) Serial monitoring of circulating tumor cells predicts outcome of induction biochemotherapy plus maintenance biotherapy for metastatic melanoma. *Clin Cancer Res Off J Am Assoc Cancer Res*, **16**, 2402–2408.

20. Carden, C.P., Sarker, D., Postel-Vinay, S. *et al.* (2010) Can molecular biomarker-based patient selection in Phase I trials accelerate anticancer drug development? *Drug Discovery Today*, **15**, 88–97.

21. DiMasi, J.A. and Grabowski, H.G. (2007) Economics of new oncology drug development. *Journal of Clinical Oncology: Official Journal of the American Society of Clinical Oncology*, **25**, 209–216.

22. Tan, D.S.W., Thomas, G.V., Garrett, M.D. *et al.* (2009) Biomarker-driven early clinical trials in oncology: a paradigm shift in drug development. *Cancer Journal*, **15**, 406–420.

23. Sarker, D. and Workman, P. (2007) Pharmacodynamic biomarkers for molecular cancer therapeutics. *Advances in Cancer Research*, **96**, 213–268.

24. Devriese, L.A., Voest, E.E., Beijnen, J.H., and Schellens, J.H.M. (2011) Circulating tumor cells as pharmacodynamic biomarker in early clinical oncological trials. *Cancer Treatment Reviews*, **37**, 579–589.

25. Lim, E., Tay, A., Von Der Thusen, J. *et al.* (2014) Clinical results of microfluidic antibody-independent peripheral blood circulating tumor cell capture for the diagnosis of lung cancer. *Journal of Thoracic and Cardiovascular Surgery*, **147**, 1936–1938.

26. Qi, F., Liu, Y., Zhao, R. *et al.* (2014) Quantitation of rare circulating tumor cells by folate receptor α ligand-targeted PCR in bladder transitional cell carcinoma and its potential diagnostic significance. *Tumour Biology*, **26**, 7217–7223.

27. Jin, T., Peng, H., and Wu, H. (2013) Clinical value of circulating liver cancer cells for the diagnosis of hepatocellular carcinoma: a meta-analysis. *Biomed Reports*, **1**, 731–736.

28. Lowes, L.E. and Allan, A.L. (2014) Recent advances in the molecular characterization of circulating tumor cells. *Cancers*, **6**, 595–624.

29. Becker, T.M., Caixeiro, N.J., Lim, S.H. *et al.* (2014) New frontiers in circulating tumor cell analysis: A reference guide for biomolecular profiling toward translational clinical use. *International Journal of Cancer*, **134**, 2523–2533.

30. Raimondi, C., Gradilone, A., Naso, G. *et al.* (2014) Clinical utility of circulating tumor cell counting through CellSearch(®): the dilemma of a concept suspended in Limbo. *OncoTargets Therapy*, **7**, 619–625.
31. Attard, G. and de Bono, J.S. (2011) Utilizing circulating tumor cells: challenges and pitfalls. *Current Opinion in Genetics and Development*, **21**, 50–58.
32. Alix-Panabières, C. and Pantel, K. (2014) Technologies for detection of circulating tumor cells: facts and vision. *Lab on a Chip*, **14**, 57–62.
33. Thiery, J.P. and Sleeman, J.P. (2006) Complex networks orchestrate epithelial-mesenchymal transitions. *Nature Reviews. Molecular Cell Biology*, **7**, 131–142.
34. Kalluri, R. and Weinberg, R.A. (2009) The basics of epithelial-mesenchymal transition. *Journal of Clinical Investigation*, **119**, 1420–1428.
35. Mikolajczyk, S.D., Millar, L.S., Tsinberg, P. *et al.* (2011) Detection of EpCAM-negative and cytokeratin-negative circulating tumor cells in peripheral blood. *Journal of Oncology*, **2011**, e252361.
36. Mostert, B., Kraan, J., Bolt-de Vries, J. *et al.* (2011) Detection of circulating tumor cells in breast cancer may improve through enrichment with anti-CD146. *Breast Cancer Research and Treatment*, **127**, 33–41.
37. Sieuwerts, A.M., Kraan, J., Bolt, J. *et al.* (2009) Anti-epithelial cell adhesion molecule antibodies and the detection of circulating normal-like breast tumor cells. *Journal of the National Cancer Institute*, **101**, 61–66.
38. Lu, C.-Y., Tsai, H.-L., Uen, Y.-H. *et al.* (2013) Circulating tumor cells as a surrogate marker for determining clinical outcome to mFOLFOX chemotherapy in patients with stage III colon cancer. *British Journal of Cancer*, **108**, 791–797.
39. Doyen, J., Alix-Panabières, C., Hofman, P. *et al.* (2012) Circulating tumor cells in prostate cancer: a potential surrogate marker of survival. *Critical Reviews in Oncology/Hematology*, **81**, 241–256.
40. Park, J.-M., Kim, M.S., Moon, H.-S. *et al.* (2014) Fully automated circulating tumor cell isolation platform with large-volume capacity based on lab-on-a-disc. *Analytical Chemistry*, **86**, 3735–3742.
41. Marrinucci, D., Bethel, K., Kolatkar, A. *et al.* (2012) Fluid biopsy in patients with metastatic prostate, pancreatic and breast cancers. *Physical Biology*, **9**, 016003.
42. Karabacak, N.M., Spuhler, P.S., Fachin, F. *et al.* (2014) Microfluidic, marker-free isolation of circulating tumor cells from blood samples. *Nature Protocols*, **9**, 694–710.
43. Nagrath, S., Sequist, L.V., Maheswaran, S. *et al.* (2007) Isolation of rare circulating tumour cells in cancer patients by microchip technology. *Nature*, **450**, 1235–1239.
44. Casavant, B.P., Guckenberger, D.J., Berry, S.M. *et al.* (2013) The VerIFAST: an integrated method for cell isolation and extracellular/intracellular staining. *Lab on a Chip*, **13**, 391–396.
45. Bode, W., Mocking, J.A., and van den Berg, H. (1991) Influence of age and sex on vitamin B-6 vitamer distribution and on vitamin B-6 metabolizing enzymes in Wistar rats. *Journal of Nutrition*, **121**, 318–329.
46. Sun, W., Jia, C., Huang, T. *et al.* (2013) High-performance size-based microdevice for the detection of circulating tumor cells from peripheral blood in rectal cancer patients. *PLoS One*, **8**, e75865.
47. Gupta, V., Jafferji, I., Garza, M. *et al.* (2012) ApoStream(™), a new dielectrophoretic device for antibody independent isolation and recovery of viable cancer cells from blood. *Biomicrofluidics*, **6**, 24133.

48. Hosokawa, M., Kenmotsu, H., Koh, Y. *et al.* (2013) Size-based isolation of circulating tumor cells in lung cancer patients using a microcavity array system. *PLoS One*, **8**, e67466.
49. Yusa, A., Toneri, M., Masuda, T. *et al.* (2014) Development of a new rapid isolation device for circulating tumor cells (CTCs) using 3D palladium filter and its application for genetic analysis. *PLoS One*, **9**, e88821.
50. Watanabe, M., Serizawa, M., Sawada, T. *et al.* (2014) A novel flow cytometry-based cell capture platform for the detection, capture and molecular characterization of rare tumor cells in blood. *Journal of Translational Medicine*, **12**, 143.
51. Pearl, M.L., Zhao, Q., Yang, J. *et al.* (2014) Prognostic analysis of invasive circulating tumor cells (iCTCs) in epithelial ovarian cancer. *Gynecologic Oncology*, **134** (3), 581–590.
52. Zhou, F., Ma, M., Tao, G. *et al.* (2014) Detection of circulating methylated opioid binding protein/cell adhesion molecule-like gene as a biomarker for ovarian carcinoma. *Clinical Laboratory*, **60**, 759–765.
53. Food Drug Administration Center for Drug Evaluation and Research (2014) Guidance for Industry and FDA Staff Qualification Process for Drug Development Tools. FDA Maryland.
54. Goodsaid, F. and Frueh, F. (2007) Biomarker qualification pilot process at the US Food and Drug Administration. *The AAPS Journal*, **9**, E105–E108.

PART IV

COMMERCIALIZATION

18

DEPARRAY™ TECHNOLOGY FOR SINGLE CTC ANALYSIS

FARIDEH Z. BISCHOFF
Silicon Biosystems, Inc., San Diego, CA, USA

GIANNI MEDORO AND NICOLÒ MANARESI
Silicon Biosystems, SpA, Bologna, Italy

18.1 CHALLENGES IN MOLECULAR PROFILING OF CTCs

Circulating tumor cells (CTCs) may serve as a surrogate for metastatic tumor tissue, allowing the opportunity to make biomarker-based treatment decisions through a liquid biopsy without subjecting patients to the risk of serial tumor biopsies [1]. Despite the clear importance of CTCs, clinical adoption has been challenged due to their rarity in comparison to hematologic cells (with a frequency of typically 1–10 CTC among 10^7 white blood cells (WBCs) per milliliter of blood), which makes them difficult to obtain reliably [2]. Given that there is also inherent phenotypic and genomic heterogeneity within CTCs from a single blood draw [3–5], tremendous technological challenges for their sequential detection, isolation, and analysis must be overcome.

Over a decade ago, the FDA cleared the CellSearch® system (Veridex, LLC, Raritan, NJ, USA; see Chapter 19 for details) for CTC detection, and CTC enumeration was introduced to the clinical community [6]. Despite the limitations of CellSearch technology and data (as discussed in other chapters in this book), investigators continue to use the platform in clinical trials, if only as a means of obtaining CTC counts as a factor in determining patients' prognosis during treatment (as observed on ClinicalTrials.gov). Clear interest in CTCs as a target for molecular analysis in place of

Circulating Tumor Cells: Isolation and Analysis, First Edition. Edited by Z. Hugh Fan.

tumor tissue has sparked the interest of academic and industry investigators to develop technologies, beyond enumeration, that enable actual recovery of CTCs for analysis. Recent reviews note that these technologies can be based on one or more properties of CTCs, including physical (i.e., label-free, density, size, electric charge, and deformability) and biological (i.e., surface protein expression and invasion capacity) features [7, 8]. Today, no single technique has been proven to be superior to all the others for delivering only CTCs when starting with whole blood. Perhaps this is due to the fact that a common definition based on the properties of a CTC is lacking, and until recently, efforts have been focused on enumeration.

Nevertheless, current CTC isolation technologies deliver only an enriched population of cells composed of only a few CTCs intermixed with hundreds to thousands of normal peripheral WBCs. The low CTC purity provided by these methods makes it impractical to carry out routine molecular sequence analysis beyond *in situ* hybridization [9] or to characterize CTC heterogeneity, a requirement of personalized medicine [10–13]. A method that enables highly sensitive subsequent recovery of individual CTCs from an enriched population would ultimately allow for more reliable investigation of the tumor cells. This combined, serial approach is now possible with DEPArray™ technology.

18.2 DEPARRAY™ TECHNOLOGY SOLUTION

DEPArray technology combines synergistically the power of microelectronics with the precision of microfluidics in an automatic platform to identify and isolate individual cells with high accuracy and precision.

The core of DEPArray technology is a disposable microsystem that integrates a microelectronic chip with microfluidic chambers and valves and exploits the basic working principle of negative dielectrophoresis (nDEP) to manipulate cells deterministically. According to the nDEP principle, neutral particles, such as cells, when subject to spatially nonuniform electric fields, experience a net force directed toward locations with decreasing field intensity due to the induction of a dipole moment as a result of electric polarization [14]. To exert and manage nDEP forces, the DEPArray microchip embeds a programmable array of more than 300,000 elements, each of which is 20 μm × 20 μm in size and consists of an electrode and embedded circuits. According to the theory of "*Moving DEP Cages*" [15] implemented by DEPArray, each electrode, energized appropriately, can create a closed nDEP cage (corresponding to a local minimum of the electric field) in suspension in the microchamber, where a single cell can be attracted and kept in levitation. Accordingly, up to ~30,000 closed nDEP cages can be established over the whole array by applying an appropriate pattern of in-phase and counterphase alternating current (AC) voltages to the 300,000 electrodes and to the lid (which is conductive and transparent). In this way, thousands of cells can be suspended in stable levitation and controlled independently. In fact, because the array is programmable, the position of each nDEP cage can be reconfigured by software, making possible the selective movement of each individual cell along any path in the whole array.

With this approach, any cell of interest can be isolated from the bulk deterministically. The selection of cells with specific desired attributes – including both physical characteristics such as size and shape and biological characteristics such as surface protein expression – is based on image analysis. The DEPArray system embeds a multichannel fluorescence microscope with six excitation filters and up to eight emission filters, in addition to bright-field, 10× and 20× magnification, solid-state LED light sources, and a high-resolution CMOS camera. A gallery of high-quality images is captured for each individual cell in the sample and processed by the system to enable accurate analysis and selection of target cells, which are easily distinguished from spurious events by the system's CellBrowser™ software (Figure 18.1).

Once identified, selected cells can be separated from the sample matrix to produce a purified preparation. The process is deterministic and automatic. The selected nDEP cages containing the target cells are moved stepwise, concurrently, and independently along trajectories calculated by the computer, dragging each target cell from its original location in the main chamber (filled with the suspension of enriched cells) into the parking chamber (filled with clean buffer). Once parked, each target cell can be moved from the parking chamber into the exit chamber, again by stepwise transfer between adjacent nDEP cages. From the exit chamber, target cells are ejected gently from the device directly into a recovery support (multiwell, tube, slide, etc.) by flowing clean buffer solution under microfluidic control. The recovery procedure can be repeated up to 35 times to obtain 35 separate recoveries of individual target cells or group of cells from the same microchamber. Recovered cell preparations are 100% pure, and the individual cells are intact and viable for further analysis.

The DEPArray system can analyze samples containing up to 40,000 cells and is compatible with a wide variety of sample suspensions, including both fixed and live cells, small cell loads from fine needle aspirates, formalin-fixed paraffin-embedded tissue biopsies, and enriched cell suspensions from tissues including whole blood, bone marrow, and cerebral spinal fluid. Cells are moved gently without contact or friction, overcoming potential problems with cell adhesion or damage. No shear forces or strain on the cells are exerted by fluidic movements, and the low-voltage high-frequency electric fields are tuned to avoid stresses to the cells.

18.3 DEPARRAY™ FOR SINGLE TUMOR CELL ANALYSIS

By enabling the isolation of pure, viable rare cells, the DEPArray system enables new solutions for applications in a variety of areas such as oncology research, drug response, and fetal cell biology. From simple spike experiments, we and others [16] have demonstrated a reliable workflow using the DEPArray system to overcome limitations associated with an impure and heterogeneous sample by delivering 100% pure tumor cells from a mixed population of tumor and normal cells after enrichment using the CellSearch system (Figure 18.2). Efficiency in single-cell recovery followed by molecular analysis was first demonstrated with healthy-donor peripheral blood samples spiked with viable KRAS-mutated tumor cell lines (SW480 colorectal cancer cell line; A549 lung cancer cell line) and enriched using the CellSearch Epithelial

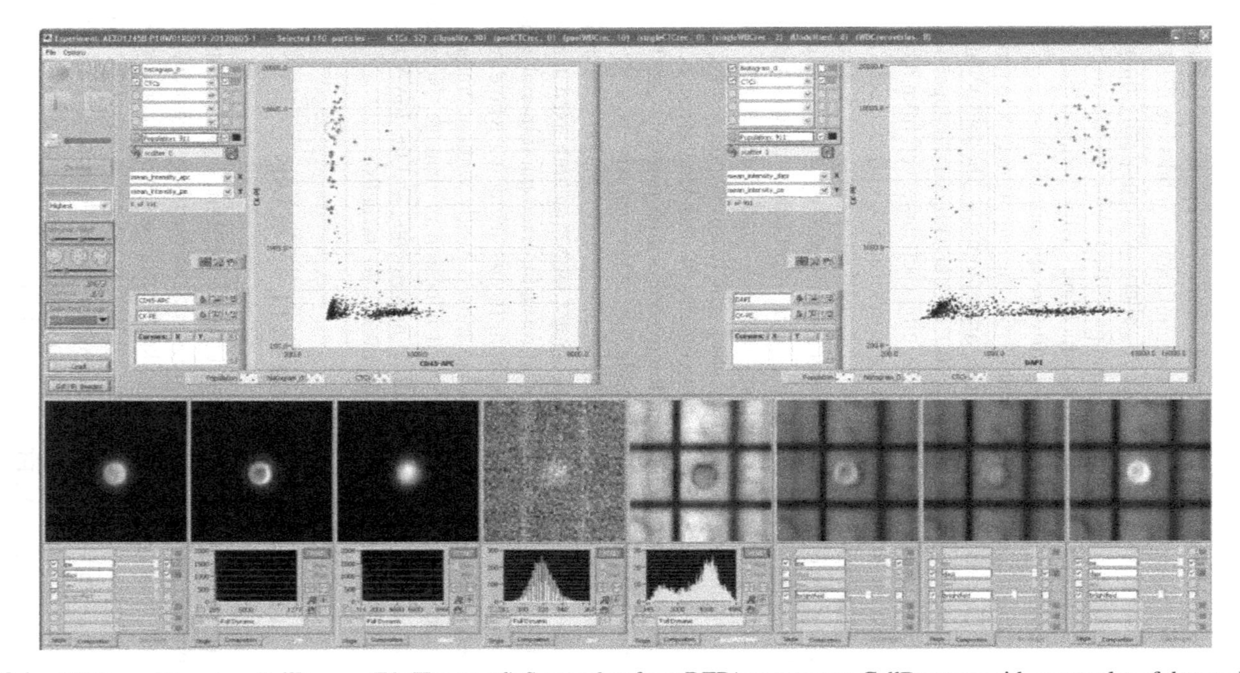

Figure 18.1 *DEPArray™ system CellBrowser™.* (Top panel) Screenshot from DEPArray system CellBrowser with scatterplot of detected cells and an image bar displaying an individual tumor cell image as identified by presence of cytokeratin (green signal), absence of CD45, and nuclear staining (DAPI, magenta signal) and individual channels in grayscale, CK-PE, DAPI, CD45-APC; bright field (BF) is also displayed along with BF/CK, BF/DAPI and BF/CK/DAPI overlay in false colors (gray/green, gray/magenta, gray/green/magenta). (Bottom panel) Gallery of tumor cell images detected (DAPI/magenta and CK-PE/green) using CellBrowser. *(See color plate section for the color representation of this figure.)*

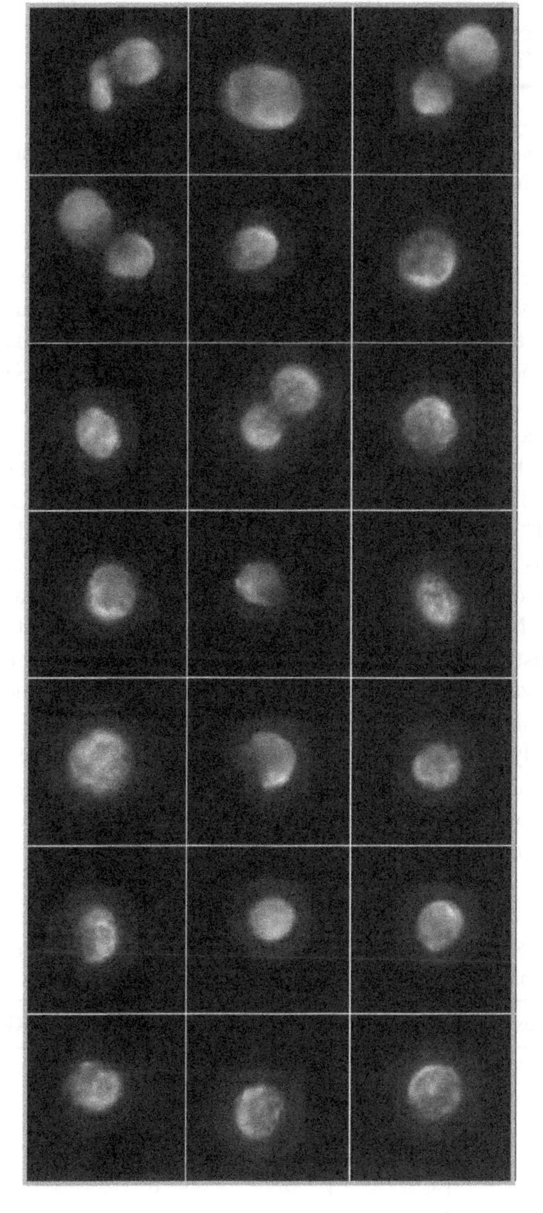

Figure 18.1 (*continued*)

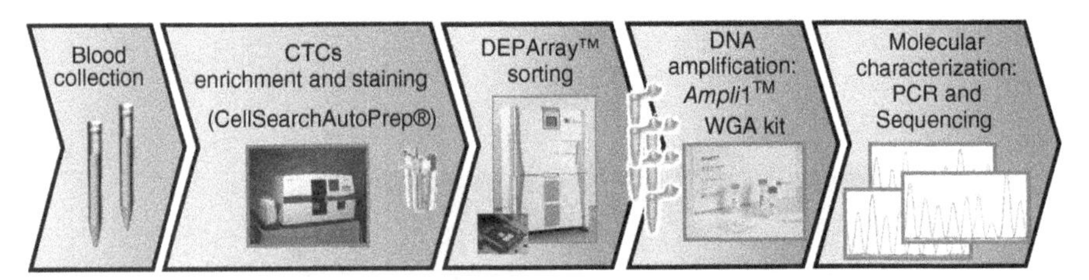

Figure 18.2 *DEPArray™ CTC workflow.* Illustration of workflow based on the DEPArray platform for CTCs separation and molecular characterization from blood samples enriched with Veridex CellSearch CTC Kit.

Cell Kit. Enriched cells were extracted from the CellSearch cartridge and prepared for sorting and analysis on a DEPArray system. Multiple individual pure collections of single tumor cells and WBC controls were recovered along with small cell pools. Following whole-genome amplification (WGA) using the *Ampli*1™ Kit (from Silicon Biosystems), DNA libraries were fingerprinted by short tandem repeat (STR) analysis and then sequenced for the KRAS gene-specific mutations. Image-based analysis of all events by the DEPArray system allowed clear identification and selection of desired "ideal" tumor cells ($CK^+/CD45^-/DAPI^+$) (Figure 18.1) and WBCs ($CK^-/CD45^+/DAPI^+$). Overall, better than 90% recovery of targeted single tumor cells (Table 18.1) as well as detection of tumor heterogeneity was achieved. All successfully amplified cells matched 100% with KRAS mutational status and DNA fingerprint analysis. No alleles from the donor (WBCs) were present in the tumor cell profiles, demonstrating high sensitivity. Similarly, Peeters *et al.* demonstrated molecular characterization of breast cancer tumor cells for the assessment of predictive biomarker analysis and heterogeneity using the DEPArray system [17]. Expected mutations in clinically relevant markers were obtained for 60% of the single recovered tumor cells and all groups of up to 10 tumor cells, demonstrating reliable gene expression profiles for spiked cultured breast cancer cells. More recently, simultaneous targeted copy number variation (CNV) and single nucleotide variant (SNV)

TABLE 18.1 KRAS Sequencing Results Following Cell Sorting Using the DEPArray™ and *Ampli*1™ WGA

DEPArray™ Recoveries	*N*	KRAS Call	KRAS No Call	KRAS Call Rate
Single cells	56	51	5	91%
TC	38	36	2	95%
WBC	18	15	3	83%
5 cells	5	5	0	100%
TC	1	1	0	100%
WBC	4	4	0	100%
Negative controls	8	0	0	NA

Note: TC, tumor cell; WBC, white blood cells; 5 cells, five cells pooled together; *N*, number of cells.

detection on DEPArray isolated single CTCs using the *Ampli*1 WGA and IonTorrent® AmpliSeq™ custom panel kits have been demonstrated. In four out of four CTCs from one breast cancer patient, they observed a common somatic heterozygous mutation in PIK3CA (not found in any of the patients' WBCs) but PTPN11 somatic homozygous mutations in only two of the four CTCs.

Fabbri *et al.* first demonstrated the use of the DEPArray platform to detect and sort pure CTCs in patient samples [18]. Analyses were conducted on peripheral blood samples from 21 patients with metastatic colon cancer initially shown to have CTCs by CellSearch. After 100% pure cell recovery and WGA, the KRAS gene mutations of the CTCs screened were compared to gene status in the primary tumor tissue. A 50% mutational concordance between CTCs and primary tumor was observed, demonstrating feasibility in performing single-CTC molecular analysis, thereby avoiding lymphocyte contamination and potentially negative results. More recently, Pestrin *et al.* reported concordance of *PIK3CA* (exon 20) mutation in five of six breast cancer cases when comparing primary tumor to matched patient CTCs recovered using the DEPArray [19].

18.4 CLINICAL SIGNIFICANCE IN SINGLE CTC PROFILING

Until recently, the clinical significance of CTCs for treatment decisions has been unclear. Several studies now suggest the clinical relevance of the DEPArray platform for single CTC profiling. Small-cell lung cancer (SCLC), an aggressive, often inoperable and not biopsied neuroendocrine tumor with poor prognosis, accounts for 15–20% of lung cancer cases. Given that CTCs are prevalent in SCLC, patients would clearly benefit from a readily accessible "liquid biopsy." Hodgkinson *et al.* have now shown that CTCs from patients with either chemosensitive or chemorefractory SCLC are tumorigenic in immunocompromised mice and that the resultant circulating tumor cell-derived explants (CDXs) mirror the donor patient's response to platinum and etoposide chemotherapy [20]. Genomic analysis of DEPArray isolated CTCs not only revealed considerable similarity to the corresponding CDX but that the most marked differences were observed between CDXs from patients with different clinical outcomes. The authors clearly demonstrate that CTC molecular analysis via serial blood sampling could facilitate delivery of personalized medicine for SCLC.

In another study, Fernandez *et al.* reported that molecular characterization of CTCs may be critical to identifying key drivers of cancer metastasis and determining the best therapeutic approach for patients [21]. CTCs from two triple-negative breast cancer patients were enriched using CellSearch and selected as single cells using the DEPArray system to investigate for the presence of *TP53* R110 fs*13 mutations by next-generation sequencing (NGS) as observed in the breast and chest skin biopsies of two patients. From six single CTCs isolated from one patient, one CTC had TP53 R110 delC, another CTC showed the TP53 R110 delG mutation, and the remaining four single CTCs showed the wild-type p53 sequence. A pool of 14 CTCs isolated from the same patient also showed TP53 R110 delC mutation. In the tumor breast tissue of this patient, only the TP53 R110 delG mutation was detected. In the second patient, a TP53 R110 delC mutation was detected in the chest wall skin biopsy. From the peripheral blood of this patient, five single CTCs and six clusters of two to six CTCs were isolated. Three of the five single CTCs showed the TP53 R110 delC mutation, and two CTCs showed the wild-type TP53 allele; from the clusters, five showed the TP53 R110 delC mutation, and one cluster showed the wild-type TP53 allele. Single WBCs isolated as controls from both patients only showed the wild-type TP53 allele. These results indicate that CTCs could represent a noninvasive source of cancer cells from which to determine genetic markers of the disease progression and potential therapeutic targets.

Polzer *et al.* also combined a workflow for enrichment with DEPArray isolation of pure CTCs and the nonrandom WGA of single cells using the *Ampli*1 method [22]. They defined a genome integrity index (GII) to identify single cells suited for molecular characterization by different molecular assays, such as diagnostic profiling of point mutations, gene amplifications, and whole genomes of single cells. Following evaluation of 510 single CTCs and 189 leukocytes from 66 CTC-positive breast cancer patients, they demonstrated >90% successful molecular analysis of high-quality clinical breast cancer samples. Genomic disparities between primary tumors and CTCs were readily identified. Microheterogeneity analysis among individual CTCs uncovered pre-existing cells resistant to ERBB2-targeted therapies, suggesting ongoing microevolution at late-stage disease. Further exploration may provide essential information for personalized treatment decisions and shed light on mechanisms of acquired drug resistance.

18.5 CONCLUSION

A growing number of studies now highlight the clinical significance of isolation and molecular characterization of individual CTCs. Moreover, they demonstrate the value of combining technologies that first enable reliable enrichment followed by a second more specific means of evaluating each target cell present in the enriched cell population individually. Unlike other systems that are best suited for enrichment, the DEPArray system has proven to be a compatible and robust platform for single-cell sorting, isolation, and recovery of CTCs for downstream molecular analysis, including NGS.

REFERENCES

1. Pantel, K. and Alix-Panabieres, C. (2013) Real-time liquid biopsy in cancer patients: fact or fiction? *Cancer Research*, **73**, 6384–6388.
2. Bednarz-Knoll, N., Alix-Panabières, C., and Pantel, K. (2011) Clinical relevance and biology of circulating tumor cells. *Breast Cancer Research*, **13**, 228.
3. Powell, A.A. *et al.* (2012) Single cell profiling of circulating tumor cells: transcriptional heterogeneity and diversity from breast cancer cell lines. *PLoS One*, **7**, e33788.
4. Heitzer, E. *et al.* (2013) Complex tumor genomes inferred from single circulating tumor cells by array-CGH and next-generation sequencing. *Cancer Research*, **73**, 2965–2975.

5. Peeters, D.J. *et al.* (2013) Semiautomated isolation and molecular characterisation of single or highly purified tumour cells from CellSearch enriched blood samples using dielectrophoretic cell sorting. *British Journal of Cancer*, **108**, 1358–1367.
6. Cristofanilli, M., Budd, G.T., Ellis, M.J. *et al.* (2004) Circulating tumor cells, disease progression, and survival in metastatic breast cancer. *New England Journal of Medicine*, **351**, 781–791.
7. Alix-Panabieres, C. and Pantel, K. (2014) Challenges in circulating tumour cell research. *Nature Reviews Cancer*, **14**, 623–631.
8. Krebs, M.G. *et al.* (2014) Molecular analysis of circulating tumour cells – biology and biomarkers. *Nature Reviews. Clinical Oncology*, **11**, 129–144.
9. Pecot, C.V., Bischoff, F.Z. *et al.* (2011) Novel platform for detection of CK+ and CK- CTCs. *Cancer Discovery*, **1**, 580–586.
10. Yu, M. *et al.* (2013) Circulating breast tumor cells exhibit dynamic changes in epithelial and mesenchymal composition. *Science*, **339**, 580–584.
11. Lustberg, M.B. *et al.* (2014) Heterogeneous atypical cell populations are present in blood of metastatic breast cancer patients. *Breast Cancer Research*, **16**, R23.
12. Hodgkinson, C.L. *et al.* (2014) Tumorigenicity and genetic profiling of circulating tumor cells in small-cell lung cancer. *Nature Medicine*, **20**, 897–903 10.1038/nm.3600.
13. Zhang, L. *et al.* (2013) The identification and characterization of breast cancer CTCs competent for brain metastasis. *Science Translational Medicine*, **5**, 180ra48.
14. Pohl, H.A. (1978) *Dielectrophoresis*, Cambridge University Press, Cambridge.
15. G. Medoro et al., CMOS-only sensor and manipulator for microorganisms, International Electron Device Meeting (IEDM), S. Francisco, USA, pp. 415-418 (2000).
16. Carpenter, E.L. *et al.* (2014) Dielectrophoretic capture and genetic analysis of single neuroblastoma tumor cells. *Frontiers in Oncology*, **4**, 1–14.
17. Peeters, D.J.E. *et al.* (2013) Semiautomated isolation and molecular characterization of single or highly purified tumour cells from CellSearch enriched blood samples using dielectrophoretic cell sorting. *British Journal of Cancer*, **108**, 1358–1367.
18. Fabbri, F. *et al.* (2013) Detection and recovery of circulating colon cancer cells using a dielectrophoresis-based device: KRAS mutation status in pure CTCs. *Cancer Letters*, **335**, 225–231.
19. Pestrin, M. *et al.* (2015) Heterogeneity of PIK3CA mutational status at the single cell level in circulating tumor cells from metastatic breast cancer patients. *Molecular Oncology*, **9**, 749–757.
20. Hodgkinson, C.L. *et al.* (2014) Tumorigenicity and genetic profiling of circulating tumor cells in small-cell lung cancer. *Nature Medicine*, **20**, 897–905.
21. Fernandez, S.V. *et al.* (2014) TP53 mutations detected in circulating tumor cells present in the blood of metastatic triple negative breast cancer patients. *Breast Cancer Research*, **16**, 445–456.
22. Polzer, B. *et al.* (2014) Molecular profiling of single circulating tumor cells with diagnostic intention. *EMBO Molecular Medicine*, **6**, 1371–1386.

19

CELLSEARCH® INSTRUMENT, FEATURES, AND USAGE

DENIS A. SMIRNOV AND BRAD W. FOULK
Oncology Biomarkers, Janssen R&D, Oncology Biomarkers, Spring House, PA, USA
MARK C. CONNELLY
Oncology Biomarkers, Janssen R&D, Huntingdon Valley, PA, USA
ROBERT T. MCCORMACK
Oncology Biomarkers, Janssen R&D, Raritan, NJ, USA

19.1 INTRODUCTION

While complete cures for specific cancers remain elusive, the number of the therapies available to treat particular cancers continues to increase. In 2013 alone, 12 new therapies gained regulatory approval in the United States (19 in 2012). The rise in the number of available therapies increases the need for optimal selection between the therapeutic regimens available to a given patient at a specific point of disease progression. Selection of optimal therapy is particularly important because cancer is a constantly evolving heterogeneous disease that has to be treated with drugs that attack its current vulnerabilities. Metastatic progression remains the cause of death in the majority of patients with solid tumors [1]. Inability to sample all metastatic sites, challenges with biopsies of primary sites, and the cost and morbidity associated with repeat biopsies clash with this increased need to monitor and characterize the dynamic changes in a patient's cancer. Circulating biomarkers such as circulating tumor cells (CTCs), circulating tumor DNA, exosomes, and microparticles are

Circulating Tumor Cells: Isolation and Analysis, First Edition. Edited by Z. Hugh Fan.

getting increased attention as a way to noninvasively obtain access to tumor material for continuous monitoring.

Of all circulating biomarkers, CTCs have the longest history. First discovered in 1869 by Thomas Ashworth, CTCs were originally believed to be important for cancer progression but have remained poorly understood due to technical challenges in their isolation and detection. Nevertheless, even as early as 1962, the prognostic value of CTCs in breast cancer patients was being considered [2, 3]. However, these early studies of CTCs suffered from inconsistencies between methods of CTC isolation and identification. For example, the percentage of breast cancer patients with CTCs varied from 3% to 50% in several studies summarized in 1963 [4]. Development of uniform, standard CTC isolation methods were hampered by the lack of available and affordable options for automated blood processing and imaging. Since the mid-1970s, development of monoclonal antibodies, flow cytometry, polymerase chain reaction (PCR), immunomagnetic solutions, and computer-controlled imaging options opened the door to a variety of novel strategies to isolate and study CTCs [5].

Flow cytometry clearly emerged as a method of choice for quantitative cellular analysis of biologic fluids such as blood. Identifying epithelium-derived tumor cells in blood by flow cytometry requires using a combination of parameters that uniquely identify tumor cells from the normal hematopoietic cells present. This can be done by simultaneously assessing light scatter parameters, fluorescent staining with a nucleic-acid-binding dye, lack of expression of the leukocyte antigen CD45, and positive expression of epithelial cell specific markers (such as epithelial cell adhesion molecule (EpCAM) or cytokeratin). Using flow cytometry, it was determined that CTCs, defined as described earlier, are rare events, with numbers as low as 1 CTC in 10^6–10^7 leukocytes in the peripheral blood of cancer patients [6–9]. It is interesting to note that these flow cytometric estimates were obtained without positive or negative cell enrichment techniques; all the cells in the blood were being interrogated. It was also estimated that, if the frequency of CTCs is 1 in 10^6 leukocytes, at least 15,705,214 nucleated events must be acquired by a flow cytometer to pinpoint 10 cells of interest with 95% confidence. This corresponds to approximately 13 ml of blood, based on the assumption that 1 ml of blood has roughly 1.2 million nucleated events per milliliter. Twenty milliliter of whole blood would have to be assessed for the lower frequency at 1 CTC in 10^7 leukocytes. It was determined that the routine limit of detection for rare cells using a flow cytometer is approximately 10^5 [6–10]. Nevertheless, since flow cytometry analysis does not require any enrichment of cells, it is possible to use this method to enumerate rare cells of a given phenotype in some patients without introducing technological or methodological bias. Thus, at least in subset of patients with high CTC numbers, flow cytometry can be used as an unbiased performance comparator for any rare cell enrichment technique. Consequently, for any rare cell analytical method if the number of cells detected in a given biological sample routinely exceeds the number of cells detected in an identical sample by flow cytometry, when identical cellular characteristics are used as identifiers, the reason for this discrepancy must be carefully scrutinized.

Since CTCs are very rare cells, their accurate and reliable detection and analysis can be difficult and confounded by small changes in methods, reagents, or technique.

In order to ensure that results can be compared across laboratories, particularly in the clinical setting, detection and characterization of CTCs must be highly reproducible. Currently, all CTC analysis methods have two basic components: enrichment and detection. The enrichment may be based on positive or negative selection. Positive selection uses markers/characteristics that are unique to the target cells (i.e., expression of epithelial markers for tumor cell selection, differences in size and deformability, electrical properties, etc.). Negative selection uses markers/characteristics that are unique for nontarget cells (i.e., CD45 for leukocyte depletion). Positive selection has the advantage of typically high target cell specificity and purity but has the disadvantage of a limited cellular range. Only those cells targeted are enriched and may or may not represent the full complement of tumor cells present. Negative depletion methods actively remove nontumor cells and have the advantage of possibly enriching the entire complement of tumor cells present in the sample. However, they generally suffer from poor specificity, being unable to remove all normal components, and subsequently require complex cell identification algorithms be applied to the depleted cell population(s) that remains. Since efficient implementation of negative selection may be difficult, and the antigenic expression of the full complement of CTCs that may be circulating were not known, it is not surprising that CELLSEARCH® and most initial attempts to develop reliable methods for CTC enrichment were focused on positive selection.

19.2 PRINCIPLES OF CELLSEARCH®

The CELLSEARCH system uses positive selection to enrich tumor cells of epithelial origin out of whole blood. It does this by use of small magnetic particles (150–180 nM) that are coated with antibody to EpCAM. The magnetic particles have an iron core of Fe_3O_4 and are small enough to be colloidally stable. As such, the reagent is referred to as ferrofluid (FF). The cell specificity for the FF is determined by the specificity of the antibody (or antibodies) bound to the particle. The CELLSEARCH system incubates anti-EpCAM coated FF with 7.5 ml of blood from a patient. It magnetically pulls the FF particles back and forth through the sample to maximize collisions between the particles and target cells. The AutoPrep also adds capture enhancement reagent (see the following text) to maximize retention of low antigen density cells. Target cells are magnetically pulled to the side of the tube, and nontarget cells, red blood cells, and white blood cells are removed by aspiration. The remaining cells are washed, and the total volume is reduced. Cells are then permeabilized and stained with the nuclear dye DAPI, and antibodies to cytokeratin 8, 18, and 19 conjugated to phycoerythrin (PE) and with antibody to CD45 conjugated to allophycocyanin (APC). After incubation, the cells are washed again and the volume is reduced to a final volume of 380 µl. The enriched and stained cell suspension is transferred to a CELLSEARCH cartridge within a Magnest® where the stained cells are magnetically mounted against a glass coverslip and then scanned and imaged on the CellTracks Analyzer II® (CTA II). The CTA II autofocuses on DAPI signals within the Magnest and then images the entire surface of the cartridge chamber in four colors. At each location in the cartridge, an image is recorded for DAPI (blue fluorescence), fluorescein isothiocyanate (green fluorescence), PE (orange fluorescence), and APC (red fluorescence). Green fluorescence channel can be used to track expression of additional markers. The software then presents images of events to the user showing the event in all four colors and a composite overly that the operator can use to identify CTCs from non-CTC events. For a cell to be scored as a CTC, it must be at least 4 μm in size, EpCAM-positive, must have a nucleus within a cytoplasm that is positive for cytokeratin, and must be negative for CD45. Clusters of cells where all or most of the cells within the cluster meet these criteria are often seen. Because it is often difficult to determine the number of cells in a cluster, whenever there are two or more cells stuck together in a single imaged event, the cluster is counted as one CTC regardless of how many cells may be present in that cluster. Events that do not contain a nucleus are often present, but not counted as CTCs, and events that are CD45-positive are not counted as CTCs regardless of other reactivities. The AutoPrep system can process up to eight samples and/or controls at a time in approximately 3.5 h, and it then takes 11 min to scan each cartridge on the CTA II.

EpCAM-positive, Cytokeratin-positive, CD45-negative cell with a nucleus is not the only possible definition for a CTC. However, it is one of the least controversial, most conservative, and highest specificity definitions for a cell of epithelial tumor origin because none of the formed elements of blood are of epithelial origin and do not express these markers [11]. This definition was chosen therefore because of its high specificity and minimum ambiguity compared to other cells and other rare cell populations in blood. As will be seen, CTCs may express other antigens, but often these other antigens are shared to a lesser or greater extent with other cells normally found in blood.

CELLSEARCH is the first and still the only standardized CTC enrichment system with regulatory approval in the United States. The CELLSEARCH system consists of two automated instruments; the CellTracks AutoPrep® system automates sample processing. The AutoPrep enriches the CTCs from blood and stains them for immunofluorescent interrogation by the CTA II. The CTA II is a computer-controlled four-color fluorescence microscope that scans the prepared samples. In addition, CELLSEARCH includes a special blood tube (CellSave tube) that standardizes blood collection and stabilizes CTCs for subsequent detection. Blood collected in CellSave can be stored at room temperature up to 96 h after collection without significant loss in the number of tumor cells. The CellSave tube is an absolutely essential component of the CELLSEARCH system that enables patient's blood to be collected for CTC analysis in many clinical settings and locations and then shipped to a laboratory for analysis. The CELLSEARCH system also includes reagent kits as well as image analysis software that ensure that CTC detection and enumeration is highly reproducible.

19.3 EpCAM DENSITY AND CTC CAPTURE

Although EpCAM is widely expressed on the cells of epithelial solid tumors (11, 12), the level of EpCAM expression may vary dramatically. Heterogeneous

EpCAM expression is seen within a tumor, between tumors, and across different tumor microenvironments. This heterogeneity is reflected in the density of EpCAM found on the cell surface of various cell lines. Some cell lines express hundreds of thousands of EpCAM molecules per cell, and others only a couple of thousand. The quantitative expression of EpCAM has been investigated [11] and reviewed [12]. Rao *et al.* investigated EpCAM expression on cell lines and also on CTCs from patient samples [11]. Using quantitative fluorescence to measure the binding of anti-EpCAM to cells and calculate the number of molecules per cell, Rao *et al.* showed the assay is able to detect a broad range of EpCAM expressions [11]. The commercially available cell line PC3 exhibited a bimodal distribution of EpCAM expression, with some cells expressing relatively high EpCAM and others expressing much lower EpCAM. PC3 cells were then subcloned to isolate a stable cell line from the low EpCAM expressing population. The PC3-9 cell line was created as a result of this subcloning effort. Table 19.1 shows expression levels on various cell lines and, importantly, on CTCs from 200 patients' samples [11].

The cell line T24 expresses 235 times fewer EpCAM molecules per cell on its surface than does MCF-7 and between 20 and 28 times less EpCAM than was found on patient sample CTCs. About 20 to 235 times fewer capture molecules available would be expected to lead to very poor recovery of T24s spiked into donor blood compared to the recovery of MCF-7 or SKBr-3s. This is indeed exactly what one finds when it is not possible to amplify the magnetic load on a target cell above the number of molecules expressed by that cell. However, because of the controlled aggregation process used by CELLSEARCH, 37% of T24 cells are recovered from spiked donor blood. Although T24 expresses 205 times fewer EpCAM molecules per cell than SKBr-3s and 24 times fewer EpCAM molecules than PC3-9s, their recovery drops by a factor of only 2 when controlled aggregation is used (Table 19.2).

Controlled aggregation allows CELLSEARCH to bind many magnetic particles to the surface of a cell for every EpCAM molecule present. This ability to amplify

TABLE 19.1 Mean Number of EpCAM Molecules Per Cell on Carcinoma Cell Lines and CTCs from the Blood of Patients with Various Metastatic Carcinomas

Cell Line or Patient CTC Number	Number of Samples	Number of EpCAM Molecules Per Cell		
		Mean	Median	% CV
MCF-7	6	509,500	508,000	40
COLO205	2	480,000	480,000	39
SKBr-3	3	445,000	455,000	43
LnCaP	3	336,667	355,000	37
PC3-9	3	51,667	54,000	78
T24	3	2,167	2,100	55
2–5 CTC	58	43,412	30,073	96
>5 CTC	42	58,404	47,535	73
>2 CTC	100	49,708	39,358	86

TABLE 19.2 Cell Recovery from Blood with or without the Use of Controlled Aggregation

Cell Line	Percent Recovery without Controlled Aggregation Mean ± CV ($n = 6$)	Percent Recovery with Controlled Aggregation Mean ± CV ($n = 6$)
SKBr-3	71 ± 7	79 ± 8
PC3-9	24 ± 19	88 ± 5
T24	1 ± 29	37 ± 14

the magnetic load on the cell following antigen–antibody binding means that the magnetic load per cell is less dependent on antigen expression than it would otherwise be if controlled aggregation were not used. Table 19.2 shows the importance of this enhanced magnetic loading for the efficient capture of low antigen expressing cells. Controlled aggregation will become critical to understanding the ability of CELLSEARCH to capture patient sample CTCs and during our discussion of cells undergoing epithelial mesenchymal transition (EMT).

CELLSEARCH is the only system to use controlled aggregation process. During manufacturing the magnetic particles, or FF, are coated with antibodies that confer the binding specificity to the particle (i.e., anti-EpCAM, anti-CD146, or combinations of antibody specificities). After the antibodies are bound, the antibody coated FF is further functionalized by coating with a covalently conjugated desthiobiotin–bovine serum albumin (dbBSA) conjugate. Desthiobiotin is a structural analog of biotin. It will bind to streptavidin at the biotin binding site, but it binds with much lower affinity than does biotin. The dbBSA imparts an important new function to the FF particle, the ability to reversibly aggregate in the presence of free streptavidin.

As a sample is processed on the AutoPrep, FF binds to target cells by virtue of the antibody present on the FF. If the cell expresses a lot of the target antigen (i.e., EpCAM), it will bind a lot of magnetic particles. If the cell expresses only a few target molecules, it will bind a few magnetic particles. The number of FF particles bound may be too few to enable efficient magnetic selection and retention during the capture and subsequent staining and washing steps. However, free streptavidin is also added to this sample–reagent mix. The multivalent streptavidin binds the desthiobiotin on the FF bound to the surface of target cells and also cross-links additional free FF to the cell surface by reacting with the desthiobiotin on the free FF. This essentially grows a crystal of iron particles onto the cell and increases the effective size of each FF particle from 150 nM to approximately 800 nM or more. Therefore, even a cell with very low EpCAM expression will bind a disproportionately large amount of iron onto its surface. This affects efficient capture of both high and low antigen expressing cells because the magnetic moment of low antigen expressing cells is greatly amplified, making their capture only loosely related to quantitative antigen expression.

As the sample continues through the processing steps, the volume is reduced from 7.5 ml to 800 ul or less, and by then, wash steps and aspirations have removed most of the nontargeted cells and all of the plasma of the original sample. The smaller volume and reduced viscosity means that the magnets are physically closer to the target cells and less iron is needed to magnetically retain target cells. As a consequence, buffers containing biotin are added to the sample. The biotin in the buffers competes effectively with desthiobiotin and disrupts the streptavidin–desthiobiotin bonds holding together aggregated FF. This effectively "melts" the FF aggregates and returns the FF back to its original 150–200 nm size, which facilitates the ability to image cells after they have been interrogate with fluorescent antibodies.

19.4 CLINICAL APPLICATIONS OF CELLSEARCH® CTCs

19.4.1 CTC Enumeration

Ever since Ashworth first speculated that cells found in a patient's blood that resembled cells from a patient's tumor might explain the existence of multiple tumors in the patient's body, many reports have speculated on the relevance of finding tumor cells in blood [2]. However, it was not until the introduction of the CELLSEARCH technology that researchers were fully able to understand the clinical relevance of CTCs. The first and only clinically and analytically valid CTC technology assured researchers and clinicians that detecting CTCs and noting changes in CTC numbers over time were not artifacts of instrumentation or technician variability.

To enter into clinical decision-making, a technology and its associated reagents must be extensively tested and validated for analytical and clinical performance. For CELLSEARCH, this was a three-step process accomplished over several years. First, a definition of CTC was determined, which has been unchanged since the early days of technology development. The definition, EpCAM-positive (by virtue of the capture antibody), cytokeratin (CK)-positive, CD45-negative (to exclude white blood cells), DAPI-positive (intact nucleus), and >4 μm in size is a very conservative but accurate definition of an epithelial cell, which is the origin for most adult cancers [13]. Coincident to development of CELLSEARCH, it was observed that tumor cells prior to entering into the blood transition to a mesenchymal phenotype (EMT), and it was proposed that this transition is necessary for successful seeding of distant sites [14]. The implication is that with such a transition, these cells would no longer be detected in the blood by CELLSEARCH. We and others have confirmed this event, but we have shown that the transition is not necessarily absolute and that most transitioning cells are epithelial–mesenchymal hybrids, readily detected by CELLSEARCH [15]. Nonetheless, the product claims and all of the studies published to date are based on the clinical significance of CELLSEARCH-defined tumor cells only.

Second, the technology and reagents were analytically validated at three clinical laboratory testing sites, using three CELLSEARCH systems, three manufactured lots of reagents, multiple technicians, and several months of testing. Shortly after commercial distribution, all analytical claims were extensively validated in academic labs located in Germany and the United States, which remains the most extensive independent validation of CTC technology to date and a key event in the acceptance of the technology [16].

Finally, the technology and reagents were clinically validated across three prospective, outcome-based clinical trials of over 600 total patients with metastatic breast [17], prostate [18, 19] or colorectal cancer (CRC) [20]. Moreover, the three trials consisted of over 400 total subjects, consisting of patients with benign disorders that might confound the interpretation of results, as well as normal, healthy individuals to establish the reference range. Collectively, results from these 1000 subjects provided the basis for the technology's first three clinical indications from the FDA and clearly delineated the test's performance characteristics.

In these studies, CTCs were analyzed as a dichotomous variable and a cutoff of ≥5 CTC was established for metastatic breast and prostate cancers and ≥3 CTCs for CRC. It has since been proposed that CTCs might best be analyzed as a continuous variable [21] since increasing CTC numbers are inversely correlated with decreasing progression-free survivor (PFS) and overall survival (OS). However, to date, most clinical and clinical research applications use CTCs with the proposed cutoffs.

All three clinical validation studies yielded highly similar clinical performance results for CTCs in the three metastatic cancers. This is best explained by two basic observations related to the biology of CTCs first shown in these three studies. First, not all patients have CELLSEARCH-defined CTCs in their blood, even during overt disease progression. Conversely, patients demonstrating RESIST-defined response to a therapy can have elevated CTCs in their blood. Interestingly, it has been shown that the CTC counts are more accurate predictors of therapy response than imaging [20], emphasizing the proposal that CTCs more strongly reflect the biology of the cancer than the amount of cancer within an individual [22, 23].

The second observation is that patients with elevated CTCs demonstrate an early and sometimes appreciable drop in CTC numbers coincident with the start of therapy. This was first seen in metastatic breast cancer patients, and it was proposed that patients who do not show such a drop in CTCs within 3–5 weeks after the start of therapy are probably on ineffective therapy [17]. Scher *et al.* introduced the concept of conversion (as a fold change) in CRPC if a patient went from ≥5 CTC to <5 CTC within 4–8 weeks of therapy initiation, indicating a favorable response to treatment [19]. These observations formed the basis for most of the clinical research of CTCs in the past 10 years.

Together, these basic observations from the first three clinical validation studies form the basis of the intended use for CELLSEARCH and its application in clinical practice: CTCs are prognostic at any time point during the course of therapy, and changes in CTC numbers, reflecting a change in a patient's prognosis, accurately predict progression-free and overall survival in patients being treated for these cancers [17–20].

19.4.2 Expanding Enumeration

Establishing clinical and analytical validation of CELLSEARCH accelerated the use of CTCs in multiple directions. Internally, CELLSEARCH reagents were reconfigured to capture and enumerate endothelial cells [24, 25], melanocytes [26], and myeloma cells [27], enabling exploration into the utility of blood tumor cells in these patients (see the following text).

Simultaneously, all clinical findings from the original three registration trials were validated by others, establishing a path toward clinical utility [28]. For example, since CTCs are one of the strongest independent predictors of a patient's prognosis prior to the start of therapy, this application has found its way into early-phase drug study entry criteria and patient treatment algorithms [18, 29]. In metastatic breast cancer, multiple studies have pursued and validated the ability of CTCs to accurately predict imaging [20, 30]. Moreover, since patients with bone-only or bone-predominant disease have higher numbers of CTCs than patients with visceral mets, several studies focused on CTCs as a suitable biomarker for patients with bone disease, and this application remains a highly attractive use of CTCs [18, 19].

A logical extension of the original CTC findings in metastatic cancer is translation to primary cancers. These studies have met with mixed results. For the most part, this is more difficult to do owing to the paucity of CTCs in patients with primary cancer: to find sufficient numbers for analysis, it is not uncommon to draw 20–30 ml of blood. However, good success was found in early-stage breast cancer, especially in Europe, investigating patients in both neoadjuvant [31] and adjuvant disease settings [32]. Regarding the latter, although the number of CTCs detected in this setting were far fewer than in the metastatic setting (most studies use a cutoff ≥1 CTC), the clinical implications were the same: patients with elevated CTCs had a shorter disease-free survivor (DFS) and OS than did patients without CTCs [32].

Additionally, CTC clinical research in other metastatic cancers accelerated in an effort to demonstrate clinical applications to these patients. Similar to the findings in early-stage breast cancer patients, there were fewer CTCs and fewer patients with CTCs in these other cancers, including bladder, gastrointestinal, non-small-cell lung cancer (NSCLC), liver, pancreatic, and others [33]. Nonetheless, using a cutoff of ≥2 CTCs, the clinical implications of a shorter PFS and OS in patients with elevated CTCs were similar to the original findings in metastatic breast, prostate, or CRC [28]. Lung cancer presents an interesting story for CTCs. Notably, small-cell lung cancer (SCLC) appears to be an excellent cancer for epithelially defined CTCs, with some studies reporting the highest detection rate of CTCs across all cancers [34]. Also, many studies have shown a low number of epithelially defined CTCs in patients with NSCLC [35] and also CRC in the peripheral blood of these patients, but the number detected increases substantially if blood is drawn from the pulmonary or portal vein, respectfully, implying a filtering mechanism for epithelial cells [36, 37].

Finally, other body fluids were also interrogated for CTCs using the CELLSEARCH technology. Two sources of diagnostic fluids that were rich in CTCs were cerebral spinal fluid [38] and pleural fluid [39], both of which have substantial applications in patient management.

19.4.3 CTC Enumeration and Clinical Utility

The extensive validation of CTCs with consistent clinical and analytical results allowed the incorporation of CTCs into clinical trials in which patients are actively managed by the CTC result. The initial observation of persistently elevated CTC as a harbinger of therapy failure [17] formed the basis for the first clinical utility trial of CELLSEARCH in women with metastatic breast cancer (SWOG S0500). The primary hypothesis in this population of patients about to start first-line chemotherapy was that women with elevated CTCs both before the start of therapy and 22 days later would demonstrate a survival benefit from an early switch of therapy. The secondary hypothesis was that CTC could accurately predict PFS and OS. Although an early switch of therapy in women with persistently elevated CTCs did not result in longer PFS or OS for these patients, CTC did accurately predict outcomes for all three groups of patients, informing the treatment decisions to continue therapy (Arm B) or change therapy (Arm C) (Figure 19.1). Additionally, patients who did not have elevated CTCs at baseline, prior to therapy start, experienced the longest OS and PFS, as predicted by CTCs (Arm A) [40].

Several other clinical utility studies are presently underway in Europe [29]. Some of these trials are making treatment decisions based entirely upon CTC number as a risk factor, while others are adding CTC phenotype to demonstrate that CTC can effectively be used to target therapy. Early explorations into shifting phenotypes during cancer progression in metastatic breast cancer patients indicated that HER2 status changes substantially, presenting an opportunity to target this receptor in patients who

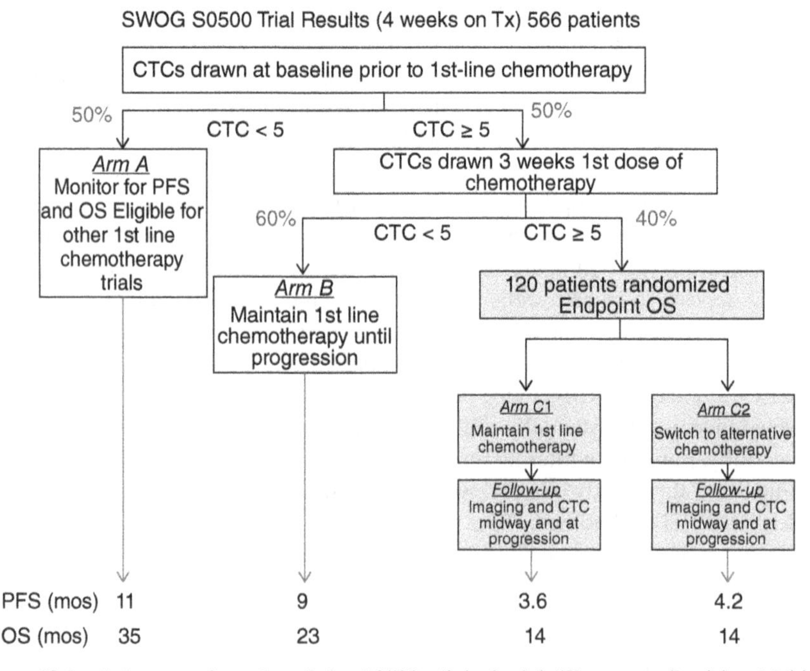

Figure 19.1 Schema and results of the S0500 clinical trial "Treatment Decision-Making Based on Blood Levels of Tumor Cells in Women with Metastatic Breast Cancer Receiving Chemotherapy."

converted to HER2-positive status [41]. However, early attempts at directing therapy based on this shift did not prove beneficial to patients [42]. The extensive clinical validation of CELLSEARCH recently culminated with a registration trial in China, in women with metastatic breast cancer that resulted in sFDA clearance in China in 2013 [43].

19.4.4 Characterization of CTCs using CELLSEARCH®

In addition to its clinical utility in enumeration of CTC, the CELLSEARCH platform has proven to be a flexible research tool for phenotypic and molecular characterization of CTC. The ability to characterize CTC may enable the assessment tumor markers, which are predictive of therapeutic response and provide insight into tumor biology in situations where tumor material is unavailable. What follows is an overview of the some of the studies that have been performed to characterize CTC using the CELLSEARCH platform. It is not intended to be comprehensive, but rather, to give an understanding of the types of analysis that have been performed using the system.

To enable phenotypic characterization of CTCs, the CELLSEARCH system allows a user-defined, fluorescent labeled tumor profiling antibody to be added to the sample concurrently with the CTC detection antibody cocktail. Presence of the tumor marker can then be detected using an extra fluorescent channel in the CellTracks scanner. Evaluation of the HER2 status in CTC has been described numerous times [41, 42, 44–46] and was included as a CTC characterization marker in a recent international study on the interoperator variably of the CELLSEARCH system [47]. Similarly, CTCs recovered from cancer patients have been stained for additional markers that are critical for monitoring during cancer progression, such as EGFR [48]. Khan *et al.* used the open channel of the CELLSEARCH system to evaluate synaptophysin and CD56 expression in CTC from neuroendocrine tumors [49]. There are numerous other examples when CTCs captured by CELLSEARCH were characterized by staining with various antibodies aiming to characterize specific cellular processes (such as apoptosis (M30), active cellular proliferation (KI67), DNA damage response (gammaH2AX)), or identify specific subpopulations of cells (cancer stem cells (CD44)) [50–54].

To further characterize samples that have been captured and stained with the CELLSEARCH system, Swennenhuis *et al.* described a method to perform fluorescent *in situ* hybridization (FISH) on CTC [55]. Following CELLSEARCH enumeration, CTCs are fixed in place in the CELLSEARCH cartridges, the liquid is removed, and the sample is dried. FISH probes are added to the cartridge, and a stringency wash is performed. Coordinates of the CTC are transferred to a modified CellTracks instrument, and individual CTC are imaged for up to four FISH probes. In 2009, Attard *et al.* used this method to simultaneously detect androgen receptor gene amplification, deletion of *PTEN*, and *ERG* rearrangements in hormone-refractory prostate cancer patients [56]. In 2010, Punnoose *et al.* described the use of the CELLSEARCH platform to characterize CTC for HER2 gene amplification status in metastatic breast cancer patients. In this study, all nine patients with evaluable CTC showed CTC FISH results that were concordant with the HER2 status of the tumor [45].

In addition to cellular analysis, the CELLSEARCH platform provides tools that enable molecular characterization of CTC. The CELLSEARCH Profile Kit isolates CTC using the same capture technology without the staining and permeabilization steps used in the CTC enumeration kits. The product of the Profile Kit is a mixture of enriched CTC and a number of carryover leukocytes that can be used for downstream molecular profiling. Samples enriched by CELLSEARCH typically contain few thousand leukocytes and variable number of CTCs per 7.5 ml of processed blood. Presence of the contaminating white cell populations requires consideration of the risk that molecular signatures arise from the leukocytes and not the CTC. Proper controls have to be employed such blood enriched for CTCs from healthy donors, CTC-depleted blood from cancer patients, whole blood from patients to facilitate proper interpretation of molecular analysis data obtained from CTC-enriched fractions. Alternatively, postenrichment isolation of individual CTCs can be used for deep molecular characterization of cancer cells.

Despite the contaminating leukocyte issue, several groups have been able to perform molecular characterization of CTC using the CELLSEARCH platform. In 2005, our group showed that CTC messenger RNA profiles of CTC could be obtained from the blood of breast, colorectal, and prostate cancer patients. This study identified cancer and epithelial specific genes such as *EpCAM*, *cytokeratin* genes, *KLK3, KLK2* that were not found in the blood of healthy donor [57]. The study also identified new genes such as *AGR2*, *S100A14*, and *FABP1* that can be used to monitor presence of CTC in the blood of cancer patients. Sieuwerts *et al.* used a similar approach to identify 55 mRNAs and 10 miRNAs more abundantly expressed in samples from 32 metastatic breast cancer patients than in blood from healthy donors or cancer patients with less than 5 CTC [58].

CTCs enriched by CELLSEARCH provide an opportunity to analyze tumor mutations, and several studies have shown that somatic mutations can be detected in the CTC from cancer patients using a variety of methods. In 2010, Jiang *et al.* showed that mutations of the androgen receptor gene can be detected in prostate cancer patients by analyzing CTC RNA with denaturing HPLC technology followed by direct sequencing [59]. Mostert *et al.* showed that KRAS and BRAF mutations can be detected in CTC from CRC patients using allele-specific PCR with a blocking reagent when as little as two CTC are present [60]. Schneck *et al.* reported a method to use SNaP-shot analysis to identify *PIK3CA* mutations in CTC enriched from metastatic breast cancer patients [61]. Using multiplex Boreal OnTarget assay, we showed that heterozygous somatic mutations can be detected in CELLSEARCH-enriched fractions when as few as five cells were present in 7.5 ml blood sample (Figure 19.2) (Patel J. *et al.*, unpublished observations).

Recent advances in molecular biology techniques have enabled more broad analysis of small numbers of cells, and these technologies have been applied to the analysis of CTCs. In separate studies, Heitzer *et al.* and Möhlendick *et al.* isolated individual CTC using a micromanipulator and flow cytometry, respectively, on samples that were enriched with the CELLSEARCH platform [62, 63]. CTC DNA was amplified by whole-genome amplification and analyzed with oligonucleotide array comparative genomic hybridization. Additionally, Heitzer used a targeted enrichment

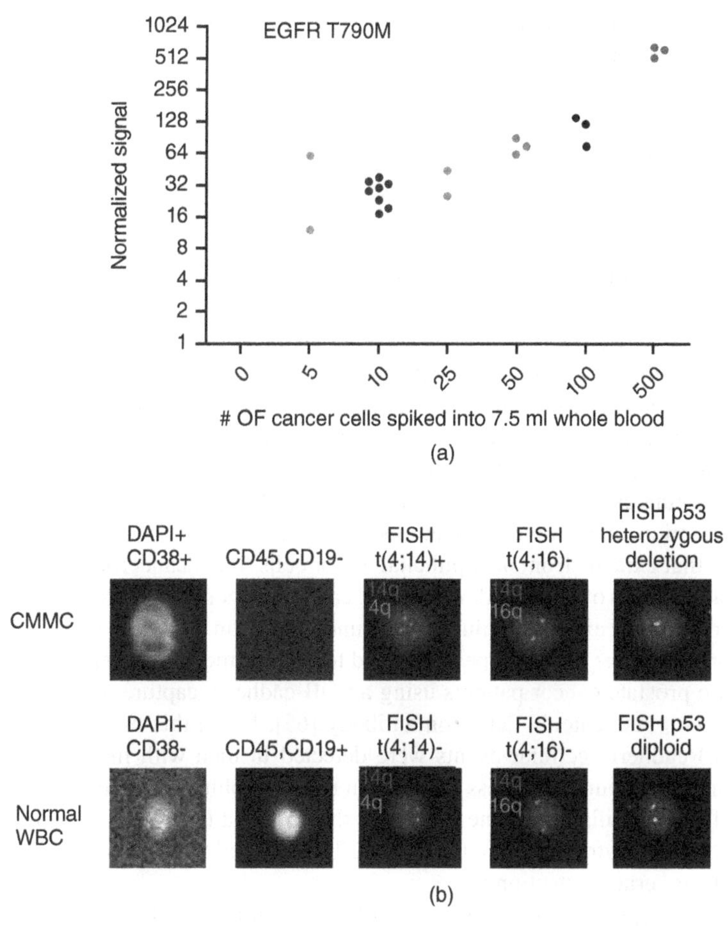

Figure 19.2 (a) Detection of EGFR T790M mutation in cancer cells H1975 spiked and recovered from 7.5 ml of healthy donor blood using CELLSEARCH® system. Mutations were detected using Boreal OnTarget™ according to manufacturer's instructions. (b) Cytogenetic analysis of circulating multiple myeloma cells. CMMCs were enriched from peripheral blood and stained as mentioned in the main text. $CD138^+/CD38^+/DAPI^+/(CD45, CD19)-$ CMMC were observed to harbor IgH rearrangements and 17p13 deletions while carryover leukocytes $(CD45, CD19)^+$ presented with wild-type IgH and diploid 17p13 signals. (*See color plate section for the color representation of this figure.*)

next-generation sequencing approach to analyze the mutational status of a panel of 68 CRC related genes in pools of CTCs collected from patients. CTC enriched by CELLSEARCH are also compatible with automatic single-cell capture tools such as DEPArray system developed by Silicon Biosystems as was demonstrated by Peeters *et al.* [64].

All of the aforementioned examples focused on analysis of genomic DNA isolated from individual CTCs. We have demonstrated that small pools of CTCs enriched from blood of cancer patients using Profile Kit can be characterized using RNAseq (Patel *et al.*, unpublished observations). Continuous improvements of protocols for single-cell RNA, DNA, and protein analysis will continue to increase utilization of CTCs for efforts to improve understanding of cancer progression. Moreover, examples presented clearly demonstrate that CELLSEARCH system is compatible with a very broad spectrum of approaches for molecular characterization of CTCs.

19.5 BEYOND EpCAM CAPTURE

The CELLSEARCH platform enables a complex characterization of target cells by employing antibodies in four general ways: (i) antibodies to capture rare cells; (ii) detection antibodies, which are specific to the target cells; (iii) counterstain antibodies, which are specific to nontarget cells; and (iv) characterization antibodies, which may or may not be specific to the target cells but are informative when used with specific capture and detection antibodies. Careful selection of the capture, detection, characterization, and counterstain antibodies enables users to study a broad range of rare cells that extends from nonepithelial tumor cells to noncancerous cells.

One application of this is to develop CTC capture reagents for cells that have down-regulated EpCAM expression due to cells undergoing an epithelial to mesenchymal transition. Bitting *et al.* developed a method to capture mesenchymal-like CTC from metastatic prostate cancer patients using an OB-cadherin capture antibody in combination with beta-catenin detection antibody [65]. Using this mesenchymal-based assay, OB-cadherin cellular events were detected in men with metastatic prostate cancer and were found to be less common in healthy volunteers. The tumor origin of these cells was confirmed in one patient by showing that the mesenchymal-like cells harbor the same cytogenetic mutations as CTC captured with the standard EpCAM capture/Cytokeratin detection kit.

Moving the platform beyond tumors of epithelial origin, Rao *et al.* described a CELLSEARCH kit configuration to capture circulating melanoma cells (CMC) using CD146 as a capture marker, HMW-MAA as a detection marker and CD45 and CD34 as counterstains [26]. In this study, it was found that patients with ≥2 CMCs per 7.5 ml of whole blood, as compared with the group with <2 CMCs, had a shorter overall survival (2.0 months vs 12.1 months, $P = 0.001$). Prognostic values were confirmed in follow-up studies in cutaneous and uveal melanoma by Khoja *et al.* and Bidard *et al.*, respectively [66, 67]. Le Rhun *et al.* described the adaptation of the CMC kit to detect melanoma cells in cerebrospinal fluid of patients and indicated a favorable comparison of the method with the current gold standard cytomorphological analysis [68].

Our internal studies have shown that circulating multiple myeloma cells (CMMCs) can be detected in the peripheral blood of patients with myeloma, smoldering myeloma, and monoclonal gammopathy of unknown significance. For this application, CD138 is used as a capture antibody, CD38 as a detection antibody, and CD45 and CD19 are used as counterstains. Circulating CMMCs were detected at a higher level in the blood of patients with multiple myeloma, smoldering myeloma, and monoclonal gammopathy of unknown significance than healthy donors. The tumor origin of these CMMC was confirmed using a subsequent FISH assay to simultaneously detect deletions of the TP53 gene (17p13) and two IgH rearrangements (t(4;14)(p16;q32) and t(14;16)(q32;q23)), which are frequently mutated in multiple myeloma [69, 70]. IgH rearrangements and 17p were detected in CMMC but not in the $CD45^+$ leukocytes (Figure 19.2)[27].

Circulating endothelial cells (CECs) are a heterogonous population of loosely defined cells detectable in the blood of healthy and diseased patients. A CELLSEARCH kit was a kit created to reproducibly capture CECs using CD146 as a capture antigen, CD105 as a detection reagent, and CD45 as a counterstain [25, 71]. Using this kit, CEC were found in the blood of healthy donors at the level of 1–20 CECs/ml blood while CEC counts were significantly higher metastatic carcinoma patients. Ilie *et al.* demonstrated that CEC counts were elevated in preoperative non-small-cell lung carcinoma patients and that elevation of CECs was significantly correlated with shorter progression-free and overall survival in this cohort [72]. Similarly to CTCs, our group generated gene expression profiles of CEC in the blood of metastatic cancer patients [73]. Many of the genes shown to be upregulated in cancer patients possess well-known association with endothelial function (e.g., *VWF*, *TIE1*, and *CDH5*). Others genes (e.g., *POSTN*, *SPARC*, and *HBEGF*) were shown to be involved in angiogenesis and tumor growth. This raises the possibility that CECs may be useful in predicting response to antiangiogenesis therapies.

Using the same kit, CECs were found to be elevated in the blood of patients undergoing ST-segment elevation myocardial infarction (STEMI) relative to blood from healthy donors and patients with peripheral vascular disease [74]. Further, the CEC detected in STEMI patients contained multicellular and multinuclear events that were not found in healthy donor or patients with peripheral vascular disease. It was suggested that these abnormal cells may arise from ruptured atherosclerotic plaques, which raises the possibility that the population of CEC may be able to predict myocardial infarction.

19.6 DISCUSSION

More than 10 years since its introduction, the CELLSEARCH system, and its reagents, remains the only widely available system that offers a highly controlled, reproducible way to enrich and analyze CTCs. It is still the only system to have gained regulatory approval in the United States and China. Such approval is significant since its opens the door for routine usage in general clinical settings. CELLSEARCH introduction and adoption contributed significantly to the establishment of CTCs as a potential biomarker with great scientific and clinical importance. It became the *de facto* gold standard that any new CTC enrichment approach must compare itself to. The introduction of CELLSEARCH was a turning point in the clinical and scientific investigation of CTCs. It enabled the first multicenter prospective clinical studies on the clinical validity of CTCs in three different human cancers. It effectively ended a century and a half of speculation on the significance of CTCs and ushered in a new era of inquiry into circulating tumor material. That CELLSEARCH was introduced as an automated system enabled not only adoption of the system in routine clinical testing laboratories, but also the rapid validation of its performance and clinical claims in numerous independent prospective clinical investigations and publication of their results in high-impact peer-reviewed journals.

The success of CELLSEARCH also lies in its extraordinary sensitivity due to the controlled aggregation technology built into the system. Having studied EpCAM density on CTCs from cancer patients, it was known that the antigen density in patient CTCs was significantly lower than that of most cell lines. To be successful, CELLSEARCH had to target these low antigen expression cells. Today, we know that even as cells undergo EMT, the level of EpCAM and cytokeratin in these transitioning cells is decreased but is often still at levels that allow for capture and detection in CELLSEARCH.

The choice of a CTC definition that was both conservative and specific also served to drive acceptance of CTCs. At a time when the existence of CTCs was questioned, and even if present, their clinical significance unknown, this highly specific unambiguous definition meant that the cells were found only in people with cancer and not in people who did not have cancer. The prognostic significance of having more than five CTCs in 7.5 ml of blood versus less than five established conclusively the importance of monitoring this CTC population in at least three human cancers. The $EpCAM^+$/$Cytokeratin^+$/$CD45^-$ CTC cell population may not be the only CTC population circulating, but after more than 10 years of research, and 15,000 publications, it is still the only CTC phenotype with proven clinical significance.

It is evident from our own research, and from the content of this book, that other CTC phenotypes exist. The CTC lexicon will undoubtedly be expanded in the coming years. Along with this, there have been, and will continue to emerge, new approaches to isolate and characterize CTCs. Several have been developed since the commercial launch of the CELLSEARCH platform. Many of the newer CTC capture approaches have been developed with the intent to minimize or eliminate a dependence on EpCAM expression for capturing CTCs. Whereas exploiting other properties of CTCs such as their size, electrical charge, compressibility, or broader antigen expression beyond EpCAM has shown a measure of success, all these properties have their own drawbacks and limitations. Tumor cell properties such as cell size can be shared by other CD45-negative circulating cells such as CECs, which are also large, and the blood of cancer patients have been shown to contain as few as 1 or 2, or often in tens of thousands of these cells [25].

New approaches for CTC enrichment often claim to capture more cells than with CELLSEARCH from a given patient's sample. These claims, although exciting and hopeful, must pass the same three critical proofs as did the CELLSEARCH system. First, are the numbers independently verifiable using other methods such as flow cytometry? Second, are the cells demonstrably of tumor origin using established methods such as FISH? Third, do the cells provide clinically significant information? In short, is the method accurate and is the result scientifically or medically significant? These are critical questions because even minor modifications or additions in the reagents used to identify CTCs (such as different clone of antibody) can have significant consequences for specificity and cross-reactivity with regard to the cells detected.

It is important to point out that these challenges are as true for CELLSEARCH as they are for the newer emerging technologies. Over the years, numerous attempts to increase the types and number of cells captured by CELLSEARCH were unsuccessful not because we did not find more cells, but because of decreased specificity of the resulting assay. The resulting assay could not pass the three proofs listed ([65], Connelly *et al.*, unpublished observations). This is perhaps not surprising since cancer is a systemic disease that affects many cell types, such as endothelial cells and immune cells. Increased number of such cells can be found in the blood of cancer patients [25, 75, 76]. Often, changes in populations of such circulating cells are affected by treatment [77]. Any lack of specificity in capture/detection methodology can lead to identification of "extra" tumor cells that may be related to the underlying cancer but are not of tumor origin.

Experience of CELLSEARCH in clinical settings is also very informative. While the system obtained regulatory clearance to be used as an aid in the monitoring of patients metastatic breast, prostate, and CRC, clinical adoption of the test has been slow. The CELLSEARCH system may be used to help physicians to predict progression-free survival and overall survival in patients with metastatic disease any time during the course of therapy. However, the use of the test has not been connected to any specific therapy or clinical decision. An attempt to examine if patients with persistently elevated CTCs following the first round of therapy would benefit from switching to another therapy sooner (such as SWOG 0500 study described) showed no benefit in OS or PFS by switching to a second line early. Although the study fully validated again the prognostic benefit of CTCs for PFS and OS in these patients, it is less clear why there was no benefit in PFS or OS for these patients, but perhaps it was due to a limited benefit of a second line of therapy in these patients, rather than a limitation on the utility of CTCs as a biomarker [78]. As a result, there has been increased pressure on utilization and reimbursement [78] for CTC tests in general, and it is likely to remain this way until specific pharmacologic or clinical decisions can be linked directly to CTCs, CTC subsets, or CTC characterization.

Part of the challenge lies with the very nature of CTCs. These cells are likely to represent universal phenomenon during cancer progression. It is clear from multiple studies in many indications that the number of CTCs detected by CELLSEARCH is related to patient survival [78, 79]. Nevertheless, it is hard to dismiss the possibility that in a particular subtype of cancer in combination with specific cancer treatment, simple CTC enumeration may not be informative. Such assumptions may lead to a need for a separate regulatory approval for every disease/treatment combination. This could put all CTC technologies on novel and very challenging economic paths.

Despite challenges for CTC technologies in clinical and regulatory settings, research-driven utilization of the CELLSEARCH platform has been impressive. Development of various methodologies for analysis of rare cells (such as methods for whole genome and transcriptome amplification, ability to measure hundreds of targets in very small samples, ultrasensitive protein detection methodologies, next-generation sequencing, etc.) has opened the door for the possibility not only to enumerate but also to fully characterize CTCs. Such characterization allows studying and understanding the role that CTCs play in the metastatic process. Fully characterizing CTCs offers the possibility of understanding and monitoring mechanisms of resistance to various cancer therapies. Such opportunities have increased the interest in CTCs for drug development purposes. Recent experiments with culturing of CTCs and creation of CTC-derived mouse models further enhance such interest [80–82]. While CELLSEARCH was designed for highly controlled enumeration of CTCs and other rare cells; the system is flexible enough to be used extensively for characterization studies. Profile kits allow for recovery of live, unstained CTCs that can be used for a variety of molecular studies, including detailed characterization of single cells by DNA and RNA sequencing. Even without isolation of single cells, the purity of CTC samples enriched by CELLSEARCH is comparable to those offered by competing approaches. When coupled with the ability to characterize CTC-enriched samples prepared with CELLSEARCH using extra protein staining markers followed by multimarker FISH, this creates a compelling case for using CELLSEARCH in a wide range of research studies. And perhaps in such research studies, clinical utilities of CTCs will be identified that eventually lead to a broad acceptance and use of CTCs in clinics around the world.

REFERENCES

1. Gupta, G.P. and Massagué, J. (2006) Cancer metastasis: building a framework. *Cell*, **127** (4), 679–695.
2. Ashworth, T.R. (1869) A case in which cells similar to those in the tumors were seen in the blood after death. *Australian Medical Journal*, **14**, 146.
3. Candar, Z., Ritchie, A.C., Hopkirk, J.F., and Long, R.C. (1962) The prognostic value of circulating tumor cells in patients with breast cancer. *Surgery, Gynecology & Obstetrics*, **115**, 291–294.
4. Fleming, J.A. (1963) Tumor cells in the blood in carcinoma of the breast. *Proceedings of the Royal Society of Medicine*, **56**, 497–500.
5. Pantel, K. (1996) Detection of minimal disease in patients with solid tumors. *Journal of Hematotherapy*, **5** (4), 359–367.
6. Terstappen, L.W., Rao, C., Gross, S. *et al.* (1998) Flow cytometry--principles and feasibility in transfusion medicine. Enumeration of epithelial derived tumor cells in peripheral blood. *Vox Sanguinis*, **74** (Suppl. 2), 269–274.
7. Hoeppener, A.E., Swennenhuis, J.F., and Terstappen, L.W. (2012) Immunomagnetic separation technologies. *Recent Results in Cancer Research*, **195**, 43–58.
8. Takao, M. and Takeda, K. (2011) Enumeration, characterization, and collection of intact circulating tumor cells by cross contamination-free flow cytometry. *Cytometry. Part A*, **79** (2), 107–117.
9. Lowes, L.E., Goodale, D., Keeney, M., and Allan, A.L. (2011) Image cytometry analysis of circulating tumor cells. *Methods in Cell Biology*, **102**, 261–290.
10. Allan, A.L. and Keeney, M. (2010) Circulating tumor cell analysis: technical and statistical considerations for application to the clinic. *Journal of Oncology*, **2010**, 426218.
11. Rao, C.G., Chianese, D., Doyle, G.V. *et al.* (2005) Expression of epithelial cell adhesion molecule in carcinoma cells present in blood and primary and metastatic tumors. *International Journal of Oncology*, **27** (1), 49–57.

12. Winter, M.J., Nagtegaal, I.D., van Krieken, J.H., and Litvinov, S.V. (2003) The epithelial cell adhesion molecule (Ep-CAM) as a morphoregulatory molecule is a tool in surgical pathology. *American Journal of Pathology*, **163** (6), 2139–2148.
13. Lenhard, R., Osteen, R., and Gansler, T. (2001) *Pathologic Evaluation of Neoplasms*, American Cancer Society's Clinical Oncology, Georgia, pp. 123–148.
14. Thiery, J.P. (2002) Epithelial-mesenchymal transitions in tumour progression. *Nature Reviews Cancer*, **2** (6), 442–454.
15. Bitting, R.L., Schaeffer, D., Somarelli, J.A. *et al.* (2014) The role of epithelial plasticity in prostate cancer dissemination and treatment resistance. *Cancer and Metastasis Review*, **33** (2), 441–468.
16. Riethdorf, S., Fritsche, H., Müller, V. *et al.* (2007) Detection of circulating tumor cells in peripheral blood of patients with metastatic breast cancer: a validation study of the CELLSEARCH® system. *Clinical Cancer Research*, **13** (3), 920–8.
17. Cristofanilli, M., Budd, G.T., Ellis, M.J. *et al.* (2004) Circulating tumor cells, disease progression, and survival in metastatic breast cancer. *New England Journal of Medicine*, **351** (8), 781–91.
18. de Bono, J.S., Scher, H.I., Montgomery, R.B. *et al.* (2008) Circulating tumor cells predict survival benefit from treatment in metastatic castration-resistant prostate cancer. *Clinical Cancer Research*, **14** (19), 6302–6309.
19. Scher, H.I., Jia, X., de Bono, J.S. *et al.* (2009) Circulating tumour cells as prognostic markers in progressive, castration-resistant prostate cancer: a reanalysis of IMMC38 trial data. *Lancet Oncology*, **10** (3), 233–239.
20. Cohen, S.J., Punt, C.J., Iannotti, N. *et al.* (2008) Relationship of circulating tumor cells to tumor response, progression-free survival, and overall survival in patients with metastatic colorectal cancer. *Journal of Clinical Oncology*, **26** (19), 3213–3221.
21. Budd, G.T., Cristofanilli, M., Ellis, M.J. *et al.* (2006) Circulating tumor cells versus imaging--predicting overall survival in metastatic breast cancer. *Clinical Cancer Research*, **12** (21), 6403–6409.
22. Cristofanilli, M., Broglio, K.R., Guarneri, V. *et al.* (2007) Circulating tumor cells in metastatic breast cancer: biologic staging beyond tumor burden. *Clinical Breast Cancer*, **7** (6), 471–479.
23. Danila, D.C., Heller, G., Gignac, G.A. *et al.* (2007) Circulating tumor cell number and prognosis in progressive castration-resistant prostate cancer. *Clinical Cancer Research*, **13** (23), 7053–7058.
24. Strijbos, M.H., Gratama, J.W., Schmitz, P.I. *et al.* (2010) Circulating endothelial cells, circulating tumour cells, tissue factor, endothelin-1 and overall survival in prostate cancer patients treated with docetaxel. *European Journal of Cancer*, **46** (11), 2027–2035.
25. Rowand, J.L., Martin, G., Doyle, G.V. *et al.* (2007) Endothelial cells in peripheral blood of healthy subjects and patients with metastatic carcinomas. *Cytometry. Part A*, **71** (2), 105–113.
26. Rao, C., Bui, T., Connelly, M. *et al.* (2011) Circulating melanoma cells and survival in metastatic melanoma. *International Journal of Oncology*, **38** (3), 755–760.
27. Gross, S., Foulk, B., Nielsen, K., Patel, J., Connelly, M., and Mata, M. (2011) 1825 Automated Enumeration and Characterization of Circulating multiple Myeloma Cells in Blood. American Society Of Hematology Abstract.

28. Yap, T.A., Lorente, D., Omlin, A. *et al.* (2014) Circulating tumor cells: a multifunctional biomarker. *Clinical Cancer Research*, **20** (10), 2553–2568.
29. Bidard, F.C., Fehm, T., Ignatiadis, M. *et al.* (2013) Clinical application of circulating tumor cells in breast cancer: overview of the current interventional trials. *Cancer and Metastasis Reviews*, **32** (1–2), 179–188.
30. Liu, M.C., Shields, P.G., Warren, R.D. *et al.* (2009) Circulating tumor cells: a useful predictor of treatment efficacy in metastatic breast cancer. *Journal of Clinical Oncology*, **27** (31), 5153–5159.
31. Pierga, J.Y., Bidard, F.C., Mathiot, C. *et al.* (2008) Circulating tumor cell detection predicts early metastatic relapse after neoadjuvant chemotherapy in large operable and locally advanced breast cancer in a phase II randomized trial. *Clinical Cancer Research*, **14** (21), 7004–7010.
32. Rack, B., Schindlbeck, C., Jückstock, J. *et al.* (2014) Circulating tumor cells predict survival in early average-to-high risk breast cancer patients. *Journal of the National Cancer Institute*, **106** (5), 1–11.
33. Allard, W.J., Matera, J., Miller, M.C. *et al.* (2004) Tumor cells circulate in the peripheral blood of all major carcinomas but not in healthy subjects or patients with nonmalignant diseases. *Clinical Cancer Research*, **10** (20), 6897–6904.
34. Hou, J.M., Krebs, M.G., Lancashire, L. *et al.* (2012) Clinical significance and molecular characteristics of circulating tumor cells and circulating tumor microemboli in patients with small-cell lung cancer. *Journal of Clinical Oncology*, **30** (5), 525–532.
35. Krebs, M.G., Sloane, R., Priest, L. *et al.* (2011) Evaluation and prognostic significance of circulating tumor cells in patients with non-small-cell lung cancer. *Journal of Clinical Oncology*, **29** (12), 1556–1563.
36. Okumura, Y., Tanaka, F., Yoneda, K. *et al.* (2009) Circulating tumor cells in pulmonary venous blood of primary lung cancer patients. *Annals of Thoracic Surgery*, **87** (6), 1669–1675.
37. Wind, J., Tuynman, J.B., Tibbe, A.G. *et al.* (2009) Circulating tumour cells during laparoscopic and open surgery for primary colonic cancer in portal and peripheral blood. *European Journal of Surgical Oncology*, **35** (9), 942–950.
38. Patel, A.S., Allen, J.E., Dicker, D.T. *et al.* (2011) Identification and enumeration of circulating tumor cells in the cerebrospinal fluid of breast cancer patients with central nervous system metastases. *Oncotarget*, **2** (10), 752–760.
39. Schwed Lustgarten, D.E., Thompson, J., Yu, G. *et al.* (2013) Use of circulating tumor cell technology (CELLSEARCH®) for the diagnosis of malignant pleural effusions. *Annals of the American Thoracic Society*, **10** (6), 582–589.
40. Smerage, J.B., Barlow, W.E., Hortobagyi, G.N. *et al.* (2014) Circulating tumor cells and response to chemotherapy in metastatic breast cancer: SWOG S0500. *Journal of Clinical Oncology*, **32** (31), 3483–3489.
41. Pestrin, M., Bessi, S., Galardi, F. *et al.* (2009) Correlation of HER2 status between primary tumors and corresponding circulating tumor cells in advanced breast cancer patients. *Breast Cancer Research and Treatment*, **118** (3), 523–530.
42. Pestrin, M., Bessi, S., Puglisi, F. *et al.* (2012) Final results of a multicenter phase II clinical trial evaluating the activity of single-agent lapatinib in patients with HER2-negative metastatic breast cancer and HER2-positive circulating tumor cells. A proof-of-concept study. *Breast Cancer Research and Treatment*, **134** (1), 283–289.

43. Jiang, Z., Cristofanilli, M., Shao, Z. *et al.* (2013) Circulating tumor cells predict progression-free and overall survival in Chinese patients with metastatic breast cancer, HER2-positive or triple-negative (CBCSG004): a multicenter, double-blind, prospective trial. *Ann Oncology*, **24** (11), 2766–2772.
44. Riethdorf, S., Müller, V., Zhang, L. *et al.* (2010) Detection and HER2 expression of circulating tumor cells: prospective monitoring in breast cancer patients treated in the neoadjuvant GeparQuattro trial. *Clinical Cancer Research*, **16** (9), 2634–2645.
45. Punnoose, E.A., Atwal, S.K., Spoerke, J.M. *et al.* (2010) Molecular biomarker analyses using circulating tumor cells. *PLoS One*, **5** (9), e12517.
46. Ignatiadis, M., Rothé, F., Chaboteaux, C. *et al.* (2011) HER2-positive circulating tumor cells in breast cancer. *PLoS One*, **6** (1), e15624.
47. Ignatiadis, M., Riethdorf, S., Bidard, F.C. *et al.* (2014) International study on inter-reader variability for circulating tumor cells in breast cancer. *Breast Cancer Research*, **16** (2), 1–8.
48. Payne, R.E., Yagüe, E., Slade, M.J. *et al.* (2009) Measurements of EGFR expression on circulating tumor cells are reproducible over time in metastatic breast cancer patients. *Pharmacogenomics*, **10** (1), 51–57.
49. Khan, M.S., Tsigani, T., Rashid, M. *et al.* (2011) Circulating tumor cells and EpCAM expression in neuroendocrine tumors. *Clinical Cancer Research*, **17** (2), 337–345.
50. Lowes, L.E., Hedley, B.D., Keeney, M., and Allan, A.L. (2012) User-defined protein marker assay development for characterization of circulating tumor cells using the CELLSEARCH® system. *Cytometry. Part A*, **81** (11), 983–995.
51. Lowes, L.E. and Allan, A.L. (2014) Recent advances in the molecular characterization of circulating tumor cells. *Cancers (Basel)*, **6** (1), 595–624.
52. Munzone, E., Botteri, E., Sandri, M.T. *et al.* (2012) Prognostic value of circulating tumor cells according to immunohistochemically defined molecular subtypes in advanced breast cancer. *Clinical Breast Cancer*, **12** (5), 340–346.
53. Larson, C.J., Moreno, J.G., Pienta, K.J. *et al.* (2004) Apoptosis of circulating tumor cells in prostate cancer patients. *Cytometry. Part A*, **62** (1), 46–53.
54. Wang, L.H., Pfister, T.D., Parchment, R.E. *et al.* (2010) Monitoring drug-induced gammaH2AX as a pharmacodynamic biomarker in individual circulating tumor cells. *Clinical Cancer Research*, **16** (3), 1073–1084.
55. Swennenhuis, J.F., Tibbe, A.G., Levink, R. *et al.* (2009) Characterization of circulating tumor cells by fluorescence in situ hybridization. *Cytometry. Part A*, **75** (6), 520–527.
56. Attard, G., Swennenhuis, J.F., Olmos, D. *et al.* (2009) Characterization of ERG, AR and PTEN gene status in circulating tumor cells from patients with castration-resistant prostate cancer. *Cancer Research*, **69** (7), 2912–2918.
57. Smirnov, D.A., Zweitzig, D.R., Foulk, B.W. *et al.* (2005) Global gene expression profiling of circulating tumor cells. *Cancer Research*, **65** (12), 4993–4997.
58. Sieuwerts, A.M., Mostert, B., Bolt-de Vries, J. *et al.* (2011) mRNA and microRNA expression profiles in circulating tumor cells and primary tumors of metastatic breast cancer patients. *Clinical Cancer Research*, **17** (11), 3600–3618.
59. Jiang, Y., Palma, J.F., Agus, D.B. *et al.* (2010) Detection of androgen receptor mutations in circulating tumor cells in castration-resistant prostate cancer. *Clinical Chemistry*, **56** (9), 1492–1495.

60. Mostert, B., Jiang, Y., Sieuwerts, A.M. *et al.* (2013) KRAS and BRAF mutation status in circulating colorectal tumor cells and their correlation with primary and metastatic tumor tissue. *International Journal of Cancer*, **133** (1), 130–141.
61. Schneck, H., Blassl, C., Meier-Stiegen, F. *et al.* (2013) Analysing the mutational status of PIK3CA in circulating tumor cells from metastatic breast cancer patients. *Molecular Oncology*, **7** (5), 976–986.
62. Heitzer, E., Auer, M., Gasch, C. *et al.* (2013) Complex tumor genomes inferred from single circulating tumor cells by array-CGH and next-generation sequencing. *Cancer Research*, **73** (10), 2965–2975.
63. Möhlendick, B., Bartenhagen, C., Behrens, B. *et al.* (2013) A robust method to analyze copy number alterations of less than 100 kb in single cells using oligonucleotide array CGH. *PLoS One*, **8** (6), e67031.
64. Peeters, D.J., De Laere, B., Van den Eynden, G.G. *et al.* (2013) Semiautomated isolation and molecular characterisation of single or highly purified tumour cells from CELLSEARCH® enriched blood samples using dielectrophoretic cell sorting. *British Journal of Cancer*, **108** (6), 1358–1367.
65. Bitting, R.L., Boominathan, R., Rao, C. *et al.* (2013) Development of a method to isolate circulating tumor cells using mesenchymal-based capture. *Methods*, **64** (2), 129–136.
66. Khoja, L., Lorigan, P., Zhou, C. *et al.* (2013) Biomarker utility of circulating tumor cells in metastatic cutaneous melanoma. *Journal of Investigative Dermatology*, **133** (6), 1582–1590.
67. Bidard, F.C., Madic, J., Mariani, P. *et al.* (2014) Detection rate and prognostic value of circulating tumor cells and circulating tumor DNA in metastatic uveal melanoma. *International Journal of Cancer*, **134** (5), 1207–1213.
68. Le Rhun, E., Tu, Q., De Carvalho, B.M. *et al.* (2013) Detection and quantification of CSF malignant cells by the CELLSEARCH® technology in patients with melanoma leptomeningeal metastasis. *Medical Oncology*, **30** (2), 538.
69. Avet-Loiseau, H., Attal, M., Moreau, P. *et al.* (2007) Genetic abnormalities and survival in multiple myeloma: the experience of the Intergroupe Francophone du Myélome. *Blood*, **109** (8), 3489–3495.
70. Fonseca, R., Blood, E., Rue, M. *et al.* (2003) Clinical and biologic implications of recurrent genomic aberrations in myeloma. *Blood*, **101** (11), 4569–4575.
71. Strijbos, M.H., Rao, C., Schmitz, P.I. *et al.* (2008) Correlation between circulating endothelial cell counts and plasma thrombomodulin levels as markers for endothelial damage. *Thrombosis and Haemostasis*, **100** (4), 642–647.
72. Ilie, M., Long, E., Hofman, V. *et al.* (2014) Clinical value of circulating endothelial cells and of soluble CD146 levels in patients undergoing surgery for non-small cell lung cancer. *British Journal of Cancer*, **110** (5), 1236–1243.
73. Smirnov, D.A., Foulk, B.W., Doyle, G.V. *et al.* (2006) Global gene expression profiling of circulating endothelial cells in patients with metastatic carcinomas. *Cancer Research*, **66** (6), 2918–2922.
74. Damani, S., Bacconi, A., Libiger, O. *et al.* (2012) Characterization of circulating endothelial cells in acute myocardial infarction. *Science Translational Medicine*, **4** (126), 126ra33.
75. Khaled, Y.S., Ammori, B.J., and Elkord, E. (2014) Increased levels of granulocytic myeloid-derived suppressor cells in peripheral blood and tumour tissue of pancreatic cancer patients. *Journal of Immunology Research*, **2014**, 879897.

76. Adams, D.L., Martin, S.S., Alpaugh, R.K. *et al.* (2014) Circulating giant macrophages as a potential biomarker of solid tumors. *Proceedings of the National Academy of Sciences of the United States of America*, **111** (9), 3514–3519.
77. Pico de Coaña, Y., Poschke, I., Gentilcore, G. *et al.* (2013) Ipilimumab treatment results in an early decrease in the frequency of circulating granulocytic myeloid-derived suppressor cells as well as their Arginase1 production. *Cancer Immunology Research*, **1** (3), 158–162.
78. Raimondi, C., Gradilone, A., Naso, G. *et al.* (2014) Clinical utility of circulating tumor cell counting through CELLSEARCH®: the dilemma of a concept suspended in Limbo. *OncoTargets and Therapy*, **7**, 619–625.
79. Lowes, L.E., Hedley, B.D., Keeney, M., and Allan, A.L. (2014) Adaptation of semiautomated circulating tumor cell (CTC) assays for clinical and preclinical research applications. *Journal of Visualized Experiments*, **84**, e51248.
80. Hodgkinson, C.L., Morrow, C.J., Li, Y. *et al.* (2014) Tumorigenicity and genetic profiling of circulating tumor cells in small-cell lung cancer. *Nature Medicine*, **20** (8), 897–903.
81. Baccelli, I., Schneeweiss, A., Riethdorf, S. *et al.* (2013) Identification of a population of blood circulating tumor cells from breast cancer patients that initiates metastasis in a xenograft assay. *Nature Biotechnology*, **31** (6), 539–544.
82. Zhang, L., Ridgway, L.D., Wetzel, M.D. *et al.* (2013) The identification and characterization of breast cancer CTCs competent for brain metastasis. *Science Translational Medicine*, **5** (180), 180ra48.

PART V

GLOSSARY

CIRCULATING TUMOR CELL GLOSSARY

JOSE I. VARILLAS

J. Crayton Pruitt Family, Department of Biomedical Engineering, University of Florida, Gainesville, FL, USA

Z. HUGH FAN

Department of Mechanical and Aerospace Engineering, University of Florida, Gainesville, FL, USA; J. Crayton Pruitt Family Department of Biomedical Engineering, University of Florida, Gainesville, FL, USA; Department of Chemistry, University of Florida, Gainesville, FL, USA

A

Acoustophoresis [1] A noncontact, continuous-flow-based method of separating cells or particles. Particles and cells, in a suspension subjected to an acoustic standing-wave field, experience a radiation force that causes them to move in the sound field to either the pressure node or the antinode of the standing wave. The direction of movement is dependent on the physical properties of the particles relative to those of the surrounding medium. Acoustophoresis is a label-free method and its implementation in microfluidic devices is discussed in detail in Chapter 10.

Antibody [2–4] Antibodies are glycoproteins that all have the same overall Y-shape and belong to the immunoglobulin superfamily. Each antibody molecule has a unique structure, known as its variable region, which allows it to bind specifically to its corresponding antigen (*see Antigen*). Antibodies are secreted by B-lymphocytes (plasma cells) of the immune system in response to infection or immunization to identify and neutralize pathogens, such as viruses and bacteria, or prepare them for uptake and destruction by phagocytes.

Circulating Tumor Cells: Isolation and Analysis, First Edition. Edited by Z. Hugh Fan.

Antigen [2, 3, 5] A structural substance that can bind specifically to an antibody (*see Antibody*). Those antigens that can induce the immune system to produce antibodies against it are called immunogens. Antigens may originate from within the body or from the external environment.

Aptamers [6] Nucleic acid oligomers, or peptide molecules, that possess distinct conformational shapes that allow them to bind to target molecules (e.g., proteins, enzymes, antibodies, cell-surface receptors, small organic molecules) with high affinity and specificity. Aptamers are generally created by chemical synthesis, although natural aptamers also exist and elicit little to no immunogenicity in therapeutic applications. Aptamers offer an alternative to antibodies and are discussed in detail in Chapter 13.

B

Benign tumor [7] A noncancerous tumor that remains in the tissue in which it originated, meaning that it does not spread into the surrounding tissue or to other organs.

Biomarker [8, 9] A quantifiable biological entity that provides information about the status of a system (e.g., organism). In clinical research, it often refers to genetic or cellular substances, or molecules, that indicate a disease if found at an abnormal level in bodily fluids, such as blood. A panel of biomarkers likely provides more accurate information than one single biomarker. Tumor biomarkers, such as circulating tumor cells (CTCs), may be used for screening, risk assessment, medical diagnosis, prognosis, and monitoring the clinical course of the cancer.

Biopsy [3, 10] The process of removing living tissue from an organ or part of the body, for diagnostic examination. For cancer diagnosis, a core needle biopsy or a fine needle aspiration biopsy is often performed. A core needle biopsy removes a small cylinder of intact tissue for histologic examination. A fine needle aspiration biopsy withdraws cells and pieces of tissue that are evaluated as individual cells and clusters of cells. Other types of biopsies for cancer diagnosis include bone marrow testing and endoscopy.

Bone marrow (BM) blood [11, 12] Found within the soft tissue in the interior of bones, BM blood contains the precursors of erythrocytes, platelets, granulocytes, stromal cells, tissue mast cells, B-lymphocytes, and osteoblasts. This biofluid is often analyzed to detect DTCs (*see Disseminated tumor cells*).

C

Cancer [3, 13] Any malignant tumor, including carcinoma, sarcoma, lymphoma, and leukemia, caused by abnormal, purposeless, and uncontrolled division of cells. As cancerous cells multiply, they invade and destroy the surrounding tissues or cells. There are many causative factors for cancer, some of which are known.

Regardless of the initial cause, cancer always results ultimately from DNA mutations and epigenetic modifications. The type of tumor, its location, and the extent of the spread determine the range of treatments that can be considered.

Cancer stem cells (CSCs) [14, 15] A small subset of cells within a tumor that are believed to be crucial for tumor growth and propagation. According to the cancer stem cell hierarchical model for tumor-regenerative capacity, CSCs are the driving force of tumorigenesis due to their ability of self-renewal and multilineage differentiation through either asymmetric or symmetric cell division. Thus, not all cancer cells are equal, as only a small fraction of them possess the properties to perpetuate the disease. However, the CSC model has been demonstrated in many cancers but not all.

Capture efficiency [16–18] In CTC isolation, capture efficiency is a performance measure defined as the ratio of the number of target cells captured or isolated to the total number of target cells introduced to the system. Cell capture efficiency is an indicator of the sensitivity of a method (See Chapter 13 for the details). It is also referred to as "recovery" rate in Chapter 4 and other work.

Carcinogenesis (oncogenesis) [3, 19] Carcinogenesis, also termed oncogenesis, is the process by which normal cells transform into cancer cells via changes at the cellular, genetic, and epigenetic level. It is a multistep process of successive genetic mutations and epigenetic modifications. The progressive stages often include hyperplasia, dysplasia, metaplasia, benign tumors, and, eventually, malignant tumors.

Carcinoma [3, 20] A cancerous tumor derived from epithelium, the tissue that lines the skin and internal organs of the body. They are the most common type of tumors, possibly because epithelia retain proliferative capacity throughout their lifespan. Carcinomas are categorized into adenocarcinoma, squamous cell carcinoma, transitional cell carcinoma, undifferentiated (or anaplastic) carcinoma, and specific types such as melanoma, basal cell carcinoma, and large cell undifferentiated carcinoma.

CD44 [21, 22] (*see Cluster of differentiation*) A family of type I transmembrane glycoproteins that function as cell-surface receptors. These receptors are involved in migration, adhesion, lymphocyte activation, and lymph node homing. CD44 can accomplish adhesive reactions during initiation of an inflammatory reaction and also during seeding of cancer cells in the course of the metastatic process.

CD45 [2, 23] (*see Cluster of differentiation*) CD45 is a transmembrane protein tyrosine phosphatase (PTP) expressed on the surface of all leukocytes (*see Leukocytes*). This molecule comprises up to 10% of the cell-surface area and is a high-molecular-weight glycoprotein with a single transmembrane domain. CD45 plays a central role in mediating the process of integrating environmental signals into cellular responses. It is frequently used as a cell-surface marker for pan-leukocytes, B-lymphocytes, as well as naïve and memory T-lymphocytes. This cell-surface receptor is often targeted for negative depletion of whole blood samples (*see Negative depletion*).

CD47 [24, 25] (***see Cluster of differentiation***) CD47, also known as the integrin-associated protein (IAP), is involved in important tumor progression-associated cellular processes including migration, adhesion, angiogenesis, apoptosis, and proliferation. Overexpression of this cell-surface marker on CTCs provides them with a mechanism to evade the immune system because this protein inhibits the cytotoxic activity of activated immune cells.

CD133 [26, 27] (***see Cluster of differentiation***) CD133, also known as prominin-1, is a transmembrane glycoprotein cell-surface marker expressed in a wide range of somatic stem and progenitor cells, indicating cancer stem cells (*see Cancer stem cells*) of solid tumors and hematopoietic malignancies. The presence of this marker is often associated with chemoresistant subpopulations.

Cell heterogeneity [28] (*See Tumor heterogeneity*) For a given cancer, cell heterogeneity arises from genetic and/or epigenetic differences among the cancer cells, divergent cancer stem cell differentiation, and the diverse cell types that are recruited to the tumor site. The manifestation of these differences includes cellular morphology and motility, metabolism, gene expression, hormonal receptors, proliferation, immunogenicity, angiogenesis, and metastatic potential.

Cell purity [16–18] In CTC isolation, cell purity is a performance measure defined as the ratio of the number of target cells captured or isolated to the total number of cells captured or isolated. The isolated cells can contain both target cells and nontarget cells. Cell purity is an indicator of the specificity of a particular method (See Chapter 13 for the details).

CellSearch® [29] A standardized, US Federal Drug Administration (FDA)-approved, semiautomated system used to capture, isolate, and enumerate CTCs (*see Circulating tumor cells*) from whole blood samples. The blood sample is pretreated, and ferromagnetic beads conjugated with anti-EpCAM antibodies are added. These cells tagged with beads are magnetically sorted, and CTCs are counted according to morphological characteristics, positive expression of cytokeratins (CK8, 18, and/or 19) (*see Cytokeratin*), and positive staining with 4′,6-diamidino-2-phenylindole (DAPI) (*see DAPI*) with the absence of the leukocyte marker CD45 (*see CD45*). CellSearch® is discussed in detail in Chapter 19.

Chaotic mixing [30] A passive method for mixing streams of fluids at low Reynolds number in a microfluidic channel. Chaotic mixing is characterized by the generation of transverse flows, which cause flow folding and stretching over the cross section of the channel or pipe. The parallel-flowing layers of fluids fold over one another increasing the degree of mixing, convection, and diffusion.

Circulating cell-free DNA (cfDNA) Cell-free DNA is DNA that circulates freely in the blood stream. It includes circulating tumor DNA (*see Circulating tumor DNA*), although the majority of circulating DNA is often not of cancerous origin. A related term is circulating cell-free RNA, especially microRNA (miRNA).

Circulating epithelial cells (CECs) [31, 32] Cells of epithelial origin found circulating in peripheral blood. These cells include malignant CTCs (*see Circulating tumor cells*) as well as benign CECs, such as endothelial cells shed from the vascular walls (circulating endothelial cells) and endothelial progenitor cells (EPCs). Increased levels of CECs have been found in noncarcinoma related cases, such as patients with different thyroid diseases and benign inflammatory colon diseases.

Circulating tumor cells (CTCs) [31, 33, 34] CTCs are cancer cells that detach from the primary tumor site, enter and travel through the vasculature. CTCs have been proven as biomarkers to diagnose and prognose cancer noninvasively. These cells are extremely rare, estimated at a few CTCs among billions of normal blood cells in the peripheral blood of patients with advanced cancer. The FDA currently defines a CTC as a nucleated cell of epithelial origin containing EpCAM as well as cytokeratin 8, 18, and/or 19, lacking expression of the CD45 surface marker (i.e., $DAPI^+$, $EpCAM^+$, CK^+, and $CD45^-$) (*see DAPI, EpCAM, Cytokeratin, CD45*). However, the term CTC could be expanded to encompass all types of cells that exhibit cancerous characteristics and are considered foreign entities in the blood.

Circulating tumor DNA (ctDNA) [35–37] Cell-free fragments of DNA that are shed into the bloodstream, arising from cells that go through apoptosis or necrosis. Although healthy subjects naturally exhibit small quantities of cell-free DNA circulating in their blood, it has been evidenced that cancer patients have up to four times more cell-free DNA present in their plasma in comparison to healthy subjects. Genetic materials extracted from the plasma of cancer patients have shown characteristics of tumor DNA, such as decreased strand stability, hypermethylation of several genes, immunoglobulin rearrangements, and the presence of specific oncogenes, tumor suppressor genes, and microsatellite alterations. Genotyping of ctDNA for somatic genomic alterations has proven to be effective in tracking tumor dynamics.

Cluster of differentiation (CD) [38] Human cluster of differentiation (CD) antigens are groups of cell surface molecules (i.e., ligands and/or receptors) recognized by a set of monoclonal antibodies. For nomenclature, the number after "CD" stands for the order of discovery. CD antigens are grouped based on their effects on cell or tissue function.

Cytokeratin (CK) [39, 40] Keratins, or cytokeratins, are proteins that constitute the intermediate filaments of the cytoskeletons of epithelial cells and are essential for normal tissue structure and function. They are characteristic of epithelial tumor cells and can be identified in individual carcinoma cells by specific monoclonal antibodies. There are 20 different cytokeratins. Since the type of cytokeratin expressed is often tissue-specific, they can be used for determining histological subtypes of cancer.

Cytopathology [3, 41] A branch of pathology, the study on cellular level for disease diagnosis. It involves the microscopic examination of individual cells, as well as clusters of cells, for the diagnosis of various diseases. The term cytology can also be used to refer to cytopathology.

D

4′,6-Diamidino-2-phenylindole (DAPI) [42, 43] A DNA-specific dye molecule that forms a fluorescent complex by multiple attachment mechanisms, including minor groove binding, intercalation, and condensation. When bound to double-stranded DNA, DAPI has maximum absorption and emission at 358 nm and 461 nm, respectively.

Diagnosis [3] The process of determining the identity and nature of a disease or disorder based on the patient's symptoms, medical history, and laboratory tests. A diagnostic opinion is reached by the medical specialists at the end of the diagnostic procedure.

Dielectrophoresis (DEP) [44, 45] The motion of polarizable particles under the influence of an applied nonuniform electric field in which the force arises from the interaction between the field and the dipole moment induced in each particle. The charge strength and distribution at the interface between the particle and the electrolyte depend on the strength of the field and on the electrical properties of the particle and the electrolyte. DEP is often used as a particle manipulation and characterization technique due to its label-free and noninvasive nature. DEP is discussed in detail in Chapter 18.

Disseminated tumor cells (DTCs) [12, 46] Tumor cells that have settled at a site (e.g., secondary organs) separate from the primary tumor (i.e., disseminated from the primary tumor). Detection of DTCs in the bone marrow has been used in clinical trials to validate its efficacy as a cancer biomarker. Genome and transcriptome analysis of single DTCs have demonstrated the majority of these rare cells contain genetic aberrations compatible with malignancy.

E

Enzyme-linked immunosorbent assay (ELISA) [3, 47, 48] An immunoassay (*see Immunoassay*) used to detect and quantify proteins (or analytes) in a sample. A primary antibody bonds to the protein of interest. A secondary antibody (also called detection antibody) conjugated with an enzyme binds to the first-antibody–protein complex. Subsequently, the enzyme reacts with a chromogenic or fluorogenic substrate, which provides the signal that allows detection and quantification of the target protein. A sandwich ELISA is a technique in which the ELISA assay is often performed with the primary antibody bound to a surface.

Enzyme-linked immunospot (ELISPOT) assay [3, 49] An antibody-capture-based assay used to monitor immune response. ELISPOT assays counts antibody-secreting cells or cytokine-producing cells with high sensitivity and selectivity. Cells are placed on a membrane coated with an antibody specific for a given protein. The protein secreted by cells is captured directly by the antibodies and is detected via a secondary antibody. Cell enumeration is achieved by counting the spots produced by protein-producing cells.

Ensemble-decision aliquot ranking (eDAR) [50] A process for CTC isolation and detection from whole blood based on positive selection. Cells are labeled with fluorescent antibodies against cell-surface markers characteristic of CTCs. The blood sample is first divided into nanoliter aliquots (typically 2 nL) via microfluidics. These aliquots are then ranked based on the presence or absence of CTCs, followed by sorting and collection based on their ranking. The eDAR method can be used for other rare cells and is discussed in detail in Chapter 3.

Epithelial cell-adhesion molecule (EpCAM) [51–53] An epithelial transmembrane glycoprotein encoded by the *GA733-2* gene, which is located on chromosomal region 4q. EpCAM, with molecular weight of 40 kDa, mediates epithelium-specific, calcium-independent homotypic cell–cell adhesion. This molecule is detected at the basolateral cell membrane of all simple (especially glandular), pseudostratified, and transitional epithelia. It resembles "classic" adhesion molecules because it associates with the cytoskeleton and forms independent adhesion. Because EpCAM is strongly expressed by carcinomas of various origins, including lung, breast, colorectal, prostate, head and neck, liver, and esophagus and is absent from hematologic cells, anti-EpCAM is useful for capturing CTCs from blood. The expression of EpCAM in carcinomas can be heterogeneous and has been thought to be affected by a shift of tumor cell differentiation to either mesenchymal (*see Epithelial-to-mesenchymal transition*) or squamous cell phenotypes. The molecule is also known by other names corresponding to chimeric and humanized monoclonal antibodies used to identify it in various tissues (e.g., EGP-40, Trop-1, 17-1A, KSA, KS1/4, AUA1).

Epidermal growth factor receptor (EGFR) [54, 55] A cell-surface marker that is found at abnormally high levels on the surface of many types of cancer cells. The EGFR signaling pathway is implicated in important tumor progression mechanisms such as proliferation, angiogenesis, metastasis, and decreased apoptosis. This receptor can be activated by various ligands, including epidermal growth factor (EGF) and transforming growth factor-alpha (TGF-α). EGFR is a member of the human epidermal growth factor receptor (HER or EGFR) family and is also known as HER1 or ErbB1.

EPISPOT (epithelial immunospot) assay [56, 57] An assay adapted from the enzyme-linked immunospot assay (*see ELISPOT*) and used to enumerate only viable epithelial tumor cells. The EPISPOT assay can detect specific proteins secreted, shed, or actively released, at an individual cell level. This allows for direct determination of viable protein-secreting cell frequencies.

Epithelial phenotype [58, 59] The epithelial phenotype refers to the observable characteristics of epithelial cells, which line cavities and the surfaces of organs, blood vessels, and other structures in the body. These types of cells are considered adherent with restricted motility and facilitated intercellular communication. Epithelial cells tend to aggregate (i.e., cluster) and function as a cohesive unit. These characteristics of epithelial cells are promoted by the presence of junctional complexes, such as tight and adherens junctions, desmosomes, and segregation of the plasma membrane into apical and basolateral domains. (see *Mesenchymal phenotype* for comparison).

Epithelial-to-mesenchymal transition (EMT) [58–61] A process for epithelial cells to change morphology and gain mobility, allowing them to move to a new site. EMT is involved in metastasis, and it is believed that during disease progression, malignant tumor cells undergo EMT in order to move from the tumor of origin and spread to a new site and revert back to an epithelial phenotype through the mesenchymal-to-epithelial transition (*see Mesenchymal-to-epithelial* *transition*). Although the EMT–MET process is often portrayed as a reversible reaction between two binary states, it has been shown that intermediate states between the two exist and play important roles in the metastasis process. However, there is still debate about whether EMT is an absolute requirement for cancer metastasis.

Epitope [21, 62] The specific region on the surface of an antigen (*see Antigen*) that is recognized by the immune system (i.e., antibodies, B-cells, or T-cells). A single antigen typically has many epitopes, each specific to different antigen receptors. Epitopes are also known as antigenic determinants.

Erythrocytes (red blood cells) [63] Red blood cells, or erythrocytes, are by far the most abundant blood cells, with roughly 5–6 billion in each milliliter of blood and about 25 trillion of them in the body's 5 L of blood. The main function of these cells is to transport oxygen, O_2. They have a biconcave structure, being thinner in the center than at the edges; this gives them a high surface-to-volume ratio, which enhances the rate of diffusion of O_2 across their plasma membranes. Erythrocytes do not have nuclei or mitochondria; they generate ATP via anaerobic metabolism, which further increases their oxygen-transporting efficiency. They range between 6–9 μm in diameter.

Extracellular matrix (ECM) [64, 65] A network of proteins and polysaccharides, including glycoproteins, glycosaminoglycans, and proteoglycans that provide a scaffolding for the adhesion and growth of cells in coherent tissues. The components of the ECM are mainly secreted by fibroblasts and smooth muscle cells. The ECM promotes cell adherence, cell migration, apoptosis, differentiation, and regulation of inflammation.

F

Flow cytometry [66, 67] A method used to count, analyze, and sort cells as they flow in a fluid stream through one or multiple beams of light. Flow cytometry allows for the measurement of a cell's relative size and shape, internal complexities, expression of proteins and other molecules, and other biological attributes based on its light-absorbing and -scattering properties.

Fluorescence *in situ* hybridization (FISH) [68, 69] A method for localizing genes and specific genomic regions on target chromosomes in metaphase and interphase cells. FISH is useful for identifying chromosomes, and parts thereof, to unravel chromosomal abnormalities and for gene mapping. Fluorescently labeled DNA or RNA probes are hybridized to chromosomes of suitably prepared cells or histological sections. The FISH technique enables microscopic visualization of genes to assess their numbers, or the expression level of mRNA or presence of other cellular nucleic acids, if applied to tissue sections.

Fluorescence-activated cell sorting (FACS) [70, 71] A specialized type of flow cytometry (*see Flow cytometry*) method for the continuous separation of mixtures of different cells. Each cell is individually placed into a droplet, using a nozzle, and passed through one or multiple laser beams at high speed. Electrostatic deflection allows for selective collection of cells into multiple containers, one cell at a time, based on the specific light-scattering and fluorescent characteristics of each cell.

H

Hematopoietic stem cells (HSCs) [72, 73] A type of unspecialized bone marrow cells that renew themselves through cell division and serve as the source for sustaining hematopoiesis throughout life. These rare cells (comprise <0.01% of normal bone marrow) are normally quiescent in the osteoblast stem cell niche of the bone marrow until cell cycle activation. HSCs undergo asymmetric and symmetric self-renewal, generating additional stem cells as well as lineage-committed progenitors, which give rise to mature cells.

Human epidermal growth factor receptor 2 (HER2) [74, 75] An epidermal growth factor receptor that plays an important role in the development and progression of certain aggressive types of breast cancer. This receptor is a member of the human epidermal growth factor receptor (HER or EGFR) family. In normal cells, HER2 helps control cell growth. It is overexpressed, or amplified, in 20–30% of invasive breast cancers and is often used as a therapeutic target molecule. The receptor CXCR4 upregulates HER2-mediated metastasis. Inhibition of this receptor suppresses metastasis.

Hydrodynamic shear [76] A force caused by the motion of different layers in a fluid when the layers move at different velocities. The force acts perpendicularly to the direction of extension of the molecule or entity of interest. For Newtonian fluids, the hydrodynamic shear force is directly proportional to the rate of deformation of the fluid.

I

Image cytometry [77, 78] A quantitative analysis tool based on microscopy and digital imaging. Image cytometry provides a method to study molecular processes as well as structural changes at the cell and tissue level. This analytical tool measures most of the same parameters as flow cytometry (*see Flow cytometry*), but with the advantage of the two- and three-dimensional imagery. Current approaches use automated microscopy and incorporate computational image processing and analysis to infer disease state based on many measurements.

Immunoassay [79, 80] A bioanalytical procedure based on the specific binding between antigens and antibodies used to quantify analytes in a sample. A number of different formats for immunoassays exist, including competitive, noncompetitive, homogeneous, and heterogeneous. ELISA (*see ELISA*) is one variation of an immunoassay.

Immunocytochemistry (ICC) [21, 81, 82] A type of immunochemistry assay used to visualize the distribution of cytoplasmic or nuclear proteins within cells fixed in fixatives, such as paraformaldehyde (PFA). This assay uses antibodies to provide highly specific binding to unique sequences of amino acids in proteins. The ICC method is widely used in cooperation with cytopathology (*see Cytopathology*) and immunohistochemistry (*see Immunohistochemistry*) to diagnose abnormal cells as well as to differentiate and subclassify types of cancer.

Immunofluorescence (IF) [3, 83] A type of immunostaining used to identify a protein, typically an antigen, in cells or tissue sections. The direct method uses a primary antibody to detect the presence of the target molecule. The indirect method uses an unlabeled primary antibody that binds to the target molecules. A secondary antibody, with an attached fluorochrome, targets and binds to the primary antibody. The fluorescence from the fluorochrome allows for the visualization of the location and quantity of the target molecules.

Immunohistochemistry (IHC) [84, 85] A type of immunochemistry assay used to detect and localize specific proteins in biological tissue sections. This assay uses antibodies with specific affinity to the antigen of interest, while retaining cellular and tissue structure. The IHC method is widely used in cooperation with cytopathology (*see Cytopathology*) and immunocytochemistry (*see Immunocytochemistry*) to diagnose abnormal cells as well as to differentiate and subclassify types of cancer.

Inertial focusing [86, 87] The use of fluid inertia effects in microchannels to manipulate particles and cells across streamlines and order them deterministically at equilibrium positions. Three forces are involved: secondary-flow drag forces, wall-induced lift forces, and shear-gradient-induced lift forces. The key parameters to control the magnitude and direction of the two types of lift forces include the channel aspect ratio and dimensions, particle diameter, and flow rate. Inertial microfluidics is discussed in more detail in Chapter 5.

Isolation by size of epithelial tumor cells (ISET) [88] A method for isolating CTCs from whole blood based on the physical properties of cells. Filters with a pore size of ~ 6–8 μm are used to trap tumor cells larger than a certain diameter threshold while allowing smaller blood cells to pass through. The filter pore size used depends on the type of cancer cells of interest. This approach is often paired up with downstream CTC analysis such as reverse transcriptase polymerization chain reaction (RT-PCR) and immunofluorescence (*see Immunofluorescence*). One version of the size-based CTC isolation methods is discussed in detail in Chapter 7.

L

Laminar flow [89, 90] A fluid flow regime in which layers of liquid slide by one another, smoothly, in the same direction and all flow elements follow a path, parallel to the surface containing the liquid. It generally occurs when dealing with small pipes and low flow velocities, at a Reynolds number (*see Reynolds number*) lower than 2300. In microfluidic devices, most fluid flows are laminar due to the size of the channels and inability of the fluids to reach high enough velocities; thus, viscous (friction) forces are much larger than inertial forces.

Leukocytes (white blood cells) [63] There are five major types (neutrophils, eosinophils, basophils, lymphocytes, and monocytes) of white blood cells, or leukocytes, whose main function is to fight infections. There are roughly 5–10 million white blood cells in each milliliter of blood, but their numbers increase temporarily when fighting an infection. Some leukocytes are phagocytic, surrounding and digesting microorganisms as well as debris from the body's own dead cells. Leukocytes are also found outside the circulatory system, where they perform immune responses in interstitial fluids and the lymphatic system.

Liquid biopsy [91,92] The process of removing blood, peripheral or directly from the bone marrow, or other biofluids, for diagnostic examination (in comparison with tissue *Biopsy*). For cancer diagnosis, CTCs (*see Circulating tumor cells*), DTCs (*see Disseminated tumor cells*), and ctDNA (*see ctDNA*) are being investigated for use as cancer biomarkers. These real-time, minimally invasive liquid biopsies can be performed at multiple time points to monitor disease and personalize cancer therapy.

M

Magnetic-activated cell sorting (MACS) [71] A passive continuous method for separating cell populations. Antibody-conjugated magnetic beads are used to bind to specific proteins on the target cells. A mechanical magnetic sweeper or an external high gradient magnetic field is then applied to isolate the labeled cells free from all other cells.

Malignant tumor [93] A tumor that invades and destroys adjacent tissues and spreads to other organs. Cancer malignancy can be characterized by anaplasia, invasiveness, metastasis (*see Metastasis*), and genetic instability.

Mesenchymal phenotype [58,59] The mesenchymal phenotype refers to the observable characteristics of mesenchymal cells; some common traits of these cells are that they are spindle-shaped, or multipolar, with fibroblast-like morphology. Mesenchymal cells lack junctional complexes, adhesion properties, and specialization of the plasma membrane into apical and basolateral domains. They only interact with other cells via focal points and are highly motile, migratory cells (see *Epithelial phenotype* for comparison).

Mesenchymal-to-epithelial transition (MET) [58–61] A process for mesenchymal cells to change morphology and become stationary in a new tissue site. Cells that undergo this transition gain characteristics of the epithelial phenotype (*see Epithelial phenotype*) and lose those of the mesenchymal phenotype (*see Mesenchymal phenotype*). It is the reverse process of EMT (*see Epithelial-to-mesenchymal transition*).

Metastasis [21,94] The spread of a malignant tumor from its site of origin to a distant site. The three main routes this occurs by are the following: through the bloodstream (hematogenous), through the lymphatic system, and across body cavities. Individual tumors may spread through one or more of the mentioned routes.

Microfluidics [95, 96] The study of fluid manipulation, with respect to the transport of samples and reagents, in microchannels on a common substrate. The channels typically have dimensions ranging from hundreds of nanometers to hundreds of micrometers. Microfluidics involves three aspects: microfabrication, microflows, and applications. Microfluidics is discussed in more detail in Chapter 2.

N

Negative depletion A cell or particle sorting technique used to isolate and enrich targets of interest from a mixture by specifically eliminating undesired portions, according to physical (e.g., size, shape) or biological (e.g., surface protein expression) properties. Negative depletion is discussed in more detail in Chapter 14.

P

Péclet number (*Pé*) [97] A dimensionless number used in the study of transport phenomena. The Péclet number value relates the effectiveness of transport by advection to the effectiveness of transport by dispersion or diffusion. It is defined as the ratio of the rate of advection of a flow to its rate of diffusion, VL/D, where V is the flow velocity, L is the flow's characteristic length (often the diameter of a tube), and D is the diffusional coefficient.

Peripheral blood (PB) [12] The circulating blood of the body not enclosed within the liver, spleen, lymphatic system, or bone marrow. Peripheral blood is composed of erythrocytes (*see Erythrocytes*), leukocytes (*see Leukocytes*), and thrombocytes (*see Thrombocytes*), which are suspended in blood plasma. Peripheral blood brings nutrients to all of the organs and systems of the body, and it also plays an important role in waste excretion and immunity.

Photoacoustics [98] Photoacoustic techniques (e.g., imaging and flow cytometry) deal with periodic laser heating and indirect probing of temperature by monitoring pressure changes (mechanical fluctuations) generated in the medium surrounding the surface of the sample. These techniques are based on photoacoustic waves, which result from the conversion of the optical energy of a photon into a mechanical disturbance. This conversion produces a detectable acoustic wave. Photoacoustic techniques are discussed in more detail in Chapter 11.

Polydimethylsiloxane (PDMS) [99] A silicone elastomer with properties that make it well suited for use in microfluidics and biomedical applications. Some of these properties include optical transparency, chemical inertness, permeability to gases, nontoxicity, nonflammability, and nonfluorescence.

Prognosis [100] The assessment of the future course and progression of a disease. It also includes the likelihood of recovery or cure after diagnosis. Prognoses are based on prognostic factors, if available, and on knowledge of the course of the disease in other patients with the general health, age, and sex of the patient. Prognostic factors, such as prognostic biomarkers (*see Biomarker*), aid in the estimation of individual patient outcome.

R

Reynolds number (*Re*) [89, 101] A dimensionless number that indicates the nature of fluid flow in a system. This value is equivalent to the ratio of inertial forces to viscous forces and is symbolically defined by $\rho VL/\mu$, where ρ is the fluid density, V is the flow velocity, L is the flow's characteristic length (often the diameter of a tube), and μ is the fluid viscosity. *Re* of 2,300 is a typical number differentiating laminar flow (*see Laminar flow*) from turbulent flow.

S

Single-cell analysis [102] Single-cell analysis is a term that describes a set of techniques used for the analysis of the molecular content of individual cells. These techniques include fluorescence correlation spectroscopy, flow cytometry, whole-genome amplification, quantitative polymerase chain reaction (qPCR), RNA sequencing, array-based comparative genomic hybridization (aCGH), FISH, and many others.

Systematic evolution of ligands by exponential enrichment (SELEX) [103] The process for generating aptamers (*see Aptamers*) with high binding affinity and specificity to a ligand of interest (e.g., proteins, small molecules). It is a reiterative process that screens a large library of oligonucleotides through *in vitro* selection and amplification. SELEX has been further developed into a method using cells as the target; the method is called Cell-SELEX, which is discussed in Chapter 13.

T

Theranosis [104] A model used to create specific treatments and therapies tailored for an individual using molecular and genetic analysis. This model involves the application of the right treatment for the right patient, at the right dose, and at the right time. Theranosis applications range from personal risk profiles leading to personalized health planning for early diagnosis, genetic counseling, databases, and decision support tools.

Thrombocytes (platelets) [63] Platelets, or thrombocytes, are cytoplasmic fragments of specialized bone marrow cells derived from megakaryocytes. These cells serve structural and molecular functions in blood clotting. When a blood vessel is cut, platelets are recruited to the site, adhere to the newly exposed subendothelial tissues, release numerous secretory granules, and aggregate together. These cells contain no nuclei, have a biconvex discoid shape in their inactive state, and are about 2–3 μm in diameter.

Tumor heterogeneity [28, 105, 106] Tumor heterogeneity refers to variation in the specific phenotypic, genotypic, and morphological characteristics of tumor cells derived from either the same tumor (i.e., intertumor heterogeneity) or from different sites (i.e., intratumor heterogeneity). Tumor diversity manifests itself in the qualitative and quantitative expression of numerous genes. Heterogeneity arising from genetic differences is intensified due to genomic instabilities.

REFERENCES

1. Lenshof, A. and Laurell, T. (2012) Acoustophoresis, in *Encyclopedia of Nanotechnology* (ed B. Bhushan), Springer Netherlands.
2. Janeway, C., Travers, P., and Walport, M. (2001) Glossary, in *Immunology: The Immune System in Health and Disease*, 5th edn, Garland Science.
3. Martin, E.A. (2010) Concise medical dictionary, in *Oxford Paperback Reference*, 8th edn, Oxford University Press.
4. Schwab, M. (ed) (2012) Antibody, in *Encyclopedia of Cancer*, Springer Berlin Heidelberg.
5. Schwab, M. (ed) (2012) Antigen, in *Encyclopedia of Cancer*, Springer Berlin Heidelberg.
6. Offermanns, S. and Rosenthal, W. (eds) (2008) Aptamers, in *Encyclopedia of Molecular Pharmacology*, Springer Berlin Heidelberg.
7. Schwab, M. (ed) (2012) Benign tumor, in *Encyclopedia of Cancer*, Springer Berlin Heidelberg.
8. Hayes, D.F., Bast, R.C., Desch, C.E. *et al.* (1996) Tumor marker utility grading system: a framework to evaluate clinical utility of tumor markers. *Journal of the National Cancer Institute*, **88** (20), 1456–1466.
9. Strimbu, K. and Tavel, J.A. (2010) What are biomarkers? *Current Opinion in HIV and AIDS*, **5** (6), 463–466.
10. Schwab, M. (ed) (2012) Biopsy, in *Encyclopedia of Cancer*, Springer Berlin Heidelberg.
11. Blood and bone marrow, in *Atlas of Clinical Hematology*, Springer Berlin Heidelberg.
12. Braun, S. and Naume, B. (2005) Circulating and disseminated tumor cells. *Journal of Clinical Oncology : Official Journal of the American Society of Clinical Oncology*, **23** (8), 1623–1626.
13. Cancer, in *Encyclopedic Reference of Genomics and Proteomics in Molecular Medicine*, Springer Berlin Heidelberg.
14. Schwab, M. (ed) (2012) Cancer stem cells, in *Encyclopedia of Cancer*, Springer Berlin Heidelberg.
15. O'Connor, M.L., Xiang, D., Shigdar, S. *et al.* (2014) Cancer stem cells: a contentious hypothesis now moving forward. *Cancer Letters*, **344** (2), 180–187.

16. Barrett, L.M. and Simmons, B.A. (2008) Cell sorting, in *Encyclopedia of Microfluidics and Nanofluidics* (ed D. Li), Springer US.
17. Sheng, W., Ogunwobi, O.O., Chen, T. *et al.* (2014) Capture, release and culture of circulating tumor cells from pancreatic cancer patients using an enhanced mixing chip. *Lab on a Chip*, **14** (1), 89–98.
18. Yu, Z.T., Aw Yong, K.M., and Fu, J. (2014) Microfluidic blood cell sorting: now and beyond. *Small*, **10** (9), 1687–1703.
19. Schwab, M. (ed) (2012) Carcinogenesis, in *Encyclopedia of cancer*, Springer Berlin Heidelberg.
20. Kirkham, N. and Lemoine, N.R. (2001) *Progress in Pathology*, vol. **5**, Cambridge University Press, London.
21. Lackie, J. (2010) *A Dictionary of Biomedicine*, Oxford University Press.
22. Schwab, M. (ed) (2012) CD44, in *Encyclopedia of Cancer*, Springer Berlin Heidelberg.
23. Thomas, M.L. (1989) The leukocyte common antigen family. *Annual Review of Immunology*, **7**, 339–369.
24. Roberts, D., Soto-Pantoja, D., and Isenberg, J. (2012) CD47, in *Encyclopedia of Signaling Molecules* (ed S. Choi), Springer New York.
25. Willingham, S.B., Volkmer, J.P., Gentles, A.J. *et al.* (2012) The CD47-signal regulatory protein alpha (SIRPa) interaction is a therapeutic target for human solid tumors. *Proceedings of the National Academy of Sciences of the United States of America*, **109** (17), 6662–6667.
26. CD133 (prominin), in *Encyclopedia of Genetics, Genomics, Proteomics and Informatics*, Springer Netherlands.
27. Corbeil, D., Karbanová, J., Fargeas, C., and Jászai, J. (2013) Prominin-1 (CD133): molecular and cellular features across species, in *Prominin-1 (CD133): New Insights on Stem & Cancer Stem Cell Biology*, vol. **777** (ed D. Corbeil), Springer New York.
28. Meacham, C.E. and Morrison, S.J. (2013) Tumour heterogeneity and cancer cell plasticity. *Nature*, **501** (7467), 328–337.
29. Allard, W.J., Matera, J., Miller, M.C. *et al.* (2004) Tumor cells circulate in the peripheral blood of all major carcinomas but not in healthy subjects or patients with nonmalignant diseases. *Clinical Cancer Research : An Official Journal of the American Association for Cancer Research*, **10** (20), 6897–6904.
30. Stroock, A.D., Dertinger, S.K., Ajdari, A. *et al.* (2002) Chaotic mixer for microchannels. *Science*, **295** (5555), 647–651.
31. Plaks, V., Koopman, C.D., and Werb, Z. (2013) Cancer. Circulating tumor cells. *Science*, **341** (6151), 1186–1188.
32. Winkens, T., Pachmann, K., and Freesmeyer, M. (2014) The influence of radioiodine therapy on the number of circulating epithelial cells (CEC) in patients with differentiated thyroid carcinoma - a pilot study. *Experimental and Clinical Endocrinology & Diabetes : Official Journal German Society of Endocrinology [and] German Diabetes Association*, **122** (4), 246–253.
33. Racila, E., Euhus, D., Weiss, A.J. *et al.* (1998) Detection and characterization of carcinoma cells in the blood. *Proceedings of the National Academy of Sciences of the United States of America*, **95** (8), 4589–4594.
34. Yu, M., Stott, S., Toner, M. *et al.* (2011) Circulating tumor cells: approaches to isolation and characterization. *The Journal of Cell Biology*, **192** (3), 373–382.

35. Shapiro, B., Chakrabarty, M., Cohn, E.M., and Leon, S.A. (1983) Determination of circulating DNA levels in patients with benign or malignant gastrointestinal disease. *Cancer*, **51** (11), 2116–2120.
36. Anker, P., Mulcahy, H., Qi Chen, X., and Stroun, M. (1999) Detection of circulating tumour DNA in the blood (plasma/serum) of cancer patients. *Cancer and Metastasis Review*, **18** (1), 65–73.
37. Diaz, L.A. Jr. and Bardelli, A. (2014) Liquid biopsies: genotyping circulating tumor DNA. *Journal of Clinical Oncology: Official Journal of the American Society of Clinical Oncology*, **32** (6), 579–586.
38. Schwab, M. (ed) (2012) CD antigens, in *Encyclopedia of Cancer*, Springer Berlin Heidelberg.
39. Schweizer, J., Bowden, P.E., Coulombe, P.A. *et al.* (2006) New consensus nomenclature for mammalian keratins. *The Journal of Cell Biology*, **174** (2), 169–174.
40. Schwab, M. (ed) (2012) Cytokeratins, in *Encyclopedia of Cancer*, Springer Berlin Heidelberg.
41. Schwab, M. (ed) (2012) Cytology, in *Encyclopedia of Cancer*, Springer Berlin Heidelberg.
42. Tanious, F.A., Veal, J.M., Buczak, H. *et al.* (1992) DAPI (4',6-diamidino-2-phenylindole) binds differently to DNA and RNA: minor-groove binding at AT sites and intercalation at AU sites. *Biochemistry*, **31** (12), 3103–3112.
43. Kapuscinski, J. (1995) DAPI: a DNA-specific fluorescent probe. *Biotechnic & Histochemistry: Official Publication of the Biological Stain Commission*, **70** (5), 220–233.
44. Morgan, H. and Green, N. (2008) Dielectrophoresis, in *Encyclopedia of Microfluidics and Nanofluidics* (ed D. Li), Springer US.
45. Demircan, Y., Özgür, E., and Külah, H. (2013) Dielectrophoresis: applications and future outlook in point of care. *Electrophoresis*, **34** (7), 1008–1027.
46. Erez, N. (2013) Cancer: angiogenic awakening. *Nature*, **500** (7460), 37–38.
47. Schwab, M. (ed) (2012) Sandwich ELISA, in *Encyclopedia of Cancer*, Springer Berlin Heidelberg.
48. Schwab, M. (ed) (2012) ELISA, in *Encyclopedia of Cancer*, Springer Berlin Heidelberg.
49. Schwab, M. (ed) (2012) ELISPOT, in *Encyclopedia of Cancer*, Springer Berlin Heidelberg.
50. Schiro, P.G., Zhao, M., Kuo, J.S. *et al.* (2012) Sensitive and high-throughput isolation of rare cells from peripheral blood with ensemble-decision aliquot ranking. *Angewandte Chemie*, **51** (19), 4618–4622.
51. Balzar, M., Winter, M.J., de Boer, C.J., and Litvinov, S.V. (1999) The biology of the 17-1A antigen (Ep-CAM). *Journal of Molecular Medicine*, **77** (10), 699–712.
52. Armstrong, A. and Eck, S.L. (2003) EpCAM: a new therapeutic target for an old cancer antigen. *Cancer Biology & Therapy*, **2** (4), 320–326.
53. Went, P.T., Lugli, A., Meier, S. *et al.* (2004) Frequent EpCam protein expression in human carcinomas. *Human Pathology*, **35** (1), 122–128.
54. Lynch, T.J., Bell, D.W., Sordella, R. *et al.* (2004) Activating mutations in the epidermal growth factor receptor underlying responsiveness of non-small-cell lung cancer to gefitinib. *The New England Journal of Medicine*, **350** (21), 2129–2139.
55. Mooren, F. (ed) (2012) Epidermal growth factor receptor (EGFR), in *Encyclopedia of Exercise Medicine in Health and Disease*, Springer Berlin Heidelberg.

56. Alix-Panabieres, C. (2012) Minimal residual disease and circulating tumor cells in breast cancer, in *Recent Results in Cancer Research* (ed M.S. Peter) Hans-Jorg S., Springer.
57. Alix-Panabieres, C. and Pantel, K. (2012) Micrometastasis, in *Encyclopedia of Cancer* (ed M. Schwab), Springer, Berlin Heidelberg.
58. Thiery, J.P. and Sleeman, J.P. (2006) Complex networks orchestrate epithelial-mesenchymal transitions. *Nature Reviews. Molecular Cell Biology*, **7** (2), 131–142.
59. Rajasekaran, A. and Rajasekaran, S. (2012) Epithelial-to-mesenchymal transition, in *Encyclopedia of Cancer* (ed M. Schwab), Springer Berlin Heidelberg.
60. van Denderen, B.J. and Thompson, E.W. (2013) Cancer: the to and fro of tumour spread. *Nature*, **493** (7433), 487–488.
61. Yu, M., Bardia, A., Wittner, B.S. *et al.* (2013) Circulating breast tumor cells exhibit dynamic changes in epithelial and mesenchymal composition. *Science*, **339** (6119), 580–584.
62. Vohr, H.-W. (ed) (2016) Epitope, in *Encyclopedia of Immunotoxicology*, Springer Berlin Heidelberg.
63. Campbell, N.A. and Reece, J.B. (2009) *Biology*, 8th edn, Pearson Benjamin Cummings, San Francisco.
64. Thiriet, M. (2011) Extracellular matrix, in *Cell and Tissue Organization in the Circulatory and Ventilatory Systems*, vol. **1**, Springer New York.
65. Schwab, M. (ed) (2012) Extracellular matrix, in *Encyclopedia of Cancer*, Springer Berlin Heidelberg.
66. Macey, M. (2007) Principles of flow cytometry, in *Flow Cytometry* (ed M. Macey), Humana Press.
67. Saxena, A. (2012) Flow cytometry, in *Encyclopedia of Cancer* (ed M. Schwab), Springer Berlin Heidelberg.
68. Garimberti, E. and Tosi, S. (2010) Fluorescence in situ hybridization (FISH), basic principles and methodology, in *Fluorescence In Situ Hybridization (FISH)*, vol. **659** (eds J.M. Bridger and E.V. Volpi), Humana Press.
69. Schwab, M. (ed) (2012) FISH, in *Encyclopedia of Cancer*, Springer Berlin Heidelberg.
70. Hong, J. and Lukes, J. (2008) Flow cytometer lab-on-chip devices, in *Encyclopedia of Microfluidics and Nanofluidics* (ed D. Li), Springer US.
71. Autebert, J., Coudert, B., Bidard, F.C. *et al.* (2012) Microfluidic: an innovative tool for efficient cell sorting. *Methods*, **57** (3), 297–307.
72. Ema, H., Kobayashi, T., and Nakauchi, H. (2010) Principles of hematopoietic stem cell biology, in *Hematopoietic Stem Cell Biology* (ed M. Kondo), Humana Press.
73. Schwab, M. (ed) (2012) Hematopoietic stem cell, in *Encyclopedia of Cancer*, Springer Berlin Heidelberg.
74. Schwab, M. (ed) (2012) HER2, in *Encyclopedia of Cancer*, Springer Berlin Heidelberg.
75. Mitri, Z., Constantine, T., and O'Regan, R. (2012) The HER2 receptor in breast cancer: pathophysiology, clinical use, and new advances in therapy. *Chemotherapy Research and Practice*, **2012**, 743193.
76. Panton, R.L. (2005) *Incompressible flow*, 3rd edn, Hoboken, N.J, J. Wiley.
77. Schwab, M. (ed) (2012) Image cytometry, in *Encyclopedia of Cancer*, Springer Berlin Heidelberg.

78. Chieco, P., Jonker, A., De Boer, B.A. *et al.* (2013) Image cytometry: protocols for 2D and 3D quantification in microscopic images. *Progress in Histochemistry and Cytochemistry*, **47** (4), 211–333.
79. O'Sullivan, M. (2005) Immunoassays, in *Principles of Immunopharmacology* (eds F. Nijkamp and M. Parnham), Birkhäuser Basel.
80. Li, D. (ed) (2008) Immunoassay, in *Encyclopedia of Microfluidics and Nanofluidics*, Springer US.
81. Burry, R. (2010) Introduction, in *Immunocytochemistry*, Springer New York.
82. Schwab, M. (ed) (2012) Immunocytochemistry, in *Encyclopedia of Cancer*, Springer Berlin Heidelberg.
83. Immunofluorescence, in *Encyclopedia of Genetics, Genomics, Proteomics and Informatics*, Springer Netherlands.
84. Gillett, C. (2006) Immunohistochemistry, in *Breast Cancer Research Protocols*, vol. **120** (eds S. Brooks and A. Harris), Humana Press.
85. Birkhahn, M., Cote, R., and Taylor, C. (2012) Immunohistochemistry, in *Encyclopedia of Cancer* (ed M. Schwab), Springer Berlin Heidelberg.
86. Zhou, J. and Papautsky, I. (2013) Fundamentals of inertial focusing in microchannels. *Lab on a Chip*, **13** (6), 1121–1132.
87. Martel, J.M. and Toner, M. (2014) Inertial focusing in microfluidics. *Annual Review of Biomedical Engineering*, **16**, 371–396.
88. Vona, G., Sabile, A., Louha, M. *et al.* (2000) Isolation by size of epithelial tumor cells : a new method for the immunomorphological and molecular characterization of circulatingtumor cells. *The American Journal of Pathology*, **156** (1), 57–63.
89. Nguyen, N.-T. and Wereley, S.T. (2006) Fundamentals and applications of microfluidics, in *Artech House Integrated Microsystems Series*, 2nd edn, Artech House, Boston.
90. Li, D. (ed) (2008) Laminar flow, in *Encyclopedia of Microfluidics and Nanofluidics*, Springer US.
91. Heitzer, E., Auer, M., Ulz, P. *et al.* (2013) Circulating tumor cells and DNA as liquid biopsies. *Genome Medicine*, **5** (8), 73.
92. Lianidou, E. (2014) Circulating tumor cells: a noninvasive liquid biopsy in cancer, in *Molecular Testing in Cancer* (eds G.M. Yousef and S. Jothy), Springer New York.
93. Schwab, M. (ed) (2012) Malignant tumor, in *Encyclopedia of Cancer*, Springer Berlin Heidelberg.
94. Muschel, R. (2012) Metastasis, in *Encyclopedia of Cancer* (ed M. Schwab), Springer Berlin Heidelberg.
95. Whitesides, G.M. (2006) The origins and the future of microfluidics. *Nature*, **442** (7101), 368–373.
96. Li, D. (ed) (2008) Microfluidic systems, in *Encyclopedia of Microfluidics and Nanofluidics*, Springer US.
97. Li, D. (ed) (2008) Peclet number, in *Encyclopedia of Microfluidics and Nanofluidics*, Springer US.
98. Tam, A.C. (1986) Applications of photoacoustic sensing techniques. *Reviews of Modern Physics*, **58** (2), 381–431.
99. Li, D. (ed) (2008) Poly(dimethylsiloxane) (PDMS), in *Encyclopedia of Microfluidics and Nanofluidics*, Springer US.

100. Schwab, M. (ed) (2012) Prognosis, in *Encyclopedia of Cancer*, Springer Berlin Heidelberg.
101. Li, D. (ed) (2008) Reynolds number, in *Encyclopedia of Microfluidics and Nanofluidics*, Springer US.
102. Klepárník, K. and Foret, F. (2013) Recent advances in the development of single cell analysis—a review. *Analytica Chimica Acta*, **800**, 12–21.
103. Tuerk, C. and Gold, L. (1990) Systematic evolution of ligands by exponential enrichment: RNA ligands to bacteriophage T4 DNA polymerase. *Science*, **249** (4968), 505–510.
104. DeNardo, G.L. and DeNardo, S.J. (2012) Concepts, consequences, and implications of theranosis. *Seminars in Nuclear Medicine*, **42** (3), 147–150.
105. Powell, A.A., Talasaz, A.H., Zhang, H. *et al.* (2012) Single cell profiling of circulating tumor cells: transcriptional heterogeneity and diversity from breast cancer cell lines. *PloS One*, **7** (5), e33788.
106. Burrell, R.A., McGranahan, N., Bartek, J., and Swanton, C. (2013) The causes and consequences of genetic heterogeneity in cancer evolution. *Nature*, **501** (7467), 338–345.

INDEX

A549 cells, 214, 369
acoustic field, 231
acoustic radiation force, 231
acoustic standing wave, 228
acoustophoresis, 227, 239, 403
ACTB (gene), 319
active cell sorting, 59
acute myeloid leukemia (AML), 354
adhesion, 95, 104, 267–269, 271, 278, 334
affinity capture/affinity-based isolation, 148, 177
ALDH1 (gene), 13, 338
aliquot ranking, 55
allophycocyanin (APC), 57, 379
*Ampli*1™, 372
androgen deprivation therapy, 337
androgen receptor, 341
angiogenesis, 5, 391
antibody, 3, 19, 289, 403
antigen, 6, 14, 20, 69, 404
anti-idiotype, 3, 4
apoptosis, 3, 12, 277, 334, 338, 387
aptamers, 20, 87, 95, 213, 287, 288, 291, 404
array comparative genomic hybridization (arrayCGH), 16, 321, 388
aspect ratios, 86, 90, 107, 132
assay validation, 212, 358 *see also* clinical validation
atomic force microscopy (AFM), 97, 333, 335

B-cell lymphoma, 3, 4
benign tumor, 404
beta-catenin, 12
binding kinetics, 104
BioMark, 319, 332, 340
biomarker, 14, 18, 73, 322, 337, 352, 354–356, 358, 360, 404
biophysical property, 148, 153, 333
biopsy, 6, 196, 323, 404
bioorthogonal nanoparticle detection, 188
bladder cancer, 385
blood, 4, 135, 228, 236, 302, 330 *see also* peripheral blood
  test, 19
bonding, 42, 93
bone marrow, 5, 404
BRAF gene, 16, 215, 322, 388
breast cancer, 8, 10, 70, 116, 274, 307, 320, 384
breast cancer metastasis suppressor-1 (BRMS1), 17

Circulating Tumor Cells: Isolation and Analysis, First Edition. Edited by Z. Hugh Fan.

C6 glioma cells, 139
cancer, 3, 114, 116, 249, 316, 352, 404 *see also* tumor
  diagnosis, 273, 331 *see also* diagnosis
  dormancy, 3, 4, 8
  metastasis, 160, 269, 353, 373 *see also* metastasis
cancer stem cells (CSCs), 66, 86, 140, 405
capture efficiency, 46, 153, 159, 176, 207, 294, 405
carcinoembryonic antigen (CEA), 352
carcinogenesis (oncogenesis), 405
carcinoma, 5, 405
castration-resistant prostate cancer, 16, 165, 337, 384 *see also* prostate cancer
CCRF-CEM cells, 291
CD2, 303
CD3, 309
CD14, 309
CD16, 303, 309
CD19, 303, 309, 389
CD24, 13, 44, 66, 79
CD36, 303
CD38, 303, 389
CD44, 13, 44, 66, 79, 316, 322, 386, 405
CD45, 6, 65, 87, 303, 309, 389, 405
CD47, 406
CD66, 303
CD68, 310
CD133, 352, 406
CD138, 389
CD168, 10
cell
  capture, 63, 91, 176, 210, 213, 255, 271, 294 *see also* cell trap
  fixation, 73, 154, 176, 236, 258
  heterogeneity, 16, 318, 368, 406 *see* also heterogeneity
  isolation, 90, 111, 140, 214, 246, 293
  labeling, 184, 236
  purity, 406 *see also* purity
  release, 213
  rolling, 104, 268
  separation, 34, 127, 132, 136, 163, 230, 244, 271, 304
  sorting, 132, 234
  trap, 160, 162, 164 *see also* cell capture
  viability *see* viability
CellCelector, 320
CellSearch®, 7, 70, 85, 148, 175, 212, 228, 270, 304, 316, 357, 367, 377, 379, 383, 406
Cell-SELEX, *see* Systematic evolution of ligands by exponential enrichment
CellTracks Analyzer, 379
centrifugal force, 100, 136
cfDNA *see* circulating cell-free DNA (cfDNA)
chaotic mixing, 45, 292, 406
chronic myeloid leukemia (CML), 355
circulating cell-free DNA (cfDNA), 17, 406
circulating epithelial cells (CECs), 6, 7, 406
circulating endothelial cells (CECs), 391
circulating fetal nuclear cells, 217
circulating melanoma cells, 250
circulating tumor cells (CTCs), 3, 4, 7, 8, 35, 47, 53, 85, 127, 147, 174, 183, 202, 227, 250, 270, 296, 301, 316, 329, 352, 367, 377, 407
circulating tumor DNA (ctDNA), 17, 377, 407
Clinical Laboratory Improvement Amendments (CLIA), 360
clinical sensitivity, 87, 116
clinical specificity, 87, 116
clinical utility, 17, 352, 385
clinical validation, 384, 387 *see also* assay validation
cluster of differentiation (CD), 407
COLO205 cell, 276, 381
collagen adhesion matrix (CAM), 15
colorectal cancer, 8, 16, 116, 274, 384
computational fluid dynamics (CFD), 101, 137 *see also* numerical simulation
computed tomography (CT), 250, 352
context of use, 352
copy number variations, 16, 218, 317, 318, 372
CTC Chip, 14, 270
CTC-iChip, 14, 89, 203
cytokeratin (CK), 6, 65, 116, 307, 407
cytopathology, 407

Dean flow/number, 101, 132, 136
deformability, 147, 155, 334
density gradient, 303 *see also* gradient centrifugation
DEPArray™, 367, 368 *see also* dielectrophoresis
detection, 17, 54, 75, 79, 331, 334 *see also* cancer detection
deterministic lateral displacement (DLD), 303
diagnosis/diagnostic, 20, 217, 250, 331, 354, 356, 374, 408 *see also* cancer diagnosis
4′,6-diamidino-2-phenylindole (DAPI), 65, 380, 407
dielectrophoresis (DEP), 153, 331, 368, 408 *see also* DEPArray™
diffusion, 35, 38, 290
disease-free survivor (DFS), 385
disseminated tumor cells (DTCs), 9, 338, 408
DNA copy number *see* copy number variation
DRAQ5, 66

droplet, 55, 57, 254, 256, 257, 259
DU 145 cells, 237

early detection, 17, 337 *see also* detection
Eastern Cooperative Oncology Group (ECOG), 352
E-cadherin, 12, 352
enrichment, 61, 164, 227, 309
ensemble-decision aliquot ranking (eDAR), 53, 56, 408
enumeration, 65, 69, 175, 367, 368, 383, 385
enzyme-linked immunosorbent assay (ELISA), 408
enzyme-linked immunospot (ELISPOT) assay, 408
epidermal growth factor (EGF), 13, 268
epidermal growth factor receptor (EGFR), 16, 116, 192, 219, 292, 307, 323, 351, 389, 409
epithelial cell-adhesion molecule (EpCAM), 6, 55, 87, 106, 148, 192, 270, 297, 307, 381, 409
epithelial immunospot (EPISPOT) assay, 15, 409
epithelial markers, 12
epithelial phenotype, 13, 409 *see also* phenotype
epithelial-to-mesenchymal transition (EMT), 12, 270, 320, 337, 357, 382, 409
epitope, 4, 410
erythrocytes (red blood cells), 54, 139, 303, 372, 410
E-selectin, 267
estrogen receptor (ER), 13, 319, 354
etching, 41, 42, 174,
extracellular matrix (ECM), 12, 206, 410
extravasate/extravasation, 268, 315
exome sequencing *see* whole exome sequencing
exosome, 377

fabrication, 39, 63, 93, 128, 153, 207 *see also* microfabrication
ferrofluid, 379
fibrosarcoma, 5
Ficoll-Hypaque/Ficoll-paque, 203, 236, 303 *see also* gradient centrifugation
flow cytometry, 6, 54, 89, 135, 203, 251, 276, 306, 308, 378, 410
flow profile *see* velocity profile
flow rate, 37, 57, 60, 63, 67, 102, 111, 128, 139, 140, 158, 229, 237, 253, 294
fluid, 34, 135, 255
fluid dynamics, 97
fluorescein isothiocyanate, 380
fluorescence-activated cell sorting (FACS), 127, 410
fluorescence in situ hybridization (FISH), 7, 10, 179, 217, 318, 352, 386, 389, 392, 410
focusing length, 131
FOXC1, 13, 320

gastrointestinal cancer, 5, 385
gene expression analysis, 179, 318, 319, 332, 392
genomic analysis, 178, 207, 373
genotypic analysis, 15
glyceraldehyde-3-phosphate dehydrogenase (*GAPDH*), 114, 319
gradient centrifugation, 203 *see also* Ficoll–Hypaque

H1975 cells, 219, 389
halloysite, 271
hematopoietic stem cells, 271, 411
Herceptin/Trastuzumab, 10
herringbone mixer or chip (HB-Chip), 14, 46, 88, 293
heterogeneity, 6, 72, 296, 330, 353, 367, 372, 381 *see also* tumor heterogeneity
high-throughput, 75, 106, 129, 203, 228, 332 *see also* throughput
HL60 cell, 275, 291
HT29 cell, 118
human epidermal growth factor receptor 2 (HER2), 10, 58, 179, 192, 307, 319, 386, 411
human prostate epithelial (HPET) cells, 135
hydraulic diameter, 36, 101, 129
hydrodynamic chromatography, 150
hydrodynamic shear or force, 95, 128, 411

ICAM-1 (protein), 271
image cytometry, 411
immunoassay, 411
immunocytochemistry (ICC), 6, 74, 209, 412
immunofluorescence, 140, 175, 412
immunoglobulin, 3
immunohistochemistry (IHC), 5, 175, 260, 412
immunological "cocktail", 19
immunomagnetic separation/method, 58, 70, 158, 202, 304, 319
immunostaining, 73, 74
inertial focusing, 129, 412
inertial microfluidics, 127
intertumor heterogeneity, 316 *see also* tumor heterogeneity
intratumor heterogeneity, 316 *see also* tumor heterogeneity
invasive assay, 89
isolation by size of epithelial tumor cells (ISET), 177, 412 *see also* size-based isolation

Knudsen number, 37
KRAS (gene), 16, 116, 179, 215, 322, 354, 369, 372, 388

L-selectin, 267
lab-on-a-chip, 34, 139
laminar flow, 35, 36, 97, 105, 136, 233, 412
laser capture microdissection (LCM), 179, 214
leukocytes (white blood cells), 63, 139, 372, 413
lift forces, 103, 129
ligase detection reaction (LDR), 117
liposome, 276
liquid biopsy, 18, 202, 323, 352, 367, 373, 413
liver cancer, 385
LNCaP cells, 237, 381
LS180 cell line, 118
lung cancer, 8, 385

magnetic-activated cell sorting (MACS), 304, 413
magnetic nanoparticle, 184
magnetic resonance imaging (MRI), 250
magnetic separation, 303
MagSweeper, 15, 203, 319, 338
malignant tumor, 15, 413
MAPK signaling pathways, 12
MART-1 (protein), 260
matrix metalloproteinases (MMPs), 12
MCF-7 breast cancer cells, 11, 60, 100, 140, 381
MDA-MB-231, 65, 140, 160, 320
melanoma, 8, 250
mesenchymal markers, 12
mesenchymal phenotype, 12, 383, 413 *see also* phenotype
mesenchymal-to-epithelial transition (MET), 13, 322, 413
metastasis, 5, 69, 72, 149, 177, 215, 250, 278, 322, 330, 413 *see also* cancer metastasis
metastatic process, 17, 393
microbeads, 243 *see also* polystyrene beads
microelectromechanical system (MEMS), 34
microfabrication, 34, 92, 175 *see also* fabrication
microfilter, 64, 89, 154, 175, 178, 203
microfluidics, 33, 34, 56, 86, 135, 155, 203, 234, 290, 414
micropillar, 88, 209, 293
micromixers, 45
micropumps, 43, 44
microvalves, 44
miniaturization, 35, 188
miniaturized total chemical analysis system (μTAS), 33
miRNA, 388
molding, 42, 92, 119
molecular profiling, 116, 191, 329 *see also* single-cell molecular profiling
morphological difference, 149
Mucin-1 (Muc-1), 6, 12, 192
multivalent binding, 294
mutational analysis, 16, 215, 317, 318, 321
multiple myeloma cells, 390

nanochemical, 336
nanofiber, 214, 216
nanomechanical, 336, 341
nanoparticles, 294
nanostructured substrate, 206
NanoVelcro, 201, 205
Navier–Stokes equation, 37
N-cadherin, 12, 307
negative depletion (selection), 14, 89, 202, 301, 414
next-generation sequencing (NGS), 16, 119, 218, 316, 373
non-small-cell lung cancer (NSCLC) *see* lung cancer
nuclear magnetic resonance (NMR), 183, 184
nuclear-to-cytoplasm ratio, 151
numerical simulation, 112 *see also* computational fluid dynamics

Oncoquick, 203
OnTarget™ assay, 388
ovarian cancer, 116
overall survival (OS), 8, 384, 386

P53, 389
pancreatic cancer, 71, 99, 114, 341, 385
PC3 cells, 237, 381
Péclet number (Pé), 104, 414
peripheral blood, 5, 287, 414
personalized medicine/precision medicine, 343 *see also* personalized therapy
personalized therapy/treatment, 15, 180, 273, 323
pharmacodynamic, 354, 355
phenotype, 4, 341, 392 *see also* epithelial phenotype; mesenchymal phenotype
photoacoustics, 249, 251, 414
photolithography, 40, 42
phycoerythrin (PE), 57, 379
PIK3CA, 16, 373, 388
platelets *see* thrombocytes
point mutation, 117, 374
polydimethylsiloxane (PDMS), 34, 39, 45, 64, 128, 139, 188, 207, 292, 414
poly(lactic-co-glycolic) acid (PLGA), 214
polymer nanosubstrate, 205, 206

polymerase chain reaction (PCR), 6, 34, 114, 288, 318 *see also* quantitative real-time polymerase chain reaction (qRT-PCR)
polystyrene beads, 140, 243
positive selection/enrichment, 53, 100, 114
positron emission tomography (PET), 352
predictive, 354, 355
pressure-driven flow, 241
progesterone receptor (PR), 13, 319, 354
prognosis/prognostic, 8, 250, 331, 353, 354, 415
progression-free survival (PFS), 8, 354, 384, 386
prostate cancer, 8, 114, 116, 150, 212, 274, 337, 384
prostate-specific antigen (PSA), 116, 213, 246, 337, 352
prostate-specific membrane antigen (PSMA), 116, 292
P-selectin, 267
PTPRN2 (gene), 338
purity, 87, 114, 134, 176, 204, 243, 261, 271, 289, 311 *see also* cell purity

quadrupole magnetic separator (QMS), 304
quantitative real-time polymerase chain reaction (qRT-PCR), 179, 319, 322, 331, 332, 336 *see also* polymerization chain reaction (PCR)

rare cell, 53, 54, 69, 86, 128, 192
ratchet, 155, 159
recovery, 87 *see also* capture efficiency
red blood cell, *see* erythrocytes
red blood cell lysis, 303
regulatory requirement, 358
reliability, 358
reverse transcription polymerase chain reaction (RT-PCR), 214, 262 *see also* polymerase chain reaction (PCR)
Reynolds number (Re), 36, 101, 129, 136, 415
RNA interference (RNAi), 12
RNA-seq, 218, 321, 333, 338, 394
RoboSep™, 304
RosetteSep™, 155, 303

sample preparation, 55, 128, 188, 236, 256
scaling, 36, 132
ScreenCell, 339
selectin, 267
separation, 43, 228, 250, 257 *see also* cell separation
SHH (gene) *see* Sonic Hedgehog (SHH)
short tandem repeat (STR), 372
signal-to-noise ratio, 57, 77, 188
signaling molecule, 4
signaling pathways, 9
single cell analysis, 80, 87, 205, 315, 318, 319, 330, 369, 372, 415
single-cell molecular profiling, 320, 329, 331, 336, 373 *see also* molecular profiling
single-cell transcriptional profiling, 16, 332
single nucleotide variant (SNV), 218, 320, 372
sinusoidal, 85, 90
size-based isolation, 133, 154, 174, 177 *see also* isolation by size of epithelial tumor cells (ISET)
SKBr-3 cells, 60, 381
SMART-Seq, 341
Sonic Hedgehog (SHH), 338, 340
spiral microchannel, 136
squamous cell carcinoma, 307
subpopulation, 4, 316, 330
surface-area-to-volume ratio, 35
surface treatment, 96, 210, 293
surrogate, 10, 354, 356
SW480 cell, 369
SWOG, 8, 386
syringe pump, 61, 128, 175, 253, 258, 293, 306
systematic evolution of ligands by exponential enrichment (SELEX), 213, 288, 415

T24 cell, 140, 381
TGFβ, 320, 338, 340
theranosis, 415
thermoplastics, 39, 91
thermoresponsive polymer, 219
thrombocytes (platelets), 140, 415
throughput, 35, 87, 153, 229, 252, 287 *see also* high throughput
TNF-related apoptosis-inducing ligand (TRAIL), 275
TP53 (gene), 374, 391
trajectory, 128, 134, 232
transcriptomic/transcriptional analysis, 218, 338
translational technology, 357
transmission electron microscope (TEM), 186
Trastuzumab *see* Herceptin
treatment monitoring, 71, 191
triple-negative breast cancer (TNBC), 12
tumor, 3, 250 *see also* cancer
tumor heterogeneity, 217, 315, 372, 415 *see also* cell heterogeneity
two-phase flow, 255

UBB (gene), 319
ultrasound, 234

UM-UC13 cells, 158
urokinase plasminogen activator (UPA), 11
urokinase plasminogen activator receptor (uPAR), 11

VCaP cells, 244
velocity profile, 37, 38, 131, 233
viability, 15, 54, 87, 139, 159, 176, 219, 245, 262
vimentin, 12, 65, 307, 352

white blood cell *see* leukocytes
whole exome sequencing, 16, 215, 320
whole genome amplification, 217, 320, 372
WNT (gene), 338, 340

Z-configuration network, 109

CHEMICAL ANALYSIS

A SERIES OF MONOGRAPHS ON ANALYTICAL CHEMISTRY AND ITS APPLICATIONS

Vol. 1 **The Analytical Chemistry of Industrial Poisons, Hazards, and Solvents**. *Second Edition.* By the late Morris B. Jacobs
Vol. 2 **Chromatographic Adsorption Analysis**. By Harold H. Strain (*out of print*)
Vol. 3 **Photometric Determination of Traces of Metals**. *Fourth Edition*
Part I: General Aspects. By E. B. Sandell and Hiroshi Onishi
Part IIA: Individual Metals, Aluminum to Lithium. By Hiroshi Onishi
Part IIB: Individual Metals, Magnesium to Zirconium. By Hiroshi Onishi
Vol. 4 **Organic Reagents Used in Gravimetric and Volumetric Analysis**. By John F. Flagg (*out of print*)
Vol. 5 **Aquametry: A Treatise on Methods for the Determination of Water**. *Second Edition* (*in three parts*). By John Mitchell, Jr. and Donald Milton Smith
Vol. 6 **Analysis of Insecticides and Acaricides**. By Francis A. Gunther and Roger C. Blinn (*out of print*)
Vol. 7 **Chemical Analysis of Industrial Solvents**. By the late Morris B. Jacobs and Leopold Schetlan
Vol. 8 **Colorimetric Determination of Nonmetals**. *Second Edition.* Edited by the late David F. Boltz and James A. Howell
Vol. 9 **Analytical Chemistry of Titanium Metals and Compounds**. By Maurice Codell
Vol. 10 **The Chemical Analysis of Air Pollutants**. By the late Morris B. Jacobs
Vol. 11 **X-Ray Spectrochemical Analysis**. *Second Edition.* By L. S. Birks
Vol. 12 **Systematic Analysis of Surface-Active Agents**. *Second Edition.* By Milton J. Rosen and Henry A. Goldsmith
Vol. 13 **Alternating Current Polarography and Tensammetry**. By B. Breyer and H.H.Bauer
Vol. 14 **Flame Photometry**. By R. Herrmann and J. Alkemade
Vol. 15 **The Titration of Organic Compounds** (*in two parts*). By M. R. F. Ashworth
Vol. 16 **Complexation in Analytical Chemistry: A Guide for the Critical Selection of Analytical Methods Based on Complexation Reactions**. By the late Anders Ringbom
Vol. 17 **Electron Probe Microanalysis**. *Second Edition.* By L. S. Birks
Vol. 18 **Organic Complexing Reagents: Structure, Behavior, and Application to Inorganic Analysis**. By D. D. Perrin
Vol. 19 **Thermal Analysis**. *Third Edition.* By Wesley Wm.Wendlandt
Vol. 20 **Amperometric Titrations**. By John T. Stock
Vol. 21 **Reflectance Spectroscopy**. By Wesley Wm.Wendlandt and Harry G. Hecht
Vol. 22 **The Analytical Toxicology of Industrial Inorganic Poisons**. By the late Morris B. Jacobs
Vol. 23 **The Formation and Properties of Precipitates**. By Alan G.Walton
Vol. 24 **Kinetics in Analytical Chemistry**. By Harry B. Mark, Jr. and Garry A. Rechnitz
Vol. 25 **Atomic Absorption Spectroscopy**. *Second Edition.* By Morris Slavin
Vol. 26 **Characterization of Organometallic Compounds** (*in two parts*). Edited by Minoru Tsutsui
Vol. 27 **Rock and Mineral Analysis**. *Second Edition.* By Wesley M. Johnson and John A. Maxwell
Vol. 28 **The Analytical Chemistry of Nitrogen and Its Compounds** (*in two parts*). Edited by C. A. Streuli and Philip R.Averell
Vol. 29 **The Analytical Chemistry of Sulfur and Its Compounds** (*in three parts*). By J. H. Karchmer
Vol. 30 **Ultramicro Elemental Analysis**. By Güther Toölg
Vol. 31 **Photometric Organic Analysis** (*in two parts*). By Eugene Sawicki
Vol. 32 **Determination of Organic Compounds: Methods and Procedures**. By Frederick T. Weiss
Vol. 33 **Masking and Demasking of Chemical Reactions**. By D. D. Perrin

Vol. 34 **Neutron Activation Analysis**. By D. De Soete, R. Gijbels, and J. Hoste
Vol. 35 **Laser Raman Spectroscopy**. By Marvin C. Tobin
Vol. 36 **Emission Spectrochemical Analysis**. By Morris Slavin
Vol. 37 **Analytical Chemistry of Phosphorus Compounds**. Edited by M. Halmann
Vol. 38 **Luminescence Spectrometry in Analytical Chemistry**. By J. D.Winefordner, S. G. Schulman, and T. C. O'Haver
Vol. 39 **Activation Analysis with Neutron Generators**. By Sam S. Nargolwalla and Edwin P. Przybylowicz
Vol. 40 **Determination of Gaseous Elements in Metals**. Edited by Lynn L. Lewis, Laben M. Melnick, and Ben D. Holt
Vol. 41 **Analysis of Silicones**. Edited by A. Lee Smith
Vol. 42 **Foundations of Ultracentrifugal Analysis**. By H. Fujita
Vol. 43 **Chemical Infrared Fourier Transform Spectroscopy**. By Peter R. Griffiths
Vol. 44 **Microscale Manipulations in Chemistry**. By T. S. Ma and V. Horak
Vol. 45 **Thermometric Titrations**. By J. Barthel
Vol. 46 **Trace Analysis: Spectroscopic Methods for Elements**. Edited by J. D.Winefordner
Vol. 47 **Contamination Control in Trace Element Analysis**. By Morris Zief and James W. Mitchell
Vol. 48 **Analytical Applications of NMR**. By D. E. Leyden and R. H. Cox
Vol. 49 **Measurement of Dissolved Oxygen**. By Michael L. Hitchman
Vol. 50 **Analytical Laser Spectroscopy**. Edited by Nicolo Omenetto
Vol. 51 **Trace Element Analysis of Geological Materials**. By Roger D. Reeves and Robert R. Brooks
Vol. 52 **Chemical Analysis by Microwave Rotational Spectroscopy**. By Ravi Varma and Lawrence W. Hrubesh
Vol. 53 **Information Theory as Applied to Chemical Analysis**. By Karl Eckschlager and Vladimir Stepanek
Vol. 54 **Applied Infrared Spectroscopy: Fundamentals, Techniques, and Analytical Problemsolving**. By A. Lee Smith
Vol. 55 **Archaeological Chemistry**. By Zvi Goffer
Vol. 56 **Immobilized Enzymes in Analytical and Clinical Chemistry**. By P. W. Carr and L. D. Bowers
Vol. 57 **Photoacoustics and Photoacoustic Spectroscopy**. By Allan Rosencwaig
Vol. 58 **Analysis of Pesticide Residues**. Edited by H. Anson Moye
Vol. 59 **Affinity Chromatography**. By William H. Scouten
Vol. 60 **Quality Control in Analytical Chemistry**. *Second Edition.* By G. Kateman and L. Buydens
Vol. 61 **Direct Characterization of Fineparticles**. By Brian H. Kaye
Vol. 62 **Flow Injection Analysis**. By J. Ruzicka and E. H. Hansen
Vol. 63 **Applied Electron Spectroscopy for Chemical Analysis**. Edited by Hassan Windawi and Floyd Ho
Vol. 64 **Analytical Aspects of Environmental Chemistry**. Edited by David F. S. Natusch and Philip K. Hopke
Vol. 65 **The Interpretation of Analytical Chemical Data by the Use of Cluster Analysis**. By D. Luc Massart and Leonard Kaufman
Vol. 66 **Solid Phase Biochemistry: Analytical and Synthetic Aspects**. Edited by William H. Scouten
Vol. 67 **An Introduction to Photoelectron Spectroscopy**. By Pradip K. Ghosh
Vol. 68 **Room Temperature Phosphorimetry for Chemical Analysis**. By Tuan Vo-Dinh
Vol. 69 **Potentiometry and Potentiometric Titrations**. By E. P. Serjeant
Vol. 70 **Design and Application of Process Analyzer Systems**. By Paul E. Mix
Vol. 71 **Analysis of Organic and Biological Surfaces**. Edited by Patrick Echlin
Vol. 72 **Small Bore Liquid Chromatography Columns: Their Properties and Uses**. Edited by Raymond P.W. Scott
Vol. 73 **Modern Methods of Particle Size Analysis**. Edited by Howard G. Barth
Vol. 74 **Auger Electron Spectroscopy**. By Michael Thompson, M. D. Baker, Alec Christie, and J. F. Tyson
Vol. 75 **Spot Test Analysis: Clinical, Environmental, Forensic and Geochemical Applications**. By Ervin Jungreis

Vol. 76 **Receptor Modeling in Environmental Chemistry**. By Philip K. Hopke
Vol. 77 **Molecular Luminescence Spectroscopy: Methods and Applications** (*in three parts*). Edited by Stephen G. Schulman
Vol. 78 **Inorganic Chromatographic Analysis**. Edited by John C. MacDonald
Vol. 79 **Analytical Solution Calorimetry**. Edited by J. K. Grime
Vol. 80 **Selected Methods of Trace Metal Analysis: Biological and Environmental Samples**. By Jon C.VanLoon
Vol. 81 **The Analysis of Extraterrestrial Materials**. By Isidore Adler
Vol. 82 **Chemometrics**. By Muhammad A. Sharaf, Deborah L. Illman, and Bruce R. Kowalski
Vol. 83 **Fourier Transform Infrared Spectrometry**. By Peter R. Griffiths and James A. de Haseth
Vol. 84 **Trace Analysis: Spectroscopic Methods for Molecules**. Edited by Gary Christian and James B. Callis
Vol. 85 **Ultratrace Analysis of Pharmaceuticals and Other Compounds of Interest**. Edited by S. Ahuja
Vol. 86 **Secondary Ion Mass Spectrometry: Basic Concepts, Instrumental Aspects, Applications and Trends**. By A. Benninghoven, F. G. Rüenauer, and H.W.Werner
Vol. 87 **Analytical Applications of Lasers**. Edited by Edward H. Piepmeier
Vol. 88 **Applied Geochemical Analysis**. By C. O. Ingamells and F. F. Pitard
Vol. 89 **Detectors for Liquid Chromatography**. Edited by Edward S.Yeung
Vol. 90 **Inductively Coupled Plasma Emission Spectroscopy: Part 1: Methodology, Instrumentation, and Performance; Part II: Applications and Fundamentals**. Edited by J. M. Boumans
Vol. 91 **Applications of New Mass Spectrometry Techniques in Pesticide Chemistry**. Edited by Joseph Rosen
Vol. 92 **X-Ray Absorption: Principles,Applications,Techniques of EXAFS, SEXAFS, and XANES**. Edited by D. C. Konnigsberger
Vol. 93 **Quantitative Structure-Chromatographic Retention Relationships**. By Roman Kaliszan
Vol. 94 **Laser Remote Chemical Analysis**. Edited by Raymond M. Measures
Vol. 95 **Inorganic Mass Spectrometry**. Edited by F.Adams,R.Gijbels, and R.Van Grieken
Vol. 96 **Kinetic Aspects of Analytical Chemistry**. By Horacio A. Mottola
Vol. 97 **Two-Dimensional NMR Spectroscopy**. By Jan Schraml and Jon M. Bellama
Vol. 98 **High Performance Liquid Chromatography**. Edited by Phyllis R. Brown and Richard A. Hartwick
Vol. 99 **X-Ray Fluorescence Spectrometry**. By Ron Jenkins
Vol. 100 **Analytical Aspects of Drug Testing**. Edited by Dale G. Deustch
Vol. 101 **Chemical Analysis of Polycyclic Aromatic Compounds**. Edited by Tuan Vo-Dinh
Vol. 102 **Quadrupole Storage Mass Spectrometry**. By Raymond E. March and Richard J. Hughes (*out of print: see Vol. 165*)
Vol. 103 **Determination of Molecular Weight**. Edited by Anthony R. Cooper
Vol. 104 **Selectivity and Detectability Optimization in HPLC**. By Satinder Ahuja
Vol. 105 **Laser Microanalysis**. By Lieselotte Moenke-Blankenburg
Vol. 106 **Clinical Chemistry**. Edited by E. Howard Taylor
Vol. 107 **Multielement Detection Systems for Spectrochemical Analysis**. By Kenneth W. Busch and Marianna A. Busch
Vol. 108 **Planar Chromatography in the Life Sciences**. Edited by Joseph C. Touchstone
Vol. 109 **Fluorometric Analysis in Biomedical Chemistry: Trends and Techniques Including HPLC Applications**. By Norio Ichinose, George Schwedt, Frank Michael Schnepel, and Kyoko Adochi
Vol. 110 **An Introduction to Laboratory Automation**. By Victor Cerdá and Guillermo Ramis
Vol. 111 **Gas Chromatography: Biochemical, Biomedical, and Clinical Applications**. Edited by Ray E. Clement
Vol. 112 **The Analytical Chemistry of Silicones**. Edited by A. Lee Smith
Vol. 113 **Modern Methods of Polymer Characterization**. Edited by Howard G. Barth and Jimmy W. Mays
Vol. 114 **Analytical Raman Spectroscopy**. Edited by Jeanette Graselli and Bernard J. Bulkin
Vol. 115 **Trace and Ultratrace Analysis by HPLC**. By Satinder Ahuja

Vol. 116 **Radiochemistry and Nuclear Methods of Analysis**. By William D. Ehmann and Diane E.Vance
Vol. 117 **Applications of Fluorescence in Immunoassays**. By Ilkka Hemmila
Vol. 118 **Principles and Practice of Spectroscopic Calibration**. By Howard Mark
Vol. 119 **Activation Spectrometry in Chemical Analysis**. By S. J. Parry
Vol. 120 **Remote Sensing by Fourier Transform Spectrometry**. By Reinhard Beer
Vol. 121 **Detectors for Capillary Chromatography**. Edited by Herbert H. Hill and Dennis McMinn
Vol. 122 **Photochemical Vapor Deposition**. By J. G. Eden
Vol. 123 **Statistical Methods in Analytical Chemistry**. By Peter C. Meier and Richard Züd
Vol. 124 **Laser Ionization Mass Analysis**. Edited by Akos Vertes, Renaat Gijbels, and Fred Adams
Vol. 125 **Physics and Chemistry of Solid State Sensor Devices**. By Andreas Mandelis and Constantinos Christofides
Vol. 126 **Electroanalytical Stripping Methods**. By Khjena Z. Brainina and E. Neyman
Vol. 127 **Air Monitoring by Spectroscopic Techniques**. Edited by Markus W. Sigrist
Vol. 128 **Information Theory in Analytical Chemistry**. By Karel Eckschlager and Klaus Danzer
Vol. 129 **Flame Chemiluminescence Analysis by Molecular Emission Cavity Detection**. Edited by David Stiles, Anthony Calokerinos, and Alan Townshend
Vol. 130 **Hydride Generation Atomic Absorption Spectrometry**. Edited by Jiri Dedina and Dimiter L. Tsalev
Vol. 131 **Selective Detectors: Environmental, Industrial, and Biomedical Applications**. Edited by Robert E. Sievers
Vol. 132 **High-Speed Countercurrent Chromatography**. Edited by Yoichiro Ito and Walter D. Conway
Vol. 133 **Particle-Induced X-Ray Emission Spectrometry**. By Sven A. E. Johansson, John L. Campbell, and Klas G. Malmqvist
Vol. 134 **Photothermal Spectroscopy Methods for Chemical Analysis**. By Stephen E. Bialkowski
Vol. 135 **Element Speciation in Bioinorganic Chemistry**. Edited by Sergio Caroli
Vol. 136 **Laser-Enhanced Ionization Spectrometry**. Edited by John C. Travis and Gregory C. Turk
Vol. 137 **Fluorescence Imaging Spectroscopy and Microscopy**. Edited by Xue Feng Wang and Brian Herman
Vol. 138 **Introduction to X-Ray Powder Diffractometry**. By Ron Jenkins and Robert L. Snyder
Vol. 139 **Modern Techniques in Electroanalysis**. Edited by Peter Vanysek
Vol. 140 **Total-Reflction X-Ray Fluorescence Analysis**. By Reinhold Klockenkamper
Vol. 141 **Spot Test Analysis: Clinical, Environmental, Forensic, and Geochemical Applications**. *Second Edition.* By Ervin Jungreis
Vol. 142 **The Impact of Stereochemistry on Drug Development and Use**. Edited by Hassan Y. Aboul-Enein and Irving W.Wainer
Vol. 143 **Macrocyclic Compounds in Analytical Chemistry**. Edited by Yury A. Zolotov
Vol. 144 **Surface-Launched Acoustic Wave Sensors: Chemical Sensing and Thin-Film Characterization**. By Michael Thompson and David Stone
Vol. 145 **Modern Isotope Ratio Mass Spectrometry**. Edited by T. J. Platzner
Vol. 146 **High Performance Capillary Electrophoresis: Theory, Techniques, and Applications**. Edited by Morteza G. Khaledi
Vol. 147 **Solid Phase Extraction: Principles and Practice**. By E. M. Thurman
Vol. 148 **Commercial Biosensors: Applications to Clinical, Bioprocess and Environmental Samples**. Edited by Graham Ramsay
Vol. 149 **A Practical Guide to Graphite Furnace Atomic Absorption Spectrometry**. By David J. Butcher and Joseph Sneddon
Vol. 150 **Principles of Chemical and Biological Sensors**. Edited by Dermot Diamond
Vol. 151 **Pesticide Residue in Foods: Methods, Technologies, and Regulations**. By W. George Fong, H. Anson Moye, James N. Seiber, and John P. Toth
Vol. 152 **X-Ray Fluorescence Spectrometry**. *Second Edition.* By Ron Jenkins

Vol. 153 **Statistical Methods in Analytical Chemistry**. *Second Edition.* By Peter C. Meier and Richard E. Züd
Vol. 154 **Modern Analytical Methodologies in Fat- and Water-Soluble Vitamins**. Edited by Won O. Song, Gary R. Beecher, and Ronald R. Eitenmiller
Vol. 155 **Modern Analytical Methods in Art and Archaeology**. Edited by Enrico Ciliberto and Guiseppe Spoto
Vol. 156 **Shpol'skii Spectroscopy and Other Site Selection Methods: Applications in Environmental Analysis, Bioanalytical Chemistry and Chemical Physics**. Edited by C. Gooijer, F. Ariese and J.W. Hofstraat
Vol. 157 **Raman Spectroscopy for Chemical Analysis**. By Richard L. McCreery
Vol. 158 **Large (C> = 24) Polycyclic Aromatic Hydrocarbons: Chemistry and Analysis**. By John C. Fetzer
Vol. 159 **Handbook of Petroleum Analysis**. By James G. Speight
Vol. 160 **Handbook of Petroleum Product Analysis**. By James G. Speight
Vol. 161 **Photoacoustic Infrared Spectroscopy**. By Kirk H. Michaelian
Vol. 162 **Sample Preparation Techniques in Analytical Chemistry**. Edited by Somenath Mitra
Vol. 163 **Analysis and Purification Methods in Combination Chemistry**. Edited by Bing Yan
Vol. 164 **Chemometrics: From Basics to Wavelet Transform**. By Foo-tim Chau, Yi-Zeng Liang, Junbin Gao, and Xue-guang Shao
Vol. 165 **Quadrupole Ion Trap Mass Spectrometry**. *Second Edition.* By Raymond E. March and John F. J. Todd
Vol. 166 **Handbook of Coal Analysis**. By James G. Speight
Vol. 167 **Introduction to Soil Chemistry: Analysis and Instrumentation**. By Alfred R. Conklin, Jr.
Vol. 168 **Environmental Analysis and Technology for the Refining Industry**. By James G. Speight
Vol. 169 **Identification of Microorganisms by Mass Spectrometry**. Edited by Charles L. Wilkins and Jackson O. Lay, Jr.
Vol. 170 **Archaeological Chemistry**. *Second Edition.* By Zvi Goffer
Vol. 171 **Fourier Transform Infrared Spectrometry**. *Second Edition.* By Peter R. Griffiths and James A. de Haseth
Vol. 172 **New Frontiers in Ultrasensitive Bioanalysis: Advanced Analytical Chemistry Applications in Nanobiotechnology, Single Molecule Detection, and Single Cell Analysis**. Edited by Xiao-Hong Nancy Xu
Vol. 173 **Liquid Chromatography Time-of-Flight Mass Spectrometry: Principles, Tools, and Applications for Accurate Mass Analysis.** Edited by Imma Ferrer and E. Michael Thurman
Vol. 174 **In Vivo Glucose Sensing.** Edited by David O. Cunningham and Julie A. Stenken
Vol. 175 **MALDI Mass Spectrometry for Synthetic Polymer Analysis.** By Liang Li
Vol. 176 **Internal Reflection and ATR Spectroscopy** By Milan Milosevic
Vol. 177 **Hydrophilic Interaction Chromatography: A Guide for Practitioners.** Edited by Bernard A. Olsen and Brian W. Pack
Vol. 178 **Introduction to Soil Chemistry: Analysis and Instrumentation.** *Second Edition.* By Alfred R. Conklin, Jr.
Vol. 179 **High-Throughput Analysis for Food Safety.** Edited by Perry G. Wang, Mark F. Vitha, and Jack F. Kay
Vol. 180 **Handbook of Petroleum Product Analysis.** *Second Edition.* By James G. Speight
Vol. 181 **Total-Reflection X-Ray Fluorescence Analysis**. *Second Edition.* By Reinhold Klockenkamper
Vol. 182 **Handbook of Coal Analysis.** *Second Edition.* By James G. Speight
Vol. 183 **Pumps, Channels, and Transporters: Methods of Functional Analysis.** Edited by Ronald J. Clarke and Mohammed A. A. Khalid
Vol. 184 **Circulating Tumor Cells: Isolation and Analysis.** Edited by Z. Hugh Fan.